THE OXFORD HANI
TECHNOLOGY AND MUSIC EDUCATION

THE OXFORD HANDBOOK OF

TECHNOLOGY AND MUSIC EDUCATION

Edited by
S. ALEX RUTHMANN
and
ROGER MANTIE

Oxford University Press is a department of the University of Oxford. It furthers
the University's objective of excellence in research, scholarship, and education
by publishing worldwide. Oxford is a registered trade mark of Oxford University
Press in the UK and certain other countries.

Published in the United States of America by Oxford University Press
198 Madison Avenue, New York, NY 10016, United States of America.

© Oxford University Press 2017

First issued as an Oxford University Press paperback, 2020

All rights reserved. No part of this publication may be reproduced, stored in
a retrieval system, or transmitted, in any form or by any means, without the
prior permission in writing of Oxford University Press, or as expressly permitted
by law, by license, or under terms agreed with the appropriate reproduction
rights organization. Inquiries concerning reproduction outside the scope of the
above should be sent to the Rights Department, Oxford University Press, at the
address above.

You must not circulate this work in any other form
and you must impose this same condition on any acquirer.

Library of Congress Cataloging-in-Publication Data
Names: Ruthmann, Alex. | Mantie, Roger.
Title: The Oxford handbook of technology and music education /
edited by Alex Ruthmann and Roger Mantie.
Description: New York, NY : Oxford University Press, [2017]
Identifiers: LCCN 2016045074 | ISBN 9780199372133 (hardcover : alk. paper) |
ISBN 9780197502983 (paperback : alk. paper)
Subjects: LCSH: Music—Instruction and study—Technological innovations.
Classification: LCC MT1 .O97 2017 | DDC 780.71—dc23
LC record available at https://lccn.loc.gov/2016045074

Contents

Foreword   xiii
    Peter Richard Webster and David Brian Williams
List of Contributors   xxi
About the Companion Website   xxxi

Editors' Introduction: Narrating the Landscape of Technology and Music Education   1
    Roger Mantie and S. Alex Ruthmann

PART 1   EMERGENCE AND EVOLUTION

1A   CORE PERSPECTIVES

1. Thinking about Music and Technology   15
    Roger Mantie

2. Technology in Music Education in England and across Europe   31
    Marina Gall

1A   FURTHER PERSPECTIVES

3. Savoring the Artistic Experience in an Age of Commodification   51
    Chee-Hoo Lum

4. Music Technology in Ethnomusicology   57
    Gabriel Solis

5. The Role of "Place" and Context   65
    Janice Waldron

6. Slow Music   71
    Rena Upitis

7. Then and Now   81
    David A. Williams

1B CORE PERSPECTIVES

8. Globalization and Technology in Twenty-First-Century Education 89
   SAMUEL LEONG

9. Technology in the Music Classroom—Navigating through a Dense Forest: The Case of Greece 105
   SMARAGDA CHRYSOSTOMOU

1B FURTHER PERSPECTIVES

10. Building a Broad View of Technology in Music Teacher Education 123
    HEIDI PARTTI

11. Technology in the Music Classroom in Kenya 129
    EMILY ACHIENG' AKUNO AND DONALD OTOYO ONDIEKI

12. Pondering an End to Technology in Music Education 137
    JOSEPH MICHAEL PIGNATO

13. A Software Creator's Perspective 143
    JOE BERKOVITZ

14. Where Might We Be Going? 149
    JONATHAN SAVAGE

15. Loaded Questions for an Emerging World of Music Education Technology 157
    JOHN-MORGAN BUSH

16. Mobile Learning in Music Education 163
    JASON CHEN

PART 2 LOCATIONS AND CONTEXTS: SOCIAL AND CULTURAL ISSUES

2A CORE PERSPECTIVES

17. Critical Perspectives from Africa 175
    BENON KIGOZI

18. Interest-Driven Music Education: Youth, Technology, and Music Making Today — 191
    KYLIE PEPPLER

19. Situating Technology within and without Music Education — 203
    JOSEPH MICHAEL PIGNATO

2A FURTHER PERSPECTIVES

20. Human Potential, Technology, and Music Education — 219
    SMARAGDA CHRYSOSTOMOU

21. "Placing" Technology within Music Education Communities — 225
    AILBHE KENNY

22. The Promise and Pitfalls of the Digital Studio — 233
    ETHAN HEIN

23. Musicking and Technology: A Further Swedish Perspective — 241
    BO NILSSON

24. Exploring Intersections of Technology, Play, Informality, and Innovation — 249
    GILLIAN HOWELL

2B CORE PERSPECTIVES

25. Pedagogical Fundamentalism versus Radical Pedagogy in Music — 257
    HEIDI PARTTI

26. The Impact of Technologies on Society, Schools, and Music Learning — 277
    VALERIE PETERS

27. Re-situating Technology in Music Education — 291
    EVAN S. TOBIAS

2B FURTHER PERSPECTIVES

28. Technology in Perspective: Who is in Control? — 311
    PATRICIA A. GONZÁLEZ-MORENO

29. The Curious Musician 317
    LEAH KARDOS

30. On Becoming Musical: Technology, Possibilities, and Transformation 323
    GENA R. GREHER

31. Ebola Songs: Exploring the Role of Music in Public Health Education 329
    CARLOS CHIRINOS-ESPIN

32. Thinking and Talking about Change in Music Education 339
    ROGER MANTIE

33. A Sociological Perspective on Technology and Music Education 345
    RUTH WRIGHT

PART 3 EXPERIENCING, EXPRESSING, LEARNING, AND TEACHING

3A CORE PERSPECTIVES

34. Power and Choice in the Teaching and Learning of Music 355
    CHEE-HOO LUM

35. Music Fluency: How Technology Refocuses Music Creation and Composition 367
    BARBARA FREEDMAN

36. Playing (in) the Digital Studio 383
    ETHAN HEIN

3A FURTHER PERSPECTIVES

37. Considering Music Technology and Literacy 399
    JAY DORFMAN

38. Technology and Music Collaboration for People with Significant Disabilities 405
    DONALD DeVITO

39. Prosumer Learners and Digital Arts Pedagogy 413
    Samuel Leong

40. A Pluralist Approach to Music Education 421
    James Humberstone

41. Augmenting Music Teaching and Learning with Technology and Digital Media 431
    Evan S. Tobias

42. Possibilities for Inclusion with Music Technologies 439
    Deborah VanderLinde

3B CORE PERSPECTIVES

43. Getting in the Way? Limitations of Technology in Community Music 449
    Gillian Howell

44. Meaningful and Relevant Technology Integration 463
    Michael Medvinsky

45. The Convergence of Networked Technologies in Music Teaching and Learning 475
    Janice Waldron

3B FURTHER PERSPECTIVES

46. Narcissism, Romanticism, and Technology 489
    Evangelos Himonides

47. Pedagogical Decision-Making 495
    Ryan Bledsoe

48. Equity and Access in Out-of-School Music Making 503
    Kylie Peppler

49. Technology, Sound, and the Tuning of Place 511
    Sandra Stauffer

PART 4 COMPETENCE, CREDENTIALING, AND PROFESSIONAL DEVELOPMENT

4A CORE PERSPECTIVES

50. Traditions and Ways Forward in the United States — 523
    Jay Dorfman

51. Technology and Invisibility in Music Teacher Education — 539
    Gena R. Greher

52. Authentic Approaches to Music Education with Technology — 555
    Jonathan Savage

4A FURTHER PERSPECTIVES

53. Technology in Music Initial Teacher Education — 569
    Marina Gall

54. Using Mobile Technologies and Problem-Seeking Pedagogies to Bridge Universities and Workplaces — 575
    Julie Ballantyne

55. Application of Technology in Music Education in Selected African Countries — 581
    Benon Kigozi

56. Defining and Acknowledging Music Education Technology in Music Teacher Training — 587
    Lauri Väkevä

57. Learner Engagement and Technology Integration — 595
    Michael Medvinsky

4B CORE PERSPECTIVES

58. Faculty Development in and through the Use of Information and Communication Technology — 603
    Patricia A. González-Moreno

59. Educators' Roles and Professional Development 619
    EVANGELOS HIMONIDES

60. Music Technology Pedagogy and Curricula 633
    DAVID A. WILLIAMS

4B FURTHER PERSPECTIVES

61. Why Isn't Music Education in the United States
    More Twenty-First-Century PC? 649
    BARBARA FREEDMAN

62. Generating Intersections between Music and Technology 655
    MATTHEW HITCHCOCK

63. Preparing for Change and Uncertainty 665
    VALERIE PETERS

Index 671

Foreword

The publication of this handbook marks a significant addition to continued dialogue on music teaching and learning and the role of "technology." The editors must be complimented on the questions posed for consideration by the distinguished group of international professionals engaged in theory, research, and practice represented here. The handbook presents a set of core chapters along with chapters by authors and others who have been encouraged to comment on these "Core Perspectives" in the form of "Further Perspectives." One imagines that the companion website will encourage dialogue long after the printed version of the handbook is released, adding still more to the impact of the project.

This book is something of a landmark. This is not because of its attention to the history of technology in music education or to the latest advances in hardware and software, although some of that is referenced. It is not a collection of writings that formally praise or condemn technology in reductionist ways, although there is praise and condemnation of sorts throughout. Rather, we have here—perhaps for first time—carefully reasoned and at times passionate chapters that critically examine the uses of technology in the ways we teach music in elementary, secondary, and tertiary settings from a multinational, global perspective.

The reader finds music learning addressed outside the formal settings of "school" and in so doing is encouraged to take an even wider view of technology's role in music education. Furthermore, as a secondary contribution, the chapters provide a wealth of research references that provide supportive data from international sources within music education and from such ancillary fields as educational technology, sociology, psychology, medicine, and more.

How best do we engineer the educational environments for music as an art form that touches us on so many personal levels? As we carefully read each chapter and commentary, we were reminded that the profession is dramatically challenged by "big" topics that are framed by our social and technological landscape. Among these are: (1) the tradeoffs between constructionist and direct instruction, (2) the encouragement of creative and critical thinking, (3) the role of interdisciplinary engagement, both within music and between music and other fields, (4) the most effective ways to assess both teaching *and* learning, (5) the consideration of who we teach, both in formal schooling and outside, (6) what we teach, in terms of both the music itself and the many social issues surrounding it, and (7) where we teach, ranging from a single person to millions and from a single location to every place where our voices, images, and sounds are encountered.

Never before has our field collected such an array of perspectives from such a global constituency of music educators, a collection that offers a worldview of melding technology with classroom music practice. Never before has our field been challenged so globally and profoundly and in such an exciting way.

The Handbook's Themes

In this foreword, we offer our collective understanding of a set of focused themes in this handbook that relate to the big topics already noted. We encourage the reader to consider each of these. Some are more conceptual, others are causes for celebration or concern, and still others relate directly to the milieu of schools and teaching. All are interrelated.

The Meaning of "Technology"

Many of the writers here directly or indirectly address a traditional understanding of the word "technology" within the music education context. Here the term might suggest "objects," "tools," or "things"—hardware and software, particularly as it might relate to computers, tablets, phones, and the potpourri of electronic music devices. Yet other writers help to provide a deeper and richer understanding of the topic. Technology is seen as integral to the species-specific need for humans to communicate and express themselves through music. Considering the history of music making from the start of civilization, music expression begins with both human voices and hands as tools or instruments, and extends to the creation of instruments readily constructed from native materials (drums, bone flutes, and monochords come to mind). The extension, or shall we say evolution, continues through the more sophisticated instruments of a culture (examples might be zithers and kotos, sitars and tablas, dulcimers and harps, violins and bassoons), then arriving at the electronic music tools of today. The very soul of what it is to be human seeks expression through whatever technology is on hand for music making.

Those who think deeply about the implications of "technology" epistemologically extend this line of thought to music learning. As we consider the "craft" and "art" of music, the writers remind us that the term "technology" might include how best we learn and what we hold as important conceptually to teach. Implied here is the notion that "technology" embraces what one does and what any given culture cherishes in its teaching. So our conversations about technology should not be centered on "things" only but also on far more fundamental patterns of thought and behavior about teaching and learning—for example, discussions about mediated music versus live performance, music creation as music production, acoustic instruments blended with electronic ones, collaborative improvisation as a staple of music learning, the role of the teacher as arbiter of what is "good" music, and other similar concerns. Nothing of importance is as simple as which ear training or notation software is best or what tablet or computer to buy—the topic is generously far broader, deeper, and richer.

Technology as Tool, Music as Central

Since the very first writings and professional work we have coauthored, spanning now some 25 years, we have considered technology not for its own sake but in terms of its potential for learning about and creating and performing music as art. This can be seen in our support of Hypercard in the late 1980s, our textbook revisions since the 1990s, and the recent work we have done on core technology competencies for music students. So many of the essays crafted in this handbook speak to this fundamental point, and it delights us that this notion continues to be celebrated and endorsed here. If one experiences a dehumanizing, boring, unorganized, and distasteful music performance or teaching sequence created by musicians using what might be thought of as impressive "cool" technology devoid of substance, fault not the technology but rather the performer or teacher.

Technological Determinism: Reductionist Thought

Perhaps one of the most profound themes that emerges in these chapters and commentary is the clear warning against thinking simplistically about technology and music education—for example, notions that (1) technology and its use in music education will somehow fundamentally change learning for the best (or the worst), (2) popular music using technology is the ruin (or the savior) of music education in the schools, or (3) music teachers are no longer of value in schools because students can create music on their own in their bedrooms, as opposed to the view that music teachers are supremely equipped to teach real music and all other music creation pathways are wasted energy; all these and thousands more binaries of this sort are a combination of myth and fallacy. If we have learned anything in the past several years about music teaching, it is that the enterprise is deeply complex, in part due to the rise of technological affordances; but at core, technology is not as much the issue as is learning in our political, social, economic, and multicultural times. That said, the best approach for teaching and learning music is to be open to pluralistic thinking and multiple ways for music to be talked about and created within well-designed structures.

The Role of Social Context

Another critical conceptual theme that is at the root of so many writings in this handbook is the importance of teaching music—or shall we say guiding music learning—while engaging students in issues of social concern. As we emphasized earlier, music has been forever tied to the times and places in which it has been made and consumed. In fact, the notions of "time" and "place" emerge as important concepts in the handbook. In addition to qualities of the music itself, many support the idea that we have a moral responsibility as teachers and scholars to engage students in the social concerns of the day with music as a central vehicle. For example, a powerful idea for us was that turntables and microphones were not the reason for hip-hop; feelings of social injustice in urban cities were more the driving force. The same might be said of a great deal of music and its history. The life work of artists like Pete Seeger, Paul Simon, Brian Wilson, David Bowie, and Keith Emerson come to mind. Or, from a different perspective, think about the work of Dmitri Shostakovich, Igor Stravinsky, Kurt Weill, Peter Gabriel, Pierre Boulez, and Yo Yo Ma.

Engaging students in music technology projects that involve music and the consideration of social context is a way for students to understand their own musical identities and develop a deeper understanding of social injustices. Such activities serve to bridge the moat between the music they experience within and without the classroom. Social media sites, collaborative networks for music distribution and performance, and other technical means for students to interrogate the music and the technologies themselves are all important for a richer music education that transcends the classroom.

Many of the authors here note that this critical reflection about music and culture, mitigated by technology in its larger context, is not common in traditional, top-down school structures. It is more likely to be found in more informal settings of the sort enhanced by social networks. Those born as "analog" and not as "digital natives" tethered to screens and earbuds may find this way of thinking challenging, but of course this kind of distinction between formal and informal leaning is its own kind of reductionist thinking that assumes that age defines who we are.

Changing Views of Composer, Performer, and Listener

Technology has afforded a change in how we view musical experiences. We were struck with the notion that critical listening is perhaps the ultimate purpose of music education, since it plays such an important part in music making. The diversity of technology tools provides a teacher with the ability to stress the aural experience as a special kind of musicianship vital for performance, composition, and improvisation. What-you-hear-is-what-you-get (WYHIWYG) music applications offer ways for aural skills and creative expression to be combined without the prerequisites of reading traditional notation and playing traditional instruments. (GarageBand, Mixcraft, Groovy Music, and Subotnick's Music Academy are cases in point.) Traditional views of the composer and performer as separate from the producer of music are also challenged at every turn. Several contributors to the handbook note that notions of a dedicated producer in an expensive recording studio as part of a master distribution scheme are anachronistic. Music production has become democratized through virtual digital audio workstations, minimal and inexpensive hardware, and social networking tools. The writings herein show this playing out globally, whether the production be in Africa, the Pacific Rim, Europe, or the United States. Lines are blurred, and music is often created and distributed in more direct ways that shake up older views of ownership and marketing rights; national boundaries of commerce disappear into the global market of the Internet. All of this becomes sources of both great opportunity and a massive challenge for those in charge of pedagogy.

Disadvantaged Populations, Remote Locations, and the Digital Divide

Some contributors offer perspectives with a more cautionary tone. Among the causes for concern are the difficulties that educators face with the costs of music technology and the division of affordance between those with resources and those without. Remote locations around the world without access to reliable supplies of electricity, healthcare, schools, and other basic necessities suffer. Even countries like the United States continue to struggle with distribution of wealth, which makes resources for education institutions and community music programs that are funded by state and local taxes a difficult problem of equity.

One is struck by the incongruity of Internet access and bandwidth across nations with very different political and economic make-ups. Many regions have far better Internet bandwidth than the United States, while others have none at all. New solutions for providing Internet access through global networks of satellites and drones promise more universal access. For some locales, urban environments provide more sophisticated music recording and production tools than rural settings. More affordable software programs designed for less expensive devices are a positive development of course, but the digital divide continues to be both a local-global and rural-urban contrast and challenge for music education. Although these concerns were noted by a few contributors, we were surprised that this was not as completely addressed as it might have been.

Expanding Research and Global Understanding

Practice and theory provide useful information, but ultimately systematic research provides empirical evidence to guide best practice. One is impressed reading through this collection of articles by the grounding of so many in research from within our field of music education, and from fields both closely aligned and far-reaching. The basic premise of imperfect induction is that we generalize from a small sample of the whole as to the characteristics of the whole. Confirmation through replication of results is critical to this process.

The handbook makes a significant contribution to this end. Too often we defer myopically to studies within our immediate realm of colleagues and academic specialization. Too often we have research results using small-scale studies that show nonsignificance and low effect sizes when attempting to measure differences in educational technology practice and performance. With this diverse set of international articles, we benefit from the global wealth of studies and results focused on a common thread: technology integration within music education.

The new model on the horizon for research is "big data." Where current research methodology is based on sampling from larger populations, "big data" refers to data sets that continue to grow in size and complexity such that traditional sampling theory is no longer relevant. Various tools are used to find predictive trends within these big data sets. Focusing our research endeavors on a global scale, as does some of the research offered here, offer a hint, a "nudge," at what might lie ahead for music education and music technology research.

The Ubiquity of Screens

We were delighted to see addressed the problems of technology overuse. Music is such a powerful and innate form of artistic expression, playing a role in the lives of so many youth, and the enabling technology delivers it into the hands of children as young as toddlers. The "information age" has made the access of data of all kinds so commonplace that one's waking hours can be completely dominated by computers, phones, tablets, televisions—screens and screens of images accompanied by sounds of all dimensions.

Many authors in the handbook offer a cautionary and balanced perspective to technology and its use in education. Concerns about slowing down our pace, cutting the links to screens and their attendant sounds, and enjoying our natural surroundings in more contemplative ways are welcomed perspectives that can balance our work with technologically-mediated activities. For example, reading a paperback book on a sailboat trip with only the sounds of the water under the transom and the sun's light dancing on the sea can be a way to inspire music teaching in ways that time in front of screens cannot.

Creative and Critical Thinking

We were thrilled with the constant celebration of music technology's role in the encouragement of creative and critical thinking. Another staple in our writing over the years has been this important part of working with sound. Contributors address the importance of revision in all aspects of music experience, of active engagement with aesthetic decision-making that is not dominated by teaching authority but guided and enhanced by the more experienced whole. If used wisely, technology becomes a tool for meaning making and developing agency for learners. Reflective thought is encouraged in this kind of collaborative learning environment that can be effectively supported by technology.

The Culture of School and of Teachers Themselves

In its closing part, the handbook includes work on the nature of school and teacher. This is a lasting and complicated question that deserves our best thinking and action. Formal schooling changes slowly, and often the pace of teachers' own attitudes toward what is of value and toward their roles change even more glacially. Despair about traditional schooling seems to be apparent in this handbook and in many other contexts. Schools are often portrayed as conservative places where teachers support linear thinking and noncreative skill building. Music made by technology is seen as undesirable and not compatible with "real" music. Smartphones and tablets in the hands of learners during school hours are seen as a negative, often not for instructional goals but for administrative, security, and disciplinary reasons.

Some contend that the virtues of more informal learning networks outside school is where the real "action is" for music learning. Many contributors touch on the topic of participatory culture as encouraging more democratic opportunities for self-expression across varied musical styles and practices. Learning communities and communities of practice are celebrated as important pathways not common in traditional schooling.

We understand these views and applaud the global discourse expressed in the handbook about the concern for formal education. We find it consonant with our own personal "wish list" for change in music education. But what we believe, too, is that these dispositions are at times overstated and overlook important work happening in hundreds of schools today. We believe that schools can and will change in coming years. Teachers will see their roles differently in times to come. This is fueled in part by teacher education programs at colleges and universities that are creating more effective curricula that support varied ensembles and concerts, more examples of multimedia performance and of student-centered learning in more constructivist ways, and more embedded assessments of skills that matter. Better scholarship on technology competencies is emerging, and an awareness of the need to consider technology skills embedded in coursework as well as dedicated courses is becoming clearer. We see "methods" classes changing in their content, adjusting to a more varied musical landscape. Younger teachers are entering the workplace with more confidence and better preparation, academically and musically, to use refined pedagogies that include technology.

Lines between formal and informal learning are starting to blur. We are finding ways to engage more students, especially at the secondary level, with music performing and creating beyond the traditional performing ensembles. Teachers are enriching the classroom music experience with activities that embrace social music skills, informal music practice, and participatory music making. Through technology tools, the learning curve for aural and performance skills and expressive and engaging music activities can be greatly accelerated. For us, the contents of this handbook provide a basis for optimism and will help propel the momentum of change internationally.

In sum, we feel the work ahead in advancing the cause of technology and music teaching and learning viewed more holistically will be hard, and perhaps circumstances outside our control will need to change for there to be dramatic progress. But the important conversations are now starting in many quarters of the globe, and the profession is starting to focus on the ideas for change that, after all, are the first and most important steps to a brighter future. The multicultural perspective of these collective writings reminds us that music expression—in all its many forms—begins with the human need for communication through music. Technology is a tool to this end, beginning with clapping, dancing, and singing and extending through any expressive device our imagination can build, any music instrument yet created.

Peter Richard Webster
Scholar-in-Residence, Thornton School of Music
University of Southern California

David Brian Williams
Emeritus Professor, School of Music
Illinois State University
March 15, 2016

List of Contributors

Emily Achieng' Akuno holds a PhD from Kingston University in Surrey, UK; a Master of Music in performance from Northwestern State University of Louisiana in the USA, and a Bachelor of Education (Arts) degree from Kenyatta University in Nairobi, Kenya. Trained as a performer-educator, Emily is Professor of Music and Executive Dean of the Faculty of Social Sciences and Technology at the Technical University of Kenya in Nairobi, Kenya. A past board member of the International Society for Music Education and chair of the Music in Schools and Teacher Education Commission (MISTEC), her research and publications focus on cultural relevance and its implications for music education, and music making in enhancing children's literacy skill development.

Julie Ballantyne is known for her work in the areas of music teacher identities, social justice, music teacher education, and the social and psychological impacts of musical engagement. An Associate Professor in music education in the School of Music at the University of Queensland, Australia, she has won commendations and fellowships for her teaching, and she holds leadership positions with organizations such as the International Society for Music Education. She has published work in key journals and has coedited the book *Navigating Music and Sound Education*. She enjoys teaching pre-service and in-service teachers at the bachelor's and master's levels, as well as supervising several PhD students.

Joe Berkovitz is the founder of Noteflight, which launched the first web-based application for creating, sharing, and consuming traditional music notation. He also cochairs the W3C Web Audio Working Group and the Music Notation Community Group. With over 35 years of experience in the software industry, he served as Senior Architect at pioneering web application company ATG. At the Education Development Corporation, he led the software design team for IBM's constructivist "Math and More" math curriculum. He is a frequent and sought-after speaker at conferences on the web platform. He studied piano and composition at New England Conservatory of Music.

Ryan Bledsoe is a doctoral student in music education at Arizona State University. She specializes in early childhood music education and has worked with infants through middle school students in Texas, Arizona, and Florida. Her articles have been published in *General Music Today* and the *Arizona Music News*, and she is an active conference presenter. Her research interests include creativity, makerspaces, and designing electronic instruments with young children. She shares her work with other educators on her website: ryanbledsoe.wordpress.com.

John-Morgan Bush is the Executive Director of the String Project at the University of Massachusetts, Lowell, and is a music education faculty member in the university's Music Department. He also serves on the executive board of the National String Project Consortium. Before joining the faculty at UMass Lowell, he served as Director of education at the Little Orchestra Society and managed music education engagement for thousands of New York City public school children, parents, and senior citizens. In 2012 he founded the Tuxedo Revolt, an arts consultancy and think tank that reimagines how to connect today's audiences with live music.

Jason Chen graduated from the University of Missouri-Kansas City Conservatory of Music in 1995 with a BMus degree, majoring in composition and piano with a 4-year scholarship, and received his MA and PGDE from Hong Kong Baptist University. Jason is currently Assistant Professor at the Education University of Hong Kong. He has a PhD in music technology from the Royal Melbourne Institute of Technology University, Australia, where he researched composition for film and media. His articles have been published by top-ranked journals, including the *International Journal of Music Education, Research Studies in Music Education*, and *Music Education Research*.

Carlos Chirinos-Espin is Clinical Assistant Professor of music business at New York University. He has been a social entrepreneur, researcher, and consultant in music and media in emerging markets in Africa and Latin America. He works in the fields of music, communication, and social change, and creates public health awareness campaigns using music, radio, and TV. In 2015 he received the Ebola Grand Challenge Award from the White House Office of Science and Technology Policy, Centers for Disease Control, and the U.S. Department of Defense for the development of Africa Stop Ebola, a global fundraising and awareness campaign about Ebola in West Africa. Before relocating to New York University, he was based at the School of Oriental and African Studies (SOAS), University of London.

Smaragda Chrysostomou is a Professor of Music Pedagogy and Didactics, Department of Music Studies, National and Kapodistrian University of Athens, Greece, where she teaches undergraduate and postgraduate courses. She also teaches at the Distance Learning MA Course on Music Education, University of Nicosia, Cyprus. She was one of the key experts responsible for the new National Curricula for Music in both Greece and Cyprus. She leads the Aesthetic Education Team in the nationwide project for media-enriched textbooks and learning objects' repository. She is a Board Member of the International Society for Music Education (ISME). More information and publications can be found at http://scholar.uoa.gr/schrysos and http://publicationslist.org/schrysos.

Donald DeVito is a music and special education teacher at the Sidney Lanier Center School for students with disabilities in Gainesville, Florida. He was a 2014–16 International Society for Music Education board member and 2011 National Council for Exceptional Children Teacher of the Year (Special Education). Dr. DeVito publishes extensively on networking universities, schools, and community-based music programs for the benefit of children with special needs throughout the world. He is developing research at the Notre Maison Orphanage in Haiti for children with disabilities. He received his PhD from the University of Florida and is an online facilitator at Boston University.

Jay Dorfman is Associate Professor and Coordinator of Music Education at Kent State University. His writing has been published in several scholarly journals. He is the author of *Theory and Practice of Technology-based Music Instruction* (Oxford University Press, 2013). He is a former president of the Technology Institute for Music Educators (TI:ME). His research interests include the uses of technology in music teaching and learning; instrumental music teacher education; and comprehensive and interdisciplinary approaches to music education. He holds BM and MM degrees from the University of Miami (Florida) and a PhD from Northwestern University.

Barbara Freedman, named the 2012 TI:ME Technology Teacher of the Year, has been teaching Electronic Music and Audio Engineering at Greenwich High School in Connecticut since 2001 and is the author of *Teaching Music Through Composition: A Curriculum Using Technology* (2013). She is also a technology trainer, leads professional development workshops around the country and is a consultant to schools and districts on building technology labs and integrating technology into the curriculum. Barbara is the Co-President of the Music Educator Technologists Association/Technology Institute for Music Educators (META/TI:ME), Connecticut Chapter. She holds a Bachelor of Science and Master of Music in Performance from Brooklyn College Conservatory of Music City University of New York and Professional Studies Diploma from the Mannes College of Music. Barbara performs regularly with the Ridgefield and Bridgeport Symphonies. She studied conducting at the Hartt School of Music, Westminster Choir College, and The Juilliard School. Barbara's motto, "Teach music. The technology will follow." has become the rallying cry for music technology teachers around the world.

Marina Gall has served as Lecturer at the University of Bristol since 1999. For 16 years prior, she taught music at both primary and secondary levels. Currently she coordinates the one-year initial teacher education program for secondary school music teachers. She is also a board member of the European Association for Music in Schools. Her research focuses on children's use of music technologies in and outside the classroom. Her current work forms part of a large-scale project called Getting Things Changed -Tackling Disabling Practices: Co-production and Change (see http://www.bristol.ac.uk/sps/gettingthingschanged/). In this project she is exploring the ways in which new music technologies are helping disabled young people make music and express themselves.

Patricia A. González-Moreno is Professor of Music Education at the Autonomous University of Chihuahua, Mexico, where she teaches courses in music, music education, educational psychology, and philosophy (Faculty of Arts and Faculty of Literature and Philosophy). She also supervises graduate student work at Boston University and the National Autonomous University of Mexico. Before earning her PhD in Music Education from the University of Illinois, she taught general music in basic education for seven years. She also holds degrees in arts and administration. Her published research includes studies on motivation, creativity, teacher education, higher music education, knowledge mobilization, and professional development. In 2013, she was acknowledged as National Researcher, by the National Council for Science and Technology in Mexico (2013–2019). She has served as a Board Member of the International Society for Music Education and chair of the ISME Advocacy Standing Committee (2012–16).

Gena R. Greher is Professor of music education at the University of Massachusetts, Lowell. Her research interests focus on creativity and listening skill development and the influence of integrating multimedia technology in urban music classrooms. Projects include Performamatics, funded by a National Science Foundation grant linking computer science to the arts; Soundscapes, a music technology intervention program for teenagers with Autism Spectrum Disorder; Making Music Count for the Thelonious Monk Institute's Math Science Music Initiative. Recent awards include being named the 2014–15 Donahue Endowed Professor of the Arts, and a University of Massachusetts Creative Economy Grant - *Discovering Cultural Identity & Self Identity: Creating Spaces for Cambodian American Adolescents to Explore Their Cultural and Artistic Heritage*. She received her EdD from Teachers College, Columbia University. Previously, she spent 20 years in advertising as a jingle producer and music director.

Ethan Hein is a PhD student at New York University, and adjunct professor of music technology and music education at New York University and Montclair State University. He maintains a widely followed music blog at ethanhein.com and has also written for *NewMusicBox, Quartz*, and *Slate*. He is an active producer and composer, and you can listen to his recent work at soundcloud.com/ethanhein. The Groove Pizza project grew out of his MA thesis for the New York University music technology program. A founding member of the university's Music Experience Design Lab (MusEDLab.org), he has also contributed to Play With Your Music, the aQWERTYon, and the IMPACT NYU impact conference and workshop. He looks forward to continuing to grow the lab's suite of online music creation and learning tools.

Evangelos Himonides held the University of London's first-ever lectureship in music technology education and is now Reader in technology, education, and music at University College London (UCL) Institute of Education. He currently leads the MA in Music Education program at UCL Institute of Education. He is a Chartered Fellow with the British Computer Society. As a musician, technologist, and educator, he has had an ongoing career in experimental research in the fields of psychoacoustics, music perception, music cognition, information technology, human–computer interaction, special needs, the singing voice and singing development. He has developed the Sounds of Intent free online resource: soundsofintent.org.

Matthew Hitchcock is Senior Lecturer in Music Technology and Program Convenor for Bachelor of Music Technology at Queensland Conservatorium Griffith University. He has worked in the music industry as a performing multi-instrumentalist, artistic manager, studio owner, recording engineer, music producer, composer, acoustic consultant, web designer and software engineer. He has received awards for excellence in teaching from Griffith University and the Carrick Institute, and has a long track record of successes in the music industry prior to joining Queensland Conservatorium.

Gillian Howell is a PhD candidate at Griffith University and adjunct lecturer in Community Music Leadership at Melbourne Polytechnic. Her research investigates music participation in war-affected countries, among newly-arrived refugees in Australia, and in intercultural collaborations. She has worked as a music leader and researcher in post-conflict settings in the Balkans, the Caucasus, South Asia, and South-East Asia, most recently as a 2016 Endeavour Research Fellow investigating music and reconciliation partnerships in Sri Lanka and Norway. She is also an award-winning musician and teaching artist, working with many of Australia's flagship arts organizations, and was the founding creative director of the Melbourne Symphony Orchestra's Community Engagement Program. She is a Commissioner of the Community Music Activity Commission of the International Society for Music Education. Further information can be found at www.gillianhowell.com.au.

James Humberstone is Senior Lecturer in music education at the Sydney Conservatorium of Music, University of Sydney. As well as focusing on the teaching of composition and technology, James's research interests cover a broad range of fields, including technology and innovation in education, musicology (experimental music and music composed for children), and nontraditional research outputs as a regularly commissioned composer.

Leah Kardos is a composer-producer making eclectic music that often combines live instrumental performance with technology, location recordings, and found sounds. She enjoys music that explores the communicative powers of timbre and psychoacoustic phenomena, memory and pattern recognition, and the beauty of spaces. She has collaborated with performers and ensembles such as Ben Dawson, R. Andrew Lee, Laura Wolk-Lewanowicz, the Ukulele Orchestra of Great Britain, and Australian chamber orchestra Ruthless Jabiru. She is currently lecturing in music at Kingston University, London, and is a signed artist with contemporary music label Bigo & Twigetti.

Ailbhe Kenny is Lecturer in Music Education at Mary Immaculate College, University of Limerick, Ireland. She holds a PhD from the University of Cambridge, is a Fulbright Scholar and European Institute for Advance Study (EURIAS) fellow. Previous positions held include Research Fellow at Dublin City University, Primary Teacher, and Arts and Education Officer at 'The Ark'. Ailbhe has led numerous professional development courses and is actively involved in university-community projects, including directing the Mary Immaculate College Children's Choir. She regularly publishes in international journals, handbooks and edited volumes on music, arts and teacher education. Her first monograph, *Communities of Musical Practice*, was published by Routledge in 2016.

Benon Kigozi holds a Doctor of Music degree from the University of Pretoria. He is a senior faculty member at the Department of Performing Arts and Film of

Makerere University, having previously served as Head of Music at Africa University in Zimbabwe. He is President of the Pan African Society for Musical Arts Education, National President of the Uganda Society for Musical Arts Education, and Chair of the *Music in Africa* Foundation on Education and Content. He is former Board member for the International Society for Music Education, a member of the National Association for Study and Performance of African American Music and member of the Association of International Schools in Africa. His current research is on information and communication technology in music education. He serves on editorial boards and has extensively published articles, books, book chapters, and conference papers.

Samuel Leong (PhD) is Deputy Director of Academic Programs and Educational Innovation at the Hong Kong Academy for Performing Arts. He was Director of the UNESCO Observatory for Research in Local Cultures and Creativity in Education and Associate Dean of Quality Assurance and Enhancement and Head of Cultural and Creative Arts at the Education University of Hong Kong. He has contributed to over 100 publications, including the *Journal of Computer Assisted Learning, Educational Psychology* and *Routledge International Handbook of Creative Learning*. His recent research projects focus on Chinese creativity, multisensory arts learning, and innovative digitally enhanced pedagogy for the performing arts.

Chee-Hoo Lum is Associate Professor of music education at the National Institute of Education, Nanyang Technological University, Singapore. He is also Head of the UNESCO-NIE National Institute of Education Centre for Arts Research in Education, part of a region-wide network of observatories stemming from the UNESCO Asia-Pacific Action Plan. His research interests include examining issues around identity, cultural diversity and multiculturalism, technology, and globalization in music education; children's musical cultures; creativity and improvisation; and elementary music methods. Currently coeditor of the *International Journal of Music Education*, he has published two edited books and numerous chapters and referred journal articles.

Roger Mantie is currently Associate Professor in the Department of Arts, Culture and Media at University of Toronto Scarborough, Roger Mantie previously held positions in music education at Arizona State University and Boston University. Prior to his university career, Roger was a high school band director in Manitoba, directed jazz ensembles at Brandon University and the University of Manitoba, directed the Royal Conservatory of Music Community School Jazz Ensemble in Toronto, and conducted the Hart House Symphonic Band at the University of Toronto. These days Roger's professional work centers on lifelong music making as an integral part of healthy living. Roger is co-editor of the *The Oxford Handbook of Music Making and Leisure* (2016).

Michael Medvinsky has engaged learner-musicians in creating and expressing through music for 10 years. He earned his Masters of Music in music education from Oakland University, where his focus was general music and music technology. He is currently an instructional technology integrator advising Pre-K–5 teachers on designing learning experiences where experiential learning encourages thinkers to look closely, think deeply, and wonder incessantly. Working in education with an engineering background, he advocates the integration of technology and global collaboration into learning environments. He actively shares ideas with practitioners. He coauthored a chapter in *Composing Our Future: Preparing Music Educators to Teach Composition* (Oxford University Press, 2012).

Bo Nilsson received his doctorate in music education at Malmö Academy of Music, Lund University, Sweden, in 2002. He is Reader in Music Education at the Faculty of Fine and Performing Arts, Lund University, Sweden and Associate Professor at Kristianstad University, Sweden, where he teaches aesthetics and pedagogy in the teacher education program and the public health program. His teaching and research interests are children's musical creativity, music in special education, popular culture and information and communication technology in music education, and public health. He was a member of the expert group that wrote the current music curriculum for Swedish compulsory school. Bo is also a member of the International Society for Music Educaton (ISME) commission for Music in Special Education, Music Therapy and Music Medicine.

Donald Otoyo Ondieki holds a PhD in Music Performance and Education, a Master of Music in Performance and a Bachelor of Education in Music from Kenyatta University, Nairobi, Kenya. Donald enjoys a wide experience as a performer, educator, researcher and music industry consultant in Kenya. Currently, Donald is the Director of the Permanent Presidential Music Commission, the state department in charge of music in Kenya, the Vice-President of the Pan African Society of Musical Arts Education (PASMAE), and, member of the Kenyan Creative Economy Working Group. Prior to that, he was Senior Lecturer and Chairman of the Department of Music and Performing Arts at the Technical University of Kenya and the Coordinator of the Kenyan Creative Arts National Working Group. His research and publications have focused on the music industry, popular music and contemporary, socio-cultural and technological issues in music education.

Heidi Partti is Acting Professor of music education at University of the Arts Helsinki, Sibelius Academy. Her research interests are initiated by a need to better understand the surrounding culture of music making, learning, and teaching, so as to help the music education profession adapt and understand the rapid changes transpiring in today's popular and participatory cultures. Her articles and book chapters on topics such as music-related learning communities, digital technology, peer learning, collective creativity, and the development of intercultural competencies in music teacher education have been published in numerous peer-reviewed journals. She is also a coauthor of a book on composing pedagogy.

Kylie Peppler, an artist by training, is an Associate Professor of learning sciences at Indiana University. She engages in research that focuses on the intersection of the visual and performing arts, computation, and out-of-school learning. She is director of the Creativity Labs at Indiana University as well as lead of the MacArthur Foundation's Make-to-Learn Initiative. Her research has been supported by the National Science Foundation, the U.S. Department of Education, the Wallace Foundation, the Spencer

Foundation, the Moore Foundation, and the MacArthur Foundation's Digital Media and Learning Initiative.

Valerie Peters holds bachelor's degrees in music and education from the University of Manitoba, a master's degree in music education from the University of Northern Colorado, and a doctoral degree in music education from Northwestern University. She is currently Full Professor of music education at Université Laval, Quebec City. She taught high school music in Montreal for 11 years. She is the recipient of a research grant to study intercultural music education, conducts research on music teacher working conditions, has been awarded an Social Sciences and Humanities Research Council (SSHRC) Insight Grant to study artistic learning and youth arts engagement in a digital age, and has received a Social Innovation Grant to implement a community music curriculum.

Joseph Michael Pignato is a composer, improviser, and music education scholar. He currently serves as Associate Professor in the Music Department at the State University of New York, Oneonta, where he teaches music industry and technology courses and directs two ensembles that perform experimental music and improvised rock. His research interests include improvisation, alternative music education, and music technology. Additional details are available at joepignato.com.

S. Alex Ruthmann is Associate Professor of Music Education & Music Technology, and Director of the NYU Music Experience Design Lab (MusEDLab) at New York University. Alex and his students research and design new technologies and experiences for music making, creative learning, and engagement together with industry and community partners including the New York Philharmonic, Shanghai Symphony, Peter Gabriel, Herbie Hancock, Yungu School, Portfolio School, Guitar Mash, Tinkamo, Peer 2 Peer University, and the Rock and Roll Forever Foundation. MusEDLab projects are in active use by over 600,000 people across the world. Alex's current research focuses on the participatory design of creative learning tools and curricula in collaboration with the Program on Creativity and Innovation at NYU Shanghai, and IRCAM in Paris.

Jonathan Savage is a Reader in education at the Faculty of Education, Manchester Metropolitan University. He is Managing Director of UCan Play, a not-for-profit company (ucanplay.org.uk) that provides consultancy, research, and training as well as a point of sale for musical instruments and audio and video technologies. He runs an active blog at jsavage.org.uk and can be followed on Twitter as @jpjsavage.

Gabriel Solis is Professor of music, African-American studies, and anthropology at the University of Illinois, Urbana-Champaign. A scholar of historical ethnomusicology, he has done research in the United States, Australia, and Papua New Guinea. His work focuses on musical racialization as a component of global modernity. In addition to the books *Monk's Music: Thelonious Monk and Jazz History in the Making* (2008), *Thelonious Monk Quartet with John Coltrane* (2014), and *Musical Improvisation: Art, Education, and Society* (2009, coedited with Bruno Nettl), he is the author of articles and book chapters that have appeared in such journals as *Ethnomusicology, Popular Music and Society, Musical Quarterly, Musicultures,* and *Critical Sociology.*

Sandra Stauffer is Professor of music education at Arizona State University. Her research and writing focus on musical creating, place philosophy, and narrative. She is coauthor and editor with Margaret Barrett, University of Queensland, of *Narrative Inquiry in Music Education: Troubling Certainty* (2009) and *Narrative Soundings: An Anthology of Narrative Inquiry in Music Education* (2012). She is also an author for K–8 music texts and online music learning platforms, and she has collaborated with composer Morton Subotnick in the development of his music-creating software for children.

Evan S. Tobias is Associate Professor of music education at Arizona State University, where his research interests and teaching include creative integration of digital media and technology, curricular inquiry, issues of social justice and equity, and integrating popular culture and music in music classrooms. He leads the Consortium for Innovation and Transformation in Music Education (CITME) (citme.asu.edu) to address how music learning and teaching can impact communities and society and to explore imaginative possibilities. His work on connected learning and creative youth development through the *Sound Explorations: Creating, Expressing, Improving Communities* project was funded by a 6th Digital Media and Learning Competition grant with support from the MacArthur Foundation. He is active on Twitter as @etobias_musiced and maintains a professional blog at evantobias.net.

Rena Upitis earned her doctorate at Harvard University after completing degrees in psychology, law, music, and education. Before securing her current position at Queen's University, where she has been Full Professor since 1995, she was a Postdoctoral Fellow at MIT. She has secured over $8 million in research funding from government, foundations, and businesses. She has authored or coauthored seven books and has published over 60 peer-reviewed articles. Her most recent book was *Raising a School* (Wintergreen Studios Press, 2010). She is currently Principal Investigator for the Music Education in the Digital Age project, Transforming Music Education with Digital Tools.

Lauri Väkevä is Professor of music education at the Sibelius Academy of the University of Arts, Helsinki. Coauthor of three books, he has also published book chapters and articles in peer-reviewed journals, as well as presenting papers at international conferences in the fields of music education, musicology, music history, and popular music studies. His main research interests cover African-American music, popular music pedagogy, history of popular music, pragmatist aesthetics, philosophy of music education, informal learning, and digital music culture. Aside from his academic career, his work assignments have covered working as a musician, music journalist, general music teacher, and instrumental teacher.

Deborah VanderLinde is Associate Professor of music education and Director of the Music Program at Oakland University. She is coeditor *of Exceptional Pedagogy for Children with Exceptionalities: International Perspectives* (2015). She has also produced "Songs for You and Me," *Illustrated Songs for Learners of All Abilities* (iBooks). She teaches graduate and undergraduate educational psychology, elementary and choral methods, and graduate qualitative research. Her research interests include the application of a constructivist approach to teaching and learning in preservice and in-service music teacher education, particularly in classrooms for learners with exceptionalities.

Janice Waldron is Associate Professor of music education at the University of Windsor, teaching music education and ethnomusicology courses. Her research interests include informal music learning practices, social media and music learning, online music communities, vernacular music, and participatory cultures. She has published in *Music Education Research*, the *International Journal of Music Education*, and *Action, Criticism, and Theory in Music Education*, among others, and chapters in the *Oxford Handbook of Community Music*, *The Routledge Companion of Music, Technology, and Education*, and *CMEA Edition 2015: Music and Media Infused Lives*.

David A. Williams is Associate Professor of music education and technology and Associate Director of the School of Music at the University of South Florida. His research interests center on the enhancement of teaching and learning situations in music education using learner-centered and informal learning pedagogies.

Ruth Wright is Associate Professor in the Don Wright Faculty of Music, Western University. She has served as Chair of the Music Education Department (2009–13) and Assistant Dean of research (2013–15) at this institution, where she received her PhD in education in 2006. She views access to socially and culturally inclusive music education as a basic human right for all young people. She is a cofounder, with Dr. Betty Anne Younker and Dr. Carol Beynon, of Musical Futures Canada, an informal learning music program. Her edited book *Sociology and Music Education* was published by Ashgate in 2010.

About the Companion Website

http://www.oup.com/us/ohtme

Included on the Oxford website are abstracts for each chapter of this handbook and a link to an interactive site specially constructed by the Volume Editors. Readers and contributors are encouraged to discuss concepts presented in the handbook with one another through this online forum, and contribute their own further perspectives.

THE OXFORD HANDBOOK OF

TECHNOLOGY AND MUSIC EDUCATION

EDITORS' INTRODUCTION
Narrating the Landscape of Technology and Music Education

ROGER MANTIE AND S. ALEX RUTHMANN

THE field of research in technology and music education is maturing. The first and second *MENC Handbooks* (Colwell, 1992; Colwell & Richardson, 2002) included only discrete chapters on technology, while the *Oxford Handbook of Music Education* (McPherson & Welch, 2012) presents two related parts focusing on technology and new media. The *Journal of Music, Technology and Education* now sustains three to four issues per year, and the new *Routledge Companion to Music, Technology, and Education* (King, Himonides, & Ruthmann, 2017) canvasses the applications of music, technology, and education across the broad landscape of music performance, creation, and research. Open almost any professional academic journal issue in music education today, and you find at least one article with technology playing a central role.

The proliferation of academic scholarship on technology and music education must be set against the historical backdrop of ever-present trade publications that are often motivated, understandably, more by consumption than education. The lines between consumption and education, however, are rarely distinct. Education in the modern age almost always involves commercial components and interests. Although technology's (and commerce's) "role" in and relation to music education will undoubtedly continue to evolve as technology and education evolve, major discourses have, we argue, plateaued and reached a point of critical mass historically, a condition that suggests longer staying power for a handbook such as this one.

In order to tease out and elucidate some of the problems, interests, and issues, we conceptualized a volume that sought to critically situate technology in relation to music education from a variety of perspectives: historical, philosophical, sociocultural, pedagogical, commercial, musical, economic, policy, and so on. We solicited essays from authors based on their potential to contribute a diversity of perspectives on technology and music education in terms of gender, theoretical perspective, geographical distribution, and relationship to the field. The overall thrust was to provide a place to stimulate and present contrasting perspectives and conversational voices rather than reinforce traditional narratives and prevailing discourses.

The handbook is organized into four parts. Each part is separated into two sets of Provocation Questions, resulting in eight subparts. We solicited twenty-two authors (two to three per subpart) to write a Core Perspective of approximately 5,000 words. These Core Perspective authors were then asked to write a 2,500-word Further Perspective for another subpart in the handbook. We selected these authors to maximize a diversity of critical perspectives. Core Perspective authors consist of ten females and twelve males across six continents that include the countries of Australia, Canada, China, Mexico, Singapore, Uganda, United Kingdom, and the United States. Selected authors include school and community music practitioners, industry members, higher education researchers, and teacher educators, embracing theoretical frames that include philosophy, history, sound studies, ethnomusicology, social and cultural psychology, and critical theory.

To underscore the perspectival nature of the handbook, we then solicited another nineteen authors to provide additional Further Perspectives to various subparts. This collection of perspectives is, by nature, not comprehensive and necessarily incomplete. All interested readers, scholars and students of technology and music education are encouraged to engage and contribute their own perspectives on the companion website[1] sharing their own critical perspectives, continuing the conversation online.

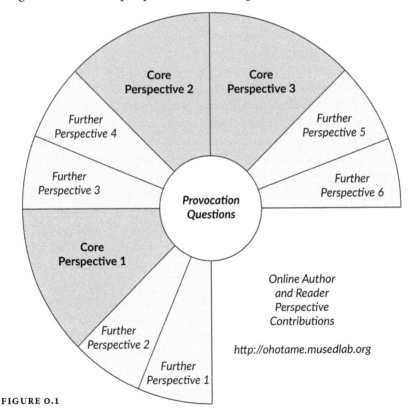

FIGURE 0.1

The chapters in this volume help vivify many of the debates involving technology and music education. Consistent with the general framing of the book, the Core and Further Perspectives highlight, we think, that there can be no ultimate or last word on matters of technology and music education. By engaging the perspectives of a diverse, international author pool of academics, K-12 practitioners, and industry leaders, one begins to develop an appreciation for the complexity of various problems and issues. The authors were tasked with providing situated, personalized perspectives, not authoritative summaries, a challenge intended to highlight the perspectival and local nature of educational matters, and our social world generally. Taken together, the chapters in this volume hopefully remind us that our relationship with technology and music education in the twenty-first century is far messier than simple narratives of technology-as-savior or naysaying technological determinism.

Narrating the Landscape

The authors in part 1 were asked to address the orienting question *Where are we now and how did we get here?* Part 1A's authors considered (1) *What constitutes a technology in music and music education?* and (2) *In what ways has technology been used, and how has technology affected music education in different times and places?* Marina Gall and Roger Mantie take very different approaches to these questions. Gall answers more directly, referencing her lifetime of experiences in the United Kingdom as a child, secondary school teacher, teacher educator, and researcher of music technology in education, and describing how the long list of technological innovations that have occurred over the past 50–60 years have impacted on education in the United Kingdom and in Europe. Mantie, by contrast, addresses the questions more obliquely, exploring the broader context of technology in our lives through a consideration of Martin Heidegger's famous 1954 essay "The Question Concerning Technology." By way of Heidegger, Mantie asks how concepts such as techné might help us to better understand our ourselves as musical agents and how this might inform our practices as music educators.

Further Perspective authors Rena Upitis, Gabriel Solis, and Emily Akuno (whose chapters appear in subpart 2B) serve up three very different takes on the question of how technology has affected our music learning and teaching at various points and in various places. Writing from a musicological perspective, Solis' views resonate well when placed against the two Core Perspectives. Is technology simply an object of our attention, or has it in fact tacitly constructed our very understandings of what it means to be musical? Similarly, the chapters by Upitis and Akuno serve as a wake-up call for those drunk on the elixir of technology. Upitis cautions us, through the metaphor of "slow food" that faster is not always better, and Akuno (like Benon Kigozi in subpart 2A) reminds us that there is more to the world that just the West—and that Western assumptions about technology may not sufficiently account for many realities around the globe. Additional Further Perspectives are provided by Chee-Hoo Lum, Janice Waldron, and David A. Williams.

Lum, without knowledge of Upitis's chapter, coincidentally employs a similar metaphor: savoring. Lum, like Upitis, encourages us to not let the technology prevent us from enjoying the artistic and the patient. Waldron, not unlike Akuno, draws attention to the importance of place, while Williams draws our attention to time: *then* and *now*.

In subpart 1B, Smaragda Chrysostomou and Samuel Leong consider (1) *How have music educators negotiated the role of technology within the broader terrain of educational policy and practice?* and (2) *What is the role and what are the effects of commerce and industry on music learning, teaching, and technology within schools?* Chrysostomou, writing from a Greek context, problematizes, with reference to international test scores, the "digital divide." Her chapter complements those by Leong and Kigozi, each of whom, in his own way, draws attention to global, neoliberal discourses that push for information and communication technology (ICT) and its associated "technocentric rhetoric" where all tech is good tech. As Kigozi rightly points out: no matter what the technology involved, the most critical element in any learning-teaching encounter is not technology or government policy but the individual teacher.

John-Morgan Bush's Further Perspective echoes the somewhat cautionary (even dystopian) tone of the Core Perspectives. Bush points out, among other things, that ICT can, if one is not careful, overemphasize individual learning to the detriment of learning communities. Joe Berkovitz and Jason Chen on the other hand offer more optimistic, hopeful outlooks in their Further Perspectives. Chen points to the possibilities of mobile technologies, and how these have created opportunities not possible in traditional paradigms of technology as student sitting in front of a computer, while Berkovitz provides a commercial software developer's viewpoint on the creativity that lies behind the invention of various hardwares and softwares. Although he does not reference the French philosopher Gilles Deleuze directly, Berkovitz makes arguments that bear a striking resemblance to Deleuze's thoughts on the productiveness of capitalism. As Berkovitz reminds us, inventors are not always motivated by profit, and capitalism is not always about the evil ogre who devours the weak. People invent for creative purposes and to fill perceived needs. There is a play element associated with invention that is often overlooked when technology is reduced to just programming or "code." Additional Further Perspectives in subpart 1B are provided by Heidi Partti, Joseph Pignato, and Jonathan Savage. Partti focuses on education and teacher training. While Pignato and Savage do not address this directly, each tackles similar issues: Pignato, drawing on Aristotle's "that for the sake of which," emphasizes the ends or goals; Savage echoes the importance of this focus on directionality in education by quoting from *Alice in Wonderland*: if you don't care (or know) where you're going, "it doesn't matter which way you go."

Part 2 is guided by the primary questions *Where is technology within music education? Who is affected and how?* In their Core Perspectives, subpart 2A authors Joseph Pignato, Kylie Peppler, and Benon Kigozi consider (1) *What are the impacts of technology (positive and negative) on different communities (rural, urban, suburban), different socioeconomic areas, and different parts of the world?* and (2) *What can be done to mitigate the negative effects of technology while accentuating the positive?* All three authors draw attention to issues often overlooked (or perhaps dismissed) by the mainstream music education establishment. Playing on the archaic sense of "without" (i.e., outside of or external to), Pignato draws attention to the fact that technology (and music technology) is a part of the social world, irrespective of music education. Too often, we in music education fail to create what Pignato calls, after Jean-Jacques Nattiez, authentic *poietic* spaces. Approaching the issue from a slightly different angle, Peppler discusses research into people's engagement with music through the video game Rock Band. Why, Peppler asks, is learning music through Rock Band not as accepted (or acceptable) as learning through conventional means? Kigozi, writing from Uganda, takes an entirely different approach to the question prompts, emphasizing that, for students and their teachers in many countries in Africa, "software and hardware need to be as reliable and as easy to use as mbira, ngoma, hosho, endeere or a Compact Disk player so that precious time is not lost due to technology glitches."

The five Further Perspectives for subpart 2A share a critical edge regarding optimism over wholesale technological embrace. Perhaps most critical is Gillian Howell, who, as a community musician who often works in impoverished settings, has little desire for things that usually need to be plugged in. Smaragda Chrysostomu and Ethan Hein are likewise critical, though in ways different from Howell. Hein points out the costs and distractions of technology, while Chrysostomu highlights how technology often exacerbates socioeconomic inequalities. Bo Nilsson and Ailbhe Kenny, critical but perhaps more optimistic in outlook, point to the importance of community. Nilsson draws on Christopher Small's concept of *musicking*, emphasizing the play element and social nature of music making, while Kenny elaborates on the concept of *communities of practice*, emphasizing the potential of online technologies to create and sustain communities not previously possible.

Guiding subpart 2B were the questions *How are technology and music technology changing us? How are we changing music technology?* and *How are music educators responding to social, cultural, and economic issues? How should they?* Heidi Partti, Evan Tobias, and Valerie Peters provide Core Perspectives that draw attention to our conceptual vocabulary. Partii focuses on "pedagogical fundamentalism" and "pedagogical populism," Tobias, drawing on the work of N. Katherine Hayles, invokes *technogenesis*, and Peters draws attention to the French term, *passeur culturel*. Taken together, the Core Perspectives raise challenging questions about how teachers wittingly or unwittingly influence what it means to be musical and what it means to be literate.

The subpart 2B Further Perspectives of Carlos Chirinos-Espin, Patricia A. González-Moreno, Gina Greher, Leah Kardos, Roger Mantie, and Ruth Wright illustrate, as well as any subpart in the book, the value of diverse (and necessarily incomplete perspectives). González-Moreno, Mantie, and Wright respond to the Core Perspectives most conventionally. González-Moreno interrogates the implications of the Core Perspectives, such as determinism, affordance and constraint, and so on, for music teachers education; Mantie leverages the theory of linguistic relativity to argue that "real" change in practices will only occur when we generate new ways of thinking and talking about what we do in music education; Wright, springboarding off the work of Basil Bernstein and Jacques Rancière, highlights issues of inequality in today's "totally technologized society" and our "musical technotopia" in music education. The Further Perspectives of Kardos, Chirinos-Espin, and Greher serve as a reminder that technology and music education is not just about conventional K-12 music learning and teaching. Kardos brings her experiences as a composer and teacher to bear on discussions of what music is and can be. Sharing his experiences coproducing the song, "Africa Stop Ebola" — an effort highlighting issues of public health and music — Chirinos-Espin reminds those of us in the West that technology isn't just about gadgets and gear, but about scale and impact related to serious crises in the world. Finally, Greher demonstates the power a technologically-mediated composition project involving students with behavioral, cognitive and physical challenges can have on preservice music educators.

Part 3 shifts from broader historical, social, and cultural issues to the direct pedagogical relationship. This part's guiding questions were *How has technology impacted the learning and teaching of music? How should (or shouldn't it)?* Core Perspective authors Barbara Freedman, Ethan Hein, and Chee-Hoo Lum focused on the following subpart 3A questions: *What are the ramifications of technology and technological change on music teaching and learning in the classroom? What can technology do for music education? In what ways has technology forced us to reevaluate definitions of musicality? Of musicianship? Of who is and is not a musician? In what ways has technology transformed our understandings of creativity? What are some of the untapped potentials in this area?* Following in the tradition of technology-as-teaching assistant, Freedman encourages us to take advantage of the efficiencies of technology in helping students become more fluent with the materials of musical sounds through the keyboard, thereby opening doors to traditional forms of symbolically based composition and creativity. Hein on the other hand emphasizes the ever-increasing possibilities for sonic manipulation through the digital audio workstation (DAW). Production, mixing, and remixing thus present ways of being musical that were unimaginable (or unattainable) prior to the advent of today's low-cost, high-power music technologies. When set against these two Core Perspectives, Lum's chapter presents an interesting commentary. Lum problematizes learner agency and "the power of institutionalized musicality," asking us to consider how technology has altered the traditional asymmetrical power differential between music teacher and learner.

The problems of learner musical agency and understandings of what it means to be musical are explored in the Further Perspectives. Donald DeVito, Jay Dorfman, James Humberstone, Samuel Leong, Evan Tobias, and Deborah VanderLinde help us to better understand and appreciate how technology is so often associated with the notion of "breakdown." Both Humberstone and Dorfman speak to how traditional roles and understandings are open to question. Humberstone, for example, points out how, historically, music technology as music education technology has been associated more with composition (per Freedman) than with performance, and implies, like Dorfman, that the classical training of most music educators might help explain their biases toward technology in/as performance. Dorfman encourages us to question our own evolving roles and understandings as music teachers. Tobias furthers the charge, using descriptive vignettes to illustrate how technology can "augment" our conceptions and understandings of what it means to be musical. Leong, taking a slightly different approach, invokes the agency implied by the term "prosumer," emphasizing the liberation that technology affords. Nowhere is this liberation felt more impactfully, perhaps, than in contexts that involve learners with special needs or "exceptionalities." VanderLinde provides several examples of how mobile technologies have the capacity to build more inclusive environments, while DeVito helps us understand how the skillful and imaginative use of technology with people with "significant disabilities" can lead to breakdowns in traditional definitions of who is and is not a musician.

Part 3B's authors considered a related set of questions: *What are examples of effective uses of technology? Under what conditions might technology be inappropriate or ineffective? What are familiar challenges to implementation and what strategies have thus far proven effective? Are there limits to what technology affords?* Gillian Howell, Michael Medvinsky, and Janice Waldron provide three diverse Core Perspectives in response to these questions. Waldron, drawing on the work of Sherry Turkle, problematizes how technology has changed the way we think, both in music and beyond, and how technology has complicated our sense of "presence." This is especially true when we think of how traditional social networks in musical communities of practice have changed with the rise of online music communities, creating, in some cases, divisions between so-called digital natives, for whom such participation comes relatively easily, and those for whom the learning curve is sometimes too much to overcome. For Medvinsky, the pedagogical possibilities of various technologies to circumvent and transcend traditional roadblocks and barriers in music points to its liberating potential. Through the integration and scaffolding of various technologies, new ways of being musical can be facilitated. Community musician Howell provides a contrarian view, highlighting how, based on her experiences in East Timor (Timor-Leste), technology more often than not gets in the way of the musical experience. Technology is in many cases expensive and prone to crashing, delays, and so on, hindering music making rather than facilitating it. Howell reminds us that technology involves thorny issues of sustainability that are too often glossed over in celebratory discourses, and that inevitably "technology begets technology." Her descriptions of the direct experience afforded by making acoustic instruments out of local and found materials offer a compelling case for those advocating "constructivism" in learning.

The Further Perspectives of Evangelos Himonides, Ryan Bledsoe, Kylie Peppler, and Sandra Stauffer build upon many of the themes presented in the Core Perspectives. Stauffer's, for example, is intriguing when read in light of Howell's Core Perspective. Reflecting on place and sound through attention to "tuning of place" and R. Murray Schafer's groundbreaking work on soundscapes helps to remind us that "knowing" in music demands an ongoing criticality. When set against the Core Perspectives of Waldron and Howell, Peppler's Further Perspective raises issues of equity, access, and inclusion, especially with respect to youth culture, that may be ignored or overlooked by traditional school music offerings (in this case those of the United States). Peppler, Waldron, and Howell all remind us that there is a big world of music learning that occurs outside the bounds of formal schooling. Himonides and Bledsoe on the other hand remind us that resistance to technology in the music classroom may be the result of teacher confidence, or even more dramatically, what Himonides calls "narcissism" on the part of musicians and teachers. The attitudes of conservatory-trained musicians and teachers may be preventing the kinds of open exploration and play described by Bledsoe—the kinds that focus on creativity and engagement without lines between acoustic and digital.

The broad overview questions in part 4 ask *How has the profession dealt with challenges to training and certifying music teachers in the area of technology? How could it? How should it?* Jay Dorfman, Jonathan Savage, and Gena Greher provide Core Perspectives in subpart 4A that look at the questions *Should music technology be taught as an independent subject, or should it serve existing curriculum and instruction? If the latter, How might technology best serve the needs of students and teachers? How can music educators develop and maintain skills beyond their teacher education programs? What sorts of canonized practices have emerged, and how might these impact music teacher education?*

Dorfman and Greher, both situated in the United States, focus primarily on aspects of music teacher training and certification. Dorfman, for example, frames the discussion in terms of the responsibility of music teacher educators to develop competencies among pre- and in-service teachers. He challenges the myth of digital natives (or "immigrants") and suggests that many preservice teachers lack the skills needed to include and incorporate technology in their teaching. Drawing on Lee Shulman's model of Pedagogical Content Knowledge, Dorfman draws attention to a distributed practice model centered on technological pedagogical and content knowledge (TPACK). Greher, bringing an industry perspective to her current role as a music teacher educator, emphasizes instead the need for university partnerships, media literacy, and an awareness of Seymour Papert's distinction between constructionism and instructionism. Greher cleverly contrasts Papert's argument that technology should ideally become invisible in the teaching-learning process with policy practices evident in the National Core Arts Standards in the United States and the accreditation practices of the National Association of Schools of Music, both of which evince a kind of invisibility that belies a lack of critical awareness. In his chapter, Savage presents an entirely different approach to the subpart 4A question prompts, focusing not on teachers but on national policy and practices in the United Kingdom that cleave a difference between music and technology by treating them as independent entities. To make his case, Savage interrogates Andrew Hugill's "digital musician," arguing that such a notion is fundamentally flawed. He also points out that the independent study of music technology has resulted in significant gender imbalances.

Further Perspective authors Marina Gall, Julie Ballantyne, Benon Kigozi, Lauri Väkevä, and Michael Medvinsky represent the most geographically dispersed group of authors in the book. As much as any part of the book, this one serves to illustrate the value that diverse perspectives bring to conversations about technology and music education. Ballantyne, for example, describes the Mobile Technologies Project, which was designed to help preservice students better cope with the workplace challenges of becoming a teacher. Gall on the other hand builds on Dorfman's teacher training concerns, drawing attention to how the 3+1 teacher training model in the United Kingdom creates challenges when trying to include some version of TPACK within the span of one year. On the other hand she points out that the teaching of technology in schools is more common in UK schools (and she echoes Savage's concerns about gendered implications), hence providing slightly greater motivation for teachers to develop technological competencies. Medvinsky, too, largely echoes many of Dorfman's concerns about the need to better prepare music teachers, but preciously adds: "if we use technology to do old things in new ways, we are still doing old things." This sentiment can be seen also in Väkevä's contribution. Writing from and about Finland, an early adopter country in the world of ICT (e.g., Nokia), Väkevä points out that, despite the prevalence of technology in Finland, digital technologies tend to be viewed in more of an additive or supportive role than "something worthy of being studied as a creative, expressive, artistic, and aesthetic field of its own," where social contexts might bring about new meanings. Finally, Kigozi reminds those of us writing primarily from westernized perspectives that technology is perceived very differently in much of the African continent. Offering perspectives on several African countries, Kigozi suggests there are sensitivities that surround African cultural practices, and that some teachers believe technology may potentially result in a kind of "deculturing."

The authors in subpart 4B responded to the following question prompts: *What are the ramifications of technology and technological change on music teacher education programs? What possible tensions exist in terms of credentialing and accreditation? Who should bear responsibility for professional development and certification?* Core Perspective authors Patricia A. González-Moreno, Evangelos Himonides, and David A. Williams offer distinct approaches to these questions. González-Moreno, for example, raises the issue—not found elsewhere in this handbook—of faculty professional development. Whereas most contributors to this volume focus on technology as it relates to K–12 students and their music learning, preservice preparation, or inservice professional development, González-Moreno suggests that a primary problem standing in the way of more fruitful engagement with and through technology is resistance by university faculty. Williams presents a variation on this theme, but points the finger at K–12 music practices and the entrenched values of American school music, which has been and continues to be dominated by a large ensemble model that emphasizes size, diminished musician autonomy, the privileging of ears over ears due to overreliance on notation, and a concert culture that privileges product over process and situates the teacher at the center of the musical encounter. Provocatively, Williams proposes new curricular and instructional models to ameliorate the damaging effects of "tradition." Himonides echoes some of the competency and certification themes found in subpart 4A, but also asks if a one-size-fits-all approach is appropriate when it comes to professional development and the potential needs of students.

Rounding out the handbook are the Further Perspectives of Barbara Freedman, Matthew Hitchcock, and Valerie Peters. Freedman begins by addressing the challenge of tradition and change posed by the Core Perspectives authors by suggesting that staying current is only a challenge for music educators (and music teacher educators) if they conceptualize competence in terms of training for a specific device or piece of software. Instead, she proposes that the teacher need only be an expert facilitator, granting students license to become experts in a community of learners. Freedman then shifts gears, however, proposing that, in the United States at least, music technology classes hold the potential to attract economically disadvantaged and visible minority students typically not found in traditional school music ensembles. Hitchcock, writing from an Australian perspective, suggests that technology and teaching have not always been "easy bedfellows," although he acknowledges that this is beginning to change, albeit slowly, as old habits apparently die hard. Notably, Hitchcock categorizes technologies as educational, administrative, social, and musical. While all four impact the lives of music teachers, he concentrates in his contribution on the ways in which technologies continue to impact on musical practices. Cautioning music educators about the ubiquity of copy-and-paste music making, he encourages us to emphasize "musically generative" ways that technology can be used—hopefully to counteract what he describes as the "catch-22" of "perpetual traditionalism." In her Further Perspective, Peters responds to the charges of all three Core Perspective authors, suggesting that we be patient as we accept the time necessary to respond to change. Differences and disparities will continue to exist, she reminds us, but by engaging with creative collaborative models, those in underresourced areas with a desire to effect change can hopefully create partnerships to compensate.

In summary, the perspectives presented in this handbook serve as the incomplete beginning to a broader and deeper conversation we hope our readers will continue to have, and build upon. We especially encourage the reader to consider the provocation questions presented throughout this handbook, and contribute their own perspectives at our forum at http://ohtme.musedlab.org/.[1] The issues, challenges, and possibilities presented to the field of music education and those who engage with it through use of technology are rich and complex. We hope this handbook provides a forum for the expression of diverse perspectives from around the world and continued dialogue.

Note

1. http://ohtme.musedlab.org/

References

Colwell, R. (Ed.). (1992). *Handbook of research on music teaching and learning.* Reston, VA: Music Educators National Conference.

Colwell, R., & Richardson, C. (Eds.). (2002). *The new handbook of research on music teaching and learning.* New York: Oxford University Press.

King, A., Himonides, E., & Ruthmann, S. A. (Eds.). (2017). *Routledge companion to music, technology, and education.* New York: Routledge.

McPherson, G., & Welch, G. (Eds.). (2012). *Oxford handbook of music education* (Vols. 1 & 2). New York: Oxford University Press.

PART 1

EMERGENCE AND EVOLUTION

1 A

Core Perspectives

CHAPTER 1

THINKING ABOUT MUSIC AND TECHNOLOGY

ROGER MANTIE

> [The famous cornet soloist Jules] Levy returned to the charge, and played his cornet fiercely upon the much indented strip [of Edison's cylinder]. But the phonograph was equal to any attempts to take unfair advantage of it. . . . [I]ts victory was unanimously conceded, and amid hilarious crowing from the triumphant cylinder the cornet was ignominiously shut up in its box.
> —"The Phonograph Wins a Victory," *Scientific American* (1878)

> There is something peculiarly weird and attractive in the idea of calling up at will some old friend or famous character and listening to an exact reproduction of his or her voice. . . . A favorite phase of the "phonogram craze" is the collection of specimens. . . . [T]he owner can give a six-hours' entertainment in his own house at any time, presenting the different artists, whose voice he has "bottled up," so to speak.
> —"Before the Phonograph," *New York Times* (1890)

MY experience with and interest in technology derives from my years teaching music in K–12 environments.[1] I still recall my initial excitement and awe in witnessing a demonstration of Hypercard by Peter Webster, who had traveled to the northern climes of Winnipeg, Canada, at some point in the late 1980s to share his expertise with an enthusiastic group of young music teachers at Minnetonka Junior High. Oh, the possibilities! From that point on I attempted to navigate the often choppy waters of technology and the music classroom. Although always intrigued by what classroom music teachers might do with technology, I have never professed to be a "power user" and, in truth, have always maintained a healthy skepticism toward technology.

My own career as a school music teacher tracks the rise of digital technologies flowing out of the introduction of MIDI and the proliferation of home computers. Technology in the music classroom in the 1990s was, for me, marked by a number of concerns: it was too expensive, it went out of date too quickly, the learning curve was often too steep for students, available spaces were too often inappropriate or inadequate, and the educational benefits were, to me, often suspect. More important, I began to question how technology was altering our most basic understandings of what it is to be a musical being in the world. Despite these misgivings, I persevered, trying to slip in CAI, notation software, sequencing hardware and software, play-along software, and any other aspects of music technology I thought might somehow help my students become more musical.

Despite having had a hand in crafting the Provocation Questions for this book, the complexity of adequately addressing the concerns of this part of the handbook only hit home while attempting to write this chapter, for one can hardly go about the task of examining technology in music and music education without first clarifying what is and is not to be considered technology.[2] Mediated music, for example, is so ubiquitous that many readers may find the opening epigraphs merely quaint rather than dire commentary. Rather than duplicate existing work on the history of technology in music and music education or the work of other authors in this volume, I have pursued a different direction, starting with a consideration of the Greek root of "technology," *techné*, which refers, in loose terms, to some combination of craft and art. This should resonate strongly with music educators, given that both the *art* of music making and the *art* of teaching require an integration of developed skill and knowledge that, at least ideally, cannot be reduced to mere formula, routine, or "training." I have also taken a cue, somewhat loosely, from the French philosopher Michel Foucault, who articulated what he described as four human "technologies": production, sign systems, power, and "the self" (Foucault, Martin, Gutman, & Hutton, 1988, p. 18). The first of these (production) is what we commonly think of when we hear the word "technology." By describing other technologies, however, Foucault reminds us of the broader meaning that lies behind the word. While I do not use Foucault in this chapter, I do take as a starting point that technology invariably impacts how we understand ourselves both as human beings and as musical beings.

I begin with German philosopher Martin Heidegger's landmark discussion of how our relationship with technology is a "revealing" of the human condition. Next, I briefly examine how technology has impacted our understandings of what it means to be human, and, by extension, how any technologically influenced change to music making practices holds the potential to alter our self-understanding because music, historically, has been so inextricably linked to humanness. I then link this idea to how technology has altered our understandings of what it means to be "musically educated." I conclude by considering a few ways in which we might begin to think about technology's potential impact on our music teaching and learning practices, with a reminder that the classroom is perhaps the ultimate technology at our disposal.

Considering Technology

One of the definitive sources for thinking about technology is Heidegger's essay "The Question Concerning Technology." The scholarly literature (not to mention the Web at large) is filled with references to and discussions of Heidegger's very provocative, if somewhat opaque, critique. For readers less familiar with the essay, its basic message is that we need to think more carefully about technology. Heidegger writes: "We shall never experience our relationship to the essence of technology so long as we merely conceive and push forward the technological, put up with it, or evade it. Everywhere we remain unfree and chained to technology, whether we passionately affirm or deny it" (1977, p. 4). Instead of focusing on the applied issues of technology use, Heidegger argues that the *instrumental* (i.e., means to an end) and *anthropological* (i.e., human activity) views of technology miss the critical point: in order to have a "free relationship to technology" we need to attend to the kinds of thinking that lie behind our relationship with it.

Many of the specifics of Heidegger's essay are likely of interest primarily to Heideggerians. Even among them, the reception is far from unanimous. Waddington (2005), for example, concludes: "Heidegger's theory of technology contains so many implausible notions that it is very difficult to defend it on a correspondence basis" (p. 578).[3] Despite various criticisms, however, Heidegger's essay provides considerable food for thought, not just for a general audience but for musicians and music educators. Rather than merely accepting the reality of technology, Heidegger argues that we need to carefully consider what it and its use reveals about us. David Lines (2013) points out that, in music education, we "have traditionally examined music technology from an instrumental-functional perspective and have not questioned it to any great degree or established creative, musical and culturally adept ways of moving forward. . . . [The] lack of criticality has become even more disabling for the profession and for music students" (p. 10). Many of Heidegger's ideas, Lines suggests, hold great potential for helping us think more carefully about what we do in music education.

Being a classical philosopher, Heidegger emphasizes many of the ancient Greek origins of the concept of technology in his essay, such as *techné* and *poiësis*. For musicians and music educators, the concepts that lie behind these esoteric-sounding Greek words are highly salient.[4] Indeed, part of Heidegger's overall argument relates to the connection between art and technology. For the ancient Greeks, *techné* referred to both the craft of making and to the artistic techniques needed as part of *poiësis*, or "bringing forth." Because the particular skill involved in bringing forth involves artistic decision-making, *techné* is in fact a way of knowing (*episteme*, "know-how"): "what is decisive in techné does not lie at all in making and manipulating nor in the using of means, but rather in the . . .

revealing" (Heidegger, 1977, p. 13). That is, what is important about techné lies not in the production but in the "truth" that is revealed in and through the process: "bringing-forth brings hither out of concealment forth into unconcealment" (p. 11). This "artistic" precept of the ancients, claims Heidegger, is no longer the dominant mode of thinking. Instead, modern technology is about mastery and what he calls "standing reserve," a phenomenon driven by a thirst for exploitation, accumulation, and profit.

Lines summarizes the potential implications of these two contrasting technological relationships for music educators, writing: "our learning through technology can be gentle or unobstructed in the techné sense of learning, in a way that supports the learner to use skills and techniques to bring about a new kind of learning experience. On the other hand technological learning can be a kind of non-learning, a 'learning' that is manipulated and imposed by dominant patterns of cultural manipulation" (2013, p. 2). When emphasis is placed on artistic decision-making it is not a simple matter of technology being subjugated as "a tool" in service of more noble goals. Rather, it is that technology is used thoughtfully and purposefully and, importantly, freely. When we are compelled to use technology, it ends up using us more than we use it. When we use technology because we feel we must because it is simply a "fact" of contemporary life or because we believe students will somehow be left behind if they fail to use it, we engage inauthentically. Phrasing it somewhat more nihilistically, Jacques Attali summarizes how music, commoditized via technology, reflects what Heidegger calls "standing reserve": "fetishized as a commodity, music is illustrative of the evolution of our entire society: deritualize a social form, repress an activity of the body, specialize its practice, sell it as a spectacle, generalize its consumption, then see to it that it is stockpiled until it loses its meaning" (1985, p. 5).

HUMAN SUBJECTIVITY AND TECHNOLOGY

Concerns about human subjectivity—our individual and collective experience of being in the world—have become more acute over the course of the twentieth century and into the twenty-first in response to the perceived impact of technology on our "human-ness."[5] Heidegger's concerns about our relationship with technology are erudite and eloquent but hardly represent a unique perspective. Ominous prognostications about the dangers of technology have been proffered practically since the time of the ancient Greeks (and likely before). The advent of the Industrial Revolution, in particular, helped to usher in a set of new and growing fears related to *technological determinism*. Fritz Lang's *Metropolis* and Charlie Chaplin's *Modern Times*, for instance, offer early examples of biting, dystopian critiques of how industrial technologies threaten humanity. Such fears have been fueled in more recent times by science fiction's portrayal of such things as cyborgs and androids that help to raise doubts about what truly separates us from machines and what separates reality from virtuality. Are we becoming less human thanks to technology's integration (or for some, intrusion) into our daily lives?

Rooting their critiques in Marx's concept of alienation and propelled by such mass society (culture, media) critics as those of the Frankfurt School (e.g., Adorno, Benjamin, Horkheimer, Marcuse), many maintain that mass media and technology represent a dehumanizing force in society. Media theorist Marshall McLuhan is probably the best known of the critics of technology's impact on who and what we are becoming, but many since have heeded his clarion call. Following on the heels of McLuhan, Alvin Toffler's *Future Shock* (1970) warned of the social problems brought on by the disorientation accompanying rapid technological change ("information overload," as he put it). Since that time there have been many theorizations and discussions related to technology and what some call "posthumanism"; in some cases, technology is actually said to enfeeble us![6] Others have written about such things as "the technological personality" (Stivers, 2004) and the "techno-human condition" (Allenby & Sarewitz, 2011). Echoing Heidegger, MIT professor Sherry Turkle argues that we need to put technology in its place by thinking about "what really matters" (2011, p. 295).[7]

Concerns about human subjectivity are relevant to music educators because music, historically at least, has been considered so central to what it means to be human. If this is so, then any "technology" that alters what music means also alters how we understand ourselves as human beings. Generally speaking, anything beyond engaging with musical sounds in the "pure" sense of the natural voice or standard acoustic string, wind, and percussion instruments has been considered by many to be a less "human" form of music making. A 1989 advertisement, for example, proclaims: "Aphex Humanizes MIDI" (Theberge, 1997, p. 123). It was asked at the "First National Assembly on Music and Technology" at Ball State University: "will human elements be lost or covered up by technology?" (Vincent & Karjala, 1985, p. 2). What is striking to consider, however, is that, as is evident in the epigraphs at the beginning of this chapter, concerns about "human elements" and their relationship to music making have a very long history.

Famously, the "canned" versus live music debates of the early twentieth century brought into sharper relief some of the ambiguities surrounding the question of what the musical experience really is. From Sousa's often-quoted essay "The Menace of Mechanical Music" to Walter Benjamin's famous essay "The Work of Art in the Age of Mechanical Reproduction," notable thinkers and personalities of the early twentieth century weighed in on the pros and cons of the then new phenomenon of mediated music. Although deserving of a much deeper discussion, suffice it to say that, among some (then and now), mediated musical experience is somehow inauthentic or less real than direct experience, harkening back to Plato's distinction between the real and the simulacrum where the copy (or reproduction) is always understood as an inferior version of the original. Closer to our own time we continue to hear similar refrains: "we've lost the ability to distinguish between essence and appearance, between live and Memorex. There is now really only music mediated by technology, only music a far cry from its reality," writes J. M. van der Laan (2007, p. 141). Significantly, however, the "live or Memorex" refrain implies not only regret over the potential loss of "the real," but the fear of losing the ability to distinguish between the real and the copy. The human ability to discern the original from the imitator or forgery has arguably been a defining feature of the Western mind and its quest for certainty. Technological change thus challenges not only many of the material practices surrounding the musical experience, but many of the metaphysical foundations upon which Western society has constructed notions of truth, reality, perception, and value.

Closely connected to the idea of original and copy is that "captured sound" (Katz, 2004) reduces the specialness and uniqueness of what had previously been music's distinctive ephemeral nature. Piano rolls and recordings, for example, brought forth the "miracle of the repeatable experience" (Roell, 1989, p. 46), challenging the dictum of Heraclitus that one cannot step twice into the same river. Through recordings one can relive a musical experience over and over again (at least to a point). While music, like other sensory experience, possesses a remarkable capacity for "indexing" our experience of the world, the mediated version of re-creation is always but a shadow of the original. Or at least this is the argument of those who proclaim the superiority of live performance as the more human experience. As Dewey points out, however, "one mode of experience is as real as any other" (1929, p. 219). Hence, while captured sound has undeniably changed the nature of the our relationship with and to music, and one can acknowledge that live performance is a different kind of musical experience, live sound cannot be proven to be superior to captured sound. What captured sound, or even studio-produced sound, does is bring into sharper relief the question of just what distinguishes the live musical experience. Is it simply hearing music being performed, or is "musicking" a more multifaceted phenomenon, as Christopher Small suggests?

Many aspects of commodification—treating music, especially recordings, as objects—are hugely important culturally but fall outside the direct concerns of this chapter. The relevance of commodification vis-à-vis music teaching and learning is that the emergence of recordings and radio brought people—regardless of economic means—into contact with more music from more places than at any previous time in human history, thus exacerbating the issue of just what was meant by the concept of being "musically educated." Writing about radio in 1935, the American music educator Peter Dykema summarized the issue well:

> [T]hose who a quarter of a century ago satisfied their love of music with the simple forms which could be produced in the home, the modest offerings of the churches, and the occasional concerts which were available in the larger centers of population, now have the resources not only of our own country, but, in an increasing degree, of the world, at their beck and call. Neither travel nor riches at the opening of this century could command what today the turn of a dial brings to humble but discriminating music lovers. ("Music as Presented by the Radio," in Taylor, Katz, & Grajeda, 2012, p. 361)

In other words, if the goal of music learning is only to be brought into contact with "good" music, then direct experience with making music is unnecessary; all one needs is to listen to the musical efforts of experts, who, presumably, only record *good* music, in order to *know* music. As famed American music educator Frances Elliott Clark, an early proponent of using recordings in schools for educational purposes, put it in 1920, "if America is ever to become a great nation musically . . . it must come through educating everybody to know and love good music" (Katz, 2004, p. 50).

Recordings and radio, then, did not just alter perceptions of the musical experience, they brought into question both what it meant to be a musically educated human being and what the goal of musical training should be by ushering in the then new concept of *music appreciation*. Whereas formerly people had no choice but to make music themselves or, when access permitted, attend live concerts, the rapid expansion of mediated musical engagement meant that music—*good* music—could be omnipresent. This, however, introduced a major problem that continues to haunt music teaching and learning: who gets to determine what people will hear, and on what bases shall music be considered "good"? On the one hand both recordings and radio were viewed as an expression of freedom and democracy based on the principle of choice and "giving the people what they want" (Theberge, 1997, p. 72).[8] On the other hand those who controlled and influenced the means of music commodification ultimately determined the options from which people could choose.[9] When a member of the American Federal Radio Commission said in 1928 "let us use radio to create a vast new army of music lovers in America" (Biocca, 1990, p. 6), he was presumably thinking of Walter Damrosch, not Louis Armstrong or Duke Ellington. The dangers of the "standardization of aesthetic experience" trumpeted by Benjamin, Adorno, and the Frankfurt School were apparently only of concern when applied to the lowbrow tastes of mass culture.[10]

As a result of commodification, the aims and purposes of music education took on new importance and meaning. People needed to be "educated" into making the right musical choices. Although the result of audio recordings and radio was that almost everyone had access to high art, there was no guarantee that people would choose it over lowbrow options. Thus, while the "democratization" of music was heralded as empowering by some, it exacerbated issues of manipulation. Technology made possible a phenomenon that continues to this day in music education: the widespread inculcation of taste by a minority under the banner of *education*. To be a musical person was no longer defined strictly according to one's ability to make music; being musical was about *knowing* music—especially knowing how to distinguish good music from bad. Viewed positively, the kinds of elitist "conspicuous consumption" practices noted by Thorstein Veblen that once served as forms of what Pierre Bourdieu labeled "distinction" were muted by the new-found abilities of the masses to enjoy the same musical performances as the cultural elite. The lowliest manual worker in the remotest town in the 1930s and 1940s might not be able to attend *La Traviata* live, but she or he could enjoy a mediated form of the music (stripped of much context, of course) just as much. Technology thus helped to promote egalitarian, democratic sentiments that rationalized high art taste as appropriate for universal education.

Interestingly—and for many feminists indicative of omnipresent patriarchal bias in society—in the 1920s and 1930s music technology was also viewed as empowering men's involvement with music, thus beginning a trend that has continued to this day (e.g., Armstrong, 2011; Shibazaki & Marshall, 2013).[11] Although music in the home had historically been considered the domain of women, and, as Katz (2004) points out, in the early days women were more than twice as likely to make decisions about phonograph purchases, the machine aspect of the phonograph served to legitimize men's interest in music. Katz notes that men could "experience the classics without the self-consciousness that might follow them into public forums such as concert halls and opera houses. The phonograph ... [permitted] them to do in private what they could not or would not otherwise do" (p. 59).

In addition to altering our conception of the musical experience, "captured sound" has helped to *fix* or standardize musical models for instruction. Whereas previously all one had was one's memory of a fine performance, recordings provide aural models that can serve as an ongoing source of comparison. Although there are arguably pedagogical merits to direct emulation, the risk is that performance becomes less about the intersubjective experience of the contextualized musical event and more about direct and canonical comparisons, leading to standardized, universal performances irrespective of audience or event. Performances of Bach's cello suites, for example, have become exercises (certainly among cellists) in how any individual rendition compares to those of Pablo Casals, Pierre Fournier, Mstislav Rostropovich, Janos Starker, Yo-Yo Ma, and so on, rather than as celebrated musical experiences in their own right. Aural models can motivate and teach, but they can also highlight failings and shortcomings; better to listen to Glenn Gould's *Goldberg Variations* than to one's own inferior efforts. Mediated music thus served, and serves, to raise questions about the nature of performance, highlighting, among many other things, the role of effort and the significance of "achievement."

Notably, the word *techné* is closely related to the word "technique." But whereas recordings and radio helped to celebrate virtuosity in music—both classical and other—modern technologies also help to challenge its preeminence as a marker of superiority and exclusion. Many vernacular musics, or at least participatory ones, are rooted in an ethic of nonvirtuosity; today's digital technologies make it possible for relative novices to produce rather sophisticated music in the absence of traditional forms of musical technique, thus undermining one of the perennial mantras of traditional music lessons: hard work. This is not to suggest that those engaged in digital music making do not work hard. They do, of course, but the work is of a different kind: the digital realm involves both creative engagement and technical know-how, something often requiring a great amount of time to develop; the hard work of daily practice, however, is, at least traditionally, grounded in drudgery aimed not at participation for the sake of participation but at being able to execute music with increasingly difficult technical demands. Digital technology thus opens up conceptions of music learning and teaching that aren't predicated primarily upon physical mastery and the perfection of the body (Foucault, 1995).

Finally, technology possesses the potential to alter the fundamental schema upon which our understandings of music are based. For example, while alternative digital input methods exist, MIDI has, for those interested in digital forms of music making at least, placed the piano keyboard at the center of human-computer interactions. The piano keyboard, it must be remembered, is but one way of interfacing with sonic production—one that differs markedly from alternatives such as the guitar, the violin, the tablet computer, or any other production mechanism. The way sound and music production is conceptualized dramatically alters what music is and how we teach and learn it. For example, not unlike staff notation, GarageBand and many other programs entrain a horizontal linearity to time and a vertical stacking to texture in their visual representations of music. Imagine for a moment how we might think differently about time and texture if the x- and y-axes of such programs were reversed. To a certain extent this what Guitar Hero and Rock Band do, but now imagine variations, such as instruments moving toward the notes (in the manner of many race car video games) rather than the notes approaching the instrument; or imagine the visual representation of time as a circle, as some looping apps do (or like the LP record perhaps?); and so on. Suffice it to say, new modes of interaction hold exciting possibilities for reconsidering the ways we understand music and how we might better help our students to learn and conceptualize music beyond the traditional staff notation model of visual representation of sound.[12]

Thoughts on Music, Teaching, and Technology

> The presence of these [pianos, violins, guitars, mandolins, and banjos] in the homes *has given employment* to enormous numbers of teachers who have patiently taught the children and inculcated a love for music throughout the various communities.
>
> Right here is the menace in machine-made music! ... The cheaper of these instruments of the home *are no longer being purchased* as formerly, and all because the automatic music devices are usurping their places.
>
> And what is the result? The child becomes indifferent to practice, for when music can be heard in the homes without the labor of study and close application, and without the slow process of acquiring a technic, it will be simply a question of time when the amateur disappears entirely, and with him *a host of vocal and instrumental teachers, who will be without field or calling.*
>
> —John Philip Sousa, "The Menace of Mechanical Music"
> (emphasis added)

Jacques Attali reminds us that "wherever there is music, there is money" (1985, p. 3). A close reading reveals that Sousa's objection to mechanical music may have had less to do with its impact on the experience of music or concern for the musical amateur and more to do with the political economy of music: instrument sales and loss of employment for music teachers. It cannot escape notice that music factors so prominently in so much technological innovation—and that there is a lot of money to be made in and through music. Where would Apple be today without the iPod? Perhaps more than anything else, might this be Heidegger's "revealing" of which we should take notice?

Finale, Sibelius, Noteflight, GarageBand, SmartMusic—these are all market-driven *products*. As with the example of the instrument manufacturers and music publishing companies that used to constitute the lion's share of our educational purchases, our endorsement of digital hardware and software invariably leads to profits for private companies. No matter how much we feign the disinterestedness of the classroom, it is folly to believe that music educators can escape Homo economicus. We never teach just music; technology usage is never just using technology.

To refer again to Sousa, music educators need to accept not only the political and economic ramifications of our instructional choices, but that our own livelihoods as music teachers are embroiled in how and what we teach. Ironically, whereas technology used to be viewed as a threat to music as an art form, a threat to music teachers and music teaching, and a threat to the humanness of music, technology is now embraced for its potential to save music education from a slow and ignominious death. As Jonathan Savage writes in "Reconstructing Music Education through ICT": "if educators fail to grasp this major cultural shift [toward technology], music as a curriculum subject will become increasingly alienated from young people's lives and they will find their music education elsewhere" (2007, p. 75).[13] Savage only relates half the story, however, as it is not just that young people will find their music education elsewhere, but that music will no longer be able to sustain its place in the curriculum in a world that sees technology as the elixir for international competitiveness. Not only do all school subjects become fair game for colonization by technology, but the neoliberal mindset considers expendable any subject that is not based in or infused with the aims of technological literacy or that fails to serve them.[14] As a result, music educators have embraced technology not necessarily because of any real benefit for students but out of occupational preservation.[15]

The danger here lies not only in failing to heed the motivations underlying those who proselytize technology, but in failing to recognize the loss of freedom to which Heidegger alludes. As Himonides and Purves point out, most discussions about music technology consist of "advice and technical tips on how to 'use' specific software packages.... [Few authors] offer[ed] critical evaluations of how the suggested tools had been used; nor discussed why they had been used, nor conferred what educational principles the authors had been celebrating" (2010, p. 126).[16] Indeed, my literature review revealed no shortage of "strategies" for teaching with or using technology. In many cases, technology is considered strictly "as a music tool" (Williams & Webster, 1996) or "in the service of music" (Hunt & Kirk, 1997). At the extreme end, however, technology is the goal, and music is the means. When technology is accepted uncritically in music education, poiësis (bringing forth) no longer involves the artistic aspects of techné. Instead, educators (and their students) are reduced to executors—cogs in the iron cage of Max Weber's technical rationality; the "freedom" to learn cannot exist in an educational environment of compulsion.

None of this is to discount the exciting possibilities opened up by music technology, however. If engaged with freely and authentically, technology can liberate and wrest control away from those with monopolies on knowledge and access. Do-it-yourself (DIY) has arguably been the poster child for music technology proponents.[17] Contra

Sousa's soothsaying about the disappearance of the amateur, many observers have pointed to the explosive growth of DIY amateuring in music. Others have pointed to how technology has helped to blur traditional boundaries between professionals and amateurs (e.g., Leong, 2003, p. 161). While radio and recordings may have created the paternalistic phenomenon of musical appreciation, further entrenching particular forms of cultural capital and "taste," technology has today arguably empowered the consumer as producer (or *prosumer*). While this undeniably has still left the greatest clout in the hands of those with the greatest economic muscle and has done little to dent what Michael Apple (2000) calls "official knowledge" (i.e., the kind taught in public schools), economies of scale have paved the way for most classes in society to exercise more of their own musical desires than at previous times in history. In research I conducted in the mid-2000s, for example, I witnessed disenfranchised youth—the kind usually abandoned by the education system—who, in this case, had also never found a spot in "traditional" music classes, produce original, very polished "urban contemporary" popular music in what was then a rather novel studio recording program at an inner-city high school. As one of my interviewees put it, "being in the studio program, you're able to do things that you never really thought you could do. When [students] record a song and they listen to it they're like, 'Man, never in a million years did I think I'd be able to do that,' you know. It boosts up your self-esteem. It makes you feel better about yourself—that you're doing something worthwhile." As a result of their music making, which was inconceivable without the assistance of modern music technology, these young people developed strong voices and strong character. My interviews with them left little doubt in my mind (then and now) about the potential benefits of music technology![18]

Many other examples could be cited, of course. Low-cost, easy-to-use music technology helps facilitate not only popular music production, but all manner of individual and collective creative enterprises that previously would have required years of specialized training and/or access to expensive equipment. Nevertheless, we need be mindful that the mere existence of music technology is no panacea in and of itself. Media critics like Tarleton Gillespie and Ted Striphas, for example, argue that what matters is not just the technologies themselves (or how we hope to use them), but the policies and arrangements that govern their implementation and practice. In an age of "algorithmic culture," confidence in our own desires is undermined. Did I freely choose that YouTube video or did I watch it because an algorithm decided my tastes for me?[19]

My final point is that, in the midst of the deluge of discussions about music technology and music education, we must not forget that the classroom is arguably the most powerful technology of all—one that, to a large extent, is within our direct control. In the "ballet of technology and music" (Williams & Webster, 1996, p. 3), we cannot afford, as Janet Mansfield points out, to simply assume "the value of 'truths' emanating from 'technological literacy,'" keeping students "unaware of the politics of musical knowledge" and the connections between technological musical literacy and "commodifying communities in the global context" (2005, pp. 139, 145). More than any technological literacy or fluency, the lack of criticality about technology may indeed be the most pressing issue with which music educators should be grappling. This need not imply rejection or hostility, however. Adapting ideas from Feenberg's (2010) "10 paradoxes of technology," for example, might offer teachers and their students new possibilities for thinking about the ways in which technology enables and constrains, affords and challenges. Bringing a historical sociocultural eye to the classroom might also help teachers and students better recognize the ways in which definitions of musicality continue to shift. As Mansfield says, technology "may bring musicality—*being musical* into unconcealment" (2005, p. 145; emphasis in original).

As educators, we have no choice but to be in the world. As music educators, this means confronting the realities of musical subjectivity as they exist in our classrooms, our communities, and our world. To imagine "technology-free" music making is to fundamentally misunderstand that we are unavoidably technological beings. Homo faber brought forth Homo technologicus in very short order. Our professional and ethical obligations must thus involve transcending naïve efforts aimed at mere competence with technology and music technology and should strive to engender critical engagement that sees students continually evaluating if and how various technologies can help them live richer and more rewarding lives in and through music. This cannot happen, however, unless music educators themselves continually question the ways in which technologies affect the musical subjectivity of our students so that we may engage freely. Just because something is possible does not make it desirable or appropriate. Conversely, failing to provide our students with critical tools that might allow them to better negotiate different the modalities of musicality made possible by various technologies would seem to make us professionally negligent as educators. Poiësis that lacks the artistic elements of techné reduces us, and our music making efforts, to mere technique. If and when that happens, the phonograph has indeed won a victory.

Notes

1. The epigraphs to this chapter are from Bill Faucett and Michael Budds, *Music in America 1860–1918: Essays, Reviews, and Remarks on Critical Issues* (Hillsdale, NY: Pendragon Press, 2008), 165–166, 168.
2. Strong arguments might be made for interrogating potential differences in meaning between technology and technologies. For the purposes of this chapter I use a general meaning of "technology."
3. On the application of Heidegger to an examination of musical instruments, Ihde writes: "it simply seems perverse to apply *Bestand, Gestell,* and so forth to this class of 'technologies'" (2010, p. 135). Ihde suggests we need to "deromanticize" Heidegger.
4. MayDay Group authors such as Philip Alperson, Wayne Bowman, David Elliott, and Thomas Regelski have written extensively on praxis, techné, poiësis, and other ancient Greek (primarily Aristotelian) thought.
5. Subjectivity in a philosophical sense is enormously complex. My point is simply that thinking about who we are and how we should live are age-old concerns.

6. Psychologist Jane Healy (1990), for example, goes so far as to suggest that overexposure to computers can negatively impact the development of the child's brain—a claim echoed in a 2008 feature in the *Atlantic* titled "Is Google Making Us Stupid?" (July/August). Retrieved from http://www.theatlantic.com/magazine/archive/2008/07/is-google-making-us-stupid/306868.
7. Following up on her highly popular critique *The Second Self* (1984), Turkle's *Life on the Screen* (1995) and *Alone Together: Why We Expect More from Technology and Less from Each Other* (2011) focus on identity and interconnectedness in a hyperconnected world. "At the extreme," she writes, "we are so enmeshed in our connections that we neglect each other" (2011, p. 294).
8. A 1907 *Current Literature* article, for example, was titled "The Democracy of Music Achieved by Invention" (Taylor, Katz, & Grajeda, 2012, p. 55).
9. I am thinking here of, for example, the phenomenon of "race records."
10. This is not quite true, of course. Adorno and the Frankfurt School were alarmed by pretty much all massification. My point is that proponents are frequently selective in their deployment of "big ideas."
11. On the basis of her research, Kip Pegley writes: "if we are to be effective educators, responsible to the needs of students under the constant pressure of technological change, we must listen more closely to the girls" (2006, n.p.).
12. One can think of similar examples outside music. Languages like Hebrew or Mandarin, for example, generate fundamentally different schema of language. Or consider alternative maps of the world, such as those that reverse the poles (with south on top) or represent countries by population rather than land mass.
13. This is but one of countless examples.
14. Janet Mansfield writes, for example: "the accountability so revered by rapidly globalizing audit culture under neo-liberal policies of 'growth and development' locks music educators into attempting to describe how music knowledge fits a 'knowledge economy.' How can such knowledge procedurally and theoretically, be conceived of in terms of economic commodity? Will it 'innovate' or contribute to national wealth?" (2005, p. 134).
15. On the theme of preservation, I could not but laugh (cry?) upon my discovery of the music technology website http://www.savedbytechnology.com.
16. In their explanation of why they wrote *Experiencing Music Technology*, Williams and Webster cite the *Handbook* of the (U.S.) National Association of Schools of Music, which stated: "through study and laboratory experience, students should be made familiar with the capabilities of technology as they relate to composition, performance, analysis, teaching, and research" (Williams & Webster, 1996, p. xv). There is no mention in the NASM *Handbook* of why this should be so other than, apparently, the fact that technology exists.
17. For a discussion of open source and music education, see *Journal of Music, Technology and Education* 5(12) (2012).
18. See Mantie (2008). One should also not overlook here the advantages afforded to those with various forms of limitation ("disability") who have benefited immeasurably from various technologies.
19. In the words of Ted Striphas (2010), "in the old cultural paradigm, you could question authorities about their reasons for selecting particular cultural artifacts as worthy, while dismissing or neglecting others. Not so with algorithmic culture, which wraps abstraction inside of secrecy and sells it back to you as, 'the people have spoken.'" Retrieved from http://www.thelateageofprint.org/2010/06/14/how-to-have-culture-in-an-algorithmic-age. Matthew Thibeault has been particularly helpful in bringing these ideas to the attention of music educators.

References

Allenby, B. R., & Sarewitz, D. (2011). *The techno-human condition*. Cambridge, MA: MIT Press.
Apple, M. W. (2000). *Official knowledge: Democratic education in a conservative age* (2nd ed.). New York: Routledge.
Armstrong, V. (2011). *Technology and the gendering of music education*. Burlington, VT: Ashgate.
Attali, J. (1985). *Noise: The political economy of music*. Minneapolis: University of Minnesota Press.
Biocca, F. (1990). Media and perceptual shifts: Early radio and the clash of musical cultures. *Journal of Popular Culture, 24*(2), 1–15. doi: 10.1111/j.0022-3840.1990.2402_1.x.
Dewey, J. (1929). *The quest for certainty: A study of the relation of knowledge and action*. New York: Minton, Balch.
Feenberg, A. (2010). Ten paradoxes of technology. *Techné, 14*(1), 3–15.
Foucault, M. (1995). *Discipline and punish: The birth of the prison*. New York: Vintage Books.
Foucault, M., Martin, L. H., Gutman, H., & Hutton, P. H. (1988). *Technologies of the self: A seminar with Michel Foucault*. Amherst, MA: University of Massachusetts Press.
Healy, J. M. (1990). *Endangered minds: Why our children don't think*. New York: Simon and Schuster.
Heidegger, M. (1977). *The question concerning technology, and other essays*. New York: Garland.
Himonides, E., & Purves, R. (2010). The role of technology. In S. Hallam & A. Creech (Eds.), *Music education in the 21st century in the United Kingdom: Achievements, analysis and aspirations* (pp. 123–140). London: Institute of Education, University of London.
Hunt, A., & Kirk, R. (1997). Technology and music: Incompatible subjects? *British Journal of Music Education, 14*(2), 151–161.
Ihde, D. (2010). *Heidegger's technologies: Postphenomenological perspectives* (1st ed.). New York: Fordham University Press.
Katz, M. (2004). *Capturing sound: How technology has changed music*. Berkeley: University of California Press.
Leong, S. (Ed.). (2003). *Musicianship in the 21st century: Issues, trends and possibilities*. The Rocks, N.S.W.: Australian Music Centre.
Lines, D. (2013). *Music education, Heidegger and emerging technologies*. Paper presented at the Eighth International Conference for Research in Music Education, Exeter, UK.
Mansfield, J. (2005). The global musical subject, curriculum and Heidegger's questioning concerning technology. *Educational Philosophy and Theory, 37*(1), 133–148.
Mantie, R. (2008). Getting unstuck: The One World Youth Arts Project, the music education paradigm, and youth without advantage. *Music Education Research, 10*(4), 473–483. doi: 10.1080/14613800802547706.
Pegley, K. (2006). "Like horses to water": Reconsidering gender and technology within music education discourses. *Women and Music: A Journal of Gender and Culture, 10*(1), 60–70.
Roell, C. H. (1989). *The piano in America, 1890–1940*. Chapel Hill: University of North Carolina Press.
Savage, J. (2007). Reconstructing music education through ICT. *Research in Education, 78*, 65–77.
Shibazaki, K., & Marshall, N. A. (2013). Gender differences in computer- and instrumental-based musical composition. *Educational Research, 55*(4), 347–360. doi: 10.1080/00131881.2013.844937.

Stivers, R. (2004). The technological personality. *Bulletin of Science, Technology & Society*, 24(6), 488–499. doi: 10.1177/0270467604269802.

Taylor, T. D., Katz, M., & Grajeda, T. (2012). Music, sound, and technology in America: A documentary history of early phonograph, cinema, and radio. Durham, NC: Duke University Press.

Theberge, P. (1997). *Any sound you can imagine: Making music/consuming technology*. Hanover, NH: University Press of New England for Wesleyan University Press.

Toffler, A. (1970). *Future shock*. New York: Random House.

Turkle, S. (1995). *Life on the screen: Identity in the age of the Internet*. New York: Simon & Schuster.

Turkle, S. (2011). Alone together: Why we expect more from technology and less from each other. New York: Basic Books.

van der Laan, J. M. (2007). "Is it live, or is it Memorex?" *Bulletin of Science, Technology & Society*, 27(2), 136–141. doi: 10.1177/0270467606298221.

Vincent, M., & Karjala, H. E. (Eds.). (1985). *Music and technology: A report*. Muncie, IN: Ball State University Press.

Waddington, D. I. (2005). A field guide to Heidegger: Understanding "The Question Concerning Technology." *Educational Philosophy and Theory*, 37(4), 567–583. doi: 10.1111/j.1469-5812.2005.00141.x.

Williams, D. B., & Webster, P. (1996). *Experiencing music technology: Software, data, and hardware*. New York: Prentice Hall International.

CHAPTER 2

TECHNOLOGY IN MUSIC EDUCATION IN ENGLAND AND ACROSS EUROPE

MARINA GALL

> I almost think that in the new great music, machines will also be necessary and will be assigned a share in it. Perhaps industry, too, will bring forth her share in the artistic ascent.
>
> Busoni, cited in H. Russcol, *The Liberation of Sound* (1972)

AMPLIFIER, Apple Lossless Audio Codec, compact disc (CD), CD-ROM, commercial music streaming service, crossover, deck, driver unit, drum machine, DVD-audio effects unit, electric guitar, electric keyboard, equalizer, file-sharing service, FX box, genres: (Balkan beat, dancehall, drum & bass, dubstep, dub regga, garage, jungle, ragga, techno, trip-hop, two-step), headphones, Internet, intranet, iPad, iPhone, iPod, loudspeaker, mediaplayer, microphone, MIDI (Musical Instrument Digital Interface), minidisk recorder, mixer, mp3, multitrack tape recorder, music video games, online audio distribution platform (NuMu), online social networking and microblogging service, open source file-sharing application, portable audio and video cassette player; preamp, sampler, sequencer, social networking website with an emphasis on music (MySpace), sound processor, speaker, super audio CD (SACD); synthesizer, tape recorder, turntable, video-sharing website, Web 2.0, website, World Wide Web.[1]

The above list of instruments, devices, and systems illustrates advancements in music and associated technologies over the last half century through which I have lived as a child, a secondary school teacher, a teacher educator, and a researcher of music technology in education. In this chapter, adopting an autobiographical perspective, I will reflect upon music and technology within English school classrooms during the last 50 years. This will include reference to summative governmental reports on school music, which, in England, derive from regular school inspections by the Office for Standards in Education.[2] I will also draw upon two of my own recent publications (Gall, Sammer, & de Vugt, 2012; Sammer, Gall, & Breeze, 2009), created in collaboration with educators in other European countries, to present a brief picture of the use of music and information and communication technology (ICT) in schools across Europe. The chapter ends with a discussion of key factors that inhibit the effective use of music technology in formal education in my own country. Throughout the sections about England it will become evident that, despite the government's constant acknowledgment of the importance of music technology within classroom work (Office for Standards in Education, 2004a, 2013), education policies and/or lack of support from school head teachers have meant that that the potential of music to enhance and transform learning has not been realized, particularly in music lessons for students aged 5–14. The section that follows provides a backdrop to my personal reflections.

Technology, Music, and Society

From the very earliest days of human existence, men and women have made use of and crafted instruments to support daily life and music making (Sachs, 2006); "technical" modifications to instruments—to keyboards between the fourteenth and nineteenth centuries, for example—have enabled new forms of musical expression (Dolge, 1972). As such, one might posit that there have always been technological developments in music, and that all music instruments are technologies. However, during the period from the 1920s to the 1960s the term "music technology," as associated with musical instruments, referred to those that were electronically powered. Examples in the popular music domain included electric keyboards, the electric guitar, and the drum machine, all of which strongly impacted contemporary styles and practices of the time (Majeski, 1990, McSwain, 1995; Schrader, 1982). During this time, music developments went beyond the creation of electronic forms of conventional musical instruments. In both classical and popular spheres, new sound-making technologies, such as the synthesizer, used extensively in the pop world, and the theremin and martenot, exploited within the works of experimental classical composers such as Varese (Risset, 2004), enabled new creative possibilities; as well as composing with new electronic sounds, musicians could actually generate new sounds themselves.

Since the early twentieth century, the term "music technology" has also related to devices used within the recording industry (Bloustein, Peters, & Luckman, 2008). In this sphere, a music technology is something that enables sounds not only to be generated but also to be manipulated in all manner of ways, and to be stored and distributed easily, thus impacting the musical processes of composition, performance, and listening

(Katz, 2004; Théberge, 1997). Indeed, in the late twentieth century new musical genres such as techno, house, and jungle emerged that are completely reliant on technology for their realization (Richard & Kruger, 2005); as such, in some cases, the technology is the music.

Ever since the advent of the radio (Juniper, 2004) and music within films (Dickinson, 2003), music has been associated with other media (Cook, 1998). However, the binding together of popular music and visuals was particularly significant in the 1980s, as this led to the creation of a new distinctive cultural (technological) art form: the music video (Kaplan, 1987; Mundy, 1999). Though not specifically classified as music technologies, other new developments are very much associated with music; for example, sound is an important aspect of all video games (Zehnder & Lipscomb, 2006) as well as the focus of some (J. Smith, 2004; Tobias, 2012). The Internet has also given rise to significant changes in music and society over the last half century: ease of access to a wide range of music, including the facilities for downloading of music and file sharing (Werner, 2009) for use on portable personal technological listening devices (Lenhart, Purcell, Smith, & Zickuhr, 2010); the availability of free composing and recording software, enabling one person to be composer, producer, and marketer (Théberge, 1997); new ways to learn musical/instrumental skills (Baxter, 2013); and new means of exhibiting one's own music, communicating ideas about music, and presenting one's own personal (musical) image within interactive music websites and social networking systems (Cayari, 2011; Raacke & Bonds-Raacke, 2008). Within these new cultural and social practices, many people's experiences of music are within multimedia settings (Pfeil, Arjan, & Zaphiris, 2008), and often involve considerable collaborative activity (Salavuo, 2006).

Music Technology in English Schools, 1960–1999

So how have such technological tools, and new practices mediated by technology, impacted young people and upon music classrooms? I was born in England in 1960. My earliest contact with music technology was undoubtedly unconscious: my parents owned a radiogram and, like many adults of the time (Cook, 2010), this formed an important part of our daily home life.[3] My first conscious experiences of music technology were of my mother playing Jim Reeves on 78 rpm vinyl records, and of listening excitedly to the Beatles on the 45-inch single bought by my elder brother.

At primary school, I was lucky to have a specialist music teacher who engaged me in a variety of mainly practical music activities, such as singing accompanied by percussion and recorder playing. At the time, technology was important in this school phase since the British Broadcasting Company's Schools Music Department, set up in the 1940s, broadcast the radio program *Music and Movement* twice weekly (Cox, 1996). Through this, schools lacking a specialist teacher or wanting to enhance their arts curriculum were able to offer dance and drama exercises to music.[4]

At secondary school, technology supported one of the two main aspects of my classroom music diet: appreciating classical music (played on vinyl).[5] Technology was also a means by which teachers decided on students' musical aptitude. The Bentley Test (Bentley, 1969), then only available on vinyl, required students to carry out such exercises as determining how many notes were playing in a chord, and which of a number of "bleepy" noises was the highest. At the time, there was disquiet about judgments being made concerning students' musical potential based on one set of tests, for a number of reasons, including the fact that, over time, these analog sources become warped and distorted slightly, thereby placing a question mark over the results (Sergeant & Boyle, 1980). Having gained high marks in these tests, I was offered what was then quite common: the opportunity to borrow a (nonelectronic) musical instrument and of have weekly lessons, all free of charge (Vulliamy, 1976).

As a teenager, although I was highly involved in classical music through instrumental lessons, "pop" music was at the center of my life and of my being, as it is for most young people. However, in the 1970s, there were almost no opportunities for any engagement with the technology of the time (keyboards, lead and bass guitars), neither in the classroom nor in school extracurricular instrumental lessons, despite the fact that these "pop" instruments were most desired by many teenagers (Vulliamy, 1976). This was largely because of the strong emphasis that was then placed upon European classical music and because of the knowledge-based, as opposed to practical, skills focus of school leaving examinations of the time (Spruce, 2002). Thus, like so many of my friends, it was only outside school that I had opportunities to engage with pop music. Luckily, by this time transistor radios were available and cheap (Christenson, DeBenedittis, & Lindlof, 1985), and so again, like many other teenagers, I was able to take control of what I listened to, and when and where this took place (Fitzgerald Joseph, Hayes, & O'Regan, 1995; Larson, Kubey, & Colletti, 1989). As a consequence, I spent a huge amount of time engrossed in the latest pop music, including late at night when Radio Luxembourg broadcast their Top 40 charts. My keen interest in the performing arts also led to my membership in the "sound team" for school productions, for which I created backing music on a reel-to-reel tape recorder, editing by cutting and splicing tape with a razor blade.

A personal dual musical identity continued throughout my years as a university music undergraduate. Owing to the near nonexistence of joint arts courses, or courses with a focus upon anything other than classical music, my three-year undergraduate course was almost exclusively focused upon "traditional" music skills, such as aural harmony and keyboard harmony;[6] in the evenings I pogoed to punk music. It was only as a trainee teacher, in 1981, when I had the good fortune to be allocated as my personal tutor George Odam—a dynamic music educator who was strongly committed to practical classroom music, including pop and, later, technology (Odam, 2000)—that I was actively encouraged to consider the importance of contemporary popular music and students' musical preferences in relation to formal music education. As a newly qualified teacher in a tough, multicultural, inner-city London school, offering the often disaffected students opportunities to perform popular music in the classroom was hugely motivating. However, practical problems associated with electronic instruments, such as lack of space and good soundproofing, mitigated against their use with anything other than my smaller examination classes. Nevertheless, in the mid-1980s, technology enabled me to make a significant music curriculum change. At this time, teachers in all subjects were permitted to develop their own alternative examination syllabi,[7] aimed at lower-attaining students (Tattersall, 1994), and so I joined with a number of other local teachers in offering my 14- to 16-year-old students a course in which they created and recorded a pop music radio program (on cassette tape) as part of their examination work.

More significant curriculum changes, which directly and indirectly impacted the use of music technology in secondary schools, came about in 1986 and 1992 with the publication of the new General Certificate of Secondary Education (GCSE) examination syllabus for all 14- to 16-year-olds (Philpott & Carden-Price, 2001) and the introduction of the orders for the first ever music National Curriculum for England and Wales (Department of Education and Science, 1991). Within both, there was a new emphasis on "music for all"; practical music, including composition (which was rare in some schools to this point), and breadth of style and genre within studies, such that world music, jazz, and pop then became commonplace within the curriculum.[8] I welcomed this, heartily, although there was a broad framework within which to work, with no detailed government prescription of content we teachers were free to design lessons of our choice.[9] Despite the growing availability of commercial classroom materials, many teachers created their own units of work and resources; in my case I included projects specifically designed with input from the students themselves, often related to popular music, and including "free" composition.

By the 1990s, music keyboards were ubiquitous within English secondary music classrooms and were the main instrument used for group composition and performance work, regardless of the style or genre (Wright, 2002).[10] Indeed, at the turn of the century the government inspectorate expressed concern that staff teaching 11- to 14-year-old students "overemphasized the use of keyboards to the exclusion of acoustic or vocal resources" (Office for Standards in Education, 2004b, p. 11), yet this was at a time when the revised National Curriculum for Music had included a new requirement for students to use "ICT to create, manipulate and refine sounds" (Department for Education and Employment/Qualifications and Curriculum Authority, 1999, p. 30).

I first saw a school computer being used for musical purposes in 1989. At an interview for my third secondary school job I was shown around the music department. Opening the door of a tiny music cupboard mainly used for storage of books, I came upon one 17-year-old music student—whom I was later to teach—squeezed in at a small desk using Cubase to compose on an Atari. Since the 1980s, the Edexcel Board had permitted students to create compositions using computer or other technologies, most typically electronic keyboards and stand-alone multitrack tape recorders.[11] In 1995, it offered a completely new Advanced level examination (for 17- to 18-year-olds) titled Music Technology. The first syllabus included a number of "traditional" elements but nevertheless gave students the opportunity to focus their work on technological aspects of music making related to sequencing and recording (Carden-Price, Philpott, & Lewis, 2002).[12] Despite this major curriculum change and the gradual development of a wide range of university undergraduate degree courses in Britain related to music technology,[13] teachers were slow to incorporate technology in classroom music lessons. This was unsurprising since, like most music teachers, I experienced the complete lack of availability of subject-specific training in music (Office for Standards in Education, 2004a), even though large funds had been designated for in-service teacher development of technological skills between 1999 and 2002 (Leask, 2002). Furthermore, lack of computer hardware and software was characteristic of most music departments throughout the 1990s (Office for Standards in Education, 2004a) despite the massive investment in computers in secondary schools at that time.[14] This meant that, for many years, most schools had insufficient equipment to work with for a whole class of 11- to 14-year-olds that often numbered up to 30 (Office for Standards in Education, 2004a). Those school that did have sufficient equipment were hampered by the absence of music-specific technical help (Office for Standards in Education, 2004a).

Music and Information Communications Technology in English Schools 2000–2013

So what changes have taken place in the 15 years since I left secondary school teaching and became involved in teacher education? Little in terms of technology and music for children aged five to ten. In primary schools, most teachers are generalists and teach all of the curriculum subjects. As such, they often lack music expertise and confidence in the subject, partly owing to the limited teacher training time typically allocated to music (Hennessy, 2012). These factors, together with the constant emphasis on literacy and numeracy in the primary school, have resulted in reduced time allocations for music within the curriculum (Office for Standards in Education, 2013). As such, particularly since 2007, when new governmental financial support for singing and for learning acoustic instruments in the classroom was allotted to primary schools (J. Evans, 2009), less of a focus has been placed upon music technology, especially for creating music; this is despite the long-standing (outgoing) National Curriculum requirement that pupils should use ICT to "capture, change and to combine sounds" (Department for Education and Employment/Qualifications and Curriculum Authority, 1999, p. 2/5d). It is difficult not to be muddled by governmental policy when one reads its inspectors' criticism that between 2008 and 2011 "the use of music technology was inadequate or non-existent in three fifths of primary schools" (Office for Standards in Education, 2012, p. 6) and then notes that the requirements for technology within the primary curriculum were removed from the 2014, slimmed-down version of the Music National Curriculum except in relation to listening to music (Gov.UK, n.d.b).

The change in focus on technology and erosion of time for music within the wider curriculum are not features of primary education only. The only statement about ICT for 11- to 14-year-olds in the most recent Music National Curriculum (2014) is that students "should use technology appropriately" (Gov.UK, n.d.b). More worryingly, following the recent introduction of the new examination framework for students aged 11–16—the English Baccalaureate, which excludes music[15]—fewer students are opting for the subject (Office for Standards in Education, 2013) and some schools are completely dropping music from their post-14 curriculum (Forgan, 2013). Additionally, in a number of secondary schools local to me the subsequent impact is a reduction of time for music within the curriculum for 11- to 14-year-olds (Office for Standards in Education, 2012, p. 38). As such, in many secondary schools today, music teachers' energy is focused upon preserving the place of music in schools, a more urgent necessity than a consideration of music technology in the curriculum.

Furthermore, lack of availability of computer hardware for whole class use in the lower secondary school music lessons with children aged 11–14 is still an issue (Gall, 2013). A recent government report noted that music technology "was inadequate or non-existent in . . . over a third of the secondary schools inspected" (Office for Standards in Education, 2012, p. 6). Yet head teachers and school governors, who nowadays have full control of their budget, have evidently not responded with financial support to help change the situation since, over the last decade, the Office for Standards in Education has constantly highlighted the same concerns (2004a, 2004b, 2009).

So what are teachers who are able to work with whole classes of 11- to 14-year-olds doing with computer music technology? The answer is mainly using it to support practical music making. Over the last 10 years, loop-based music creation software has been a favorite of many teachers owing to ease of use (Cooper, 2007) and its potential to enable the creation of music in popular styles (Gall & Breeze, 2005). I have also seen students offered the chance to develop programmatic music using sequencing software, for example sequencing a walk around an art gallery, stimulated by Mussorgsky's "Pictures at an Exhibition" (Breeze, 2009). Another local teacher encourages his 11-year-olds to create their own sounds, which are recorded into sequencing software and become a bank for pairs to use in designing their own soundscapes, which are then discussed in relation to *musique concrète*. Composing music for an advertisement or creating music to accompany a film are also popular activities for students of this age (Gall Lazarus, Tidmarsh, & Breeze, 2009). However, in England, aside from projects that I have read about in journal articles (Baxter, 2013; Challis, 2007; Savage, 2012; Savage & Butcher, 2007; Savage & Challis, 2001, 2002), I have seen very limited use of technologies for creative activities with 11- to 14-year-olds other than paired work at computer workstations using sequencing or sample-sequencing software. That this is a national issue is confirmed by the latest triennial government music report, which suggests that technology should be better used "to promote creativity" (as well as to widen participation and make assessment more musical[16]) (Office for Standards in Education, 2012, p. 8).

While the use of computers with 11- to 14-year-olds has not been widespread in English schools, other technologies have become more commonplace: at the turn of the century, the development of Musical Futures, a new pedagogical approach to musical teaching and learning in schools, has led to a wider use of electric guitars and drum kits/pads in the classroom. The primary aim of the approach is to enable students to take more ownership of their music making and to engage with the contemporary music they invariably enjoy outside school. In groups they are given a song on CD and a range of (preferably) rock instruments, and then, through self-directed learning, they produce their own version of the piece (Paul Hamlyn Foundation, n.d.).[17] More recently, the Musical Futures approach has been extended to include music other than contemporary pop, and the website now indicates ways to integrate technology into projects.[18] One local school has fully embedded rock music practical work within the curriculum for 11- to 14-year-olds but in a different way: through the purchase of five JamPods, each of which has a central student mixer linked to up to six "rock" instruments, most often used with headphones, thus enabling the whole class to work in small groups, but in a "silent" classroom (Bristol Post, 2013). Despite the expense of the JamPod, heads of departments with whom I work are very keen to acquire at least one or two sets in order to offer opportunities for more students to engage in "band" work within classroom lessons.

Interestingly, over the last decade there have been many positive developments in the use of technology in music classrooms for students aged 15–18. In the 1990s, changes to the Edexcel A level syllabus (for 17- to 18-year-olds), discussed above, also impacted the board's syllabus for the school music examination for 14- to 16-year-olds (GCSE) and, by 2000, sequencers, multitrack recorders, samplers, and record decks (for DJ-ing) were permitted for composing and for solo and ensemble performing within both examination syllabi (Terry, 2007). This further impacted other examination boards, which followed suit with respect to both GCSE and A level syllabi, although no other copied the creation of a separate music technology A level. At some of the schools with whom I work, the music technology A level has become so popular that the traditional music syllabus is not offered. This accords with governmental findings on the popularity of the subject: in 2011, a third of entries for A level music courses were for music technology (Office for Standards in Education, 2012). Furthermore, during the early 2000s, alongside these traditional courses most commonly found in schools, a number of significant alternatives, more vocational in nature, have also emerged such as Business and Technology Education Council BTEC courses (Pearson, 2017), which enable technology to be included in a range of ways within work-related/contemporary music projects (K. Evans, 2009).

These curriculum changes have been radical in offering opportunities for students with a broad range of musical skills/interests to engage in formal school examination courses. For years, 14- to 16-year-olds have been keen to utilize music technologies for composing. Indeed, in 2004, a government monitoring project for music reported that ICT was used most "to be creative; with 57% of key stage 4 [14- to 16-year-old] pupils using ICT frequently for this purpose" (Qualifications and

Curriculum Authority/ 2003/4, cited in Mellor, 2008, p. 452) and in 2005 the Edexcel GCSE examination board's annual report on composing stated: "those [students] writing pieces that featured technology heavily, such as electronic music and club dance remix, often deserved full marks" (Edexcel, 2005, p. 14).[19] However, recently staff from the 20 school music departments with whom I am currently partnered for teacher education work expressed concern that they have never been able to support students in offering DJ-ing as a performance option because no one in the department has knowledge/skills in this area.

Technology in Schools in Other European Countries, 2008–2011

Thus far, the focus of this chapter has been on classroom work in England. However, as part of my work within the European Association for Music in Schools, I have engaged in research on the use of music technology in schools across Europe. In the first project in 2008, music educationists from 24 countries completed questionnaires on attitudes toward, and uses of, music technology in schools in their countries (Sammer, Gall, & Breeze, 2009).[20] In the second, authors from 14 European countries each contributed a chapter to a volume in which they reflected upon a wide range of issues related to ICT and music from their countries' perspectives (Gall, Sammer & de Vugt, 2012).[21] In this next section, I will draw upon these two publications to discuss similarities and differences across countries, including England.

In all countries the significance of the Internet was apparent, with some noting its use as "a complementary textbook and teaching aid" (Sweden: Scheid & Stranberg, 2012, p. 245; Norway: Grønsdal, 2012), accessible at home as well as in school (Poland: Konkol & Kierzkowski, 2012). Some countries even relied solely on online curricula, which were seen to enable more student autonomy (Belgium: de Baets & Meyer, 2012; Netherlands: Arnold, Overmars, & van de Putte, 2012). The Internet was also viewed as a quick means of obtaining a wide range of music for teaching purposes, particularly through the use of YouTube, in which listeners' experience is enhanced by visual stimulation (Poland: Sammer, Gall, & Breeze, 2009; Cyprus: Savvidou, 2012) and a place for uploading students' music for display purposes (Sweden: Scheid & Stranberg, 2012). From the chapter writings it would seem that access to these resources using mobile devices is only common practice in Sweden (Sweden: Scheid & Stranberg, 2012). Recording technologies were cited as a means of capturing student performances (Germany: Sammer, Gall, & Breeze, 2009; Belgium: de Baets & Meyer, 2012), mainly for the purpose of self-evaluation (Cyprus: Savvidou, 2012; Poland: Konkol & Kierzkowski, 2012). Interestingly, the most recent English governmental report on music in schools 2008–2011 expressed concern that there was underuse of the free online platform NuMu to share and appraise students' composing and performing work, and insufficient use of audio recording by teachers as a means of assessing and aiding the improvement of students' work (Office for Standards in Education, 2012).

Another suggested use of technology was to support the development of what I have previously termed "traditional skills," including: ear training (Poland & Switzerland: Sammer, Gall, & Breeze, 2009; Cyprus: Savvidou, 2012; Italy: Biasutti, 2012); musical listening and appreciation (Estonia & Slovenia: Sammer, Gall, & Breeze, 2009; Belgium: de Baets & Meyer, 2012); analysis (Lithuania: Jautakyte, 2012); information retrieval from the Internet (Estonia:Sammer, Gall, & Breeze, 2009; Austria: Höfer & Reubenz, 2012; Italy: Biasutti, 2012) especially in relation to the history of music (Slovakia: Sammer, Gall, & Breeze, 2009; Norway: Grønsdal, 2012); notation (Germany & Poland: Sammer, Gall, & Breeze, 2009); and score writing skills (Cyprus: Savvidou; Italy: Biasutti, 2012). One of the most novel uses of technology to support musical learning was cited in relation to work in Finland, where video conferencing is used to support classroom and individual instrumental teaching within schools situated in remote areas of the country (Finland: Myllykoski, 2012).

As might be expected, in every country the use of technology in relation to the development of musical skills was seen to vary according to the perceived significance of each skill within the cultural context. For example, technology supports vocal work in Slovakia and dance in Poland (Sammer, Gall, & Breeze, 2009) and is used in relation to performing popular music in Sweden (Scheid & Stranberg, 2012). Indeed, music educationists from three countries expressed their unease that technology could undermine the "core" classroom activities of singing and group music making (Estonia, Slovakia, & Slovenia: Sammer, Gall, & Breeze, 2009).

In both publications the focus on composing that exists in England was not apparent in many countries, despite the fact that, within the research by Sammer, Gall, & Breeze (2009), composing and arranging were cited most often as skills developed through using technology. Data from almost all countries suggested that, as in England, the chief inhibitors to creative work in the classroom in the period 2008–2011 were the shortage of computer hardware, especially for use by nonexamination (larger) classes, and the lack of training for teachers to feel confident and competent in supporting learning with technology in the classroom.

And the Future? 2014 Onward

While a number of the points that I raise in this final section may also relate to ICT and music in schools across Europe, I now return to some overarching thoughts about music technology in formal education in England and my concerns about the future. It feels as if music technology in schools has been caught up in a vicious cycle: inadequate pre- and in-service training has resulted in many school music teachers lacking confidence or competence and, as a consequence, interest in ICT and music; a shortage of equipment for whole class use has then led to further hesitation in bringing technology into music lessons and a disinclination to develop new skills and awareness in this area of classroom work, especially with children under the age of 15, that is, in nonexamination classes. The result is that many teachers remain unskilled and underconfident (Gall, 2013).

Sadly, music's low status position within the curriculum (Forgan, 2013) means that we are unlikely to see more money being allocated to music departments for new technological devices, despite the government's awareness of the likely impact: "music technology is changing rapidly and the schools found it difficult to develop their own resources in line with the quality of equipment which students were seeing—and sometimes using themselves—outside school. Consequently, ICT in school could appear dated to them" (Office for Standards in Education, 2009, p. 34). This should be a matter of concern for all music educationists in a country where student disengagement in school music has been recognized for years (Lamont, 2002; Office for Standards in Education, 2013).

Paradoxically, sophisticated mobile phones, which enable a range of multimedia activities, are invariably in great supply within most English classrooms as many young people own them. Yet, while certain teachers have exploited these in school music lessons (Baxter, 2007), this is rare, since most schools' policies completely forbid mobile usage within lessons (Stowell & Dixon, 2014). Furthermore, although many students own or have access to iPads at home, in many secondary schools with which I work there is not even a central supply, let alone a set accessible for regular music classroom work. The availability of such a resource would overcome a key problem that pertains only to music and ICT: the typical lack of space to enable work with acoustic instruments (or drum kits and electric guitars, which are often housed in annexed practice rooms) at the same time as with computers and/or other technologies (Gall, 2013). Furthermore, one of the government's key foci for more effective use of music technology in schools is "as a tool for inclusion" (Office for Standards in Education, 2012, p. 42); iPads nowadays not only offer a wide range of musical programs and software but also act as an easily-accessed sound palette for exploration by students, including those with severe physical disabilities (DM Drake Music, n.d.).

Effective use of ICT to support musical learning is not, however, merely an issue of equipment availability and/or teacher expertise. The above discussion has touched upon the significance of students' use of technology outside school. It is axiomatic that extracurricular work in music is important in skills development, yet, as I have discussed earlier in relation to pop music in schools in the 1970s, children's musical activities outside formal education often not only differ from those in school classrooms but are at odds with them. First, in children's lives outside school, work on one task often includes engagement with a variety of modes of learning, involving multitasking "hands-on" approaches with, often, rapid movement between different modes and tools; this is at odds with the more linear learning traditionally experienced and expected in school. Second, staff interested in considering new approaches to student learning more in sympathy with children's out-of-school practices may be hindered by examination requirements that not only tend to be conservative in nature, but, in the main, focus upon the achievements of the individual; this is inconsistent with the networking, sharing, and co-construction of knowledge that typify not only children's engagement with technology in their own time but also professional practice in music and the arts (Sexton, 2007). Moreover, outside school, children often experience music within multimedia settings (Pfeil et al., 2008), yet most English secondary schools organize learning in relation to separate "subjects." Therefore, at a time when the latest English National Curriculum (2014) (Gov.UK, n.d.a) makes no mention of media or multimedia contexts and "English schools are hounded by league tables of exam results" (K. Smith, 2004, p. 90), music activities within authentic multimedia settings and other classroom music innovations are less likely to be seen than established practice that teachers feel confident will support the attainment of high grades.

In light of the above discussion, it is ironic that recent governmental reports have remarked that "best practice" regarding technology in music classrooms occurs when teachers relate students' work to how ICT is used in the real world so that students can explore processes used professionally (Office for Standards in Education, 2009). Many of the underlying structures and policies that underpin formal education in the twenty-first century are the chief inhibitors to these "real world" practices. If there is a true concern to help teachers overcome the problems of inadequate or ineffective use of ICT and music in classrooms noted regularly in school inspection reports (Office for Standards in Education, 2004a, 2004b, 2009, 2012), a major shift in school culture and established practice will be required. This is unlikely at a time when the place of music in schools is itself in question. However, it is my personal hope that the popularity of music technology as a school examination subject, the huge rise in numbers of students taking music technology–related undergraduate courses (Boehm, 2007), and the importance of the music business—much of which revolves around technology—as a chief revenue provider for the United Kingdom (Anderton, Dubber, & James, 2013) might leverage educators' attempts to retain music's place within the secondary school curriculum.

Notes

1. The epigraph is from Russcol (1972, p. 33).
2. Reports, made available to the public via a website, follow visits to schools, made with very short notice. Failure to gain either a Good or Outstanding grade leads to further visits; continuing low standards can result in the dismissal of school staff.
3. Throughout I write "English," although I call myself "British," because the education systems and school music are not the same in all countries of the United Kingdom.
4.. See Historical Boys' Clothing (2004) for more information on the broadcast and Giles, A. (n.d.) for a sound recording of one of the programs,
5. The other was singing British folk songs.
6. Practical activities happened "outside" the course, and it was only during my time at the university that composition was introduced as a possible area of study.
7. For the Certificate of Secondary Education Mode 3.
8. At this time, a typical lesson for students aged 11–14 focused on performing or composing tasks, with students working in groups of three to six using music keyboards and/or voices and/or acoustic (including Orff) instruments.
9. This continues to date.

10. The National Music Council reported that, in the United Kingdom in 1997, more money was spent on keyboards than on any other instrument (National Music Council, 1999).
11. Schools choose from three government-regulated examination boards (Assessment and Qualifications Alliance (ACA), Edexcel, and Oxford, Cambridge, and Royal Arts Society (OCR), who set their own syllabi and examinations; the requirements of these vary, but in each there is the same focus upon active music making that is found within the National Curriculum for younger pupils.
12. At the same time, computer music notation software was being used by students on the traditional A level Music course, who were required to submit a score of their compositions.
13. Boehm (2007, p.10) notes that, in 2007, there were 351 in the category of music technology (although only 131 incldued "music technology" in the title).
14. Most schools created centralized computer suites that were primarily used by teachers of subjects other than the arts, partly because they did not include the associated equipment, such as high-quality sound cards and music (MIDI) keyboards.
15. Students must achieve in mathematics, English, the sciences, a language, and history or geography.
16. By: significantly improving the use of music technology to record, store, listen to and assess pupils' work; placing greater emphasis on pupils' *musical* development through the use of technology—with the acquisition of technical skills and knowledge supporting, rather than driving, musical learning [and] making more creative and effective use of music technology to support performing and listening work (Ofsted 2012, p.8).
17. This methodology is now widely known in England. It is difficult to know how widespread its use is, although many teachers (including those from countries outside England) network within their website, and a number of the teachers with whom I work include one or two Musical Futures projects within their curriculum for 11- to 14-year-olds.
18. See Paul Hamlyn Foundation (n.d.) for a range of resources.
19. This was in relation to one assessment criterion: "use of technology"; grades were also allocated to a number of other criteria for each piece.
20. Austria, Belgium, Croatia, the Czech Republic, Denmark, England, Estonia, Finland, France, Germany, Greece, Hungary, Iceland, Ireland, Italy, Kosovo, Norway, Portugal, Scotland, Slovenia, Spain, Sweden, Switzerland, and the Netherlands.
21. Austria, Belgium, Cyprus, England, Finland, Germany, Italy, Lithuania, Norway, Poland, Slovenia, Spain, Sweden, and the Netherlands.

REFERENCES

Anderton, C., Dubber, A., & James, M. (2013). *Understanding the music industries*. London/California: Sage.

Arnold, J., Overmars, J., & van de Putte, R. (2012). ICT and teaching music in schools in Netherlands: The intro-model education. In M. Gall, G. Sammer, & A. de Vugt (Eds.), *European perspectives on music education: New media in the classroom* (pp. 163–176). Innsbruck: Helbling.

Baxter, A. (2007). The Mobile Phone and Class Music: A Teacher's Perspective. In J. Finney & P. Burnard (Eds.), *Music Education with Digital Technology* (pp. 52–62). London/New York: Continuum.

Baxter, A. (2013). Playstation power: Removing performance barriers to musical composition. In J. Finney & F. Laurence (Eds.), *MasterClass in music education: Transforming teaching and learning* (pp. 173–181). London/New York: Bloomsbury.

Bentley, A. (1969). Measurement and development of musical abilities: Some research interests and findings. *Journal of Research in Music Education*, 17(1), 41–46.

Biasutti, M. (2012). ICT in music education in Italy: Contexts, potentialities and perspectives. In M. Gall, G. Sammer, & A. de Vugt (Eds.), *European perspectives on music education: New media in the classroom* (pp. 135–147). Innsbruck: Helbling.

Bloustein, G., Peters, M., & Luckman, S. (2008). 'Be not afeard; the isle is full of noises': Reflections on the synergies of music in the creative knowledge economy. In G. Bloustein, M. Peters, & S. Luckman (Eds.), *Sonic synergies: Music, technology, community, identity* (pp. xxi–xxviii). Aldershot, U.K.: Ashgate Publishing.

Boehm, C. (2007). The discipline that never was: Current developments in music technology in higher education in Britain. *Journal of Music Technology and Education*, 1(1), 7–21.

Breeze. (2009). Learning design and proscription: How generative activity was promoted in music composing. *International Journal of Music Education*, 27(3), 204–219.

Bristol Post. (2013, January 21). Kick out the jams! Student rock out with music system. *Bristol Post*. Retrieved January 31, 2017, from http://www.bristolpost.co.uk/Kick-jams-Students-rock-music/story-17919091-detail/story.html.

Carden-Price, C., Philpott, S., & Lewis, M. (2002). Public examinations 2: A Level and Post-16 examinations. In C. Philpott (Ed.), *Learning to teach music in the secondary school: A companion to school experience* (pp. 201–221). London/New York: RoutledgeFalmer.

Cayari, C. (2011). The YouTube effect: How YouTube has provided new ways to consume, create, and share music. *International Journal of Education & the Arts*, 12(6).

Challis, M. (2007). The DJ factor: Teaching performance and composition from back to front. In J. Finney & P. Burnard (Eds.), *Music education with digital technology* (pp. 65–75). London/New York: Continuum.

Christenson, P. G., DeBenedittis, P., & Lindlof, T. R. (1985). Children's use of audio media. *Communication Research*, 12(3), 327–343.

Cook, C. (2010). Entertainment in a box: Domestic design and the radiogram and television, music in art. *Rethinking Music in Art: New Directions in Music Iconography*, 35(1/2), 261–270.

Cook, N. (1998). *Analysing musical multimedia*. Oxford: Clarendon Press.

Cooper, L. (2007). The gender factor: Teaching composition in music technology to boys and girls in year 9. In J. Finney & P. Burnard (Eds.), *Music education with digital technology* (pp. 30–41). London/New York: Continuum.

Cox, G. (1996). School music broadcasts and the BBC 1924–47. *History of Education: Journal of the History of Education Society*, 25(4), 363–371.

de Baets, T., & Meyer, H. (2012). Embedding ICT in music education: A Belgian perspective. In M. Gall, G. Sammer, & A. de Vugt (Eds.), *European perspectives on music education: New media in the classroom* (pp. 45–54). Innsbruck: Helbling.

Department for Education and Employment Qualifications and Curriculum Authority. (1999). *Music: The National Curriculum for England*. London: HMSO.

Department of Education and Science. (1991). *National Curriculum: Music working group interim report*. London: DES.

Dickinson, K. (Ed.). (2003). *Movie music: The film reader*. London/New York: Routledge.

DM Drake Music. (n.d.). *Using iPads for the first time in a SEN school*. Retrieved January 31, 2017, from http://www.drakemusic.org/dm-education/experiences/using-ipads-first-time-sen-school.

Dolge, A. (1972). *Pianos and their makers: A comprehensive history of the development of the piano*. New York: Dover.

Edexcel. (2005). *GCSE Examiner's Report: Music (1426)*. Mansfield: Edexcel.

Evans, J. (2009). Continuing pupils' experiences of singing and instrumental learning from Key Stage 2 to Key Stage 3. In J. Evans & C. Philpott (Eds.), *A practical guide to teaching music in the secondary school* (pp. 74–81). London/New York: Routledge.

Evans, K. (2009). Musical teaching and learning in 14–19 education. In J. Evans & C. Philpott (Eds.), *A practical guide to teaching music in the secondary school* (pp. 17–26). London/New York: Routledge.

Fitzgerald, M., Joseph, A., Hayes, M., & O'Regan, M. (1995). Leisure activities of adolescent schoolchildren. *Journal of Adolescence, 18*(3), 349–358.

Forgan, L. (2013, January 13). Arts Council chief accuses Gove of abandoning cultural education. *The Guardian*. Retrieved January 31, 2017, from http://www.guardian.co.uk/culture/2013/jan/15/arts-council-chief-gove-education.

Gall, M. (2013). Trainee teachers' perceptions: Factors that constrain the use of music technology in teaching placements. *Journal of Music, Technology & Education, 6*(1), 5–27.

Gall, M., & Breeze, N. (2005). Music composition lessons: The multimodal affordances of technology. *Educational Review, 57*(4), 415–433.

Gall, M., Lazarus, E., Tidmarsh, C., & Breeze, N. (2009). Creative designs for learning. In R. Sutherland, S. Robertson, & P. John (Eds.), *Improving classroom learning with ICT* (pp. 88–114). London: Routledge.

Gall, M., Sammer, G., & de Vugt, A. (Eds.). (2012). *European perspectives on music education: New media in the classroom*. Innsbruck: Helbling.

Giles, A. (n.d.). *Whirligig Snippets*. Retrieved January 28, 2017, from http://www.whirligig-tv.co.uk/tv/memories/snippets/snippets7.htm

Gov.UK. (n.d.a). *Collection: National Curriculum*. Retrieved January 31, 2017, from https://www.gov.uk/government/collections/national-curriculum.

Gov.UK (n.d.b). *National curriculum in England: Music programmes of study*. Retrieved January 31, 2017, from *https://www.gov.uk/government/publications/national-curriculum-in-england-music-programmes-of-study*

Grønsdal, I. (2012). ICT in music education in Norwegian secondary schools. In M. Gall, G. Sammer, & A. de Vugt (Eds.), *European perspectives on music education: New media in the classroom* (pp. 177–191). Innsbruck: Helbling.

Hennessy, S. (2012). Improving primary teaching: Minding the gap. In G. McPherson & G. Welch (Eds.), *The Oxford handbook of music education* (Vol. 2, pp. 625–628). Oxford: Oxford University Press.

Historical Boys' Clothing (2004). *English Schools Dance Program: Music and Movement*. Retrieved January 28, 2017, from http://histclo.com/act/dance/les/cou/eng/sdle-mam.html

Höfer, F., & Reubenz, J. (2012). New media in Austrian music teaching in school. In M. Gall, G. Sammer, & A. de Vugt (Eds.), *European perspectives on music education: New media in the classroom* (pp. 31–43). Innsbruck: Helbling.

Jautakyte, Z. (2012). ICT in Music: A Case of Lithuanian upper school education. In M. Gall, G. Sammer, & A. de Vugt (Eds.), *European perspectives on music education: New media in the classroom* (pp. 149–162). Innsbruck: Helbling.

Juniper, D. (2004). The First World War and radio development. *History Today, 54*(5), 32–38.

Kaplan, E. A. (1987). *Rocking around the clock: Music television, postmodernism and consumer culture*. London: Methuen.

Katz, M. (2004). *Capturing sound: How technology has changed music*. Berkeley: University of California Press.

Konkol, G. K., & Kierzkowski, M. (2012). ICT in music teaching in the context of a new education system in Poland. In M. Gall, G. Sammer, & A. de Vugt (Eds.), *European perspectives on music education: New media in the classroom* (pp. 193–204). Innsbruck: Helbling.

Leask, M. (2002). *The new opportunities fund: Training for teachers and school librarians in the use of ICT.* London: Teacher Training Agency.

Lamont, A. (2002). Musical identities and the school environment. In R. A. R. MacDonald, D. J. Hargreaves, & D. E. Meill (Eds.), *Musical identities* (pp. 41–59). Oxford: Oxford University Press.

Larson, R., Kubey, R., & Colletti, J. (1989). Changing channels: Early adolescent media choices and shifting investments in families and friends. *Journal of Youth and Adolescents, 18*(6), 583–599.

Lenhart, A., Purcell, K., Smith, A., & Zickuhr, K. (2010). *Social media and mobile internet use among teens and young adults.* Washington, DC: Pew Internet and American Life Project.

Majeski, B. T. (1990). *A history of the US music industry.* Englewood, NJ: Music Trades.

McSwain, R. (1995). The power of the electric guitar. *Popular Music and Society, 19*(4), 21–40.

Mellor, L. (2008). Creativity, originality, identity: Investigating computer-based composition in the secondary school. *Music Education Research, 10*(4), 451–472.

Mundy, J. (1999). *Popular music on screen: From Hollywood musical to music video.* Manchester: Manchester University Press.

Myllykoski, M. (2012). The use of ICT and music technology in Finnish music education. In M. Gall, G. Sammer, & A. de Vugt (Eds.), *European perspectives on music education: New media in the classroom* (pp. 115–124). Innsbruck: Helbling.

National Music Council. (1999). *A sound performance: The economic value of music to the United Kingdom.* London: NMC/KPMG.

Odam, G. (2000). Teaching composing in the secondary school: The creative dream. *British Journal of Music Education, 17*(2), 109–127.

Office for Standards in Education. (2004a). *2004 report: ICT in schools—the impact of government initiatives: Secondary music (HMI 2189).* Manchester: Ofsted.

Office for Standards in Education. (2004b). *Ofsted subject reports 2002/3: Music education in secondary schools (HMI 1981).* Manchester: Ofsted.

Office for Standards in Education. (2009). *Making more of music: An evaluation of music in schools 2005–08. (HMI 080235).* London: Ofsted.

Office for Standards in Education. (2012). *Music in schools: Wider still, and wider.* Retrieved January 31, 2017, from http://www.ofsted.gov.uk/resources/music-schools-wider-still-and-wider.

Office for Standards in Education. (2013). *Music in schools: What hubs must do (Ref 130231).* Retrieved January 31, 2017, from http://www.ofsted.gov.uk/resources/music-schools-what-hubs-must-do.

Paul Hamlyn Foundation. (n.d.). *Music futures: Resources.* Retrieved January 31, 2017, from https://www.musicalfutures.org/Resources.

Pearson (2017). *Qualifications. Music: Overview.* Retrieved January 31, 2017, from https://qualifications.pearson.com/en/subjects/music.html

Pfeil, U., Arjan, R., & Zaphiris, P. (2008). Age differences in online social networking—A study of user profiles and the social capital divide among teenagers and older users in MySpace. *Computers in Human Behaviour, 5*(3), 643–654.

Philpott, C. & Carden-Price, C. (2001). Public examinations 1: the General Certificate of secondary Education (GCSE) in music. In C. Philpott (Ed.) *Learning to Teach Music in the Secondary School* (pp. 177–200). London/New York: RoutledgeFalmer.

Raacke, J., & Bonds-Raacke, J. (2008). MySpace and Facebook: Applying the uses and gratifications theory to exploring friend-networking sites. *Cyberpsychology & Behavior, 11*(2), 169–174.

Richard, B., & Kruger, H. H. (2005). Ravers' paradise: German youth cultures in the 1990s. In T. Skelton & G. Valentine (Eds.), *Cool places: Geographies of youth culture* (pp. 162–175). London: Routledge.

Risset, J-C. (2004). The liberation of sound, art-science and the digital domain: Contacts with Edgard Varèse. *Contemporary Music Review, 23*(2), 27–54.

Russcol, H. (1972). *The liberation of sound*. London: Prentice Hall.

Sachs, C. (2006). *The history of musical instruments*. New York: Dover.

Salavuo, M. (2006). Open and informal online communities as forums of collaborative musical activities and learning. *British Journal of Music Education, 23*(3), 253–271.

Sammer, G., Gall, M., & Breeze, N. (2009). Using music software at school: The European framework. In G. Fiocchetta & F. Ballanti (Eds.), *NET Music Project 01: New education technology in music field* (pp. 155–177). Rome: Anicia srl.

Savage, J. (2012). Moving beyond subject boundaries: Four case studies of cross-curricular pedagogy in secondary schools. *International Journal of Educational Research, 55*, 79–88.

Savage, J., & Butcher, J. (2007). DubDubDub: Improvisation using the sounds of the Worldwide Web. *Journal of Music, Education and Technology, 1*(1), 83–96.

Savage, J., & Challis, M. (2001). Dunwich revisited: Collaborative composition and performance with new technologies. *British Journal of Music Education, 18*(2), 139–149.

Savage, J., & Challis, M. (2002). A digital arts curriculum? Practical ways forward. *Music Education Research, 4*(1), 7–23.

Savvidou, D. (2012). The use of ICT in music education: The Cypriot perspective. In M. Gall, G. Sammer, & A. de Vugt (Eds.), *European perspectives on music education: New media in the classroom* (pp. 55–76). Innsbruck: Helbling.

Scheid, M., & Stranberg, T. (2012). Schools' permeable walls and media cultures: An example of new prerequisites for music education in Sweden. In M. Gall, G. Sammer, & A. de Vugt (Eds.), *European perspectives on music education: New media in the classroom* (pp. 237–251). Innsbruck: Helbling.

Schrader, B. (1982). *Introduction to electro-acoustic music*. Englewood Cliffs, NJ: Prentice Hall.

Sergeant, D., & Boyle, J. D. (1980). Contextual influences on pitch judgement. *Psychology of Music, 8*(2), 3–15.

Sexton, J. (Ed.). (2007) *Music sound and multi-media: From the live to the virtual.* Edinburgh: Edinburgh University Press.

Smith, J. (2004). I can see tomorrow in your face: A study of "Dance Dance Revolution" and music video games. *Journal of Popular Music Studies, 16*(1), 58–64.

Smith, K. (2004). An investigation into the experience of first year students at British universities. *Arts and Humanities in Higher Education, 3*(1), 81–93.

Spruce, G. (2002). Ways of thinking about music: Political dimensions and educational consequences. In G. Spruce (Ed.), *Teaching music in secondary schools: A reader* (pp. 3–24). London/New York: RoutledgeFalmer.

Stowell, D., & Dixon, D. (2014). Integration of informal music technologies in secondary school music lessons. *British Journal of Music Education, 31*(1), 19–39.

Tattersall, K. (1994). The role and functions of public examinations, assessment in education. *Principles, Policy & Practice, 1*(3), 293–304.

Terry, P. (2009). *GCSE music study guide* (3rd ed.). London: Routledge.

Théberge, P. (1997). *Any sound you can imagine: Making music/consuming technology.* Middletown, CT: Wesleyan University Press.

Tobias, E. (2012). Let's play: Learning music through video games and virtual worlds. In G. McPherson & G. Welch (Eds.), *The Oxford handbook of music education* (Vol. 2, pp. 531–548). Oxford: Oxford University Press.

Vulliamy, G. (1976). Pupil-centred music teaching. In G. Vulliamy & E. Lee (Eds.), *Pop music in school* (pp. 49–61). Cambridge: Cambridge University Press.

Werner, A. (2009). Girls consuming music at home: Gender and the exchange of music through new media. *European Journal of Cultural Studies, 12*(3), 269–284.

Wright, P. (2002). ICT and the music curriculum. In G. Spruce (Ed.), *Teaching music in secondary schools: A reader* (pp. 143–165). London/New York: RoutledgeFalmer.

Zehnder, S. M., & Lipscomb, S. D. (2006). The role of music in video games. In P. Vorderer & J. Bryant (Eds.), *Playing video games: Motives, responses, and consequences* (pp. 241–258). Marwah, NJ: Lawrence Erlbaum.

1 A

Further Perspectives

CHAPTER 3

SAVORING THE ARTISTIC EXPERIENCE IN AN AGE OF COMMODIFICATION

CHEE-HOO LUM

The advancement, availability, and proliferation of technology have broadened both the collaborative and personal learning spaces for music learners. The digital learner now has a plethora of options for engaging through technology with music as a listener, composer/improviser, and performer that one could hardly have imagined possible just a few decades ago. Pedagogies in the teaching of general music to medium and large class sizes have also transformed because of technology, as more student-centered approaches are made possible through self-directed technological music learning tools and the ability of the teacher to multitask and capture individual musical learning processes.

The lines between amateur and professional musician are also gradually blurring through the aid of technology. Pitch, rhythm, volume, and timbre are only a few examples of musical elements that can easily be explored and manipulated through the aid of technological tools. The digital learner's musical palette thus broadens beyond and is not restricted to the confines of skill and technique that the physical body can manage. Mediated music has also opened up a limitless selection and access for the listener.

Peppered with a broad range of music choices through technology, music educators can have a challenging time trying to cater to the musical appetites, curiosity, and creativity of the learner. The ease with which musical skills and knowledge can literally be had at the fingertips (with efficiency, accuracy, and speed) also presents a monumental challenge for music educators in reflecting on their rapidly changing roles as teachers and facilitators in the music classroom amid learners' unabated pursuit and celebration of technology in music.

Savoring

In suggesting what education can learn from the arts, Eisner (2002) pointed out that a key characteristic in an artistic experience is

> the importance of taking one's time to relish the experience that one seeks ... [of] paying close attention to what is at hand, of slowing down perception so that efficiency is put on a back burner and the quest for experience is made dominant. There is so much in life that pushes us toward the short term, toward the cursory, toward what is efficient and what can be handled in the briefest amount of time. The arts are about savoring. ... Enabling individuals to learn how to attend with an eye toward the aesthetic and with time to undergo its flavors is a nontrivial outcome of education that is perhaps most acutely emphasized in effective art education. (p. 207)

Perhaps this is a timely reminder that notions of slowing down perception, paying close attention, and savoring can sometimes be polar opposites of the efficiency and speed that technology brings into the music learning environment. Depending on the musical context, if savoring is deemed significant in the development of the music learner toward becoming a feelingful embodied being, refining sensitivities and subtleties toward the aesthetic experience, then music educators need to focus careful attention on (1) critically scaffolding "savoring" processes in their facilitation of music digital learners, and (2) engaging them musically and aesthetically while recognizing their growing hunger for ever-alluring technological music commodities.

Taking the live engagement with instruments as a case in point, physically playing on a guzheng is quite different from playing the guzheng on an app. The intensity and articulation of body movements and vibrations felt on the body as it engages with the instrument involved and the aural acuity and sensitivity needed to articulate particular dynamics and expressions through the instrument are never quite equivalent to its technological counterpart.

So, while a good way to get students engaged musically in a general music class is by watching YouTube video clips of performances or listening to good audio tracks of guzheng repertoire, or by using a music app to enable a listening, composing, or even performing experience, music educators also need to engage students directly with the acoustic instrument whenever possible so that they can get to touch, feel, and play the guzheng in order to savor the timbral qualities, the physical articulation, and the aesthetic experience of having an actual guzheng player perform for them. This kind of experience might apply to many acoustic instruments with which technology has now created diverse options for musical engagement. Strumming a guitar, hitting the cajon, and blowing on the shakuhachi are hardly the same musical experiences as tapping on the touchscreen of a tablet to simulate these instruments.

Commodification

In Singapore, as in many parts of the world, technology is seen as a crucial component in education; thus, substantial resources from the government have been devoted to the provision and upgrading of technological facilities (both hardware and software) within the formal school system. This was evident in the opening address by Dr. Ng Eng Hen (former minister for education) announcing the launch of the third Masterplan for information and communication technology (ICT) in education (2008), in which he presented four key strategies: (1) bringing ICT into the core of the education process, to integrate ICT during planning and design of lesson plans and work through implementation details of curriculum and assessment; (2) focusing on improving the capabilities and skill sets of teachers; (3) improving the sharing of best practices and successful innovations; and (4) further building up infrastructure where it is needed to upgrade the technology to maximize the potential of ICT.

In more recent years, within local music education, technological provisions have also significantly increased, ranging from the establishment of keyboard and computer labs (specific to music compositional and recording purposes) to the provision of individual tablets for students to use during music classes. Digital music collaborative spaces have also been encouraged in some schools. Inclusion of technology aimed at students' developing skill sets (like operating recording systems, creating varied commercially viable music loops, using notational software, etc.) suitable for entering the creative music industry is also becoming more pervasive, with polytechnics offering diplomas in music and audio technology and the Normal (Technical) syllabus (Singapore Examinations and Assessments Board, 2016) targeting similar skill sets: "the Normal (Technical) music syllabus aims to provide students with a broad-based music education that is relevant to new developments in the music and music-related industries. It recognizes the use of music technology in contemporary musical expression and experience, and integrates music technology in developing basic skills and creativity. It fosters students' musicianship and understanding through exposure to a wide range of music" (p. 2). Even within the stated aims and objectives of the general music program in Singapore (Ministry of Education 2008), music technology is featured to some extent in the program's key objectives of discerning and understanding music from various cultures and of various genres, and understanding the role of music in daily living through "the use of technology in creating the varied identity of contemporary music (e.g. loops in dance music); [the student is expected to] describe how music conveys ideas and communicates messages in advertisement, MTV, film, documentary" (p. 8).

These links between music education and technology have led to a suggestion by some local general music educators that general music students in the primary and secondary schools should be grounded in foundational Western music systems, both theoretically and in practical terms. After all, they say, Western classical harmonic theory is seen as the basis for acquiring practical knowledge of the construction of popular music (needed for the creative industry). The pervasiveness of media (like YouTube and other sound/video sources) in the proliferation of popular music culture cannot be denied. To develop in students a discerning ear for engaging with popular music, it is the responsibility of music educators to develop such skills and knowledge in students. This leads to the belief that particular "fundamentals" are essential, further reinforcing Western music systems' position within the local music education system as a convenient "truth." Greene warns: "we must acknowledge the fixities and corruptions of our consumer-based and technicized culture. We must take into account the languages of technology and violence, even as we do the mis-education in much that is done in schools . . . to do what we can to include within it the voices of the long silent or unheard in this country" (1995, p. 56). Discerning music educators are mindful to include popular music in their general music curriculum so that students can relate and critically engage in what they encounter in their daily musical experiences with the music classroom context, listening, performing, re-creating and creating popular music in a *glocalized* way beyond just mere replication. Music educators also need to be cognizant of the many marginalized musical cultures/voices amid the local/regional that may not have benefited as much from the power of media and commodification and remain underserved. A cautionary note needs to be sounded out to music educators so that they are mindful of the need to provide a diverse and inclusive range of musical voices and contexts in the music classroom, to offset the proliferation of commodified musics through technology and media influences, entering ever so easily into the ears, eyes, and minds of music consumers young and old.

Beyond the Singular Experience

Linked to the commodification of music is the musical event for the digital learner in the twenty-first century, which is often beyond just an aural experience. Visuals to accompany the musical experience, ranging from MTVs to advertisements and films, are commonplace. The digital learner is constantly immersed in a multimedia, multisensorial environment through technology. The politics of consumerism and materialism in popular culture notwithstanding, the reality is that digital learners are getting bombarded and surrounded by multimedia experiences all the time. Music educators need to be mindful that the singular musical experience that is provided in classroom activities may sometimes not appeal to these digital learners. A multiarts experience would seem closer to the experiences of digital learners in their daily encounters.

In Singapore's schools, music and visual arts general programs are compulsory for primary and lower secondary schools, often with dance and drama programs offered not within the main curriculum but as cocurricular options or enrichment programs. Music and visual art programs are also run as separate subjects with little overlap. The arts in local schools are thus seen as four distinct entities, making multiarts experiences uncommon in students' arts experiences in schools. Local music teachers are hardly prepared within their professional training and development to deal with other art forms, and most music teachers also are not well equipped with technological skills to facilitate multimedia and multiarts experiences for students. There is no visible indication that paradigms will shift in the near future within formal school systems, in terms of a more holistic approach to the arts, structured within a multimedia, multiarts environment. This is food for thought, in terms of future planning for music and arts education, about reflecting and engaging current realities toward technologically integrated arts environments that are already quite the norm in the daily experiences of students' lives.

References

Eisner, E. (2002). *The arts and the creation of mind.* New Haven, CT: Yale University Press.

Greene, M. (1995). *Releasing the imagination: Essays on education, the arts and social change.* San Francisco: Jossey-Bass.

Ng, E. H. (2008, August 5). *Opening address by Dr. Ng Eng Hen, Minister for Education and Second Minister for Defence, at the International Conference on Teaching and Learning with Technology (ICTLT), Suntec Convention Hall.* Retrieved July 20, 2014, from http://www.moe.gov.sg/media/speeches/2008/08/05/opening-address-by-dr-ng-eng-h-1.php.

Singapore Examinations and Assessments Board. (2016). *Music Syllabus T, GCE Normal (Technical) Level (2016) (Syllabus 6129).* Retrieved December 7, 2016, from http://www.seab.gov.sg/content/syllabus/nlevel/2016Syllabus/6129_2016.pdf.

Ministry of Education. (2008). *General music programme syllabus (primary/secondary).* Retrieved July 20, 2014 from www.moe.gov.sg/education/syllabuses/.../general-music-programme.pdf.

Ministry of Education. (2014). *Normal course curriculum.* Retrieved July 20, 2014, from http://www.moe.gov.sg/education/secondary/normal/.

CHAPTER 4

MUSIC TECHNOLOGY IN ETHNOMUSICOLOGY

GABRIEL SOLIS

Ethnomusicology has often had an ambivalent relationship with technology, as I see it. Officially, we owe our discipline to mid-twentieth-century developments in recording technology. Nevertheless, in my experience there is a strong counter-modern streak that characterizes ethnomusicologists as a group. I might not go so far as to call us Luddites as such (though that extreme may be represented in small numbers) but would say that our broad commitment to participatory music making, to modernity's others, and to artisanal music represents an antitechnological subtext in our research practice and pedagogy. My perspective on this comes as a scholar of jazz and popular music, an ethnographer of Afro- and Indigenous modernities, and a fan of mass-mediated musics. To explore this perspective I offer some reflections from my experiences teaching the history of ethnomusicology and doing research since the 1990s. In the end I call for a move from ambivalence to critique.

Teaching the Origins of Ethnomusicology

I teach a course on the history of theory and method in ethnomusicology for first-year graduate students. The course has two different origin stories embedded in it, both of which revolve around sound recording. The first of these locates the beginnings of comparative musicology as a late-nineteenth-century break with musical folklore studies. This story focuses on two things: Alexander Ellis's experiments with the measurement of pitch, and the Berlin Phonogramm-Archiv's collection of recordings from around the world under the curatorship of Carl Stumpf and Erich von Hornbostel. We read from

Ellis's article "On the Musical Scales of Various Nations" (1885) and from a number of sources on the Phonogramm-Archiv (e.g., Nettl, 2005; Sachs, 1961).

As it happens, both of these scholarly accomplishments were possible only because of contemporary advances in technology, but the narratives in which they are now embedded differ considerably in the ways they make music technology visible. Ellis relied on a tonometer to measure pitch to the nearest cent in his work, and yet the ethnomusicological literature that references him spends almost no time on this matter, focusing instead on the more human aspects of his method and conclusions. We tend to see a precedent for later work in the fact that he turned to experts in the musical systems he wanted to measure in order to get authoritative pitches for his tonometer; likewise, we appreciate his cultural relativism in determining not a single system of pitch but rather a world of distinct, incommensurable systems.

By contrast, much of the discussion of the Berlin school focuses primarily on the centrality of recordings to Stumpf and Hornbostel's work. The term "armchair ethnomusicology" gets to the heart of the sense of unease with which later ethnomusicologists have responded to the studies that comparative musicologists derived from the recordings housed in the archive. The narrative my students derive from looking at the Phonogramm-Archiv—a narrative I think represents something like a consensus view—is one in which music technology stood in for a more holistic, humane, and ultimately authentic kind of musical experience. There is a certain appreciation for Stumpf and Hornbostel's work, but in the end their authority is undercut by their reliance on recordings (Nettl, 2005; Goble, 2010, p. 46).

The second origin story describes the beginnings of contemporary ethnomusicology as a break from comparative musicology in the 1950s. While the primary explanation of ethnomusicology's differentiation from its predecessor was the interdisciplinary fusion of musicology and anthropology, what made this fusion possible in the work of a single researcher was a technological advance of the time: the development of (relatively) easy to use, (relatively) lightweight, portable reel-to-reel tape recorders. This advance facilitated two changes in research protocol. First, and perhaps most obviously, it allowed one relatively hale person to transport and operate the recording device. Second, and perhaps more important, it allowed the ethnomusicologist to make recordings in a far wider array of locations and contexts. In a sense, true musical ethnography, in which the discipline would move from the study of musics of the world's cultures to the study of music in/as culture, was made possible by this device.

That said, my teaching (and the literature in ethnomusicology more broadly) quickly forgets the remarkable impact of this music technology. There are pedagogical articles that discuss such technical issues as recording media, tape speed, microphone placement, and so on, but for the vast majority of ethnomusicology from any time since the 1960s, these issues are at most a footnote (Myers, 1992). Want to know what kind of microphones someone used to document gamelan in a Javanese village? Want to know what kind of tape (or later, what kind of file) they used or where they placed microphones? You may find some documentation in Folkways or Nonesuch recording notes, but probably not in a scholarly publication.

I wouldn't make the argument that we should always discuss the specifics of our music technology in more detail, but I do believe the fact that we don't is telling. It underscores our ambivalence. There are exceptions, of course—Steven Feld has been particularly attentive over the years to an explicit discussion of the role of technology as a mediator between himself and his interlocutors in Papua New Guinea and elsewhere, as has Simha Arom in Africa, among others (Feld & Brenneis, 2004; Arom, 1991). But as a rule, I think we have argued that the "real" ethnomusicology happens in an embodied exchange in the field.

Technology and the Interpretive Turn

The next major shift in the history of ethnomusicology came at the time I was a student: the "interpretive turn." In the late 1980s and 1990s ethnomusicologists drew on the work of the "Writing Cultures" group, a number of critically inclined anthropologists whose work was collected in James Clifford and George E. Marcus's book by that name (Clifford & Marcus, 1986). This work, which was also heavily influenced by Clifford Geertz's *Interpretation of Cultures* (1973), as well as semiotics and practice theory, reimagined both the object of ethnographic study and the nature of its texts. From Geertz it took the idea that culture was the "webs of significance [man] himself spun," and that ethnography therefore was "not an experimental science in search of law but an interpretive one in search of meaning" (Geertz, 1973, p. 5). Even further, the Writing Cultures group argued that the interpretive ethnography in search of meaning was more like fiction in its goals than like science.

At the same time that we realized that we were interpreters of the musical lives of our interlocutors, ethnomusicologists began to discover the popular music that imbued those lives. Rather than evidence of "cultural greyout"—the fear that mass media and cultural imperialism were making the world's musics uniform—we came to see a world full of culturally distinct, socially innovative mass-mediated musics. We were no longer the only ones using music technology; our interlocutors everywhere from Uganda to Mongolia, and from the arctic to the tropics, were making their own recordings, running their own studios, listening with Walkmans, and playing synthesizers.

The question all this raised for me, the ambivalence it produced, had to do with ethnologists' use of technology in the field. If the goal of ethnographic scholarship was to produce thickly interpretive, novelistic accounts of the musical lives of the community I studied, to what extent did I need to invest in high-tech sound and video recording gear? On the one hand the era saw the rise of increasingly miniature digital recording—first on tape, then on disc, and eventually solid-state. I could make high-quality recordings and edit them without the hassle of managing tape. On the other hand I never expected to make documentary films or release field recordings. In the end, I took advantage of the miniaturization of recording technology to make recording interviews as unobtrusive as possible.

Technology remained fundamental to the practice of ethnomusicology, but our own use of it remained relatively opaque. There were a small number of studies that took the technologized lives of musicians as an object of ethnography—Louise Meintjes's monograph on South African recording studios, *Sound of Africa* (2003), stands out to me as one of the best—and some that looked at the impact of particular technologies on cultures of musical listening in the world—Peter Manuel's *Cassette Culture* (1993) is perhaps the most influential. Still, our own use of technology, from recording to telecommunications, remained opaque in much of our scholarship. As before, but perhaps even more because of the language of interpretive critical theory, we saw real ethnomusicological insight coming from embodied, human interaction in the field.

Technology and the New Digital Humanities

I believe we are in another major point of articulation in the field today, one characterized by the rise of the so-called digital humanities. Digital humanities (or DH, as its leading figures call it) purports to offer fundamentally new ways of viewing the traditional subject matter not only of one or another field, but of humanistic study at large. It would see itself as a Kuhnian paradigm shift. I suggest that ethnomusicology might bring some of the same ambivalence it has had toward technology in the past to bear on its engagement with digital humanities in the present. The digital turn, as I see it, offers considerable opportunities, but its technological orientation needs to be leavened with the more traditionally humanistic concerns that have characterized ethnomusicology from the start, not least the attention to embodied, human interaction in music.

I'm not the only one to take this position. In a position piece published in *Hybrid Pedagogy*, Adeline Koh, a leading scholar of postcolonial digital humanities, calls for a renewed effort to support "work that focuses more on culture than computation," in order to make digital humanities critical in the core humanistic sense, not "simply a handmaiden to STEM." Likewise, Helle Porsdam has argued in an article in *Digital Humanities Quarterly* that we "need to find a better balance between the 'how' of the new digital methods and technologies and the 'what' of older, more traditional humanities methods" (2013, n.p.). Her reasoning stems from a concern that, so far, digital humanities has tended to produce "information" (in her terms) without turning that information "into knowledge by means of critical interpretation and contextualization." I do not yet see this playing out extensively in ethnomusicology, in large measure I think because of our historical ambivalence to technology. We have simply not been strongly represented in digital humanities, but I think we could be, and in so being could add something of exactly the nature that Koh and Porsdam are both calling for. My experiences with digital humanities may be representative of why we have not done so yet.

In 2008, I received a small grant from the Madden Foundation as part of a targeted initiative broadly conceived as the study of "technology and the arts," which would now surely be worded using the term "digital humanities." The grant supported my field work in the Northern Peninsular Area (NPA) of Cape York in North Queensland, Australia, with local Indigenous people—members of the Angamuthi and Injinoo Ikya language groups. My proposal was to use digital technology, principally GPS tagging, to help members of the local community produce maps of the relationship between Angamuthi-language songs and sacred places in the landscape of the NPA, a method that a number of Australian scholars had shown to be useful. As it turned out once I got to Cape York, the digital component of the research was only a small part of the real question. For the local community the most valuable thing was not the GPS tagging but rather the process of recording local repertoires, many of which had never been committed to tape. The archival record was important, but even more so was my ability to make dubs on CD at the moment; for the next few weeks, at least, and perhaps longer, Johnny Mack's synthesizer and vocals, Larry Woosup's guitar, and a group of community women and men's backing vocals were in regular rotation in cars and trucks in the five towns of the NPA. For me the most compelling scholarly questions couldn't be answered by the GPS data, or by the recordings at all, for that matter. They revolved around how people asserted ownership over song items, how they maintained songs over time, how they built affective ties with repertoire—through contemporary social ties and through connections between music, history, and memory. A large-scale collection of GPS-tagged song data for Indigenous music in Australia remains a worthwhile endeavor, and collaboration with digital humanities scholars with computer science backgrounds offers the potential to uncover locally valuable information from that data; but making meaningful knowledge from that information will require the critical work of local knowledge-holders and scholars with the requisite humanities background to see how to make the data speak to critical questions such as those I encountered in Cape York.

The J-DISC project, a digital jazz studies collective based at Columbia University's Center for Jazz Studies, of which I am a member, offers another model for how this can happen. This project has brought together jazz musicians, archivists, humanists, and STEM specialists to explore the production of a jazz research platform, a kind of enhanced discographical archive. As I see it, two types of music digital humanities are emerging from this research group: big data "distant listening" projects and computer-assisted, intensified "close listening" projects. The first involve such things as using a computer to derive and statistically model information about tempo from the entire collection of jazz recordings in the J-DISC archive. The second involve such things as using a computer to model timbral variation over time in avant-garde jazz recordings. Both of these kinds of projects are potentially useful for extending ethnographic and historical jazz studies, but it is important to bring the history of ethnomusicology to bear on them so as to maintain a critical ambivalence about the power of technology to provide answers—to resist the "hegemonic drive of the qualitative approach" to humanities scholarship, as Porsdam puts it (2013).

To return to the teaching I began these reflections with, I am struck by the fact that the history of ethnomusicology is peppered with both big data distant listening and technologically assisted close listening: from Mieczyslaw Kolinski's studies of "reiteration quotients" (1982) and Alan Lomax's cantometrics (1962) on the one hand to the uses of the Seeger Melograph (Hood, 2000) on the other, we have been down this path before, albeit with smaller data sets and less sophisticated signal processors. We have the opportunity to learn not only from the new technology but also from the shortcomings in research design and analysis of this history.

Conclusion: From Ambivalence to Critique

Ethnomusicology and music education have historically shared a number of things, not least a value placed on research drawn from embodied musical interaction. A look at the history of ethnomusicology makes it clear that technology has mediated that research from the beginning, even as we have maintained a kind of ambivalence about the role of that technology. It is my position, however, that ambivalence is no longer a useful stance to take. Ambivalence is, ultimately, unproductive. It precludes forthright evaluation of the opportunities and consequences of technology, and leads to a haphazard adoption of those technologies that are ready to hand. Whether we continue to pursue projects based in participant observation or newer methodologies involving the statistical analysis of big data (or both), I think we would do well to reflexively turn a critical lens on our own use of technology. We have been better at seeing the affordances and limitations music technology has had for our research subjects than we have for our own research. Understanding the history of the discipline is a good starting point for developing such a perspective

References

Arom, S. (1991). *African polyphony and polyrhythm: Musical structure and methodology.* (M. Thom et al., Trans.). Cambridge, UK: Cambridge University Press.

Clifford, J., & Marcus, G. (Eds). (1986). *Writing culture: The poetics and politics of ethnography.* Berkeley: University of California Press.

Ellis, Alexander J. (1885). On the musical scales of various nations. *Journal of the Society of Arts, 33*, 485–527.

Feld, S., & Brenneis, D. (2004). Doing anthropology in sound. *American Ethnologist, 31*(4), 461–474.

Geertz, C. (1973). *The interpretation of cultures: Selected essays.* New York: Basic Books.

Goble, J. S. (Ed.). (2010). *What's so important about music education?* New York: Routledge.

Hood, K. M. (2000). Ethnomusicology's Bronze Age in Y2K. *Ethnomusicology, 44*(3), 365–375.

Koh, A. (2015, April 19). A letter to the humanities: DH will not save you. *Hybrid Pedagogy*. Retrieved June 1, 2015, from http://www.hybridpedagogy.com/journal/a-letter-to-the-humanities-dh-will-not-save-you/.
Kolinski, M. (1982). Reiteration quotients: A Cross-cultural comparison. *Ethnomusicology*, 26(1), 85–90.
Lomax, A. (1962). Song structure and social structure. *Ethnology, 1*, 425–451.
Manuel, P. (1993). *Cassette culture: Popular music and technology in North India*. Chicago: University of Chicago Press.
Meintjes, L. (2003). *Sound of Africa! Making music Zulu in a South African studio*. Durham, NC: Duke University Press.
Myers, H. (Ed). (1992). *Ethnomusicology: An introduction*. New York: Norton.
Nettl, B. (2005). *The study of ethnomusicology: Thirty-one issues and concepts*. (2nd ed.). Urbana: University of Illinois Press.
Porsdam, H. (2013). Digital humanities: On finding the proper balance between qualitative and quantitative ways of doing research in the humanities. *Digital Humanities Quarterly, 7*(3). Retrieved June 1, 2015, from http://www.digitalhumanities.org/dhq/vol/7/3/000167/000167.html.
Sachs, C. (1961). *The wellsprings of music*. (Jaap Kunst, Ed.). The Hague: Martinus Nijhoff.

CHAPTER 5

THE ROLE OF "PLACE" AND CONTEXT

JANICE WALDRON

IMPROVING practice is all about challenging assumptions—one of the principal reasons, I think, for the existence of this book. But rocking the "habitus" boat comes with its own set of problems; primarily that, in our eagerness to embrace the virtues of the "new," we sometimes fail to critically examine the *a priori* of the very thing we are extolling. Exuberant positive narratives alone are not enough; success in changing a dominant paradigm—no matter what that may be—requires transparency and critical reflection on the part of those doing the advocating.

One primary concern to consider is the role and influence that local context and culture play in the issue under discussion—in this case, technology use and music learning and teaching. Those advocating for technology use in the field usually begin by raising relevant issues based on personal and meaningful but localized narratives. Although this is a good place to start—people rarely argue for change not grounded in their own experiences—building arguments for technology use also requires a more nuanced interpretation of what technology in music learning and teaching means to and for practitioners and researchers in the larger global context. Specifically, how does technology's evolution from "thing" to "thing and place" change our perceptions of its use(s) in music learning and teaching? How do the roles of local context, cultural assumptions, and musical genre fit into a discussion of what constitutes technology and technology in music education?

THE EVOLUTION OF SOFTWARE, HARDWARE, AND THE WEB: "THINGS" AND "PLACES"

First, the term "technology" needs unpacking and redefining within a general historical context as well as in music learning and teaching. In the 1970s, technology generally meant hardware and software; in music education contexts, technology as physical "thing" also included electronic instruments and analog recording devices (Webster, 2011). With widespread Internet availability in the 1990s, "technology" evolved to become an all-inclusive term for software, hardware, and the World Wide Web—thus designating both "place(s)" and "things"—and the ability to make and share content with anyone with an Internet connection became a reality now taken for granted (Turkle, 2011).

In music learning and teaching, "technology" now also includes digital instruments and digital recording hardware and software, along with global access to myriad music cultures, genres, and online communities. The last constitutes perhaps the biggest change in technological developments and advancements in music education because knowledge acquisition and the collaborative opportunities offered by social media, video conferencing technology, and cloud storage systems via the Internet easily transform music making and music learning and teaching from being a local activity into becoming a global one. Earlier forms of computer-mediated communication—email, listservs, instant messaging, and chat rooms, for example—were designed to function as stand-alone nonnetworked applications, but because social media, video conferencing technology, and cloud storage systems function as collaborative social networks embedded within new media systems, they are powerful enablers of what Jenkins, Ford, and Green (2013) dub "spreadability."

At the core of "spreadability" is the idea that content, whether user-generated or corporately created, can be shared by anyone and distributed across a network of participants, thus giving it the potential to reach large-scale audiences. Not only are social networks continually created and re-created during this process, but the more "spreadable" the application, the larger the network can grow (Jenkins, Ford, & Green, 2013). A similar argument comes from communications researcher Leah Lievrouw (2011); she contends that "new media" systems facilitate online collaboration and participation by allowing people to create:

1) Material *artifacts* or devices that enable and extend people's abilities to communicate and share meaning;
2) Communication activities or *practices* that people engage in as they develop and use those devices; and,
3) Larger social *arrangements* and organizational forms that people create and build around the artifacts and practices. (p. 7)

Lievrouw concludes that new media systems are, at their core, fundamentally interactive because users can pick and choose their sources of information and cultural resources, and the most important effect of interactivity is that it promotes or provides the necessary conditions for active *participation* to occur. This then further facilitates the expansion of new media networks and systems (2011). She explains: "New Media systems do not just deliver content; people must actively use them to do something i.e. search, share, recommend, link, argue, and so on. *Use is an action by definition, which may encourage new media users toward more involved social and cultural participation online and off* (p. 19; emphasis added)." Studies on online music communities and music education research have already emerged that illustrate the ideas put forth by Jenkins, Ford, and Green (2013) and Lievrouw (2011) above. Examples include research on collaborative websites chronicling (1) composition in opera (Partti, 2014, 2013; Partti & Westerlund, 2013) and composition in general (Ruthmann, n.d.); (2) digitally mediated performances, such as those found in Eric Whitacre's Virtual Choir (see ericwhitacre.com) (Talbot & Paparo, 2013) and in the digital remix community Indabamusic.com (Michielse & Partti, 2015); and (3) sharing information and resources such as YouTube for music learning and teaching in various online music communities/affinity groups (Cayari, 2011; Salavuo, 2006; Waldron, 2013).

Local Contexts, Curriculums, Expectations, Assumptions, and "Philosophies"

As Gall in her chapter rightly points out, considering how useful technology is or could be in music education contexts must also include, articulate, and address the roles that local context and cultural assumptions hold in school music teaching. Just as "music education" is not a "one-size-fits-all" "thing," "technology in music education" has different meanings to different people depending on assumed local philosophies and performance practices, both of which are not always articulated in a written curriculum.

For example, chapter author Chee-Hoo Lum writes from a Singaporean perspective, which has inherited the mores of the UK conservatory model of a "musical education" and has its own issues deeply embedded in its colonialist DNA. But I would also argue that because the conservatory model stresses and rewards individual solo work over ensemble membership it is easier, on a practical level, to implement compositional, creative musical work with acoustic, electronic, and digital mediums through various technologies because the idea of musician as "soloist" is already deeply entrenched (see Brook & Upitis, 2014). The hurdle does remain of convincing practitioners of the benefits of what technologies can offer for creative musical work—but I would argue that this is an easier task to accomplish and implement in a context where music teaching and learning has traditionally been perceived more as an individualistic than as a group endeavor.

Now, contrast the UK/Commonwealth conservatory model with the standard North American model of music education—that of large acoustic performing ensembles. Band, choir, and orchestra have taken a bit of a bad rap over the past 10–15 years for various reasons, but they are as deeply embedded in the zeitgeist as the conservatory model is in places like the United Kingdom and elsewhere that share a Commonwealth colonialist history. Although there are individual competitions (solo and small ensemble,

All-Region/All-State Band, etc.) that are important in the North American context, "music education" is mostly seen as a group activity, where it is assumed and expected that the individual will subsume her musical individuality into the group for the greater good of all.

Nowhere is this more starkly evident than in my native Texas, where "music education" at the secondary level (grades 6–12) means membership in a large acoustic performing competitive ensemble—band, choir, and orchestra. "Technology" in music education is generally assumed to refer to any number of specific "tools," such as software for writing marching band shows and smartphones used as tuners/metronomes and video recording devices, as well as information sites for sharing and communicating practical issues (e.g., the website of the Texas Music Educators' Association, tmea.org). Technology's value in the Texas context lies in how much it can improve practice in performing groups in order to meet the end goal of receiving a First Division at regional and statewide competitions. Thus technology is neither perceived nor used as a primary means of delivering music instruction, and this speaks to an underlying philosophy regarding local practice of what "music teaching" is (or is not) (Waldron, 2006).

In my recent research (Waldron, in review), my participants, all of whom were Texas high school/middle school band directors, were perplexed when I asked them if they used technology to teach composition; this wasn't because they didn't think that it was or could be a good thing but because the idea of teaching composition wasn't anywhere on (or even close to being near) their radar screens—they'd never considered it before—and, gathering from the collective looks on their faces, they thought it an odd question for me to be asking. When they did answer the question, most seemed to think that it would *perhaps* be a "nice" thing to offer students—but not particularly useful because it would not contribute to their primary raison d'être, that is, winning a First Division/Sweepstakes at University Interscholastic League marching band, concert band, and sight reading competitions (or, in the Texas vernacular, "contests"). The Texas music education culture is a *very* strongly entrenched community of practice: I graduated from a Texas high school in 1975, and in 2013, for the first time since 1987, I attended the Texas Music Educators' Association Convention. At a foundational level, not much had changed since 1975; it very eerily felt like being in a time warp.

Whether one agrees or disagrees with the practice of music education in the state of Texas is irrelevant (and a discussion for another time and place); what is significant is that it is an example of how important it is to consider local cultural context when advocating for integrating technology use in music education practice in a *meaningful* way.

Related to the above are assumptions associated with Western art music. While genre-based and thus not as "place" specific as my two examples above, Western art music also carries with it unexamined issues of hegemony directly connected to its performance practices that still exist *and* thrive. Although this has begun to change, one must be careful not to make assumptions based on Western art music practice as to what is important in technology use for music education.

For example, the idea that musical "fluency" can universally be defined is flawed because it is based on the assumption that musical fluency and appropriate performance practice in Western art music equals, and is appropriate performance practice in, other music genres. One must be careful not to equate the idea of "musical fluency" with "good" musicianship in genres where playing by ear (combined with observational learning, kinesthetic learning, and/or some form of skeletal notation, for example tab notation, etc.) is the basis of performance practice; in those genres, "musical fluency" is the ability to learn the genre within the dictates of the performance practice, whatever those might be.

For example, I am reminded of Timothy Rice's study of Bulgarian bagpipers, in which his teacher—a well-known Bulgarian bagpiper—informed Rice that "he [Rice] did not have a bagpiper's fingers" (2008, p. 502). In the performance practice of Bulgarian bagpiping, "musical fluency" is equated with the ability to "read" players' fingers in tandem with playing by ear; the ability to read Western (or any other kind) of written notation is not, and never has been, part of that performance practice. Any arguments about the value of teaching musical "fluency" through technology use are context-dependent and unless stated otherwise, rendered moot. In a postmodern world, one must always be aware of hegemonic bias when making universal statements as to what music *is*, and, by extension, *what is*—*or what is not*—of value when advocating for technology use in music teaching and learning. Context and philosophical *a priori* matter.

References

Brook, J., & Upitis, R. (2014). Can an online tool support contemporary independent music teaching and learning? *Music Education Research, 17*(1), 34–47. http://dx.doi.org/10.1080/14613808.2014.969217.

Cayari, C. (2011). The YouTube effect: How YouTube has provided new ways to consume, create, and share music. *International Journal of Education and the Arts, 12*(6), 1–28.

Jenkins, H., Ford S., & Green, J. (2013). *Spreadbale Media: Creating Value and Meaning in a Networked Culture*. NY, NY: New York University Press.

Lievrouw. L. (2011). *Alternative and activist new media*. Cambridge UK: Polity Press.

Michielse, M., & Partti, H. (2015, in press). Producing a meaningful difference: The significance of small creative acts in composing within online participatory remix practices. *International Journal of Community Music*.

Paparo, S., & Talbot, B. (2013). *Real voices, virtual performing: Phenomena of digitally-mediated singing*. Retrieved November 14, 2014, from https://prezi.com/na9osenopcmx/copy-of-real-voices-virtual-performing-phenomena-of-digitally-mediated-singing/.

Partti, H. (2013). Learning to be an opera composer: The Opera by You online community as a platform for growth of music-related expertise. *Musikki, 1*, 33–50.

Partti, H. (2014). Cosmopolitan musicianship under construction: Digital musicians illuminating emerging values in music education. *International Journal of Music Education, 32*(1), 3–18.

Partti, H., & Westerlund, H. (2013). Envisioning collaborative composing in music education: Learning and negotiation of meaning in operabyyou.com. *British Journal of Music Education, 30*(2), 207–222.

Rice, T. (2008). Music in Bulgaria: Experiencing music, expressing culture. *Journal of American Folklore, 121*(482), 501–502.

Ruthmann, A. (n.d.). *Noteflight*. Retrieved November 16, 2014, from http://www.alexruthmann.com/blog1/?page_id=130.

Salavuo, M. (2006). Open and informal online communities as forums of collaborative musical activities and learning. *British Journal of Music Education, 23*(3), 253–271.

Turkle, S. (2011). *Alone together: Why we expect more from technology and less from each other.* New York: Basic Books.

Waldron, J. (2013). User-generated content, YouTube, and participatory culture on the Web: Music learning and teaching in two contrasting online communities. *Music Education Research, 31*(1), 91–105.

Waldron, J. (in review). The more things change, the more they stay the same: Texas high school bands and music education as "contest."

Waldron, J. (2006). Adult and student perceptions of teaching and learning at the Goderich Celtic College: An ethnographic case study. Unpublished doctoral dissertation, Michigan State University.

Webster, P. (2011). Key research in music technology and music teaching and learning. *Journal of Music, Technology and Education, 4*(2-3), 115–130.

CHAPTER 6

SLOW MUSIC

RENA UPITIS

As I sit here on the train with my laptop, musing about technology in music education, I find myself thinking about the mobile apps that promise to change the face of music making. Practice record tools. Electronic tuners. Sight-reading miracle makers.

But will they? And are these technologies new? Or are they merely digitized versions of preexisting technologies? Take, for example, the apps that keep practice records and send email messages to teachers when students log the minutes and hours spent on their instruments. Is this really all that different from keeping a practice log in a notebook and sharing it with the teacher at the weekly lesson? And the sight-reading apps—well, let's face it—to learn to sight-read, you have to practice sight-reading. If an app helps with that, great. But fundamentally, the task is the same: sit with your instrument with music you have never seen before. Play it. And repeat the process, again and again, until you become a fluent reader. This is a slow and methodical undertaking.

What Is Music Technology?

One of the questions posed by the editors of this book, to shape the perspective that you are reading here, was to define what constitutes a technology in music and music education. A complex question, as it turns out. In addition to apps and software for music teaching, thinking of music technologies brings to mind recording software, mixing boards, and various multimedia tools. In this vein, Wikipedia defines "music technology" as computers, effects units, and software used to help musicians record and play back, compose, store, mix, analyze, edit, and perform music. We would also do well to examine how older music technologies have shaped music learning, performance, and composition. I consider two such technologies: acoustic instruments and devices designed to enhance music learning.

Acoustic Instrument Technology

Music history courses are filled with descriptions of the evolution of piano technology. From the time that Bartolomeo Cristofori first invented the fortepiano in the early eighteenth century, the instrument we now call a piano has evolved to become a pervasive presence in countries where the Western musical canon is embraced. The fortepiano, like the piano, allowed the performer to vary the volume of the notes by altering the touch. While the fortepiano has physical similarities to the modern piano, many changes have been made over the past several hundred years. Take the number of notes, for example. Cristofori's fortepiano was four octaves, while modern pianos are over seven octaves. While Cristofori incorporated a soft pedal, it has since evolved, as have the methods for building the body of the instrument, which now include metal frames and bracing. Along with these changes in piano technology there have also been changes in the repertoire, as composers responded according to the technological advances that were made. Mozart wrote piano works for five-octave instruments, while Beethoven, toward the end of his career, was writing for six octaves, following the changes in instrument technology. Like other music technologies, changes in the piano have led to changes in composition, technique, and music education itself.

Devices

There are an abundance of tools that enable teachers to share both their knowledge and love of music with their students as they teach them new skills. These include digital and analog recordings, YouTube videos, and devices such as tuners that students might use to accurately tune their stringed instruments (whether those tuners are digital or not).

Of these devices, perhaps one of the most humble yet extraordinary forms of music technology is that of the metronome, first developed in the early nineteenth century by Johann Maetzel. Research has repeatedly demonstrated that students who regularly use the metronome to practice are more likely to advance as musicians (Barry, 1992; Hallam et al., 2012; Pace, 1992). While many musicians use the analog metronome, others have switched to digital versions, attracted by the accuracy of the beat keeping, as well as having access to both visual and aural depictions of the underlying rhythmic structure. I would suggest that it matters not whether the metronome is analog or digital: what matters is that students learn to use the metronome effectively in their journeys as developing musicians. Tuners have a similar history: from the days of the tuning forks and pitch pipes, we have seen tuner technology evolve into digital forms that use needle, LCD, or LED outputs to indicate whether the desired pitch has been created as a string player tunes his or her instrument.

Technology for Composition, Recording, and Performance

Music technologies, like pianos and metronomes and tuners, can be easily overlooked against the backdrop of more sophisticated Internet-based tools and mobile apps. Other technologies can also be overlooked, not because of their lack of digital sophistication but because of their ubiquitous nature. In this context, it is worth recalling how many music-related technologies that we now take for granted have, like the piano, changed the face of music education.

From the time that we were first able to record music—toward the very end of the nineteenth century—the changes and advances in music technologies have been rapid and substantial. Surely the ability to broadcast music over the radio was one of the most thrilling technological advancements in the early years of the twentieth century. Many sources mark the legendary January 13, 1910, broadcast of Enrico Caruso and Emmy Destinn in concert from the Metropolitan Opera House in New York City as the birth of public radio broadcasting (Whitely, 2002). Imagine the excitement, as reported the next day in the *New York Times*, that the arias they sang could be "trapped and magnified by the dictograph directly from the stage and borne by wireless Hertzian waves over the turbulent waters of the sea to transcontinental and coastwise ships and over the mountainous peaks and undulating valleys of the country." Two decades later, the technology for recording music with magnetic tape was developed, further revolutionizing the radio broadcasting as well as music recording industries, as magnetic tape allowed musicians to record and rerecord with minimal loss of quality and, further, to edit the recordings, rearranging segments as required (Onosko, 1979).

While there are competing views regarding the dating of the first computer, many identify the Atanasoff-Berry Computer as the first digital computer. Completed in 1946, it occupied 1,800 square feet, used close to 18,000 vacuum tubes, and weighed nearly 50 tons. Hardly portable or wireless. But by the late 1970s, there were affordable home computers on the market, and by the 1980s, home computers were a common phenomenon.

At the same time that computer technologies were evolving, so too were music recording and playback technologies. Long-playing albums were introduced in 1948, followed by audio cassettes in 1962. Home multitrack systems were introduced in 1978. Digital audio was quick to follow (1979), and then a torrent of changes ensued: CDs were introduced in 1980 and MIDI files, a technology that is still well used, in 1983 (Cumberland, 2012).

Fast-forward to the mid-1990s, and we have MP3 files, DVDs, and iPod-type technology—all before the turn of the millennium. Children born in 2003 arrived along with the birth of iTunes. The growth of iTunes over the ensuing decade is staggering: in 2013, iTunes business exceeded $4 billion in the first quarter alone, more than Yahoo, Facebook, and Netflix combined. People are downloading and listening to recorded music. By the time this book is published, streaming may be usurping the downloading trend. Unimagined ways of listening to music mediated through technology are yet to come (Yarow, 2013).

To a greater or lesser extent, all of these technologies have influenced music education. They also have broad appeal outside the boundaries of our educational systems: iPods, mobile phones, and other music-playing devices are everywhere: in homes, in schools, at soccer fields, in shopping malls. Those outside influences affect what happens in the classroom and the independent music studio. They cannot be ignored.

Pushback

This litany of technologies may make it seem that I am a music technology enthusiast. To some extent, that is the case. Full disclosure: for many years, I have conducted research on music education, both in classroom settings and in private music instruction settings where independent music teachers work with students one-on-one in weekly lessons. I am currently involved in a partnership between Queen's University, Concordia University, and Canada's Royal Conservatory on a project called Music Education for the Digital Age. The work of the partnership involves the development of digital tools to support music learning for students taking weekly lessons, as well as the dissemination of those tools and of research that explores the use of tools and studio teaching practices in general (Brook & Upitis, 2014; Upitis, Varela, & Abrami, 2013; see www.musictoolsuite.ca for a detailed account of the research, tools, and knowledge mobilization activities associated with the partnership project).

Despite my active involvement with music technologies, I have deep concerns. To address those concerns, I have found it is helpful to consider whether or not the technologies are enabling. That is, does any given form of technology, digital or otherwise, actually help the teacher to teach and the student to learn? Or does it limit or constrain the learning? In terms of the latter, there are a number of mobile apps that allow teachers and students to make digital markings on musical scores. No doubt, this type of technology can be enabling when teachers and students are separated by distance, or when the teacher wishes to provide feedback to the student during the week. But at the lesson, when a teacher is working with a student and the student is reading from the score, using a traditional approach (e.g., pencil markings on the score) might be not only quicker than using an app but more effective as well. While making a marking on the score, the teacher may also be explaining something orally to the student, after, perhaps, demonstrating the point on the instrument—traditional forms of multimedia involving notation, speech, gesture, and sound.

Too Much Technology

I am not alone in suggesting that the pervasiveness of technology can be problematic. Critical concerns have been raised by others contributing to this book as well. And

I know this to be true because of the technologies we have all employed in creating the book that you hold in your hands or, I daresay, are reading on your computer screen.

While I have written several chapters for Oxford handbooks, on a number of topics, I have never been invited to make a submission to a handbook that has not only included—but required—the use of online interactions. A website. Blogs. Notifications. Dutifully, I explored the pieces that had already been uploaded before I began my work. Upon reading the work of Jonathan Savage and Chee-Hoo Lum, I found myself paralyzed; they were saying many of the things that I would have said, albeit in different ways, if I hadn't read their submissions first. But I have read their submissions, theirs being earlier on the blogging queue. I found myself in a quandary. Do I write what I would have said anyway, pretending I didn't see them? Or do I respond to the issues raised, in this virtual way, not through a conversation with either of the contributors (which, by the way, I would very much enjoy) but rather through this technology-mediated forum?

Alas, I cannot pretend that I did not read the work of Savage and Lum. Like Savage, I share the view that we are less and less "in the world and interacting with it and the people around us" and more and more in a mediate world, "one that we interact with through screens" (chapter 14 in this book). Quoting the work of Turkle (2011) and Morozov (2014), Savage concludes that while there are certainly powerful affordances associated with music technology, the loss of the "raw human part of what it means to be with each other" has, what he calls, "devastating consequences for musical expression and the process of music education." I agree.

What's more, many of us have become addicted to a life mediated through screens. Well over a decade ago, MIT professor Alan Lightman (2002) wrote a provocative essay in which he claimed that most people in Western societies live lives filled with "a sense of urgency, a vague fear of not keeping up with the world" (2005, p. 186). With this sense of urgency comes a sense of loss—loss of time to waste, loss of silence, loss of some part of one's inner self. Lightman attributes this sense of loss to the pervasiveness of technology of all kinds—computers and mobile telephones, yes, but also microwaves and televisions—making us what he calls "prisoners of the wired world."

Lightman identifies other negative outcomes of living in a wired world. In addition to the loss of silence and private time, he lists the prevalence of frenetic speed and impatience, information overload, excessive consumption, and lack of connection to the natural world. He suggests that these outcomes can be counteracted by replacing speed and impatience with deliberate and measured choices, by substituting the information overload from screens with the drama presented by nature, by fighting the obsession with consumption by living well with less, by letting the natural world speak for the virtual world, by embracing silence, and by finding private time for the inner world. And, I might add, by making music.

Parental Pushback

Earlier I explained that my own research has often focused on technology in music education—back as early as 1981, when I used an Apple IIe with a soundboard to explore music composition with young children (Upitis, 1982a, 1982b). And so, I have been asking, for over three decades, permission from students and their parents to conduct such research. When I started working on the Music Education for the Digital Age partnership project in 2011, we designed a study that would allow us to gauge the ways in which one of our tools was used during lessons and during the time between lessons, by both teachers and students. Because most of the students were under the age of consent, we approached parents, asking if their children could take part. Many of the parents agreed. But by no means did all. In my experience, this is unusual: most parents agree to research that promises to advance the field, poses no known risks to their children, and requires little time from the students themselves. Why did these parents not consent to the research? There was only one reason given: parents who did not consent claimed that music was the one place where their children were not using computers, tablets, phones, or other similar technologies, and they wished to retain that nonscreen aspect of music learning.

If Lightman (2002, 2005), Turkle (2011), and Morozov (2014)—and the parents—are right, what would a music education look like in which teachers and young musicians made deliberate and measured choices, made music with less, and were more fully connected with their inner lives through music? What might music education look like if we were to make music with less, let natural sounds speak, and make more mindful, deliberate, and measured choices? Perhaps music education would look like slow food.

Slow Food

In response to the 1986 promise of a McDonald's restaurant opening near the Spanish Steps in Rome, Carlo Petrini founded a movement that has come to be known as Slow Food. The movement has expanded globally; there are over 100,000 official members in 150 countries, in addition to the millions of local farmers, chefs, and consumers who embrace the slow food tenets without being official members (see slowfood.com). Slow food advocates are committed to enjoying the pleasures of good food in combination with a commitment to community and the environment. The slow food movement highlights traditional and regional cooking, and encourages the cultivation of plants, seeds, and livestock that are characteristic of local ecosystems.

Slow food is part of a larger movement that calls for a cultural shift in slowing down the pace of life—much like the call Lightman issued in 2002. Carl Honoré's (2004) book *In Praise of Slowness: How a Worldwide Movement Is Challenging the Cult of Speed*, suggested that the slow philosophy could be applied to all aspects of human endeavor. He writes that the slow movement "is not about doing everything at a snail's pace . . . the Slow philosophy can be summed up in a single word: balance. Be fast when it makes sense to be fast, and be slow when slowness is called for. Seek to live at what musicians call the tempo giusto—the right speed" (Honoré, 2004). I have found blogs and articles and books on slow fashion, slow cities, and slow design. All of them emphasize not only that slowing down allows us to make more environmentally conscious decisions but also that such decision-making may lead to more ethical choices while at the same time preserving local traditions. Various sources suggest that there are a growing number of vibrant slow movements including slow travel, slow gardening (isn't gardening slow by definition?), slow schools, and slow parenting (Honoré, 2004; see slowmovement.com). Oddly enough, slow music is rarely mentioned. Yet slow music strikes me as a perfect candidate for the movement that makes savoring experiences and fostering relationships paramount.

SLOW MUSIC: A RESPONSE TO TECHNOLOGY PUSHBACK

Music and food are inextricably linked. Bioevolutionary scholar Ellen Dissanayake (2000, 2003) makes the convincing claim that music and other kinds of art making are essential to human survival. Although some researchers take issue with aspects of her model (e.g., Davies, 2005), her main claim—that art is essential to human life—is difficult to dispute. Dissanayake (2003) theorizes that humans "have a specifiable biological nature that is the product of millions of years of adapting to the world in which they (and their ancestors) came into being" (p. 246). Her analyses demonstrate that art fulfills some of our biological needs, claiming that art, like language, "is inherent in human nature, and will emerge in every normal individual during normal development and socialization" (p. 246).

A powerful part of Dissanayake's argument is the set of five features that support her thesis that art making is essential to human evolution. The first of these features is universality: art making is present in every society and cultural group. Second, the investment of resources in the arts—especially in premodern societies—is disproportionately greater than one would expect for a peripheral undertaking. Third is the biological importance of the aspects of daily life that are "artified" through ritual ceremonies concerned with "safety, subsistence, prosperity, health, social harmony, and the successful traversing of birth, death, and other life stages" (2003, p. 247). Fourth, the arts are associated with pleasure—just like the other essential requirements of life such as food, sex, familiar surroundings, rest, conversation, and close relationships. Finally, Dissanayake suggests that the juvenile predisposition to the arts provides evidence that the arts are essential to the development of humanity. Other keen observers of children's musical behavior similarly claim that "all children, to a greater or lesser degree, are musical" (Campbell, 1998, p. 169). Clearly, food is also essential to survival. And interestingly, both food and music have fast versions, an idea that I will ponder in the final section of this essay.

Evidence of Slow Music

Even in this fast-paced, wireless musical world, there are many musicians who make what I would call slow music. These include the singers in community choirs, meeting once a week to sing and socialize. Or professional string quartets whose members practice for many hours every day, with pencils propped behind their ears as they make their way through new and familiar scores. There are instances of slow music in the world of music technology as well, such as the growing interest in returning to vinyl recordings. In a recent article in the *New York Times* titled "Music Technology Comes Full Circle," the owner of the Rik Stoet High End Audio store in the Hague describes how the recent surge in vinyl recordings not only is due to the market segment who still remember the feel and sound of vinyl and yearn to return to a physical medium but also is made up of young people who may never have seen a record player (Schuetze, 2014).

An entirely different approach to slow music involves a lush recording studio called Manifold, near Raleigh, North Carolina. The design for this high-tech facility was inspired by the CBS 30th Street Studio, a recording venue favored by the likes of Miles Davis and Glenn Gould. The key feature for Manifold's founders was this: Davis would bring dozens of people into the studio for a live recording session. Michael Tiemann, one of Manifold's founders, describes this recording style as a salon approach, where people are gathered in a room small enough to support intimate conversation. The Manifold model captures the intimacy of the salon and offsets the costs by having a select group of people experience the music live in the studio as it is created—a recording salon. Tiemann claims that "just as the slow food movement encourages eaters to think more holistically about how food is grown, prepared, and brought to the table, this co-producer model gives people much more access to the creative process of music" (Menconi, 2012).

Slow Music and Our Students

In a world of iTunes and earbuds, in a world where music can be performed and produced at the press of a button, would students have interest in a slow music approach? As Lum states so poignantly in his essay in this book (chapter 3): "The digital learner is constantly immersed in a multimedia, multisensorial environment through technology. The politics of consumerism and materialism in popular culture notwithstanding, the reality is that digital learners are getting bombarded and surrounded by multimedia experiences all the time. Music educators need to be mindful that the singular musical experience that is provided in classroom activities may sometimes not appeal to these digital learners. A multiarts experience would seem closer to the experiences of digital learners in their daily encounters." Perhaps here, too, there are lessons from the slow food movement. Students aren't necessarily interested in slow food, at the outset, either. For students whose diet consists primarily of fast food, it is a long journey to a slow food mentality. One of the most successful demonstrations of how that journey can be made is illustrated by the Edible Schoolyard at the King Middle School in Berkeley, California. The Edible Schoolyard was founded several decades ago with the visionary dedication of Chez Panisse chef Alice Waters (see edibleschoolyard.org). One of the driving goals of the project was to serve real food in the lunchroom, grown by the students on the land surrounding the school. Studies have shown that when students become engaged in the process of growing, harvesting, and preparing food, they are more likely to eat, enjoy, and value local food, leading to positive health outcomes as well as curricular gains in mathematics and science as compared to students in control schools (Murphy, 2003).

We might imagine a similar trajectory with slow music. A challenge for music educators is to think about how—in their local contexts—music can be savored more fully (a notion also explored by Lum in his chapter). What would slow music education look like? As I pose this question, I can imagine some possibilities, such as learning to play an acoustic instrument, with dedicated, slow, and methodical intention. Or making instruments. Drumming and singing with others, deepening relationships, as called for by Savage and the slow movement alike. Exploring new musical tastes. Or maybe just pulling the old LP player out of the basement and bringing it into the classroom as a way of pausing to listen thoughtfully, carefully.

Closing (Slow) Thoughts

So, now that I have responded to the task at hand, in this mediated world, I find myself reflecting on how I feel about it. Indeed, it would have been much easier not to know what Savage and Lum had written. Knowing that the points they made, and even the references they used, were ones that I would have called upon, pushed me to take a different perspective. Slow food. Slow music. And indeed, slow thinking. Rather than reaching for the familiar, I pondered a great deal about how to write this piece. I reread some of Savage's (2005, 2007) earlier work. I took long walks through the forest and thought about how to complement their perspectives without repeating their points. Perhaps I have done this. Perhaps I will have a real-time conversation with some of the people I have quoted in order to find out.

References

Barry, N. H. (1992). The effects of practice, individual differences in cognitive style, and gender upon technical accuracy and musicality of student instrumental performance. *Psychology of Music, 20,* 112–123.

Brook, J., & Upitis, R. (2014). How can an online tool help independent studio teachers and their students? *Music Education Research.* doi: 10.1080/14613808.2014.969217.

Campbell, P. S. (1998). *Songs in their heads.* New York: Oxford University Press.

Cumberland, R. (2012). Music business timeline. Retrieved January 20, 2017, from http://www.bemuso.com/musicbiz/musicbusinesstimeline.html

Davies, S. J. (2005). Ellen Dissanayake's evolutionary aesthetic. *Biology and Philosophy, 20,* 291–304.

Dissanayake, E. (2000). *Art and intimacy: How the arts began.* Seattle: University of Washington Press.

Dissanayake, E. (2003). Art in global context: An evolutionary/functionalist perspective for the 21st century. *International Journal of Anthropology, 18*(4), 245–258.

Hallam, S., Rinta, T., Varvarigou, M., Creech, A., Papageorgi, I., Gomes, T., & Lanipekun, J. (2012). The development of practising strategies in young people. *Psychology of Music, 40*(5), 652–680. doi: 10.1177/0305735612443868.

Honoré, C. (2004). *In praise of slowness: How a worldwide movement is challenging the cult of speed.* San Francisco: HarperOne.

Lightman, A. (2002, March 16). Prisoners of the wired world. *Globe and Mail.*

Lightman, A. (2005). *A sense of the mysterious.* New York: Pantheon.

Menconi, D. (2012, January 28). *Can music learn from the slow food movement?* Retrieved January 20, 2017, from http://www.salon.com/2012/01/28/can_music_learn_from_the_slow_food_movement/.

Morozov, E. (2014). *To save everything, click here: The folly of technological solutionism.* London: Penguin.

Murphy, J. M. (2003). *Findings from the evaluation study of the Edible Schoolyard.* Report to the Center for Ecoliteracy. Berkeley, CA: Center for Ecoliteracy.

Onosko, T. (1979). *Wasn't the future wonderful? A view of trends and technology from the 1930s.* New York: Dutton.

Pace, R. (1992). Productive practising. *Clavier, 31*(6), 17–19.

Savage, J. (2005). Working towards a theory for music technologies in the classroom: How pupils engage with and organize sounds with new technologies. *British Journal of Music Education, 22,* 167–180. doi: http://dx.doi.org/10.1017/S0265051705006133.

Savage, J. (2007). Reconstructing music education through ICT. *Research in Education, 78,* 65–77.

Schuetze, C. F. (2014, November 24). Music technology comes full circle. *New York Times.* Retrieved January 17, 2017, from http://www.nytimes.com/2014/11/24/style/international/music-technology-comes-full-circle.html?_r=0.

Turkle, S. (2011). *Alone together: Why we expect more from technology and less from each other.* New York: Basic Books.

Upitis, R. (1982a). *A computer-assisted instruction approach to music for junior-age children: Using ALF for teaching music composition.* Unpublished master's thesis. Queen's University, Kingston, Ontario, Canada.

Upitis, R. (1982b). Microcomputers and music composition: Some possibilities for elementary students. *Canadian Music Educator, 24*(1), 11–19.

Upitis, R., Varela, W., & Abrami, P. C. (2013). Enriching the time between music lessons with a digital learning portfolio. *Canadian Music Educators Association, 54*(4), 22–28.

Whitely, S. (2002). *On this date.* New York: McGraw Hill Professional.

Yarow, J. (2013, May 13). *Chart of today: Apples surprisingly steady iTunes growth.* Retrieved January 20, 2017, from http://www.businessinsider.com/chart-of-the-day-itunes-revenue-2013-5.

CHAPTER 7

THEN AND NOW

DAVID A. WILLIAMS

As a high school student, I was devoted to the school instrumental music program. I played the clarinet and the tenor saxophone and enrolled in every class offered by the school that had anything to do with band. This included three years of concert band, marching band, and jazz band, and a year of music theory. I was accepted into both All-County and All-State ensembles, and I participated in solo and ensemble festivals each year. I took part in an extracurricular pep band that performed at junior varsity football games, and I also played in the pit orchestra for every school musical. Music technology for me was the development of woodwind instruments, reeds, and printed music. I was glued to any development in the manufacture of clarinets, ligatures, mouthpieces, and reeds.

Music technology was also a part of my life through sound recording and playback. In school we were able to record rehearsals and listen to them later. We received recorded feedback from adjudicators at music festivals. In the car, I could listen to the radio. At home, I could listen to prerecorded records and cassette tapes, and even use my personal cassette player to record myself playing. These were exciting times for me, as they were for countless other teens in high schools across the United States in the 1970s.

A lot has changed in our society during the many years since I was a high school band student. Music technology has created a whole new landscape of possibilities for school music programs. But change doesn't happen quickly in music education. Music teachers tend to be creatures of habit—even other people's habits. Characteristics true of my high school music experience were true for my music teachers when they were in high school, and possibly true also for their music teachers. Most of them are still true today. The clarinet, for example, has been an important technology for music education for a very long time, and while the instrument has developed some even in my lifetime, it remains very much the same instrument it was at the beginning of the twentieth century. Experiments with new materials have resulted from the diminishment in supplies of Grenadilla wood, and some models include a new key or two, but a clarinet today is basically the clarinet of yesteryear. In the hands of the right person it accomplishes what it is intended to accomplish quite well, yet it is no longer an innovative technology.

This is not to suggest that innovation itself should be the goal of music education, nor is it necessarily essential. Avoiding innovation in technology shouldn't be a goal either, however.

Even the sounds that come out of a clarinet appear to be passing some point of being innovative. Beginning at least in the mid-1900s, composers experimented with the sound palette of the clarinet, asking performers to play multiple tones at one time, to play higher notes than ever before, to take the instrument apart and play on segments, to slap and click the keys, to hum while playing, and so forth. While every new composition is innovative in different ways, it's possible that we have reached some endpoint of innovation for the technology we call the clarinet—especially in K–12 music education, where only a fraction of these extended techniques find a home anyway. And this is not to belittle the clarinet. The same, or similar, could be said about the flute, oboe, bassoon, trumpet, horn, trombone, tuba, violin, viola, cello, string bass, snare drum, bass drum, xylophone, and marimba. Compared to these instruments the saxophone is relatively young, but it still fits into this discussion of musical instruments that can no longer be considered innovative technology.

Taken together, all these traditional wind, string, and percussion instruments form the basis of school instrumental music study in the United States—both today and 100 years ago. Our current model of instrumental music, as a technology and a pedagogy, is basically the same as that implemented in the early part of the twentieth century. With few exceptions, music teachers today use the same instruments and technologies, in the same ways, in the same teacher-led large ensembles, as was the norm at the beginning of the profession in the United States.

On the vocal side of things, while technologies have improved our understanding of the voice and how to train it, the human voice in music education is still pretty much the human voice. The ways in which the voice is used in music education have changed little. Outside our classrooms, however, technologies have added richly to the sound of the human voice, providing a wide pallet of possibilities for singers and rappers. Yet our current model of vocal music education, as a technology and a pedagogy, is also basically the same as that implemented in the early part of the twentieth century. With few exceptions, vocal music teachers today use the voice and corresponding technologies, in the same ways, in the same teacher-led large ensembles, as was the norm at the beginning of the profession in the United States.

I should pause briefly and define the term "we" as I am about to use it. I am referring to those involved in the music education profession, especially in the United States. I am generalizing and lumping the vast majority of our profession into a very big whole. I realize, however, that anytime I make a statement about "we" there are probably exceptions. There are individuals, sometimes even small groups of individuals, who are exceptions to the rule. But these exceptions tend to be few and very far between.

We are a profession of old, perhaps even ancient technology, and for the most part we are both happy with and proud of this. Certainly there are exceptions. We have individuals who regularly employ more recent technologies in music teaching and learning environments. We have nonconformists who stretch possibilities. We have outliers who use new technologies in amazing ways, but taken as a whole, the music education profession, especially in the United States, does not have much interest in adapting to instruments and technologies developed in the past 50 years.

Instead of being a profession that lacks criticality about technology, I suggest it is possible we are overly critical. So critical, that we, as a whole, will not allow new technologies to even be considered. This is not necessarily the kind of criticality to which Roger Mantie refers in chapter 1. It is more a criticality borne from fear of the unknown—a fear of change—a fear of losing control over that with which we are comfortable and confident. I would even go so far as to suggest that this is the core of the issue. Music teachers, especially those who are primarily involved with traditional secondary performing ensembles, are afraid to give up the comfort they have developed. I think one could argue that our criticality toward technology has potentially more to do with what is best for ourselves than what is best for our students.

I mentioned earlier that innovation itself should not be the goal of music education. Instead, I would suggest that increasing the musicality of all students should be our goal. For me this means helping students become independent, lifelong music makers, capable of making creative musical decisions on their own. There is little evidence our traditional model of music education has done a very good job with this. One of the arguments for the incorporation of new technologies and instruments into music education is that, together with well-conceived pedagogical approaches, we could witness an increase in students' independent music making (Green, 2008).

I must admit, at this point, that I certainly fall into the group to which Mantie is referring when he suggests that some use music technology "because we feel we must because it is simply a 'fact' of contemporary life." I do not, however, feel that this assures inauthentic engagement. I'd like to believe it is possible to both "feel we must" and to do so in authentic ways that are of great benefit to ourselves, our students, and society in general. At the same time, I recognize that the use of newer music technology does not present a utopian condition. I can't think of a technology that does. Certainly there were those during the beginning of the twentieth century who feared the birth of the automobile. They feared the unknown—feared change—feared losing control over that with which they were comfortable and confident. I am also certain there were others who were early adopters of the automobile as technology because they felt they must, "because it [was] simply a 'fact' of contemporary life." As it turned out of course, this latter group helped change the way we commute, and the automobile has changed the way we interact with each other and the world. While there are serious issues related to the automotive industry (injury, death, road-rage, pollution, etc.), the automobile is a fact of contemporary life.

I can imagine, as the automobile began to catch on, the blacksmiths who had choices to make—either to continue making horseshoes or to retool and produce car parts. Those who retooled no doubt achieved two things denied to those who didn't. First, they remained relevant to society. They produced something that was seen as useful to the vast majority of people. They adapted to the needs of the culture in which they lived instead of remaining true to an old tradition that was quickly moving to the fringes of the culture. Second, they probably sustained employment, unlike many of those who stubbornly kept on doing only what they had always done.

There are some striking similarities in all this regarding new music technology. It is a fact of contemporary life. Musical instruments and related technologies associated with popular musics have changed the way music is conceived, made, produced, performed, bought, and sold. Practically every aspect of contemporary music business is dominated by new technologies. Every aspect, of course, except for music education—especially in the United States, where we continue to ignore new music technologies. This ignoring has caused two things to occur. First, as a profession we have not remained relevant to society. We no longer produce something that is seen as useful for the vast majority of people. We have not adapted to the needs of the culture in which we live; instead, we have remained true to an old tradition that has moved to the fringes of culture. Second, we are having issues sustaining employment opportunities as music teaching positions continue to be consolidated and cut.

It is important to note, however, that pedagogical models involving new music technologies have been employed in schools in the United Kingdom, Scandinavia, and Australasia for several years. Many of these have had a positive impact on student participation and retention, student attitudes, self-esteem, on-task behavior, and the development of a greater range of musical skills and higher levels of musical understanding (Byrne & Sheridan, 2000; Finney & Philpott, 2010; Folkestad, 2006; Georgii-Hemming & Westvall, 2010; Hallam, 2005; Karlsen, 2010; Wright & Kanellopoulos, 2010). How is it that such changes haven't affected music programs in the United States to any significant degree? How can it possibly be that my musical experience as a middle school and high school student are so incredibly similar to what students experience today? How can our "then" look so much like our "now," while music as practiced in our society has changed so dramatically?

The responses I receive when asking these kinds of questions normally have to do with quality. We (this is still the same "we" I described previously) believe that what we do is the most illustrious way to do music. All other musical involvements are less worthy than what we do. We work with the very best music in ways that produce the very best type of musician who performs on the very best musical instruments or uses the voice in the very best ways. We are convinced of this. Other musics, including all popular musics, are seen as inferior to the music we use in our classes. These inferior musics tend to be too simple and lack the complexity of our quality musics. And complexity is good. Other instruments, including all those newfangled electronic and digital instruments, are also seen as inferior to the great musical instruments we use in our classes. Mastering the instruments we use takes years and years of committed study, while the inferior instruments are too simplistic to produce any music that is worthy of serious study. Learning to make music on the instruments we use is formidable. And formidable is good. We believe that being complex and formidable serves us well, but this thinking actually serves to demean the musics that students find most enjoyable and in which they find meaning. Demeaning students' interests is not an effective strategy to recruit large numbers of students.

We are a profession that resists the use of new music technologies, especially if use of these technologies would mean changing what we do in any substantial way. We are a profession that resists change, and this resistance has hurt us. This resistance is fast making us irrelevant in a musical world that is ever changing. Students currently in K–12, as well as in higher education, have grown up with new music technologies and related musical styles that are quite different from what they encounter in their schools. The vast majority of these students see no place for themselves in school music programs. Lucy Green (2008) puts it this way: "we can surmise that many children and young people who fail and drop out of formal music education, far from being either uninterested or unmusical, simply do not respond to the kind of instruction it offers" (p. 3). As a result, we concentrate on music in which few people outside our doors seem interested, and then we perform basically for ourselves (including family and friends who are most often more interested in supporting than in listening).

We are missing out on exciting opportunities that would be made possible by embracing new music technologies, especially when used in conjunction with corresponding pedagogies. With successful implementation of musical technologies we could help students gain personal experience with musical styles that interest and hold meaning for them; reach students with musical experiences that are typical in the music industry outside the schools; attract a wider audience of students into school music programs, perhaps even saving or reestablishing teaching positions in some schools; increase the likelihood that students might develop into independent music makers and continue making music without us after they leave our programs; afford musical experiences to students with various forms of limitations that currently are not well served; and better incorporate the rich resources of the Internet to reach students outside the traditional school setting. In short, we could bring music education into the twenty-first century and create new relevance for our programs. We could do all these things, or we could continue to disregard the possibilities and remain devoted to our traditions—rehearsing and performing the "best" music, the "best" way, with the "best" instruments—as we continue to drift even further away from the students in our schools.

REFERENCES

Bartel, L. (2004). *Questioning the music education paradigm*. Volume 2 of Research to Practice, a biennial series. Toronto, ON: Canadian Music Educators Association.

Byrne, C., & Sheridan, M. (2000). The long and winding road: The story of rock music in Scottish schools. *International Journal of Music Education, 36,* 46–58.

Finney, J., & Philpott, C. (2010). Informal learning and meta-pedagogy in initial teacher education in England. *British Journal of Music Education, 27,* 7–19.

Folkestad, G. (2006). Formal and informal learning situations or practices versus formal and informal ways of hearing. *British Journal of Music Education, 23,* 135–145.

Georgii-Hemming, E., & Westvall, M. (2010). Music education—a personal matter? Examining the current discourses of music education in Sweden. *British Journal of Music Education, 27,* 21–33.

Green, L. (2008). *Music, informal learning and the school: A new classroom pedagogy*. Surrey, England: Ashgate.

Hallam, S. (2005). *Survey of musical futures*. A report from the Institute of Education, University of London for the Paul Hamlyn Foundation.

Karlsen, S. (2010). BoomTown Music Education and the need for authenticity—informal learning put into practice in Swedish post-compulsory music education. *British Journal of Music Education, 27*, 21–33.

Wright, R., & Kanellopoulos, P. (2010). Informal music learning, improvisation and teacher education. *British Journal of Music Education, 27*, 71–87.

1 B

Core Perspectives

CHAPTER 8

GLOBALIZATION AND TECHNOLOGY IN TWENTY-FIRST-CENTURY EDUCATION

SAMUEL LEONG

THE 2013 Global Information Technology Report ranked the "network readiness" of 144 countries based on their capacity to benefit from new information and communication technologies (ICTs). The top 10 countries, in descending order, were Finland, Singapore, Sweden, Netherlands, Norway, Switzerland, the United Kingdom, Denmark, the United States, and Taiwan. Information and communication technology has created a "space of flows," where global interactions have been rearranged to create "a new type of space that allows distant synchronous, real-time interaction" (Castells, 2004, p. 146). This new kind of interaction has enabled the creation of a global "network society" that increasingly expands the connections and interdependency between people, territories, and organizations in the educational, cultural, economic, and political domains (Barney, 2004). The rapid advancement of technologies has facilitated and driven the spread of globalization and the growth of the global knowledge economy. These have opened up the world and facilitated the flow of information and knowledge. In the mid-1970s, it took more than two months for my book order from England to arrive in Singapore, and a photocopy then was 60 cents (US) a page. I remember that my first computer videoconference between Australia and the United States in 1998 was virtually an Internet version of walkie-talkie communication, and Hypercard and hypermedia were aiding music educators as interactive technologies then. Today's technologies have enabled people across the world to be connected educationally and socially via media and telecommunications, educationally and culturally through movements of people, economically through commerce and trade, environmentally through dependency on the planet's resources, and politically through international relations and systems of regulation. For my students in Hong Kong, these provide many new and increasingly common ways and platforms for them to share, consume, and produce information. And today's education policies and practices have connected with the globalization trend that has expanded the scope of how, when, and where learning takes place.

As the product and driver of globalization, ICT has enabled the international transfer of knowledge and the creation of a transnational private market of education provision that complements and competes against local and national education providers. It has also propelled the popularization of neoliberalism, the dominant political-economic ideology worldwide. This has impacted policies that support market mechanisms, including choice, competition and decentralization, the liberalization and privatization of the education sector, and the transplantation of management techniques from the corporate sector. Globalization has also fostered transnational attention to education as a global public good and a human right while contesting the neoliberal education agenda. UNESCO's recent global monitoring report has focused on the applications of ICT and associative technologies to empower persons with disabilities so as to widen access to information and knowledge for all (UNESCO, 2013). It calls for applying accessibility standards to the development of content, products, and services and for creating classrooms that are more inclusive, physical environments that are more accessible, teaching and learning content and techniques that are more in tune with learners' needs. The Education for All agenda of UNESCO has acknowledged the innovative use of technology to help improve learning by enriching teachers' curriculum delivery and encouraging flexibility in pupil learning but cautions that new technology is not a substitute for good teaching. UNESCO's *2013/14 Education for All Global Monitoring Report* (UNESCO, 2014) notes that ICT can be more effective as a means of improving learning and addressing learning disparities if it plays a complementary role, serving as an additional resource for teachers and students. Careful consideration should be given to students' access to technology, as those from low-income groups are less likely to have experience of ICT outside school and may thus take longer to adapt to it. It also notes that the use of mobile phones and other portable electronic devices, such as MP3 players, is a promising way of increasing the accessibility of ICT for teaching and learning. While mobile learning can increase learning opportunities, these new technologies need to tailor content and delivery to the varying needs of learners, especially weaker students. Considerations of these issues are now being integrated into the music teacher education program at my current workplace in Hong Kong.

The advent of powerful computer technologies has ushered in the so-called new information age, in which human knowledge is said to double every 13 months, nanotechnology knowledge every 2 years, and clinical knowledge every 18 months (Schilling, 2013). Digital literacies have emerged as being as essential as reading literacy; these include e-literacy, screen literacy, multimedia literacy, information literacy, ICT literacies, and new literacies. Today's literates are expected to possess the ability to interpret and write various codes, "such as icons, symbols, visuals, graphics, animation, audio and video" (Nallaya, 2010, p. 48). While passive entertainment, such as standard television, typified Web 1.0, Web 2.0 has seen a rise in audience-generated content, such as blogs and podcasts. Users are engaged at an ever deeper level of engagement in the Web 3.0 era of virtual worlds—where the content seeks out the users, and users' activities and interests determine what finds them, and it is delivered how and where they want. Since the World Wide Web was made available to the public in 1991, technological advances and concomitant services have provided new possibilities for users in a globalized and mobile world, including those with implications for music and education:

- 1991: World Wide Web launched to the public
- 1995: Amazon.com bookstore appears
- 1996: First mobile phone with Internet connectivity
- 1998: Google named the search engine of choice by *PC* magazine
- 1999: Blackberry launched
- 2001: Wikipedia opened
- 2003: Apple introduced online music service iTunes
- 2003: Intel incorporated Wi-Fi (wireless Internet receiving capability) in their Centrino chip
- 2004: Podcasting commenced
- 2004: Facebook launched
- 2005: YouTube, the first video sharing site, came online
- 2006: Twitter launched
- 2006: Google purchased YouTube
- 2007: iPhone, the first of the smartphones, introduced
- 2008: App Store opened as an update to iTunes
- 2008: OpenNebula, enhanced in the European Commission–funded project RESERVOIR (Resources and Services Virtualization without Barriers), became the first open source software for deploying private and hybrid clouds, and for the federation of clouds
- 2010: iPad tablet computer released
- 2011: Amazon released the Kindle Fire tablet
- 2011: iCloud, free online synchronization and backup service, launched by Apple
- 2012: 1.7 billion mobile phones (722 million smartphones) sold worldwide
- 2012: 5.2 billion mobile subscriptions in the developing world (76.6% of global subscriptions)
- 2013: 6.8 billion mobile subscriptions worldwide
- 2014: Apple Watch released
- 2015: 3D touchable holograms made of iodized air that pulsed at one quadrillionth of a second created
- 2016: China's Sunway TaihuLight becomes the world's fastest supercomputer at theoretical peak performance of 125 petaflops, 10,649,600 cores, and 1.31 petabytes of primary memory

New technological developments are challenging my institution's music and music teacher education programs to evaluate their impact on established education approaches. These include: (1) cloud-based services such as G-mail, Google+, iTunes, Twitter, and YouTube, which are becoming the standard repositories for educational content; (2) Open Educational Resources, such as MIT OpenCourseWare, Khan Academy, iTunes U, MOOCs (massive open online courses), and other repositories that provide massive amounts of quality online learning materials that can be leveraged to supplement and assist the classroom regardless of delivery modality; (3) social networks such as Facebook, Twitter, Pinterest, and Google+, which offer opportunities for creating a sense of connectedness among students at course, program, and institutional levels; and (4) blending/merging of classroom and online approaches, including the increasingly popular "flipped classroom," that are helping students achieve success and reduce dropout rates.

INFORMATION AND COMMUNICATION TECHNOLOGY AND EDUCATION 3.0

Information and Communication Technology and Education Policy

The global knowledge explosion has seen a proliferation of government policy statements on the role of ICT in education. Governments and education systems in a number of Western and Asian countries have attempted to transform schools and their practices through the incorporation of ICT. Many schools have been wired up and networked, fitted with hardware and software, and given sets of policies and curriculum guides, as well as some professional development courses.

Australia's *National Goals for Schooling in the Twenty First Century* (Ministerial Council for Education, Employment, Training and Youth Affairs, 1999) and *Learning in an Online World: The School Action Plan for the Information Economy* (Ministerial Council for Education, Employment, Training and Youth Affairs, 2000) recognized the role of ICT in improving student learning, offering flexible learning opportunities, and improving the efficiency of school practices. These two documents were followed by an information and competency framework for Australian teachers (Department of Education, Science, & Training, 2002). In 2005, the role of ICT in education was sealed in a "national vision" to build an Australian knowledge culture and create an innovative society where all learners could achieve their potential, efficiencies could be achieved through the sharing of resource and expertise, and education could be internationalized (Ministerial Council for Education, Employment, Training and Youth Affairs, 2005). The new Australian Curriculum has recognized the importance of ICT both as a "general capability" and as a learning area in partnership with design and technology (Australian Curriculum, Assessment and Reporting Authority, 2013). In the United Kingdom, technology is recognized in "a national plan for music education" (Department of Education [UK], 2011) for its important role in supporting, extending, and enhancing music teaching—helping to connect communities in ways that rely less on location, being used to inspire, motivate, and stretch pupils, extending musical experiences, and helping children with additional needs to further engage in music making. In the United States, the National Education Technology Plan of 2010 recognizes technology-based learning and assessment systems as pivotal in improving student learning and generating data that could be used to continuously improve the education system at all levels. Technology enables the execution of collaborative teaching strategies combined with professional learning that could better prepare and enhance educators' competencies and expertise over the course of their careers.

The Singapore government launched its initial 5-year "Masterplan for IT in Education" in 1997, with subsequent Masterplans in 2002 and 2008. Together with another education reform initiative, Thinking Schools, Learning Nation, the three IT-in-education policies worked in tandem to address both hardware-software provision and pedagogical issues related to ICT integration into a more flexible and dynamic curriculum. The "iN2015 Masterplan" (Info-communications Development Authority of Singapore, 2006) aimed to secure a digital future for everyone in an intelligent nation and global city that caters for the elderly, the less privileged, and people with disability, enabling them to enjoy connected and enriched lives for self-improvement and lifelong learning. At about the same time, the Hong Kong government launched two 5-year master plans for ICT in education, in 1998 and 2004, respectively. Skills in ICT are essential to a new culture of learning and teaching that emphasizes "learning how to learn," "integrated learning," and "integrated and flexible arrangement of learning time." Information communication technology serves as a powerful educational tool and plays a catalyst role in transforming school education from a largely teacher-centered approach to a more interactive and learner-centered approach; e-learning is considered central to the learner-centered mode of education (Chen, 2012; see fig. 8.1).

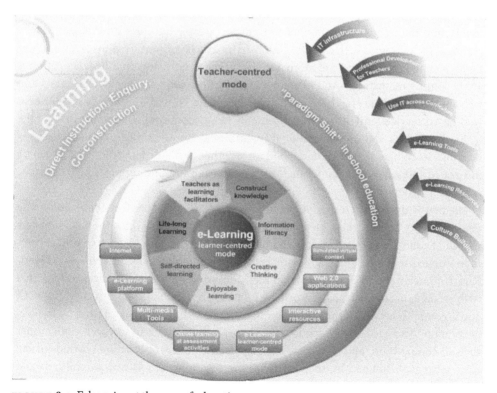

FIGURE 8.1 E-learning at the core of education

Information Communication Technology and Education Practices

Progress, convergence, and integration in ICT have driven fundamental changes in the kinds of ICT competencies that faculty, students, colleges, and universities have or might be expected to acquire. Many of today's young people are familiar with playing Nintendo games in three dimensions without 3D glasses, and the days of 3D television, 3D screens, 3D projectors, and 3D printing have arrived (see Leong, 2011). Popular video games such as Guitar Hero, DJ Hero, and Rock Band provide music-themed action games for players who push buttons in a sequence dictated on the screen in time to a soundtrack, causing the player's avatar to dance or to play a virtual instrument correctly (see Peppler, chapter 18 in this book). With programs capable of compiling and converting television broadcasts and other recordings, the traditional distinction between class time and nonclass time is disappearing; soon search engines will yield both text and spoken results ("Top ten forecasts for 2011," retrieved from http://www.wfs.org/page/futuristmagazine).

Information communication technology poses significant challenges for modern educational practices, and there are concerns about the "digital divide" between those who have ready access to ICT and those who do not. Advanced and emerging technologies are blurring and even merging boundaries that have traditionally separated disciplines, organizations, structures, and peoples. Today's digital media enables the combining of text, graphics, sound, and data in an integrated multisensory, multimedia, and multinetworked manner to transport people into new experiences of new realities. Technologies such as lecture capture, smart classrooms, response clickers, and tablet computers have become more popular, user-friendly, and seamless. E-portfolios are becoming mainstream and being integrated into course management systems and web applications. There has been an increased demand for mobile access to institutional web-based services and instructional content; and e-collaboration tools are driving the technology plans at some educational institutions. However, instructional effectiveness and learning enhancement depend on whether available technology tools can be integrated and sustained without major adoption, training, and support challenges (see Ingerman & Yang, 2011).

The new "Education 3.0" paradigm emphasizes technology as a key enabler of young people's ability to build their "personalized learning spaces," in which they can "amass a wealth of education resources, in rich multimedia format, gain access to world experts in multiple disciplines, enjoy authentic learning using online data, receive instant feedback from team mates and teachers on their ideas and their performance, and interact with students from all over the world as they collaborate on group projects" (CISCO, 2008, p. 13). Today's educators are coming to terms with technologies used by the Net generation to socialize, work, and learn. These learners prefer to work in teams rather than in competition, and to engage in collaborative forms of learning beyond formal class settings, facilitated by social networking technologies. As today's children and young people are prosumers, they engage in self-directed arts projects solely because they want to (Gauntlett, 2011). This requires "holistic transformation" in the way teachers and students learn through innovative pedagogy, increased creativity and collaboration, creation of learning communities, provision of real-time feedback and assessment, and harnessing of the potential of digital participatory cultures to transform individuals and societies so as to develop engaged citizenries.

New technologies are making it possible for college students to live wherever they want and take many (if not all) of their courses online, earning degrees that are accredited by international accrediting agencies (Dew, 2010). In the United States, university student enrollment in at least one online course has doubled between 2005 and 2010 (Allen & Seaman, 2011). The era of hyperconnectivity is enabling many professionals to weave their careers and personal lives into a blended mosaic of activity. As work and leisure are interlaced throughout the waking hours every day of the week, student life is also reflecting this trend. The social networking lifestyle has brought on a critical need for social skills such as self-discipline, responsibility, and media literacy, and curricula need to broaden to include the development of these interpersonal skills. With the explosion of information made available by ICT, new skills are also needed for accessing, evaluating, and organizing information in digital environments. Knowledge societies and economies also require people who are able to transform information into new knowledge or to apply it to the devising of new ideas, as well as people who possess research and problem solving skills for defining, searching, evaluating, selecting, organizing, analyzing, and interpreting globally accessible information.

Technology and Music Education

Technology has played a vital role in music education for a long time, and a number of useful texts to assist music teachers have been produced over the years (e.g., Leong & Robinson, 1995; Reese, McCord, & Walls, 2001; Rudolph, 2004; Williams & Webster, 2006; Burns, 2008; Rudolph & Frankel, 2009; Dillon & Hirche, 2010). From the early use of the phonograph with stylus, magnetic tape recorders, and chalk and blackboards to the more recent use of MIDI synthesizers, sound modules, sequencers, and samplers and wireless microphones, musicians and music educators have always embraced technology. In fact, all musical instruments can be considered "technology," as they are outcomes of the tools ("hardware") and the skills ("software") by means of which humans produce and use them (see Bain, 1937). Until more recently, knowledge creation and transfer have oriented around the individual and involved the oral-aural, writing-printing, and inventions-products dimensions. While music technology associated with electronic- and digital-based technology (e.g., computer, effects unit, or piece of software) used in the process of making music (e.g., recording and playback, storage, mixing, editing, analysis, and performance) is important, it is the individual who is central in playing the key roles of producer-creator, communicator-transmitter, and valuer-interpreter of knowledge. As the world becomes increasingly characterized and dominated by technology, the emphasis on the possibilities offered by the electronic-digital and cyber dimensions should not come at the expense of the human dimension.

Indeed the learner/teacher today can benefit from hardware and software that are increasingly complementing and converging to create products capable of multiple functions. This is augmenting pedagogy that emphasizes positive learning experiences and performance outcomes of individual learners. An example is the SmartMusic program, which is used by many students. As an interactive application for teachers and students, it provides instrumental and vocal students with practice opportunities for accompaniment by a piano or synthesized band or orchestra. Besides offering a less "lonely" musical practice experience, the accompaniment can adapt to the individual student's tempo changes; the keys and tempi of pieces can be changed to suit specific learning and performance purposes. SmartMusic also has a built-in tuner, metronome, and source of information that includes fingering charts, musical terms dictionary, and notes on composers and compositions. It can also serve as a recording, home practice, and assessment tool. A record of students' efforts can be made and graded using rhythm and pitch data, and the recording and accompanying grade can be emailed to the teacher and automatically entered into the teacher's database grade book. Another example is the visual feedback voice training software system Sing & See. It enables singers to actually see their voices on a computer screen—providing real-time spectrographic analysis of the quality of their voices in terms of pitch and timbre. It can help to highlight vocal issues such as overshooting or scooping, vibrato, and roughness in a nonconfrontational way, as well as providing focus for self-directed learning. The visual feedback can be obtained while singers are singing or during playback. A research study has found evidence of its usefulness among my undergraduate music and music education students (see Leong & Cheng, 2014).

Today's technological advancement has necessitated the ability to negotiate the increasingly complex and shared knowledge spaces beyond the confines of school classroom, rehearsal, and studio settings. The titles of 20 sessions for a recent music education conference focusing on using technology creatively in music learning reveal some of the changes characterizing the evolving field of music education in a fast-changing world, and show the range of professional development needs of today's music educators:

- Social networking and composition
- Remix isn't copy and paste
- Live looping on your iPad
- Connecting and learning 24/7: the importance of Twitter and Facebook
- Composing online
- Make your own website in 90 minutes
- iPad Band
- Online resources test drive
- Awesome ideas for interactive whiteboards in primary and middle school music
- Music ICT pedagogy

- Making music with new media
- Creating effective systems: How technology can benefit instrumental teaching
- Songwriting made simple using technology
- World music using technology
- Online resources for teaching theory, aural skills, music literacy, and music history
- You can be a film composer: Scoring with Sibelius
- World Café: Interconnected world
- The self-publishing teacher
- Using the cloud to teach and learn about instruments and ensembles
- QR (Quick Response) codes in music education

This list will change and grow as new music hardware and software products are released into the marketplace. New technological development has greatly expanded the possibilities of integrating technology in a variety of music learning settings and activities, including music reading, vocal, choral, and instrumental performance, musical improvisation and composition, analysis and evaluation of music, combination of music and other art forms, and music history and culture. New sounds and musical styles are being created every day, and the wide array of options that music technology provides is expanding all the time. For example, a wealth of musical styles can be created by the combination of a synthesizer, acoustic guitar, and recording software with a computer. The Internet, music technology, and open file sharing have brought about a huge change in the way people discover new music and new artists. Even children with little or no formal musical training are able to find ways to compose and record their own music, share it with the world from their own rooms, and offer judgments about music and performances by artists they have heard. These technological advances have enlarged the potential for creative individuals to find their voices and satisfaction through music either as a hobby or a profession. They have also enabled global connections between people who would otherwise never meet.

Novice musicians are now able to produce professional-sounding work with widely available programs. There are also new mobile apps that have expanded the opportunities for creating music. It is now possible to create original compositions featuring realistic virtual instruments, radio-ready beats, and audio engineering effects with a little more than an electronic keyboard and a laptop, thus blurring the traditional divide between artist, record company, and distributor. The widespread availability of inexpensive recording hardware and software (e.g., Aviary, Audacity, and Digidesign's Pro Tools), coupled with the expanding opportunities for amateurs to distribute and share their work online, has caused tremendous shifts in the music industry.

The globalization of music practices has also made if possible for social learning networks, such as Remix Learning, to empower youth, including disadvantaged middle-school youth, to publish their original music, images, and videos, as well as to post comments and maintain an embedded blog. Youth are able to work in a simulated recording industry via NuMu, a supportive online community that allows them to showcase their music, collaborate, compete, and develop their talents as performers, producers, managers, and engineers. From the NuMu homepage, youth can search for music by artist or genre, listen to tracks, review and comment on artists' work, enter into competitions, and communicate and collaborate with like-minded peers. In collaboration with Musical Futures, these youth develop skills in the process of managing their school's (or organization's) "record label" on NuMu, which may include creating music, publishing, or marketing and promoting.

Despite significant developments in globalization, education policies and technologies, the process of technology uptake in schools has been unsatisfactorily slow (Buckingham, 2011). A 2012 government report on UK schools found music technology underused across all levels. This was a significant barrier to pupils' musical progress, and an example given was the insufficient use made of audio recording to assess and improve students' work. The report noted the need for greater emphasis on using technology to nurture pupils' musical development and making more creative use of music technology to more effectively create, perform, record, appraise, and improve students' work. In the United States, a major survey of over 1,800 high schools found that only 14% of responding schools offered technology-based music classes (Dammers, 2012). Recent graduates surveyed indicated a lack of music technology classes in U.S.college music education curricula that make the connection between technology usage, enhancement of music learning, and assistance with assessment (Nielsen, 2011). The types of technology used by music teachers for assessment purposes tend to be notation software and digital recording devices. Teacher respondents described using notation software for making quizzes or tests based on music knowledge content as well as performance excerpts either for face-to-face assessment or in conjunction with SmartMusic. Implementation of new technologies, such as interactive whiteboards or web-based tools such as rubrics and e-portfolios, seems to be still developing within the music education profession, as was indicated in another survey (Williams & Webster, 2011), which identified eight music technology competencies of undergraduates rated highly by music faculty. These were not musically or pedagogically profound: use a notation program; record and mix a performance with digital audio; understand copyright and fair use; burn a CD or DVD; edit digital audio; basic understanding of acoustics and audiology; use presentation software and connect to a projector or smartboard; set up a computer music workstation and troubleshoot problems.

The situation does not appear to be better in Australia, with Southcott and Crawford (2011) reporting that ICT in the Australian music education curriculum is perceived as "a different set of tools to support music education, much as a new set of percussion instruments might do" (p. 124), or as computer-aided instruction might do (Mark & Madura, 2010). In this tool-oriented approach, the potential benefits of technology in music teaching and learning are reduced simply to assisting with aural training and the acquisition of notational and music composition skills. An earlier study of preservice and early-career music teachers found that technology skills and knowledge were not mentioned as an attribute necessary for teaching music effectively (Harrison & Ballantyne, 2005). Another study found that preservice teachers' experiences in music and technology did not develop into confidence, knowledge, and motivation to teach music (Capaldo & Bennett 2013). My study in the 1990s found that lack of confidence and lack of expertise were reasons cited by Australian music teachers for not using technology in their teaching (Leong, 1995). A Hong Kong study of BA and MA students in a tertiary institution found that only less than half the respondents would expect a music classroom to be equipped with multimedia equipment (Ho, 2009). Although nearly all of them had their own computers, their online learning was infrequent, and their use of AV media/technology and email communications was limited to assignments and presentations. While computers were considered to be useful tools for teaching musical composition in primary schools, only 14 of over 800 Hong Kong primary schools had secure government funding to set up computer-based music teaching rooms (Tang, 2009). However, the provision of IT training for music teachers by the government was found by a researcher to be "practically adequate in terms of quantity and content" as a whole (Lee, 2010, p. 25).

Sustaining a Human-Techno Future

An online survey of US-based musicians by the Future of Music Coalition found that most of the respondents used Internet-based technologies for their work, including collaboration with other musicians and producing/recording their music. The top five technology-based tools used by them to create their work were, in descending order: Finale, Pro Tools, GarageBand, Sibelius, and Logic. These reflect the types of software used in the music classroom to teach composition with technology as a tool with which students create, edit, save, and reproduce music (Freedman, 2013). My involvement as codirector of the Australian National Review for School Music Education found that in order to improve and sustain the quality and status of music education needs to work in supportive and productive partnerships and networking with the music industry, musicians, music organizations, and the community (Pascoe, Leong, MacCallum, Mackinlay, Marsh, Smith, Church, & Winterton, 2005). Indeed many music industry players, such as Roland Music, Apple, and local music suppliers and retailers, have contributed greatly to the music education community through their sponsorships of music education events and conferences and provision of professional development workshops and materials. In addition, some companies, such as Roland and Apple, provide web-based resources that support and supplement the teaching and learning of music.

As seen above, the growing array of technologies are useful as tools for teaching, learning, and assessment purposes. When they are used skillfully and implemented properly, ICT and associated technology can serve as:

- Complementary tools that can be adapted to work with various learning approaches
- Information tools for enhancing access to resources in a variety of formats

- Communication tools that facilitate communication among the community of learners, practitioners, researchers, etc.
- Assistive tools that enhance the efficiency of both teacher and learner in tasks such as, among others, searching, storing, preparing, editing, producing, capturing, transcribing, and transposing music
- Construction tools that facilitate the construction of knowledge and understanding through hands-on investigation, inquiry, and discussion
- Tutorial tools (similar to computer-assisted instruction) that cater to the various learning skill levels, providing differentiated and targeted practice, feedback, and tracking and recording of learners' achievements
- Situating tools that enable the situating of musical knowledge, skills, and activities in the context of real-world practices
- Simulation tools that provide virtual environments (including gaming) for applying and executing real-world tasks free from the constraints of time and space

Based on my experiences with technologies over the past three decades, the integration of technology in music education could be said to have undergone three stages of development:

(1) Computer-assisted learning (CAL), which emphasized drill and practice, tutorials, simulation, and instructional games that were learner centered with self-paced instruction and limited interactivity
(2) Multimedia-based instruction, which extended CAL to include multimodal instruction (such as printed text, spoken words, pictures, illustrations, or animations) and modeling, with enhanced interactivity and some level of learner's construction of knowledge.
(3) Web-based instruction, which features on-demand and just-in-time learning in asynchronous and synchronous virtual collaborative learning environments that are primarily constructive in nature

My adventures with ICT have shown how software, hardware, and connectivity are becoming more global, powerful, and sophisticated so as to allow for enhanced integration of ICT in music education, but practices in schools have not always reflected this. Educational systems and schools are often unable to adopt and execute ICT-facilitated teaching and learning, and there may be a lack of access to the expertise necessary for setting in place appropriate infrastructure and processes. As such, partnerships with universities and the music industry, as well as building up professional networks, are essential. I have learnt the importance of preservice teacher education and professional development being tailored to cater to the needs of real-world teachers and classrooms. The possibilities, advantages, and limitations of using ICT should be included as the key content focus of teacher education, which should emphasize ICT as the core technology for delivering teacher education courses. Its use should also be embedded in music teaching-learning approaches, and it should be used to facilitate professional development and networking. My involvement with a number of different systems and levels has taught me that for successful and sustainable uptake and adoption of ICT in music education to be realized, synergies need to be created at a number of levels—government, school, teacher education, music industry, and individual teacher. These include:

- Government: policy initiation, development, and implementation
  Music teacher education: ICT-infused preservice education; professional development; postgraduate education; research and knowledge transfer
- School music education: integrating ICT; supporting ICT-rich pedagogy and ICT infrastructure; industry-university linkages
- Music industry: development of products and services; support for development of music learning pedagogy; school-university linkages
- Individual music teacher: updating ICT-related pedagogies; professional development; postgraduate education; school-industry linkages

The issues of uptake, adoption, and use of ICT by educational systems, individual institutions, and individual teachers are realities to be addressed. It can play two different but overlapping roles in any system-wide development: enabling gradual change in established mechanisms and processes, and transforming or replacing those mechanisms and processes. A UNESCO study has identified four broad stages that systems would need to undergo, and these may be conceived as a continuum or series of steps (Anderson & van Weert, 2002). The first emerging stages would see teachers beginning to become aware of the potential of ICT. The second, applying stages would see teachers learning how to use ICT for teaching and learning purposes. The infusing stages would see a variety of ICT tools being integrated into the curriculum. The final transforming stages would involve new ways of approaching teaching and learning contexts and situations with specialized ICT tools to explore a variety of real-world problems through innovative learning.

With ICT becoming more accessible now than ever before, the future is certainly bright for its expanded use in music education in a globalized and hyper-connected world and for music educators to make a difference as digital citizens. I am mindful that the learning curve will always be there, but the journey is exciting, being full of surprises, challenges, gee-whiz technologies, and amazing new products. Music education is blessed with the opportunity and ability to offer learners an education that can transform how they value, think, and act. I believe it's crucial that technology should never be allowed to define who we are or drive human choices and actions in an era of incomprehensible technological complexity and change. As Einstein allegedly put it: "the human spirit must prevail over technology." Indeed, it is humans living in a globalized world who should apply and design technology that works in tandem with innovative pedagogies in open learning communities.

References

Allen, I. E., & Seaman, J. (2011). *Going the distance: Online education in the United States.* New York: Babson Survey Research Group. Retrieved January 5, 2017, from http://sloanconsortium.org/publications/survey/going_distance_2011?date=121011.

Anderson, J., & van Weert, T. (Eds). (2002). *Information and communication technology in education: A curriculum for schools and programme of teacher development.* Paris: UNESCO.

Australian Curriculum, Assessment and Reporting Authority. (2013). *General capabilities in the Australian curriculum.* Sydney: Author. Retrieved January 5, 2017, from http://www.australiancurriculum.edu.au/generalcapabilities/overview/general-capabilities-in-the-australian-curriculum.

Bain, R. (1937). Technology and state government. *American Sociological Review, 2*(6), 860–874.

Barney, D. (2004). *The network society.* Cambridge, UK: Polity Press.

Buckingham, D. (2011). *Beyond technology: Children's learning in the age of digital culture.* London: Polity Press.

Burn, A. M. (2008). *Technology integration in the elementary music classroom.* Milwaukee, WI: Hal Leonard.

Capaldo, S., & Bennett, S. (2013). Examining pre-service primary teachers' experiences of music and technology. In *Proceedings of the World Conference on Educational Media & Technology* (pp. 484–493). Chesapeake, MD: AACE.

Castells, M. (2004). An introduction to the Information Age. In F. Webster, R. Blom, E. Karvonen, H. Melin, K. Nordenstreng, and E. Puoskari (Eds.), *The Information Society Reader* (pp. 138–149). London and New York: Routledge.

Chan, K. (2012, April 18). *ICT in education: A Hong Kong perspective.* Presentation at the Open Educational Resource Symposium 2012, Open University of Hong Kong. Retrieved January 5, 2017, from http://oer.ouhk.edu.hk/ppt/oer2012_chen.ppt.

CISCO. (2008). *Equipping every learner for the 21st century.* San Jose, CA: CISCO Systems, Inc.

Dammers, R. J. (2012). Technology-based music classes in high schools in the United States. *Bulletin of the Council for Research in Music Education, 194,* 73–90.

Department of Education (UK). (2011). *The importance of music: A national plan for music education.* London: Author. Retrieved January 5, 2017, from https://www.gov.uk/government/publications/the-importance-of-music-a-national-plan-for-music-education.

Department of Education, Science, & Training. (2002). *Raising the standards: A proposal for the development of an ICT competency framework for teachers.* Canberra, Australia: Author.

Dew, J. (2010). Global, mobile, virtual, and social: The college campus of tomorrow. *Futurist,* March-April 2010. Retrieved June 11, 2011, from http://findarticles.com/p/articles/mi_g02133/is_201003/ai_n52370702/.

Dillon, S., & Hirche, K. (2010). Navigating technological contexts and experience design in music education. In J. Ballantyne & B. Bartleet (Eds.), *Navigating music and sound education* (pp. 175–192). Newcastle, UK: Cambridge Scholars.

Freedman, B. (2013). *Teaching music through composition: A curriculum using technology.* Oxford, UK: Oxford University Press.

Gauntlett, D. (2011). *Making is connecting: The social meaning of creativity, from DIY and knitting to YouTube and Web 2.0.* Cambridge, UK: Polity Press. Retrieved January 5, 2017, from http://www.makingisconnecting.org/.

Gordon, D. G., Rees, F., & Leong, S. (2000). An evolving technology: Interactive televised instruction's challenge to music education. *Australian Journal of Music Education, 1,* 40–51.

Harrison, S., & Ballantyne, J. (2005, July 6–9). *Effective teacher attributes: Perceptions of early-career and pre-service music teachers.* Paper presented at Australian Teacher Education Association Conference, Gold Coast.

Ho, W. C. (2009). The role of multimedia technology in Hong Kong higher education music programs. *Visions of Research in Music Education, 13.* Retrieved from http://www-usr.rider.edu/~vrme/.

Info-communications Development Authority of Singapore. (2006). *iN2015 Masterplan offers a digital future for everyone.* Retrieved January 5, 2017, from https://www.imda.gov.sg/infocomm-and-media-news/buzz-central/2006/9/in2015-masterplan-offers-a-digital-future-for-everyone.

Ingerman, B. L., & Yang, C. (2011, May 31). Top-ten IT issues 2011. *Educause Review,* 1–44. Retrieved January 5, 2017, from http://er.educause.edu/articles/2011/5/topten-it-issues-2011.

Lee, B. K. Y. (2010). ICT integration in primary school music education: Experience of pioneering countries and its implications for implementation in Hong Kong. *Asia-Pacific Journal for Arts Education, 8*(4). Retrieved January 5, 2017, from http://www.ied.edu.hk/cca/apjae/Vol8_No4.pdf.

Leong, S. (1995). Music technology competency and effective teacher preparation. *Australian Journal of Music Education, 1,* 21–25.

Leong, S. (2011). Navigating the emerging futures in music education. *Journal of Music, Technology and Education, 4*(2), 2–22.

Leong, S., & Cheng, L. (2014). Effects of real-time visual feedback on pre-service teachers' singing. *Journal of Computer Assisted Learning, 30*(3), 285–296. doi:10.1111/jcal.12046.

Leong, S., & Robinson, G. (1995). *Using computer technology in music education.* Perth, Australia: CIRCME.

Long, M. K. (2011). *The effectiveness of the SmartMusic® assessment tool for evaluating trombone student performance.* Unpublished Doctor of Musical Arts dissertation, University of North Carolina, Greensboro.

Mark, M., & Madura, P. (2010). *Music education in your hands.* New York: Routledge.

Ministerial Council for Education, Employment, Training and Youth Affairs. (1999). *National goals for schooling in the twenty first century.* Carlton South, Victoria: Curriculum Corporation.

Ministerial Council for Education, Employment, Training and Youth Affairs. (2000). *Learning in an online world: The school action plan for the information economy.* Canberra: Author. Retrieved January 5, 2017, from http://www.scseec.edu.au/site/DefaultSite/filesystem/documents/Reports%20and%20publications/Archive%20Publications/ICT/LOW-ContemporaryLearning.pdf

Ministerial Council for Education, Employment, Training and Youth Affairs. (2005). *An education and training action plan for the information economy.* Carlton South, Victoria: Curriculum Corporation.

Nallaya, S. (2010). *The impact of multi-modal texts on the development of English language proficiency.* Unpublished doctoral dissertation, University of Adelaide, Australia.

Nielsen, L. D. (2011). *A study of K–12 music educators' attitudes toward technology-assisted assessment tools* (Student Research, Creative Activity, and Performance Series, Paper 43). Lincoln, NE: School of Music, University of Nebraska-Lincoln.

Pascoe, R., Leong, S., MacCallum, J., Mackinlay, E., Marsh, K., Smith, B., Church, T., & Winterton, A. (2005). National review of school music education: Augmenting the diminished. *Department of Education, Science and Training.* Retrieved January 5, 2017, from http://researchrepository.murdoch.edu.au/9459/

Reese, S., McCord, K., & Walls, K. (2001). *Strategies for teaching: Technology.* Reston, VA: MENC, National Association for Music Education.

Rudolph, T., & Frankel, J. (2009). *YouTube in music education*. Milwaukee, WI: Hal Leonard Corporation.

Schilling, D. R. (2013, April 19). Knowledge doubling every 12 months, soon to be every 12 hours. Industry Taps into News. Retrieved January 5, 2017, from http://www.industrytap.com/knowledge-doubling-every-12-months-soon-to-be-every-12-hours/3950.

Southcott, J., & Crawford, R. (2011). The intersections of curriculum development: Music, ICT and Australian music education. *Australasian Journal of Educational Technology, 27*(1), 122–136.

Tang, P. H. (2009). *Teaching music composition with computer-based technology in Hong Kong primary schools: Critiques of present pedagogical practices and recommendations for future development*. Unpublished doctoral dissertation, Deakin University, Australia.

Top ten forecasts for 2011. (n.d.) *The Futurist Magazine*. Retrieved January 5, 2017, from http://www.wfs.org/page/futuristmagazine.

UNESCO. (2013). *UNESCO global report: Opening new avenues for empowerment ICTs to access information and knowledge for persons with disabilities*. Paris: Author.

UNESCO. (2014). *Global monitoring report: Education for all*. Paris: Author.

Williams, D. B., & Webster, P. R. (2006). *Experiencing music technology: Software, data, and hardware* (3rd ed.). Belmont, CA: Thomson/Schirmer.

Williams, D. B., & Webster, P. R. (2011). *Music Technology Skills and Conceptual Understanding for Undergraduate Music Students: A National Survey*. College Music Society National Conference/Association for Technology in Music Instruction, Richmond, VA.

CHAPTER 9

TECHNOLOGY IN THE MUSIC CLASSROOM—NAVIGATING THROUGH A DENSE FOREST
The Case of Greece

SMARAGDA CHRYSOSTOMOU

Introduction

It was spring of 1994, at the computer center of Bulmershe College, Reading University, and I was sitting next to my friend and fellow first-year PhD student Ros, exchanging emails, practicing our skills in this new medium for long-distance communication. I remember exploring my way around a (very dark and boring) Disc Operating System (DOS) environment interface, sending back and forth short greetings and small pieces of everyday conversation (Hello! How are you? When are you going for dinner? Did you finish the literature review?), engaged in an activity that (today, in hindsight) could be described as the prequel to instant messaging.

Fast forward to 2014. Our world today is dominated by ICT technology in ways that were unimaginable 20 years ago. We are daily submerged in activities that would not be possible without the use of digital technology and digital media. Moreover, we know that technology we use today becomes obsolete with a speed that exceeds 7% a year. This practically means that in 10 years we will be using technology that is currently unheard of.

Digital literacy and digital advancement is therefore rightly located at the top of the priorities list for all future agendas pushed forward by international and European bodies like the Organisation for Economic Co-operation and Development (OECD) and the European Union. All countries consider skills that are included under the umbrella term "digital literacy" vital for the successful citizen of the future. For example, the Digital Agenda for Europe (European Commission, 2010) introduced recently, aims to "maximize the social and economic potential of ICT" (p. 3) for Europe by contributing significantly to EU's economic growth and spreading the benefits of digital era to all sections of society. Wider and more effective use of digital technologies will "spur innovation, economic growth and improvements in daily life for both citizens and businesses" (p. 3).

At the same time, technology's use and impact on education interests many international and European institutions. Accordingly, technology and education have been an ongoing focus of OECD work since the 1980s, and recent Programme for International Student Assessment (PISA) measurements include variables related to access and use of technology in and out of schools (Kools & Istance, 2013).[1] An important conclusion that these international surveys have reached is that "access to ICT at school and at home on its own does not show a positive impact on students' educational performance" (Fuchs & Wossmann, 2004). However, these surveys have consistently uncovered a positive relationship between ICT and students' and teachers' motivation and concentration in class, as well as some confirmation of the ability of ICT to encourage independent learning and teamwork.

Of course this is not a unified picture of the world. Studies by the OECD and PISA in 41 member and partner countries reveal a gap between participating countries in relation to computer and digital technology access and frequency of use, a gap that is a reflection of unequal socioeconomic resources. In addition, the latest surveys focus on a "second-level digital divide," which refers to young people's capacity to take advantage and engage with ICT even among the countries that have similar access and frequency of use. This second gap appears to also be a reflection of unequal socioeconomic and cognitive resources (Selwyn, 2004; Claro, n.d.).

The above research and reports suggest that in order to assess whether technology enhances and advances learning we need to look beyond the numbers and percentages of schools, students, and teachers that have and use computers and digital technology, and concentrate more intently on what they are actually doing with these technologies. It is a matter of not only how much technology we have and use but also how we use it, what kind of activities we engage in, what kind of skills we acquire, and what goals we accomplish. Even though our students may be "digital natives" (Prensky, 2001) and may appear technologically "savvy"—it does not necessarily mean that they are competent at using technology in a critical, responsible, and creative way. And this inevitably brings into focus our teaching strategies and methods.

With the teacher then in focus, the OECD and PISA reports conclude that education standards need to include the types of skill and competence that can help students develop into responsible and performing users of technology and also acquire new competences required in today's knowledge society—those related to knowledge management. It is the duty of the teachers "responsible for teaching the new millennium learners . . . to be able to guide them in their educational journey through digital media." Therefore, teacher training, both initial and in-service, is crucial for equipping teachers with the essential competences (Organisation for Economic Co-operation and Development, 2006).

Apart from the types of skill needed (for both teachers and students), the educational aims proposed, and the teaching methods involved, even long-established and conventional physical resources like buildings and classrooms are disputed. Fisher (2010) points out that the recent advent of wireless broadband internet access and mobile communications has enabled the "emergence of a true synchronous/asynchronous and virtual/physical matrix of learning opportunities" (p. 3). He sees the "knowledge age" rapidly transforming into a "creative age," yet students are still being crammed into closed classrooms that represent not only "a physically outdated teaching model" and a mismatch to "the inter-connected virtual world we now live in" but also the "least creative space they can learn in" (p. 3). It appears, then, that as Savage (2004) claims, it is hard to overestimate the change taking place and determined by the arrival of digital technologies and their incorporation in the teaching and learning processes. New technologies create the imperative of transforming the way we understand education while, at the same time, representing a fraction of a much larger social and cultural change.

Research reveals that music teachers have attitudes and constraints similar to teachers from across the curriculum when it comes to the role and application of technology in their classrooms (Greher, 2011; Wise, Greenwood, & Davis, 2011). Problems created by frequent "glitches" during use of technology, like incompatible software, computer crashes, Internet unpredictability, and outdated hardware and equipment, are often mentioned by teachers. The fact that many teachers struggle to keep up with the constant advances and developments whereas their students are always many steps ahead does not help boost their confidence. Thus, music teachers end up adopting the technology that fits within their comfort zone, which usually means that they use technology to teach in the same way they already do, with the same objectives and aims and the same philosophy. Research from around the world reveals that teachers primarily employ technology in order to prepare their lessons and materials and do not seem to incorporate it in their lessons, at least not as much or as differently as we would expect (Barron, Orwig, Ivers, & Lilanois, 2002; Cuban, 2001; Greher, 2011; Savage, 2004).

The profound impact that the use of technology in music has had on musicians and their conceptualization of musical practice has been evident for a long time. Varese, commenting on electronic instruments four decades ago, exclaimed: "I have been waiting a long time for electronics to free music from the tempered scale and the limitations of musical instruments. [They] are the portentous first step toward the liberation of music" (as cited in Machlis, 1970, p. 324). This statement could be perceived as prophetic for the revolution in music today that technology has enabled. Indeed, it seems that music technologies changed musical priorities and enabled musicians to "engage with the microphenomena of musical sound itself" (Savage, 1997, as cited in Savage, 2004, p. 73), shifting the emphasis away from traditional issues such as melody, rhythm, or harmony, "to an increasing focus in dealing with the sound itself, thinking about its intrinsic value and place in a wider musical structure" (Théberge, 1997, as cited in Savage, 2004, p. 73).

New music technology has also offered freedom by making accessible to all, free through the Internet, tools and sound and music manipulation capabilities that, until recently, belonged to professional music studios. A wider crowd of young people can now create music, orchestrate their ideas, invent new sounds and assonances, add sound effects, and explore their musical creativity in ways that a few years ago were unthinkable. Whereas long hours and years were previously needed in order to master the necessary skills to start making music (learning to play an instrument, learning the theory, the notation, etc.), nowadays technological advancements have enabled a much shorter and direct process. Music intelligence is thus exhibited easily and (often) surprisingly by students with no prior formal knowledge of music.

However, our music education system is still predominantly rooted in traditional beliefs and values, conventional musical skills connected with usual musical instruments, learning to read and write in (mainly) Western notation, composing by paying attention to the specific rules and constrains of a particular musical genre (which is usually part of Western classical traditions). The majority of aims and objectives in a music curriculum spring from these skills. Although there is evidence from around the world that some current changes in the way that music is heard, created, owned, understood, and explored have filtered through the musical practices in our classrooms, it is still a slow process.

On the other hand new technology is created daily, and our students continue to outpace us. Especially in music, with the majority of young people today being high-end users and consumers of music technology, it is unfortunate that music teachers do not incorporate the types (or uses) of technology that are relevant to students' interests and manners of thinking. The paradigm of "digital native" is insufficient to cover the multifarious new intelligence that has been revealed with the "blast" of digital technology. Shifting the focus away from teaching about technology and with technology to *thinking with technology* is the only way to transform pedagogical practice and achieve the significant cultural change that is crucial in order for music education to move into the twenty-first century. Teachers need to broaden their understanding of what constitutes musical compositional and performance activity, to recognize the transformative powers of ICT for music teaching and learning, to find ways to bring into the school and the classroom knowledge and skills that students develop outside school about music and its creation (Savage, 2004; Finney & Burnard, 2009; Wise, Greenwood, & Davis, 2011; Greher, 2011). "This will be achieved as more teachers recognize the potential of new technologies to teach new musical content in new ways" (Savage, 2004, p. 74). However, if "educators fail to grasp this major cultural shift, music as a curriculum subject will become increasingly alienated from young people's lives and they will find their music education elsewhere" (p. 75).

The case of Greece

So far I have engaged in a bird's-eye view of the forest. Its density often obscures the available paths. By focusing on the specific case of Greece, it would be possible to illuminate the trail that was followed so far and recommend a future route. How do all the different choices and problematics relate to Greece? To what extent has technological transformation impacted Greek education? How did the Greek educational system respond to worldwide changes brought about by technology? In what way and to what extent have the international priorities been incorporated into Greek national policy?

Technology in Education in Greece

Information communication technology and its use in education has been a topic of interest for the last 15–20 years in Greece. Technological advances in society were filtered into education, incorporating aims and objectives relevant to basic skills of technology use. As in many countries around the world, the autonomous subject of informatics was introduced to the National Curriculum initially in secondary education, and aspects of it later in primary education as well. However, as was expected from similar research worldwide, teachers either lacked the appropriate knowledge or hesitated to accept and adopt technology in their classrooms (Drenoyianni & Selwood, 1998). In the last decade, efforts have been made to integrate ICT across the curriculum, not as a separate subject but as a tool and a means to enhance learning in all subjects.

Nonetheless, statistics and research results testify to the limited outcomes these efforts have had so far. For the period 2007–2010, research by the Greek Observatory of Digital Society () demonstrated that only 22% of Greek citizens had a satisfactory level of digital literacy compared to 44% of 27 EU countries. Even though that proportion was significantly improved during that 3-year period, it remained considerably lower than the EU average. Similarly, in education, the Greek Observatory of Digital Society for that same period discovered that very little exploitation of ICT in education was achieved.[2]

In 2010, a multilateral and multilevel framework of reforms in education called New School was initiated in order to respond to twenty-first-century challenges. One of the four pivotal areas of reform, called Digital Advancement, introduced a large number of relevant actions encompassing schools, teachers, students, processes, equipment, and so on. Following the directions of the 2020 Digital Agenda for Europe and international trends, and taking into account experiences and lessons learned from previous attempts to integrate technology in education and cultivate digital literacy, the large-scale national scheme called Digital School was launched by the Greek Ministry of Education. It introduced a series of actions organized into five key areas: Infrastructure, digital educational content, teacher training, electronic management of education, and support actions. Some of the actions and projects of this scheme were:

- Creation of interactive e-books, which are school textbooks for all subjects in primary and secondary education, in html format, enriched with multimedia content and "click-and-play interactive learning resources" (Megalou & Kaklamanis, 2014, p. 3)

- Development of open, reusable learning objects from all areas of the Greek curriculum; these were initially created as part of the previous action but were subsequently incorporated in the national learning object repository (see the following paragraphs)
- Design, development, and operation of the Greek National Digital Learning Repositories for hosting, organizing, and distributing learning resources for schools: Photodentro LOR is the cornerstone of the infrastructure, hosting learning objects (photodentro.edu.gr/lor); the Photodentro Repository ecosystem also includes Photodentro EduVideo (http://photodentro.edu.gr/video), hosting short educational videos suitable for in-class use, and Photodentro UGC (http://photodentro.edu.gr/ugc), hosting learning resources developed by teachers, thus representing the user-generated branch of the ecosystem (Megalou & Kaklamanis, 2014, p. 3)
- Design, development, and operation of the Greek National Educational Content Aggregator Photodentro, a national service for harvesting and accumulating educational metadata from various repositories and collections (museums, libraries, audiovisual archives, etc.), thus serving as the central access point to learning resources for schools in Greece
- Design, development, and operation of the Greek Digital Educational Platform for all schools, pupils, and teachers in primary and secondary education: The e-me platform—currently under development—aims at providing a safe working space for pupils and teachers, with a modern and intuitive environment, to share their content, connect, communicate, and collaborate with friends; publish their work; download useful apps; and access and utilize efficiently learning resources (Megalou & Kaklamanis, 2014; Chrysostomou, 2013; Jimoyiannis et al., 2013)

The majority of the above actions and projects can be accessed through the platform of Digital School (http://www.dschool.gr), an open source platform with Web 2.0 tools. This wide scheme is still in progress (2011–2015), together with parallel actions under the Digital Advancement agenda that include equipment provisions for primary and secondary education, in-service teacher training programs, and large-scale national campaigns in schools and mass media for the safe use of the Internet.

At the same time, a lively discourse on the philosophy and aims of technology integration in education has developed during the last 10 or more years. Greek national policies were accused of a "technocentric rhetoric" that presupposes that all and any technology is an "undisputed good." This effectively assumes that technology integration in education, regardless of the framework or process, will undoubtedly democratize education, improve cooperative teaching and learning, and raise the quality of education. This rhetoric and philosophical stance was identified as the reason for technology's failure to bring about the intended results. Into the bargain, another strong view complementing the above and widening the discourse, maintains that technology is not neutral and therefore cannot be integrated in the same way in all education systems, ignoring particular social, economic, political, cultural, and educational circumstances in different countries. Technology represents power and can lead to educational inequalities between stakeholders. According to these opinions, such a process of integrating technology in education results in a "functionalistic" application where, regardless of the novelty of the tool, teachers reproduce traditional educational models. Supporters of this critical view conclude, cynically, that the only group that seems to benefit are the companies that create and sell software and hardware (Bourantas, 2012; Grollios, 1999; Raptis & Rapti, 2004).

As mentioned earlier, the latest initiatives and actions for technological advancements in Greek education are still under way, and therefore their effect is not yet possible to assess. In addition, the economic crisis has hit Greece hard during the last few years, and all areas of the public sector (including all levels of education) have suffered considerable loss of funding, which has brought many developments to a standstill. The above reasons could establish the shortfall of technology's exploitation, which appears disproportionate to the level of its prioritization. It is not a surprise, then, that according to the latest PISA results (Chrisalita & Cretu, 2014) Greece is still at the lowest end of ICT usage and access in comparison to all 21 European countries participating in the research. Even though this percentage was raised significantly after the 2011 PISA report, it is still the lowest in Europe, and the gap between Greece and the other countries remains.

The picture becomes clearer and some of the aforementioned inequalities become apparent as we look at specific numbers and areas of interest within the PISA results. Approximately 20–25% of Greek students (15-year-olds) do not have access to a desktop or a laptop at home (almost the highest percentage among the participating countries), and a huge percentage (75%) do not have access to tablets. On access to the Internet Greece is again at the lowest end of the scale. About 82% of the population have access to the Internet, the lowest percentage among OECD countries. In the majority of participating countries, access to the Internet is well over 90%. The lack of access to the Internet is repeated in the relevant use of smartphones with or without an Internet connection.

The level of ICT equipment, access, and usage in schools is a slightly different picture. Despite the lack of equipment (the lowest in Europe), Greece, surprisingly, holds one of the highest percentages of access to tablets in schools among the 21 European countries, as well as the highest percentage of access to and use of e-books (16% compared to 5% for the rest of European countries). Although the majority of schools have access to the Internet, a significant percentage of students do not use it at school. This is particularly interesting and would need further probing, as it appears to be a direct result of the teaching methodologies used as well as lesson design. However, despite the lack of equipment, 33% of students access the Internet every day for information, schoolwork, and communication with their teachers and friends in order to discuss and submit work. It also appears that Greek students use the Internet least of all for accessing social networks, fun, and entertainment but hold the highest percentage in Europe for uploading their own content and for downloading music!

Some initial conclusions that can be drawn from the above information is that although the overall picture is dim, with Greece lagging behind regarding technology's access and use, certain initiatives seem to have already had some positive effect in specific areas. The high percentage in the use of tablets and e-books in schools is no doubt a result of the investment that was promoted in those specific devices, as well as the development of enriched textbooks (e-books) that were recently introduced in all school subjects. Another interesting result that seems promising and requires additional inquiry is the types of activity and usage that Greek students appear to engage in—more educationally concentrated than their peers around Europe, uploading their own content and downloading music in higher percentages that the rest of the European countries. The skills and attitudes that are behind this type of behavior and activities should be cultivated and built upon because it can promote the coveted critical and creative use of technology.

Music Teaching and Learning with Technology in General Music Classrooms

How did all this affect music education and music teachers in Greece? How did music teachers react? Have they embraced this change? What are their views and what do they actually do in their classrooms? Is there evidence of adaptation?

Music is taught by specialist music teachers in both primary and secondary general education in Greece.[3] Moreover, general music is a mandatory subject for all of primary school and the first three years of high school. The latest version of the National Curriculum for Music was developed in 2010–2011 as part of the latest New School reform (mentioned above). This information, combined with a highly centralized educational system where everything is instigated by the Ministry of Education, explains the importance of the large-scale schemes that I will describe in the following paragraphs. The idiosyncrasies of the Greek language and the system's centralization and bureaucracy limit the possibility of extensive change in teaching culture and practices, even within specific boroughs or counties. Individual efforts and good practices are documented around the country, as will be mentioned in the paragraphs that follow. However, in order to promote the kind of change that is needed here, as was established in the beginning of this analysis, isolated efforts with localized impact will not be enough.

Initiatives and projects during the last few years that utilize technology and can be used in teaching music in primary and secondary education in Greece include large-scale, nationwide schemes and actions, as well as smaller school-centered and teacher-centered projects. Some of these include the following.

- The CD-ROM Emmeleia is for all primary education. It was created during the previous educational reform (2001) as part of the digital educational material in the form of DVDs and CD-ROMS funded by the Ministry of Education. This original scenario is presented in the form of a virtual village called Emmeleia, where neighborhoods and buildings hold a variety of applications and multimedia content to develop, train, practice, and play with basic music concepts, different music genres, music history, musical instruments, and music civilizations. The CD-ROM utilizes sounds, images, videos, musical examples, exercises in the form of games, short presentations for music history, music theory, and so on. It can be accessed online and downloaded free from the internet (http://photodentro.edu.gr/lor/subject-search?locale=el).

- Melodisia is an interactive online digital resource focusing on the history of music. It was created by the Music Library of Greece "Lilian Voudouri" of the Athens Concert Hall (http://melodisia.mmb.org.gr/). Its focus group is secondary education students and their teachers. Music excerpts as well as other art from the Middle Ages to 20th century are organized into six historical periods. Music examples presented include music content from the wide variety of concerts held in the Athens Concert Hall during its operation.
- Euterpe is an online interactive repository designed as a resource for music teachers (http://euterpe.mmb.org.gr/euterpe/). It was the result of a collaboration between the Greek Society for Music Education and the Music Library of Greece "Lilian Voudouri" after they were awarded the Gibson Award during the thirtieth world conference of the International Society for Music Education in 2012. The content of this repository is continually being enriched and includes Greek songs for educational use (traditional and by contemporary Greek composers). Each song is catalogued, with teaching notes, suggestions for lesson plans and related activities, accompaniments, and additional information to assist music teachers in adapting the material to their particular needs and teaching styles (Chrysostomou, 2014b).

The Digital School national initiative, mentioned earlier, includes a number of relevant and parallel projects. As the coordinator in a number of them for aesthetic education (music, visual arts, and theatre), I was directly involved in their design, development, and dissemination process, and I have a personal interest in their application and assessment in the classroom. Within each of these projects, extensive digital educational content was created to support primary and secondary music education. Learning objects that were created (initially as part of the media-enriched textbooks and subsequently as autonomous, reusable applications) can be used by the music teacher in multiple ways, in various environments and teaching formats, promoting a variety of teaching strategies. Available for use by teachers are:

- E-books/interactive textbooks: all music textbooks for primary school (last four years) and secondary school (first three years) (years 3–9) were enriched with multimedia content, following the process already described (http://ebooks.edu.gr/2013/). Html versions of these textbooks were enriched with a number of original learning objects, which range from presentations with minimal interaction to exercises and edu-games for self-assessment and music improvisation activities with a high level of exploration. These html versions were enriched with additional

content in the form of links to sites with particular interest for students or teachers, for example a Greek educational TV archive, museums, orchestras, and so on.
- All autonomous learning objects created for the music textbooks were stored in Photodentro after having been completed with educational metadata. They can be downloaded or accessed directly through the Internet (http://photodentro.edu.gr/lor/subject-search?locale=el).
- Older digital educational content and projects are being reevaluated and, following the metadata authoring process, will be stored in Photodentro ready to be used freely.

In-service training for music teachers has been planned in order to introduce the whole range of possibilities that the above schemes have created and showcase the available resources that can be incorporated in the music lesson. A recent small-scale study, based on case studies of secondary education music teachers in Greece and Cyprus, revealed some of the potential of the above scheme, which is currently a national priority, as well as some of the challenges that need to be addressed (Chrysostomou, 2013, 2014a).

More specifically, in order to assess the applicability of these resources and document teachers' views and their impressions from students' reactions, a small-scale pilot application was initiated as soon as the first three music e-books (for the three first years of high school) became available online. An open invitation was sent to high school music teachers in Greece and Cyprus to voluntarily take part in this initial inquiry. Participating teachers were sent directions for self-monitoring and a questionnaire with open-ended questions. A 2- to 3-month application period led to the completion of the questionnaire, which was then followed up with an interview seeking to complete any gaps, interpret teachers' opinions, and explore their ideas further. Analysis of the music teachers' oral and written responses showed that overall their views were positive. The ease of use and versatility of the new multimedia content of the e-books were included among the assets. Students were motivated and happy to use them, both in school and at home, and this was particularly important for the teachers. At the same time, the teachers commented that they would have liked more learning objects and applications that encourage free exploration and music creativity. In addition, internet volatility and lack of equipment presented points for concern (Chrysostomou, 2013).

Apart from the aforementioned large-scale and nationwide schemes, in many schools around the country technology is being used in music classrooms in a variety of ways and with different purposes. Whether it is through European projects' applications that enable international comparisons or smaller projects, based on the creativity and zeal of individual music teachers, supported by head teachers and schools, in public and private schools, primary and secondary education, some music teachers take advantage of the opportunities, equipment, hardware, and software offered to them. They follow their interests and utilize the available technology and music technology to create applications, presentations, and enticing music lessons for their students. Evidence of these efforts is available in conferences and publications where music teachers present their work.[4] However, an official record of the range of applications and types of utilization of technology in the music classroom is not available.

An important point worth mentioning here is that many music teachers are interested in creating their own materials and applications and are eager to share them with the community of Greek music teachers. These views were apparent from discussions and written reports during the pilot application of e-books mentioned earlier, as well as during in-service training seminars that have taken place.[5] It is a tendency that was also exhibited by the students in the PISA analysis of the 2012–2013 results. This trend, combined with the overall positive attitudes toward technology in the classroom and optimism for the future of technology's use and applications that was evident in recent research, provides an important foundation upon which to build for the future (Kyridis, Drossos, & Tsakiridou, 2006; Pesmatzoglou & Papadopoulou, 2013; Chrisalita & Credu, 2014).

Coda

Our navigation so far has highlighted certain points of interest that bring into focus possible paths for our future venture. Technology and the latest web tools have had an enormous impact on education. Greater access to information, communication, and expression, as well as open access to a variety of educational material, including a large quantity of high-quality teacher-generated content, changes the relationship between instruction and learning, along with inquiry and knowledge construction. Moreover, flexible licensing schemes, like Creative Commons, that offer authors and institutions wider opportunities for creating and sharing dissolve the boundaries between classroom and community and lead the way to a more democratized and globalized education.

At the same time, digital technology in music and music education has revolutionized the teaching and learning of music. Technology enables students to develop their musical intelligence further and with less effort. Thinking in sound, processing information, constructing meaning, and showing understanding through music and sound are as immediate and instant as clicking the mouse, selecting the instrument, and manipulating the final product through software, an "app," a tablet, or a smartphone. New meanings for musicality and musical intelligence, new ways of experiencing and creating music have emerged, and music educators have the responsibility to allow their students to move forward and explore their potential in new ways, in new contents, and in new learning environments.

It is a misconception to think that technology alone can enable an improvement in the quality of education received. The real lesson we can learn from navigating through the dense forest of international research and literature on technology and music education and from looking at specific situations, countries, and contexts is that it is the teacher who matters most. The teacher remains the important factor in generating change and is "the closest thing we have to a silver bullet for the education system today" (Ferreira, 2013). The future is bright, and technology is the basic ingredient. However, we need to be cautious. In order for the future to actually become brilliant we have to harness technology's transformative powers to open up new frontiers in our music civilization while enhancing its humanistic qualities.

Notes

1. A worldwide study by the OECD in member and nonmember nations, PISA gathers data on 15-year-old school pupils' scholastic performance on mathematics, science, and reading. It was first performed in 2000 and then repeated every three years.
2. Reports of the Greek Observatory of Digital Society can be found in: http://www.infosoc.gr/infosoc/en-UK/sthnellada/operators/observation/.
3. In addition, a large number of special music schools are operating around the country where different music specializations are pursued through additional music subjects during an extended school day and a different national curriculum. This analysis, however is focusing on general schools in primary and secondary education and therefore on general music teaching.
4. This very short list of such projects and publications is only indicative of the variety of such projects and efforts:

    Digital Music Education and Training project
       (http://dmet.iema.gr/index.php?lang=0&id=1);
    Musical Interaction Relying on Reflexion project
       (http://www.mirorproject.eu/default.aspx);
    Prelude—ICT in Music Education project (http://www.ea.gr/ep/prelude/);
    NetSounds—European network project (http://netsoundsproject.eu/).

    See Chourdakis (2012); Makropoulou and Varelas (2009); Triantafyllaki, Kotsira, and Anagnostopoulou (2013).
5. As the coordinator in a number of the aforementioned actions, I have also had the opportunity to come in contact with music teachers during in-service training seminars. In the course of those seminars and the discussions that have taken place I have had the opportunity to listen to music teachers' concerns and views regarding the offered resources.

References

Balanskat, A., Blamire, R., & Kefala, S. (2006). *The ICT impact report: A review of studies of ICT impact on schools in Europe*. Retrieved August 9, 2014, from http://ec.europa.eu/education/doc/reports/doc/ictimpact.pdf.

Barron, A., Orwig, G., Ivers, K., & Lilanois, N. (2002). *Technologies for education: A practical guide*. Portsmouth, NH: Libraries Unlimited.

Bauer, W., Reese, S., & McAllister, P. (2003). Transforming music teaching via technology: The role of professional development. *Journal of Research in Music Education*, 51(4), 289–301.

Bourantas, O. (2012). ICT in Education. (In Greek). *Ta Ekpaideutika*, 103–104, 145–168.

Finney, J., & Burnard, P. (Eds.). (2009). *Music education with digital technology*. London: Continuum.

Chourdakis, M. (2012). *Automatic melody recording with musical notation: Pedagogical applications.* Unpublished doctoral dissertation, National and Kapodistrian University of Athens.

Chrisalita, O. A., & Cretu, C. (April 24–25, 2014). *What do PISA 2012 results tell us about European students' ICT access, ICT use and ICT attitudes?* Paper presented at the 10th International Scientific Conference on eLearning and Software for Education, Bucharest. Retrieved August 9, 2014, from https://www.academia.edu/6866143/What_do_PISA_2012_results_tell_us_about_European_students_ICT_access_ICT_use_and_ICT_attitudes.

Chrysostomou, S. (2013). The use of media-enriched textbooks in the music classroom: Examples from Greece and Cyprus. *Proceedings from eighth International Conference for Research in Music Education*. Exeter, U.K.: University of Exeter

Chrysostomou, S. (2014a). Interactive text books in the music classroom: Applications of an innovative project in Greece and Cyprus. *Proceedings from 31st International Society for Music Education World Conference*. Porto Alegre, Brazil: ISME

Chrysostomou, S. (July 14–19, 2014b). *Meaningful music teaching with the use of technology in Greece.* PechaKucha presentation at ISME Music in Schools and Teacher Education Commission Seminar, Curitiba, Brazil.

Claro, M. n.d. *ICT and educational performance.* Retrieved August 9, 2014, from http://www.oecd.org/edu/ceri/39485718.pdf.

Cuban, L. (2001). *Oversold and underused: Computers in the classroom* (2nd ed.). Cambridge, MA: Harvard University Press.

Drenoyianni, H., & Selwood, I. D. (1998). Conceptions or misconceptions? Primary teachers' perceptions and use of computers in the classroom. *Education and Information Technologies,* 3(, n2), 87–99.

European Commission (2010). *A Digital Agenda for Europe* [.doc file]. Retrieved from http://eur-lex.europa.eu/legal-content/EN/ALL/?uri=CELEX:52010DC0245R

Eurydice. (2004). *Key data on ICT in schools in Europe.* Brussels: Eurydice European Unit. Retrieved August 9, 2014, from http://www.eurydice.org/ressources/eurydice/pdf/0_integral/048EN.pdf#search=%22Eurydice%20Key%20data%20on%20ICT%20in%20schools%20in%20Europe%22.

Ferreira, J. (2013, June 18). *The real lesson from PISA* (blog post). Retrieved August 9, 2014, from http://www.knewton.com/blog/ceo-jose-ferreira/the-real-lesson-of-pisa/.

Fisher, K. (2010). *Technology-enabled active learning environments: An appraisal CELE exchange 2010/7.* Retrieved August 9, 2014, from www.oecd.org/publishing/corrigenda.

Fuchs, T., & Wossmann, L. (2004). Computers and student learning: Bivariate and multivariate evidence on the availability and use of computers at home and at school. *Brussels Economic Review* (Universite Libre de Bruxelles), 47(3-4), 359–386. Retrieved August 9, 2014, from http://ideas.repec.org/p/ces/ceswps/_1321.html.

Greher, G. (2011). Music technology partnerships: A context for music teacher preparation. *Arts Education Policy Review,* 112, 130–136.

Grollios, G. (1999). *Ideology, pedagogy and educational politics: Logos and praxis of European educational programs* (in Greek). Athens: Gutenberg.

Harrison, C., Comber, C., Fisher, T., Haw, K., Lewin, C., Lunzer, E., ... Watling, R. British Educational Communications and Technology Agency (BECTA), corp creator. (2002). *ImpaCT2: The impact of information and communication technologies on pupil learning and attainment.* U.K.: Becta. Retrieved August 9, 2014, from http://www.becta.org.uk/page_documents/research/ImpaCT2_strand1_report.pdf.

Jimoyiannis, A., Christopoulou, E., Paliouras, A., Petsos, A., Saridaki, A., Toukiloglou, P., Tsakonas, P. (2013). Design and development of learning objects for lower secondary education in Greece: The case of computer science e-books. *Proceedings of EDULEARN13*: Fifth International Conference on Education and New Learning Technologies, 41–49. Barcelona: IATED

Kiridis, A., Drossos, V., & Tsakiridou, H. (2006). Teachers facing information and communication technology (ICT): The case of Greece. *Journal of Technology and Teacher Education*, 14(1), 75–96.

Kools, M., & Istance, D. (2013). OECD work on technology and education: Innovative learning environments as an integrating framework. *European Journal of Education*, 48(1), 43–57.

Machin, S., McNally, S., Silva, O. (2006). New technologies in schools: Is there a pay off? Bonn, Germany: Institute for the Study of Labor. Retrieved August 9, 2014, from http://ftp.iza.org/dp2234.pdf#search=%22New%20technologies%20in%20schools%3A%20Is%20there%20a%20pay%20off%3F%20%22.

Machlis, J. (1970). *The enjoyment of music*. New York: Norton.

Makropoulou, E., & Varelas, D. (2009). *Plug and play: New technologies in the music lesson* (in Greek). Athens: Fagotto.

Megalou, E., & Kaklamanis, C. (2014). Photodentro LOR: The Greek National Learning Object Repository. In *Proceedings of INTED2014*: Eighth International Technology, Education and Development Conference, 309–319. Valencia: IATED.

Network for IT Research and Competence in Education. (2004). *Pilot: ICT and school development*. Oslo: University of Oslo. Retrieved August 9, 2014, from http://zalo.itu.no/ITU/filearchive/ENG_PILOT_FV.pdf.

Organisation for Economic Co-operation and Development (2001). *Learning to change: ICT in schools*. Paris: OECD.

Organisation for Economic Co-operation and Development. (2004). *Learning for tomorrow's world: First results from PISA 2003*. Paris: OECD.

Organisation for Economic Co-operation and Development. (2005). *Are students ready for a technology-rich world? What PISA studies tell us*. Paris: OECD.

Organisation for Economic Co-operation and Development. (2006). *Are the new millennium learners making the grade? Technology use and educational performance in PISA 2006*. Retrieved August 9, 2014, from www.oecd.org/edu/ceri/.

Pesmatzoglou, E., & Papadopoulou, A. (2013). The intentions of primary school teachers for ICT integration in the learning process: Research data (in Greek). In A. Ladias, A. Mikropoulos, X. Panagiotakopoulos, F. Paraskeva, P. Pintelas, P. Politis, … A. Halkidis (eds). *Proceedings of third PanHellenic Conference "ICT in Educational Procedures"* Piraeus: University of Piraeus.

Peter, J., & Valkenburk, P. M. (2006). Adolescents' internet use: Testing the "disappearing digital divide" versus the "emerging digital differentiation" approach. *Poetics*, 34, 293–305.

Prensky, M. (2001). *Digital natives, digital immigrants*. Retrieved August 9, 2014, from http://educationcabinet.ky.gov/NR/rdonlyres/F9E83D7C-95BA-4053-9B6F-A913A5278CF0/0/DigitalNativesPartIII.pdf.

Raptis, A., & Rapti, A. (2004). *Learning and teaching in the information era: A holistic approach* (in Greek). Athens: Aristotle Raptis.

Ravitz, J., Mergendoller, J., & Rush, W. (April, 2002). Cautionary tales about correlations between student computer use and academic achievement. Paper presented at annual meeting of the American Educational Research Association. New Orleans, LA. Retrieved August 9, 2014, from http://www.bie.org/images/uploads/general/def781d1d449c5298bd7691bdeccc1f6.pdf.

Rossiou, E. (2012). Digital natives ... are changed: An educational scenario with LAMS integration that promotes collaboration via blended learning in secondary education. In H. Beldhuis (Ed.), *Proceedings of 11th ECEL-2012* (pp. 468–479).

Savage, J. (2004). Reconstructing music education through ICT. *Research in Education, 78*, 65–77.

Selwyn, N. (2004). Reconsidering political and popular understandings of the digital divide. *New Media & Society, 6*(3), 341–362.

Triantafyllaki, A., Kotsira, L., & Anagnostopoulou, C. (2013). Learning to compose using interactive technology: A case study investigation of whole-class composition processes using the MIROR platform in a Greek primary school. *Proceedings from eighth International Conference for Research in Music Education.* Exeter, U.K.: University of Exeter

Wise, S., Greenwood, J., & Davis, N. (2011). Teachers' use of digital technology in secondary music education: Illustrations of changing classrooms. *British Journal of Music Education, 28*(2), 117–134.

1 B

Further Perspectives

CHAPTER 10

BUILDING A BROAD VIEW OF TECHNOLOGY IN MUSIC TEACHER EDUCATION

HEIDI PARTTI

What does it take to develop an educational system that caters to the needs of the next generation?

READING the perspectives from around the globe provided by the chapters of this book, I have found myself pausing to ponder this question several times. Each time, my thoughts have been drawn back to the importance of improving teacher education—the imperative task of ensuring that teachers are equipped with a wide availability of knowledge and skills needed in the rapidly changing world. These abilities, sometimes referred to as "twenty-first-century skills," include, among others, the capacity to network and collaborate with others around the world in order to engage in processes by which new meanings are generated and novel works created (see, e.g., Jenkins, Clinton, Purushotma, Robinson, & Weigel, 2006; Understanding and Providing a Developmental Approach to Technology Education, 2010). Inspired by Samuel Leong's and Smaragda Chrysostomou's contributions to this part of the handbook, I will discuss some challenges, opportunities, and visions related to the use of technology in supporting the cultivation of creative and collaborative skills in music teaching, particularly from the viewpoint of music teacher education.

Diagnosis: What is Emerging; What Is Evolving?

Samuel Leong's and Smaragda Chrysostomou's chapters center on the emergence and evolution of technology in music education. Chrysostomou and Leong take the reader on a journey to the depths of the relationship between ICT and the school. The journey is led by questions such as *Where are we now?* and *How did we get here?* The level of accuracy in answering these questions holds great significance. An effective action plan calls for an appropriate diagnosis of the situation at hand.

It is noteworthy, although hardly surprising, that both chapters address the gap between the promise of technology and the realization of that promise in schools. Although Chrysostomou and Leong use wide-angle lenses to examine the impact of ICT in global and national education policies and practices, both authors end up focusing more closely on the underdeveloped or insufficient use of technology in music teaching.

In her chapter, describing a variety of policies and development schemes introduced in Greek schools, Chrysostomou addresses the music teachers' struggle to integrate technology into their teaching practices in creative, effective, and meaningful ways. Leong, writing from an Asian perspective, discusses the lack of expertise and confidence among teachers in utilizing technology in their teaching. According to both writers, the issues highlighted in the chapters are evident across all levels of education, resulting in a considerable barrier to students' musical development.

The diagnoses provided by Chrysostomou and Leong provide various important considerations, being by no means limited to the authors' own contexts. Despite the vast differences between the aims, processes, and infrastructures of educational systems in different countries, the challenges and possibilities related to music technology in Greece and Hong Kong are easy to relate to in Finland, where I work as a researcher and music teacher educator.

Between the Potential and Reality of Technology in Music Teaching

Despite Finland being one of the leading countries in the use and availability of ICT, Finnish music teachers seem to be facing challenges similar to the ones described by Chrysostomou and Leong. In a recent survey conducted among music teachers in Finland,[1] less than one-fifth of the respondents rated their own know-how in music technology as "excellent" or "good," whereas almost half rated it as "below average" or "poor" (Partti, 2015).

Arguably, the reasons for Finnish teachers' sense of lack in technology-related expertise are diverse. The availability of tools and division of resources varies from one school to another, as do focus areas in local curricula. The results of the study suggest, however, that even when the essential resources and equipment are available, many teachers either opt out of making use of technology in their music teaching practices or use technology in extremely limited ways. Technology is mostly used to enable singing, playing, and listening to music and to search for music-related information. The possibilities that digital technology offers for composing one's own music are scarcely utilized. Songwriting, collaborative composing and improvising, remixing, mash-upping, powermixing, and other creative music-making activities highly popular in music communities outside the school (e.g., Partti, 2014; Michielse, 2015; Michielse & Partti, 2015) are only *occasionally* part of the lessons rather than implemented on a regular basis.

According to the aforementioned study (Partti, 2015), teachers may be aware of the potential benefits of technology in music teaching and learning, but feel unconfident and inept in putting the technology to work. Those respondents who rated themselves with a low grade in their know-how in music technology often mentioned that they felt they had not been equipped with adequate skills during their studies, or that those skills were out of date due to the lack of in-service training.

The Finnish teachers' limited tendency to employ music technology, mostly to enable the reproduction and appreciation of music made by other people, could be understood as a form of *tool-oriented approach* to technology, as discussed by Leong in his chapter. Rather than making the most of the full potential that technology provides for students' processes of creating new knowledge and cultural artifacts, teachers implementing the tool-oriented approach to technology settle for utilizing technologies in much the same ways they utilize any other set of tools.

A tool-oriented approach to technology is problematic in music education, as it fails to provide students with resources and environments to explore and invent music, learn in social interactions, and develop creative thinking, cooperative problem solving, and initiative-taking abilities. Furthermore, overemphasizing the role of musical reproduction by reducing the use of technological tools to merely assisting in aural training and skill acquisition is in stark contrast to the values and ideals of the reality of music making that students are likely to face outside the school classroom. At its worst, this reduced use of technology can seem confusing and alienating to the students. As Rasinen and Parikka (2012) argue, technological instruments, appliances, and interfaces are not considered mere tools by those who have grown up in technology-centered environments but rather as—at least on an emotional level—closely related to their own lifestyles.

INTEGRATING TECHNOLOGY INTO DAILY MUSIC EDUCATION PRACTICES

As illuminated by Leong and Chrysostomou, there seems to be a gap—one with worryingly extensive measures—between the *potential* that technology could provide for music teaching and learning processes and the *cultivation* of this potential in classroom practices. To bridge this gap, it seems clear that we should take a more holistic approach to technology and its use in music education. In other words, rather than understanding technology as a separate and optional part of music teaching, technology should be studied on a regular basis at all educational stages by utilizing practical and diverse methods, as recommended by the multinational European project Understanding and Providing a Developmental Approach to Technology Education (Understanding and Providing a Developmental Approach to Technology Education, 2010), among others.

Understanding technology as an integral part of any teaching situation entails a broad view of technology. According to this view, technology is systematically used to support and extend learning objectives and to encourage pupils to collaboratively identify and solve problems. In music education, the integration of hands-on technology activities into daily classroom practices refers specifically to increasingly systematic efforts to provide pupils with possibilities to engage in collaborative processes of creating their own music. An example of recent endeavors to achieve this in Finnish general upper secondary schools is ongoing design-based research examining how a compulsory music course could be pedagogically and technically redesigned in order to broaden the scope from musical reproduction and performance—still prevalent in traditional educational settings—to creative music-making activities (Ojala, 2017). The research project is conducted by a researcher-teacher who has developed a new pedagogical approach, called Learning Through Producing, in cooperation with his own students, as well as four other music teachers and their students. The Learning Through Producing approach aims to contribute in the cultivation of practical musicianship through sustained collaborative work, such as arranging, songwriting, recording, and mixing music, and the production of tracks, videos, and other artifacts that can be shared with others in formal and informal networks. The project illustrates a fruitful way to combine research and practice to advance and test theoretically inspired innovations and to generate new knowledge. By inviting the students to actively contribute to the process by choosing the aims, problems, and tools that they consider significant in terms of their musical learning, the project design can be viewed as particularly fitting for efforts to use technology to support the growth of students' creativity and the skills of collaboration and problem solving.

Integrating technology as part of music teaching at all educational stages may necessitate an immense reorganization of music teacher education. In addition to becoming fluent in their subject content, music teachers are also called upon to understand the wider culture of music making and to envision the impact of sociocultural phenomena on the values, goals, and practices within formal music education. As I discuss in chapter 25, sociocultural changes—increasingly driven by globalization and digital technology—are not phenomena *out there*, taking place in a reality outside the classroom walls. Rather, they exist inside the classroom environment. In order to facilitate opportunities for the emergence of our students' meaningful and lifelong interest in music, educators should actively engage with those phenomena.

The call for schools not only to engage but also to *reflect* social reality is not a new one. Yet when it comes to opportunities brought about by the arrival of digital technology, we, as music teaching professionals, are still far from fully engaging in and reflecting on today's social realities, as a number of contributors to this volume point out. Instead of leading the way in technology-facilitated teaching and learning, educational institutions are criticized for lagging behind in adopting and using technological opportunities. However, blaming individual teachers is neither helpful nor justified. Without the necessary pedagogical support and personal experiences in using technology to promote the development of versatile musicianship, it is understandable that teachers may fall back on the experiences and training they *do* have. Typically, their default training has to do with studying and reproducing musical works by singing and playing (electro) acoustic instruments.

One could ask whether we still expect preservice teachers to adapt themselves to the existing (and often outdated) institutional musical landscapes by practicing certain repertoires or instrumental combinations—be they choirs or rock bands—instead of being able to cultivate an experimental attitude through technology-facilitated creative music making. I have elsewhere argued (Partti 2012) that a widespread excitement over online music communities, for instance, reveals that people have an apparent *need* for experimenting with music, manifested in the ways in which the participants of music-related (online) communities are willing to make public and openly share their individual and open-ended musical learning processes.

How are we taking into account this need in music teacher education? As Leong emphasizes, the inadequacies hindering the integration of technology in music education practices should be a concern for both preservice teacher education and teachers' professional development. This also sets a pressing challenge for the international field of music education research. Apart from some recent development projects which specifically examine the possibilities of using digital technology to support university-level music teacher training, music education research on technology issues has so far focused mostly on developing basic education rather than teacher education. The need for up-to-date knowledge and understanding about the integration of technology into music teacher education is pressing; regardless of the significance of innovative policy schemes, program visions, and curricular changes in schools, it is impossible to transform the classroom without, in tandem, altering the teacher education. Importantly, the call for research includes both the projects, with their focus on generating new knowledge about technology-facilitated teaching and learning, and the studies that aim to develop new music-making practices in music teacher training.

A broad view on the use of technology in music education indicates a paradigm shift from viewing technological innovations as tools to enhance one's learning *about* music to seeing technology as a powerful way to facilitate more possibilities to *participate* in different musical practices and musical worlds and to open up new avenues for musical imagination. Pursuing an educational system that supports individual and societal successes thus necessitates the advancement of institutions in which developing teachers are purposefully supported to cultivate their creative and collaborative skills through the multifaceted use of technology; hence, they would be prepared to advance the learning of a wide variety of students in swiftly changing and increasingly digitized twenty-first-century settings.

Note

1. The respondents (n = 618) were composed of class teachers who taught music in compulsory basic education (children from the ages 7–16) as well as music subject teachers working in upper secondary schools (students aged about 16–19) all across Finland.

References

Jenkins, H., Clinton, K., Purushotma, R., Robinson, A. J., & Weigel, M. (2006). *Confronting the challenges of participatory culture: Media education for the 21st century*. Retrieved February 21, 2014, from http://www.digitallearning.macfound.org/atf/cf/%7B7E45C7E0-A3E0-4B89-AC9C-E807E1B0AE4E%7D/JENKINS_WHITE_PAPER.PDF.

Michielse, M. (2015). *Remix, cover, mash: Remediating phonographic-oral practice online*. Maastricht: Maastricht University.

Michielse, M., & Partti, H. (2015). Producing a meaningful difference: The significance of small creative acts in composing within online participatory remix practices. *International Journal of Community Music, 8*(1), 27–40.

Ojala, A. (2017). Developing learning through producing: Secondary school students' experiences of a technologically aided pedagogical intervention. In G. D. Smith, Z. Moir, M. Brennan, S. Rambarran, & (Eds.), *The Routledge Research Companion to Popular Music Education* (pp. 60–73). London & New York: Routledge.

Partti, H. (2012). *Learning from cosmopolitan digital musicians: Identity, musicianship, and changing values in (in)formal music communities*. (Studia Musica 50). Helsinki: Sibelius Academy.

Partti, H. (2014). Cosmopolitan musicianship under construction: Digital musicians illuminating emerging values in music education. *International Journal of Music Education, 32*(1), 3–18.

Partti, H. (2015, March 3–5). *The bliss and dread of creative music making: Finnish music teachers' approaches to teaching composing*. Presentation at the Nordic Network for Research in Music Education Conference, Helsinki.

Rasinen, A., & Parikka, M. (2012). Teknologiakasvatus ja tietoyhteiskunnassa pärjääminen. *Kasvatus, 43*(2), 207–213.

Understanding and Providing a Developmental Approach to Technology Education. 2010. *Understanding and providing a developmental approach to technology education* (Final activity report). Retrieved February 24, 2014, from http://update.jyu.fi/images/b/b7/UPDATE_042941_Final_report_D8.3.pdf.

CHAPTER 11

TECHNOLOGY IN THE MUSIC CLASSROOM IN KENYA

EMILY ACHIENG' AKUNO AND DONALD OTOYO ONDIEKI

A Context

For a majority of music educators in Kenya, the story of scholarship in music is a tapestry of experiences, disappointments, and revelations. For one of the authors of this chapter, these include experiencing song from childhood, an activity that was shared equally well at the playground at home and at primary school, at the numerous song activities at church, and at the formal, organized "serious" singing that marked preparations for the annual Kenya Music Festival, in which I started participating at the age of 9. At this stage, the gadgets that were used to aid the generation of music were really confined to the choir master's melodica, an end-blown instrument with a single-octave keyboard. At home, my father purchased and brought home a Philips record player, a novelty that made it possible to hear songs even when the radio was off. This was a boost to my ability to choose what and when to listen to in the form of music, though limited to my father's library of 45- and 33-speed records and long-play albums.

These experiences continued into secondary and high school. Choral music was ever present, with the junior and later senior choir, the house choir, and hymn practice. The school's repertoire of recorded music included cassettes, with the capacity to record what we did and said or sang. This opened an avenue for creativity and space to hear one's own voice, which sounded so strange and different coming from a gadget in contrast to what I heard as I spoke.

I was privileged to attend a well-endowed institution with a music practice and study tradition where I encountered music instruments of a different caliber. Each of us had to learn to play the recorder, and not even the squeaks could deter the determined class of excited girls in Form 1. But the greatest novelty came in the form of what appeared to be a piece of furniture (from color and material) and turned out to be a beautiful sound-producing item that allowed one to generate both melody and accompaniment, and to sing while playing. The presence of this piano (and the many others in the school) was a boost to music making, because many students explored its offerings, and it freed several to explore their creative and performance capacities, coming up with many memorable music moments.

The two schools mentioned above, though presented as they were several years ago, are a reflection of the state of a number of public schools in Kenya today. Though there are more schools in the first category, there are several private schools that own and utilize different types of learning materials, providing a rich learning experience for learners. At one time, music education in schools was the teaching and learning of music literacy, and the singing and performance of song and dance for the entertainment of guests and for festivals. This is still rampant, but the university has moved to a different level. Of the six public universities that offer instruction in music, the processes of knowledge transmission and skill development call for the use of a variety of resources as educators endeavor to prepare learners to develop the appropriate attitudes and behavior that are requisite for operation in the industry. My own study at Kenyatta University included the playing of several local instruments, an endeavor to ensure cultural relevance in music study.

The Story of Music Education

Music education in Kenya has transformed from the initial indigenous learning environments where one learnt music through participation in cultural activities. Today, it is a classroom-based activity (Kidula, 1996; Apudo, 2007). This evolution has seen the teaching and learning of music traverse a number of formats. With a relatively flexible curriculum structure, the indigenous knowledge system provided for a very practical and experience-based teaching and learning approach. This ensured learner engagement with technological devices that were required for the execution of necessary tasks.

Music in the classroom setting in Kenya has evolved through curriculum and policy developments over the years (Akuno, 2005, 2011). Some of these have focused on making cultural experiences available for learners, with music found to be a valuable tool toward achieving this. The provision of teaching and learning has exposed learners to music styles from near and distant cultures, with their respective bodies of materials and resources. Whereas technology may not have been the focus of music education provision, it has always been understood to be a contributing feature of the learning landscape. The nature and role of this technology are what may need to be established and evaluated.

Music in the society has also evolved from indigenous, mono-stylistic genres to hybrid contemporary popular genres. These have mainly been propagated in and through technology (Okumu, 1998; Ondieki, 2010), especially electronic devices for storage and retrieval of the musical works. This, today, makes the learner's soundscape an environment in which one learns and for which one's training is aimed.

With this continuous evolution of the music scene, alongside that of technology, the classroom teacher is faced with challenges in theory, practice, curriculum, content, approaches, and methodology. The syllabus content has been, over the years, designed according to the ideology of the administrating authority. The colonial, missionary, postindependence, and contemporary settings speak of sociocultural and economic situations that dictate what is taught, how it is taught, and subsequently the resources used for that teaching. This affects how we learn music, articulated through developments in curriculum and policy.

Today the debate on the role of indigenous music versus contemporary music, world music, and Western classical music in the music curriculum in Kenya is slowly informing policy and curriculum. Curriculum issues are discussed in relation to how much foreign content versus local content is included, how much indigenous versus contemporary styles and methods is covered, and how to balance the rural versus the urban learning and practice scenarios. From a cultural perspective, it is notable that what is relevant to a rural child may not be to the urban child. Today, methods, approaches, content, theory, and practice present challenges to the music educator and thus impact the appropriation of technology, despite its availability.

The Problem

The classroom teacher, in today's environment full of fast-changing technology, is at an advantage when it comes to availability of technological resources. This teacher is at a disadvantage, however, in that this technology may not be readily accessible. With no clear policy in Kenya regarding the use of technology for music instruction in particular, and no guide for the practice of applying technology for teaching in general in primary and high schools, we found it appropriate to interrogate teachers' experiences and responses to the technology in education debate, with reference to music study. This chapter shares the findings of a study that sought to find out how the classroom teacher manipulates the theories, approaches, and methodologies acquired in training to the practical and theoretical changes produced by technology by addressing the following questions:

- What technologies are available for music education in Kenya?
- Why use music technology in the classroom?
- How can these technologies be used in the Kenyan curricula?
- Who will use which technology?
- Does the appropriation of technology in the classroom foster skill development?
- How does this impact the learner?
- Does this scenario then call for a review of the teacher training curriculum and application of the traditional music education theories, methods, and approaches?

Methodology

The data were generated through focus group discussions conducted during the annual Kenya Music Festival's training workshop of music teachers, choral trainers, and educators by the Ministry of Education, in which over 300 teachers were involved. The teachers were a fair representative of the nation's teaching force, being drawn from all counties in the country and from both primary and secondary schools. The aim of the focus group discussions was to provide a platform where music teachers would share ideas on the use of music technology in the classroom while rethinking creativity and curriculum.

Findings

Technologies Available for Music Education in Kenya

The teachers reported the following resources for teaching music in varying degrees of availability:

- Computer labs—only available to some schools
- Mobile phone applications—only possible for some teachers
- SMART boards and tablets—only available in private schools
- Electronic keyboards and synthesizers—only available to some schools
- Internet—online information, limited access to some schools
- Access to classroom content material—only possible for some teachers
- YouTube—free online audio and video access—only possible for some teachers
- Internet radio—only possible for some teachers
- Online purchases of scores, audio, books, etc.—only possible for some teachers

Whereas previously the teachers had relied on workshops done by other experienced teachers and lecturers from universities for training in the use of technology, today they access the information directly on the Internet. This makes it possible for them to provide a wide choice of knowledge for their learners. The challenge for the country now is to develop online content for the indigenous Kenyan aspects of the curriculum, because literacy, theory, and other information is readily available online.

A number of mobile apps for phones and tablets were reportedly used by teachers. These included GarageBand, pitching apps, aural training apps, tuning apps, music score reading apps, concert repertoire apps, music score buying apps, free chords and lyrics apps, and so on. Access was, however, only possible for some teachers. Teachers further employed music writing software, with the most popular being

Finale, Sibelius, and Noteworthy. These, too, were only possible for some teachers. The music creation/production softwares cited were FruityLoops, Audio logic, and Pro Tools, once more with limited spread. Some teachers reported using music education software, and these included Auralia and Musition for aural training. These, and e-books, were only possible for some teachers. One encouraging factor is the knowledge that some teachers were generally aware of resources to use for teaching and appeared keen on employing them for knowledge transmission and skill development in learners.

Music Technology in the Classroom

Despite the large sample size, few teachers, notably those from private schools, gave rationales for the use of technology in the classroom. These included:

1. The ability of technology
   - to expand instructional time
   - to engage students in meaningful, directed instruction
   - to allow each student to work at his or her own pace
2. Technology is able to effect such amazing accomplishments due to the thoughtful pedagogy embodied within well-written software.
3. Good programs imitate good human teachers by introducing concepts and skills in a logical, understandable sequence of small steps.
4. They provide stimulating music problems, frequent checks for student understanding, meaningful and useful feedback, and added exercises to reinforce new skills and concepts.

How These Technologies Can Be Used in the Kenyan Curriculum

Teachers were asked to indicate their appreciation of technology's role in education. Their response indicated their perception that technology can be used in the delivery of the curriculum content. This would work if e-content for the curriculum is developed and would bank on the provision of hardware by government, such as phones, tablets, computers, and so on for effective application of these technologies.

To use technology successfully, it was noted that teachers needed to adopt a new way of thinking about music learning and instruction. Many strategies have been developed for organizing and integrating technology instruction into typical elementary general music settings, and each is useful for meeting specific needs. Some teachers acknowledged sending individual or small groups of students to cybercafés or computer labs to graze randomly in whatever programs interested them.

Two teachers acknowledged having developed more directed and integrated sequences using individualized or cooperative learning settings. This shows an awareness of the need to adopt appropriate learning practices to accommodate and/or benefit from technology. It was also noted that some important pedagogical principles must be considered when designing strategies for using technology.

Who Will Use the Technologies?

Most of the technologies noted in the survey were teacher-based, which raised the discussion on why teachers were not looking out for technologies that would be appropriate for the learners. It was noted that:

- Different types of programs satisfy different needs
- Technology can be used for entire classrooms to introduce or reinforce music skills and concepts
- Network systems can be used to serve many computers and hundreds of students
- Effective instruction can be provided by having all students in the class take turns running a program on a single computer (Kassner, 2000)

Does Appropriation of the Technology Foster Skill Development?

A majority of the teachers agreed that technology in the classroom fostered skill development. A smaller number, however, seemed more concerned that this appropriation would only foster limited development.

Some of the ways technology impacted the learner that were discussed included:

- Helps the learner to work independently
- Helps improve confidence
- Learners in the urban centers more privileged than those in the rural setting
- Encourages creativity
- Learner centered approach helps students own the process of their learning

Following challenges faced in teaching aural skills, such as teaching echo singing to woodwind students, especially boys aged 9–18, following the Associated Board of the Royal Schools of Music syllabus, I started using the app Aural Books, among other aural training apps, on the IPad in class to address these areas of musicianship development.

The results over the last four years have been impressive, chief among them being learners conquering their fear of aurals. Many individual instruction teachers are adopting this approach.

Conclusions

There appears to be a long-standing suspicion of technology, even among teachers. Most teachers did not view technology as an avenue for solving some of their challenges, such as:

- Limited time for delivering skills and knowledge in music
- Limited time for well-rounded music curriculum delivery
- Limited time for individual and small-group instruction
- Limited time for individual student assessment
- Limited time for delivering and pacing instruction with a variety of differently abled students
- Limited time for motivating students and helping them achieve high standards

With all these reservations pegged on limitation of time, it appears that the role of facilitating learning that has been assigned to technology was not appreciated by these teachers. Furthermore, their listing of technologies that are for teaching instead of for learning point to a need for a broader understanding of the technology available for teaching, and how it can be appropriated.

In response to the stated calls for a review of teacher training curricula and for the application of traditional music education theories, methods, and approaches, this article underscores the need to appreciate the nature of the technology available. It further seeks an understanding of the manner in which that available technology is designed to operate. It should be understood that technology is a tool that makes activities possible. Such an understanding should enable teachers to adapt their teaching to accommodate and make use of available resources.

The classroom teacher in Kenya is faced with many challenges in theory, practice, curriculum, content, approaches, methodology, and infrastructure, and needs in-service and preservice training in and exposure to:

What technologies are available for music education
Which of these technologies are better for the teacher and which are better for the student
The use of music technology in the classroom
The role of technology in fostering skill development and its impact on the learner

The challenges of an urban teacher versus those of the rural teacher need to be addressed while rethinking technology, creativity, and curriculum in music education in Kenya.

Recommendations

The following recommendations are made:

Educate classroom teachers in Kenya toward rethinking their creativity and curriculum through technology. Encourage and promote the appropriation of technology by classroom teachers in Kenya in the teaching of music. Examine how meaningfully these technologies can be used in the classroom. Develop more local content for e-learning.

References

Akuno, E. A. (2005). *Issues in music education in Kenya*. Nairobi: Emak Music Services.

Akuno, E. A. (2011). What's in a Song? Exploring the analytical-creative learning process in indigenous Kenyan children's songs. *Problems in Music Pedagogy, 8*, 49–72.

Apudo, M. A. (2007). *The role of technology in music education: A survey of computer usage in secondary schools in Nairobi Province, Kenya*. Unpublished Master of Music Education thesis, Kenyatta University, Nairobi.

Kassner, K. (2000). One computer "can" deliver whole-class instruction. *Music Educators Journal 86*(6), 36–40.

Kidula, J. N. (1996). Cultural dynamism in process: The Kenya Music Festival. *Ufahamu: A Journal of African Activist Association, 24*(2-3) 63–81.

Okumu, C. C. (1998). *The development of Kenyan popular guitar music: A study of Kiswahili songs in Nairobi*. Unpublished MA thesis, Kenyatta University, Nairobi.

Ondieki, D. O. (2010). *An analysis of Kenyan popular music of 1945–1975 for the development of instructional material for music education*. Unpublished doctoral dissertation, Kenyatta University, Nairobi.

CHAPTER 12

PONDERING AN END TO TECHNOLOGY IN MUSIC EDUCATION

JOSEPH MICHAEL PIGNATO

The Core Perspective chapters presented in part 1B of this book consider two broad issues: (1) the ways in which music educators negotiate the role of technology within the contexts of educational policy and practice, and (2) the influences of commerce and industry upon music learning, teaching, and technology within schools. Samuel Leong and Smaragda Chrysostomou address the Provocation Questions from two distinct perspectives. Leong provides an overview of technology in music and music education as an observer, and to some degree prognosticator, of global trends in policy and in practice. Chrysostomou starts in similar fashion, offering descriptions of the ubiquitous nature of technology in contemporary life, but then distinguishes between the "forest," or more general observations, and the "trees," a narrower focus on technology within music education in Greece.

Chrysostomou and Leong repeatedly refer to an international push, on the part of governments, nongovernmental organizations, and market forces, to codify, via public policy, the development of, distribution of, and applications for information and communication technology (ICT) in education. The combined powers and resources of international governments, corporations, and advocacy groups have shaped particular brands of policy that reflect and serve to reaffirm, through reproduction, particularized political economic values. Leong describes this trend quite effectively: "As the product and driver of globalization, ICT has enabled the international transfer of knowledge and the creation of a transnational private market of education provision that complements and competes against local and national education providers. It has also propelled the popularization of neoliberalism, the dominant political-economic ideology worldwide. This has impacted policies that support market mechanisms including choice, competition and decentralization, the liberalization and privatization of the education sector, and the transplantation of management techniques from the corporate sector."

Chrysostomou summarizes the somewhat abstract priorities that guide the current frenzy for technology policy by noting that "digital literacy and digital advancement is therefore rightly located at the top of the priorities list for all future agendas pushed forward by international and European bodies like the Organisation for Economic Co-operation and Development and the European Union." One wonders about such a push. What are we to make of "digital advancement"? What is it exactly, and whom or what does it advance: technology, users of technology, learning, or the economy? In this chapter, I would like to put forth Aristotle's (2014) concept of "that for the sake of which" as a framework through which we might critique, interrogate, and even counterbalance the implications of the policy momenta and the underlying assumptions, values, and forces identified by Chrysostomou and Leong from which they draw fuel.

Aristotle (2014) distinguished between the essential nature of objects, that is to say what an object is in material terms, and "that for the sake of which" we create, and subsequently use, objects; that is, the end or ultimate purpose that drives our conception, creation, and implementation of the object in the first place. Remaining cognizant of "that for the sake of which" we make, use, and politicize technology prompts us to do the following:

1. Distinguish technology itself from those who use or are served by it
2. Consider the interests of those making technological devices, products, and services
3. Consider the interests of those developing legislation and policy pertaining to the implementation of technological devices, products, and services in music and in education

In order to make this distinction or engage in such considerations, we might begin by applying Aristotle's concept to a rewording of the part 1B Provocation Questions:

What is that for the sake of which music educators negotiate the role of technology within the broader terrain of educational policy and practice?

What is that for the sake of which commerce and industry play roles within and affect music learning, teaching, and technology within schools?

As Roger Mantie notes in chapter 1, the root of the word "technology" comes from the Greek notion of *techné*, or skilled, artful doing. Although one might think of techné as it is involved in developing and manufacturing the array of technologies available to young musicians today, I am interested in techné as it is associated with how young musicians use such technologies.[1] Thus we might add a third Provocation Question: What is that for the sake of which young musicians use technology?

Understanding how young musicians use technological devices, products, and services requires considering the end users: young people seeking to learn, to make music, and to engage others in those processes. Understanding such complex phenomena raises essential questions such as *Who are those young musicians? What do they wish to do with technology? What opportunities exist for them should they use technology?*

How can technology serve their interests, aspirations, and development? How might policy governing technology in music and in education impact their lives, needs, and pursuits? Pondering questions like these compels educators and policy-makers to give foundation to otherwise abstract ideals like "digital literacy" or "digital advancement" by casting them in terms of human needs, desires, interests, and vulnerabilities.

Thus, understanding techné with regard to "that for the sake of which" people develop and, perhaps more tellingly, use technologies seems particularly relevant to the themes, questions, and phenomena addressed throughout this book and in the Core Perspectives offered by Chrysostomou and Leong. Thomas Regelski (2005) differentiates techné from the purposeful doing that reflects human music learning quite effectively:

> Techne was associated with poiesis, the making of things. Such practiced skill and craft operate in the service of uncontroversial and thus taken-for-granted ends, such as making certain useful objects or playing an instrument. Techne becomes problematic, however, to the degree that musical technique (i.e., virtuosity) becomes an end in itself, and in relation to claims that only technical expertise can properly realize the full aesthetic or artistic integrity of the music performed. Praxis, however, involved not the making of things but "doings" of various kinds that concern people. As such, praxis is governed by an ethical dimension (called phronesis) where "right results" are judged specifically in terms of the people served or affected. (p. 16)

Mantie offers a distinction similar to Regelski's, noting that "technology invariably impacts how we understand ourselves both as human beings and as musical beings." Just the same, I might add that how we understand ourselves both as human beings and as musical beings invariably impacts technology. In other words, the technologies we create, implement, and seek out reflect how we understand ourselves and how and why we gather, posture, create, share, learn, compete, and express our thoughts, our musical creativity, and our identities. Thus we might seek to understand how the usage of technology among young musicians reflects them, their needs, their aspirations, and their identities.

Chrysostomou makes a similar distinction, noting that "it is a matter of not only how much technology we have and use but also how we use it, what kind of activities we engage in, what kind of skills we acquire, and what goals we accomplish." Young musicians use technology in ways that amplify long-standing human tendencies or "doings" that "concern people." Such uses include, among others: gathering or forming communities of response, social posturing, creating, sharing, learning, and competing, and expressing thoughts, creativity, and identity.[2] Such uses should inform policy-making pertaining to technology in music and music education. Further, Chrysostomou advises that "in order to assess whether technology enhances and advances learning we need to look beyond the numbers and percentages of schools, students, and teachers that have and use computers and digital technology, and concentrate more intently on what they are actually doing with these technologies."

Developers create products based on perceptions of what people want to do. They do so largely in the interest of generating commercial devices, products, or services to build market share. Policy-makers make policy in large part for political purposes, either to affirm the prevailing sentiment or values of voters or to forward a particular political, economic, or social agenda, such as the often unquestioned push for technological progress alluded to earlier.

Users are active, have agency, and therefore act upon those products in ways that sometimes affirm market intent and in other instances subvert it. Developers make products, but people use them and give them their ultimate meaning. Users act. They perform. They use technology with the purposeful phronesis described by Regelski. In so doing, they change technological products, or at least change the conception of what they could or ought to be in their lives.

By giving technological products meaning, users claim the technology and make it their own in service of fundamentally human needs, tendencies, desires, and ends. People use technology to realize themselves, albeit in response to the perceptions of developers and policy-makers and the hegemony of the prevailing political economy and cultural contexts. Policy-makers, commercial interests, and education administrations ought to study the give and take between developers and users to understand that for the sake of which they make policy, that is, to understand who the policy services, for what purposes, and to what ends.

Discussions pertaining to policy—local, national, or international—that governs technology and its applications in education must start with considerations of human needs, aspirations, inclinations, and, as I argue in chapter 19, human vulnerabilities. Samuel Leong notes that "the rapid advancement of technologies has facilitated and driven the spread of globalization and the growth of the global knowledge economy." Of course I agree. But I would encourage education administrators, or those who must implement policy, to vigorously question marketplace models, references, and ideals when crafting policy or legislation that impacts human experiences like learning, music making, and cultural expression.

The pursuit of human needs and of human knowledge led to and drove the development of technology. The rapid advancement of technologies reflects fundamentally human tendencies of problem solving, of seeking knowledge beyond that which we already possess. Market forces seek to exploit those fundamentally human characteristics under the often paired guises of technological convenience, progress or advancement, and status, for one sole purpose: to sell products. Similarly, governments, legislators, and education policy-makers have sought to increasingly politicize technology in market terms and for market purposes. Leong notes: "the global knowledge explosion has seen a proliferation of government policy statements on the role of ICT in education." It seems fitting to apply Aristotle's concept of "that for the sake of which" as a counterbalance to prevailing assumptions that drive, shape, and champion implementation of technological devices, products, and services in educational policy, in educational institutions, and in the lives of young musicians.

Although this chapter is a Further Perspective contribution, I think it raises more questions than it provides perspective. The questions raised herein arise in response to and to further the astute observations of Smaragda Chrysostomou and Samuel Leong in their thoughtful Core Perspectives chapters. Technology developers, policy-makers, school administrators, and music educators can only begin to develop meaningful technology protocols, or policies, for education by authentically addressing such questions. Pondering such questions might help us realize what Leong called a sustainable "human-techno future," which is ultimately predicated on human and humane considerations with regard to technology policy, practice, and implementation in music and music education. By engaging such questions and keeping them at the fore of policy deliberations, we might begin to ponder the end to, or that for the sake of which, we use technology in music education.

Notes

1. In chapter 19 I note that I use the term "musicians" in much the same way education scholars refer to "readers" or "speakers," with the assumption that doing music, in one way or another, is an innate ability in need of development through socially dynamic praxis. I do not refer to musicians as specialists, an elite class of doers set apart from a majority spectator class. That usage also applies here.
2. In chapter 19, I define "communities of response" as groups of individuals who come together for a "gather round" experience, connecting around interests, content, cause, or activity. I began using the phrase in my teaching of popular music to describe the ways in which fans come together around an artist, a musical genre, or a subcultural movement. Communities of response represent a kind of precedent step to Lave and Wenger's notion of "communities of practice"; See Lave and Wenger (1991).

References

Aristotle. (2014). *The poetics*. In *Aristotle collection* (J. C. Douglas, Ed., & I. Bywater, Trans.) N.p.: Amazon Digital Editions.
Lave, J., & Wenger, E. (1991). *Situated learning: Legitimate peripheral participation*. New York: Cambridge University Press.
Regelski, T. (2005). Music and music education: Theory and praxis for "making a difference." *Educational Philosophy and Theory*, 37(1), 7–27.

CHAPTER 13

A SOFTWARE CREATOR'S PERSPECTIVE

JOE BERKOVITZ

OTHER contributors to this book bring with them experience as educators and as what one might term "metaeducators": thinkers about the nature of teaching and learning and technology. The Provocation Questions deserve responses that spring from broad knowledge and thoughtful interpretation of the many things that have already been said and done in this field.

You can read just such responses elsewhere, but perhaps not here. For better or worse, my perspective cannot be erudite in this way: I am a software creator and a musician, but not an educator. My experience of education is from the student's vantage point (in the past) and from an observer's and vendor's vantage point (in the present). Even with these deficits, though, I believe my kind of viewpoint can be useful to the reader. This work is about technologies and their impact, and the mindset of technology creators is not without some relevance.

WHY MAKE ANYTHING?

I would like to begin by confessing a few personal motivations as an inventor, since these are the lenses through which I cannot help but view the creation and use of educational technology. These are best described quite simply and shamelessly as biases; I don't ask where they come from or why I have them.

First, my wish in creating software (of all kinds, not just music related) is to provide people with some kind of medium that serves their creative and expressive impulses. In the musical sphere, in particular, I hope for users to engage with the imaginative, mystical, open-ended quality of music making that is so hard to measure with a machine but in which so much of music's human value inheres. I want to provide people with new kinds of active experience of creation, not passive ones of being entertained or repetitive ones of perfecting some particular skill.

Second, I want to leverage the emergence of information and communication technology (or, as many software developers say in a kind of metonymic shorthand, "the Web"). This is the great technological and cultural sea change of our time, possibly much greater in impact than the emergence of substantial personal computing power that necessarily preceded it. We've transformed our ability as a planet full of people to connect and collaborate with each other, and I am happiest when creating things that exploit this new-found capacity of humanity. Such technologies, to me, are bound to be transformative: they harness one of the greatest changes our species has ever unleashed on itself.

Third, I wish to create systems that support the traditional while providing space for the novel, and that allow both to flourish and to come into contact with and nourish each other. I believe that there is great value in the continuity of our musical culture: its breadth in the dimension of time as well as in geography and genre. My favorite musical inventions are those that do not posit a rejection of the past. Music has always been full of such inventions—in fact, this category includes many of the once-novel acoustic instruments that we now accept as canonical.

Finally, like many software developers, I like engaging with new techniques and tools purely for novelty's sake, in a mode that popular culture might refer to as "geeky." To name just one example, the recently developed ability of web browsers to synthesize a musical audio signal is outright fun to play with and make use of in a software program. This sort of technically driven interest is often a large piece of the motivation for the creation of many technology-based products, although not one that holds intrinsic value to the end user.

I believe that most inventors and developers have such biases and follow them, whether they acknowledge it or not. Sometimes the fruits of our labors will serve the needs of learners, sometimes the needs of school administrators or politicians, and sometimes the needs of a company's investors. But it is more common for technologists to forgo thinking about these larger outcomes (which in any case are difficult to comprehend in advance) and simply carry on with building their ideas. We make things. Their impact on our culture, which is after all enacted by the culture itself and not by the inventor, unfolds downstream. The technologies we create do not bring some imaginary purity of purpose to the table, or have some value that derives purely from their modernity or degree of innovation. Our creations need to be critically examined after the fact to understand how they mesh with our society.

Technology and Education: A Look Back at Mathematics

In some of my earlier work building educational software in the 1980s, I worked in the realm of primary and early secondary school mathematics. At the time, the teaching of mathematics was (and no doubt still is) in a state of flux in the United States. In many areas—algebra providing one key example—math was usually taught as an abstract subject lacking application to everyday life. This had the not completely unexpected effect of leading students to reach the same conclusion.

The personal computer revolution was unfolding at the same time, and a wave of technological panaceas was washing over the educational establishment. Software "solutions" for learning mathematics, though, were concentrated in those areas where their adoption best fit the current teaching model: algebra drills. At the same time, new software applications were being created that made it much easier to link algebraic thinking with real-world concerns—indeed, such tools (e.g., graphing calculators and geometric sketchpad applications) allowed students to actively think and work in a mathematical way while freeing them from mechanical effort and drudgery that had previously presented barriers. Other applications went even further and were best employed in a constructivist setting in which students built the relevant mathematics for themselves from the ground up.

The latter type of solution—call them "tools for active thinking"—arguably offered greater potential for positive change in the classroom than the more popular drills and rote skill-building programs. Yet the adoption of such "active" software tools did not fare well until more than a decade later, when new models of teaching math had gradually gained acceptance on their own merit. It is fair to say that the technology did not itself effect this change; rather, it was the shift in teaching philosophy and the increasingly obvious failure of the status quo. But if technology did not itself act as a change agent, its existence certainly aided the process. There were groups of technologists, philosophers, and educators who were aligned in their views from the start, and the availability of concrete technologies fed the growth and dissemination of the teaching ideas that could usefully employ them.

Technology and Education: A Look at the Musical Arts

In the wake of creating Noteflight, I have had an opportunity to look at many of the same issues again, this time in the area of music. What I have seen has been fascinating, and in some respects a reprise of my earlier experience with math. We again see the same spectrum of tools that originate in the status quo mindset of "skills mastery" at one end, with more open-ended technologies that provide opportunities for active musical thinking at the other.

There is of course much for technology to accomplish in the attainment of various musical skills that can be analyzed and rated, such as sight-reading, intonation, rhythmic regularity, and the knowledge of theoretical schemas to name only a few. While these can be very useful and they generate a great deal of economic activity in the industry, I personally have little interest in addressing these problems. They are in essence "better mousetraps" for skill-building rather than transformative of the context in which we approach imagining and making music. Many different repetitive and deliberate regimens (including traditional or antiquated ones) can yield the desired result of improving some musical skill or other. Those that employ computing technology are often no more than incrementally better than what they replace, although if they are more appealing to music learners purely because of their currency, that has some value.

To use the language of the first Provocation Question, I have come to believe that music educators, like their math counterparts, are disposed to "negotiate the role of technology" by favoring choices that require the minimum change in approach. They do so not out of any desire to obstruct change but because the availability of this or that technology does not alter anyone's mind about how to teach. At the moment, the fact remains that the U.S. approach to music education is rooted in skills mastery and in the desiccated remains of what was once an apprenticeship tradition. Many of the putative apprentices are not interested in what the master has to offer, since it has little connection to their everyday musical culture (which in turn is not taken very seriously as a subject of study). Like their math counterparts of yore, music teachers all too often teach music in a way that appears almost calculated to evoke disinterest except from a small fraction of "music kids." (Change agents in music education, unlike those in mathematics, have a harder time basing their arguments for change on efficacy or sufficiency with respect to the job market, STEM notwithstanding.)

This doesn't mean that technologies supporting active and creative musical thinking are doomed to a life at the margin. Such tools become seeds for change. Change from the outside is encouraged as critics use new tools to demonstrate their ideas and push their agendas. Change from the inside is encouraged as both teachers and students begin to use a tool on one obvious premise but then discover the alternative modes of use in front of them and begin to experiment with them.

Technology and Education in the Language and Visual Arts

I want to call attention to another contrast: how differently the musical arts are taught and learned in our schools in comparison to language arts and visual arts. In these other disciplines, little or no controversy attaches to the idea that students learn meaningful skills from engaging their creative impulses. We take creative writing, drawing, painting, video production, and sculpture as givens in the lives of schoolchildren today. We expect learning to take place even if their works are not sophisticated or competent. Yet composing and arranging music are still placed (as algebra once was) at the terminus of a long path of mastering skills and abstract concepts.

The picture of technology that emerges in these other disciplines is, perhaps surprisingly, quite ordinary. It is an unfancy technology world, with few grand claims. Students use word processors, image manipulators, and video production applications. These are not skill-building or knowledge-imparting technologies in a didactic sense (although using them does develop some nontrivial skills in itself). Rather, they are workaday applications that are used by teachers and society at large, in roughly the same form that they present to students.

Yet there is something magical in this ordinary digital world of everyday apps. These programs relieve the drudgery of creating and editing physical artifacts. They free the creative person to explore alternatives, undoing and redoing at will. In their newer, web-based incarnations, they now allow a student to collaborate and exchange documents with teachers and with fellow students effortlessly, and can capture and reflect edits, comments, and suggestions from multiple parties. There seems to be a convergence in this area between what the world at large finds useful and what is useful for education.

I feel that music technology can benefit from an infusion of this sensibility. Perhaps some of the most valuable things that technology can do for music are some of the simplest: take basic capabilities like recording, playback, notation, and video and endow them with simplicity, collaboration, annotation, and communication. No need to embed these in a completely unguided or unstructured setting, no more so than in any English or art class—but let's bring creativity back into the music classroom, where it deserves to be, and support it with simple everyday tools that any musician would think of using.

The Role of Commerce and Industry

Generally speaking, commerce and industry reflect the values and thinking of the society that buys the wares. Our culture already values money highly; in our commercial and financial culture, profit is paramount (as must inevitably be the case under capitalism). Most money is invested in creating new solutions that are perceived to have a ready market. In this facet of commerce and industry, conservatism reigns.

As I've mentioned, however, inventors also invent things based on their biases, which are unpredictable and manifold. They also occasionally follow the views of educators outside the mainstream and indirectly help those views gain currency. There is a kind of Darwinian aspect to commerce: there is a steady stream of products that do not reflect mainstream thinking and that reach the market and (like mutations) take root and thrive despite the odds. This facet of commerce is disruptive rather than conservative. When successful, disruptive products can be even more profitable than their conservative counterparts. (Of course, all successful disruptive inventors claim they knew it all along!)

I think it is rare for commercial visions to shift the viewpoints of educators, and I hesitate to say that commerce should or should not have any particular role in advancing change. Commerce is just the vast sum of a great deal of human activity directed at making money, without any average philosophy or concept of its own role. But the very blindness and avarice of commerce is perhaps its own saving grace when it comes to fueling change in education and supporting new models of thinking. Commerce throws off products and innovations continuously, both good and bad. Some of these products are helpful to some causes and serve to highlight the advantages of change.

For my part, I hope to build products in my lifetime that are flexible enough to provide a stage on which many different possibilities can be played out, by educators, by students, and by ordinary musicians exercising their creative will. On this stage, ideas will come and go and many lines will be spoken and sung. Out of this play of ideas will come the next generation of ways to teach, learn, and enjoy music.

CHAPTER 14

WHERE MIGHT WE BE GOING?

JONATHAN SAVAGE

When considering the emergence and evolution of music education with technology, one conversation is always at the forefront of my mind. It comes from Lewis Carroll's *Alice in Wonderland*, specifically the point in chapter 6 where Alice asks:

> "Would you tell me, please, which way I ought to go from here?"
> "That depends a good deal on where you want to get to," said the Cat.
> "I don't much care where—" said Alice.
> "Then it doesn't matter which way you go," said the Cat.
> "—so long as I get SOMEWHERE," Alice added as an explanation.
> "Oh, you're sure to do that," said the Cat, "if you only walk long enough." (1920, p. 90)

Wherever one looks across the globe, the history of music education with technology is a difficult one. At this point it in our history, when digital technology is pervasive in so many different cultures, it is worth considering three basic questions:

1. Does anyone know which way we are going?
2. Does anyone know where we want to go?
3. Does it matter where we end up?

My answers to these questions are (1) no, (2) no, and (3) yes. This does worry me. As the chapters here show, now, perhaps more than ever, the pace of technological change is rapid and wide-ranging. Music education needs a robust, principled, and defendable position. The chapters here reflect deeply on the issues associated with the emergence and evolution of technology and the impact this has had on music education. Similar reflections can be found throughout this book from many countries. Pausing, as the chapters here do, to consider the journey we have undertaken is a highly productive and useful exercise. Change for the sake of change is seldom productive. Nor is decoupling the present from the past, as Furedi so helpfully illustrates: "in the worldview of the educational establishment, change has acquired a sacred and divine-like character that determines what is taught and what is learned. It creates new "requirements" and "introduces" new ideas about learning. . . . Typically change is presented in a dramatic and mechanistic manner that exaggerates the novelty of the present moment. Educationalists frequently adopt the rhetoric of breaks and ruptures and maintain that nothing is as it was and that the present has been decoupled from the past" (2009, p. 23).

There is a strong argument within educational philosophy, particularly in the work of Hannah Arendt, that the process of familiarizing ourselves with the world as it has been and as it is today gives us the best opportunity (the "existential security") to have a chance of attempting something new. Even more powerfully, Arendt applies this thought to our roles as adults, parents, and teachers:

> Education is the point at which we decide whether we love the world enough to assume responsibility for it and by the same token save it from ruin which, except for renewal, except for the coming of the new and young, would be inevitable. And education, too, is where we [adults and teachers] decide whether we love our children enough not to expel them from our world and leave them to their own devices, nor to strike from their hands their chance of undertaking something new, something unforeseen by us, but to prepare them in advance for the task of renewing a common world. (1961, p. 196)

The educational affordances of digital technology are often accepted and celebrated without this significant degree of responsibility. Thankfully, this book is evidence of a new seriousness in addressing key issues through a collaborative approach to writing and engagement. This is a vital first step in planning for a productive future. It may result in positive responses to the first two of my key questions.

But, still, the notion of "disruptive innovation" (Christensen, 1997) is overwhelming and powerful in educational discourses. Technological innovation is conceptualized as positive change making services or products simpler, easier to access, and less expensive. These chapters provide evidence of political rhetoric that builds on this narrative. The power and commercial interests of the industries behind the technological transformation of education need to be acknowledged and interrogated more explicitly. Clearly, this is a much broader issue than music education per se, but similar issues will emerge. One of the key features of these chapters that I have noted is that recent years have seen an emergence of musical tools that are less specialist and more generalist. In other words, musical tools—software, hardware, instruments—are embedded within common platforms or formats: Internet browsers, phones, tablets. General arguments about these technologies and their affordances and limitations have begun to impact our work as music educators. The "augmentation" of what counts as a musical instrument or compositional tool into these spheres is, of course, both productive and worrisome.

The arguments of key skeptics of the world of educational technology need to be taken more seriously in music education. In *To Save Everything, Click Here*, Evgeny

Morozov looks at two discourses that have permeated the technology industry and through it shaped our lives and government policy over the last 10 years (2013). One is "Internet-centrism," the tendency to look at the Internet as the technological structure and inevitable master narrative that will shape the future of all our institutions and social practices. The second is "technological solutionism," the strategy of jumping onto simple technological solutions to problems before knowing exactly what those problems are and the ways they might be framed and understood.

Morozov does not write directly about education in this book, but it is not difficult to extrapolate some of his arguments. We are currently witnessing a model of education that was conceived in the era of industrial mass production giving way to another less top-down, more participatory, networked model where knowledge is created through the connections we make. This is a long journey. While writing this chapter I noted that the headmaster at Eton College, one of the United Kingdom's top independent schools, has been complaining about the examination system within the United Kingdom. He believes that an exam system that "obliges students to sit alone at their desks in preparation for a world in which, for much of the time, they will need to work collaboratively" is one that is failing (Ratcliffe, 2014). As music education moves forward, on whatever roads it takes and wherever it ends up, there are some key things that we need to be aware of. Nobody wants to work as a part of a failing system. These are observable in other areas of our social interaction with technology, and they will apply equally to the teaching and learning of music.

First, we must beware of the technological quick fix. Khan Academy is a classic example. It is not a substitute for a coherent curriculum in mathematics, biology, or chemistry. It is also not a substitute for a real teacher. It is not a complete educational resource in any meaningful sense, because it was not designed to be that in the first place. Robert Talbert put it this way in his blog:

> When we say that someone has "learned" a subject, we typically mean that they have shown evidence of mastery not only of basic cognitive processes like factual recall and working mechanical exercises but also higher-level tasks like applying concepts to new problems and judging between two equivalent concepts. A student learning calculus, for instance, needs to demonstrate that s/he can do things like take derivatives of polynomials and use the Chain Rule. But if this is all they can demonstrate, then it's stretching it to say that the student has "learned calculus," because calculus is a lot more than just executing mechanical processes correctly and quickly. To say that it is not—that knowledge of calculus consists in the ability to perform algorithmic processes quickly and accurately— is to adopt an impoverished definition of the subject that renders a great intellectual pursuit into a collection of party tricks. . . . Khan Academy is great for learning about lots of different subjects. But it's not really adequate for learning those subjects on a level that really makes a difference in the world. Learning at these levels requires more than watching videos (or lectures) and doing exercises. It takes hard work (by both the learner and the instructor), difficult assignments that get students to work at these higher levels, open channels of

communication that do not just go one way, and above all *a relationship between learner and instructor that engenders trust*. (Talbert, 2012; emphasis added)

Productive questions for us as we move forward could be:

- What are the quick fixes in music education?
- What are the hidden dangers that they contain?
- What is lost in the process of engaging with them?
- How can these be avoided?

Second, there is a real danger of us losing our sense of humanity. This would be devastating for us as musicians and teachers, and this danger should be taken seriously. Sherry Turkle examines the irony that in a world where young people have never been so connected, there is more loneliness than ever before. She also examines the consequences of arriving at what she calls the "robotic moment," when we are delegating more and more of our interactions to robots through technology. She has researched this issue for many years in various locations, including the use of robotic toys for children, immersive online environments such as Second Life, and the adoption of technological gadgets to "care" for elderly folk in residential homes. Rather than our being in the world and interacting with it and the people around us, her book documents a mediated world, one that we interact with through screens. In her conclusion, she describes the narrative of her study as an arc:

> We expect more from technology and less from each other. This puts us at the still center of a perfect storm. Overwhelmed, we have been drawn to connections that seem low risk and always at hand: Facebook friends, avatars, IRC chat partners. If convenience and control continue to be our priorities, we shall be tempted by sociable robots, where, like gamblers at their slot machines, we are promised excitement programmed in, just enough to keep us in the game. At the robotic moment, we have to be concerned that the simplification and reduction of relationship is no longer something we complain about. It may become what we expect, even desire. (2011, p. 295)

As with Morozov, who argues that we are losing the "narrative imagination," Turkle suggests that in increasingly defaulting to the digital we are losing the raw human part of what it means to be with each other. This has devastating consequences for musical expression and the process of music education. But the affordances, skillfully handled, are potentially incredible too.

Key questions for us to consider together here might include:

- What do we expect from the music technology that we use as teachers and students?
- How does music technology impact our relationships—with sound, the "creative moment," our fellow musicians?
- How does music technology shape our musical expectations, and even our sense of musical desire?

Third, we need to be aware of the dangers of technological use resulting in a lack of joy in learning. You don't need to look far to find evidence of this in most academic institutions. Learning management systems are often joyless environments. They are principally concerned with managing users, standardizing access and roles, and ensuring that "privileges" are allocated on a top-down basis. The result is often a learning environment that is dull, routine, and rigid. They are often inflexible, they depersonalize learning, and they rarely teach or utilize the broader range of digital literacy skills that students might have developed in their wider lives. Formalizing music education within such a management system is a recipe for disaster. It is not a road that I would want music education to go down.

Finally, we should be wary of the dangers associated with constant distraction. Here and now in the United Kingdom, the iPad is the new must-have "educational" gadget. I often joke with my students that it is hard to move in the many schools I visit across the northwest of England due to the stacks of iPads (other tablets are available) that have been bought by head teachers anxious for their next technological fix. The iPad is simply another learning management system, closed and top-down. The model of interaction that the iPad represents puts the student fair and square in the role of the consumer, a consumer who does not need to know anything about the "device." In one sense, of course, there is nothing to know; the maker movement's dictum, "if you can't open it, don't own it," should be read as a warning here. This is system control; enabling privileges to mix and modify are the preserve of the lucky few; the "politburo" of such a system lies in the iTunes store. You might be able to make an app, but you cannot distribute it without permission; you cannot distribute it to work on different platforms. Just like students in a learning management system, you are locked in.

Mark Bauerlein's book *The Dumbest Generation: How the Digital Age Stupefies Young Americans and Jeopardizes Our Future* (2008) presents a compelling alternative that we would be wise to consider as music educators. In his wonderful blog entry on this book, Peter Lawler (2013) asks whether this might be both the smartest and dumbest generation that has ever existed. Through 12 provocative points, he amplifies Bauerlein's thesis in a highly entertaining way:

1. Virtually all of our students have hours—and often many, many hours—of daily exposure to screens.
2. So they excel at multitasking and interactivity, and they have very strong spatial skills.
3. They also have remarkable visual acuity; they're ready for rushing images and updated information.
4. But these skills don't transfer well to—they don't have much to do with—the non-screen portions of their lives.
5. Their screen experiences, in fact, undermine their taste and capacity for building knowledge and developing their verbal skills.
6. They, for example, hate quiet and being alone. Because they rely so much on screens keeping them connected, they can't rely on themselves. Because they're

constantly restless or stimulated, they don't know what it is to enjoy civilized leisure. The best possible punishment for an adolescent today is to make him or her spend an evening alone in his or her room without any screens, devices, or gadgets to divert him or her. It's amazing the extent to which screens have become multidimensional diversions from what we really know about ourselves.

7. Young people today typically are too agitated and impatient to engage in concerted study. Their imaginations are impoverished when they're visually unstimulated. So their eros is too. They can't experience anxiety as a prelude to wonder, and they too rarely become seekers and searchers.
8. They have trouble comprehending or being moved by the linear, sequential analysis of texts.
9. So they find it virtually impossible to spend an idle afternoon with a detective story and nothing more.
10. That's why they can be both so mentally agile and culturally ignorant. That's even why they know little to nothing about how to live well with love and death, as well as why their relational lives are so impoverished.
11. And that's why higher education—or liberal education—has to be about giving students experiences that they can't get on screen. That's even why liberal education has to have as little as possible to do with screens.
12. Everywhere and at all times, liberal education is countercultural. And so today it's necessarily somewhat anti-technology, especially anti-screen. That's one reason among many I'm so hard on MOOCs, online courses, PowerPoint, and anyone who uses the word "disrupting" without subversive irony.

Lawler highlights, simultaneously, the potential affordances and limitations of screens and the experiences they present. What are the consequences of these ways of thinking for the practice and process of music education? As researchers and teachers, we need a serious debate around these issues. Winston Churchill once quipped that "it is always wise to look ahead, but difficult to look further than you can see." This might be true. Through these chapters, this book, and our work together as teachers and researchers, we are developing a good understanding of where we have come from. However, the way forward is still unclear. There are various roads we could take, but we need some clarity for our direction of travel. "Somewhere" or "anywhere" is not good enough. Just "walking long enough" will only make us tired. This chapter has raised some key questions for us to consider together. But the debate is only just beginning.

References

Arendt, H. (1961). *Between past and future*. London: Faber & Faber.
Bauerlein, M. (2008). *The dumbest generation*. London: Penguin.
Carroll, L. (1920). *Alice's adventures in Wonderland*. London: MacMillan and Company.

Christensen, C. M. (1997). *The innovator's dilemma: When new technologies cause great firms to fail.* Boston: Harvard Business School Press.

Furedi, F. (2009). *Wasted: Why education isn't educating.* London: Continuum.

Lawler, P. (2013). Is this both the smartest and the dumbest generation? Retrieved January 14, 2014, from http://bigthink.com/rightly-understood/is-this-both-the-smartest-and-dumbest-generation-3.

Morozov, E. (2013). *To save everything, click here: The folly of technological solutionism.* London: Penguin.

Ratcliffe, R. (2014, August 5). Eton headmaster: England's exam system unimaginative and outdated. *Guardian.* Retrieved May 5, 2014, from http://www.theguardian.com/education/2014/aug/05/eton-headmaster-exam-system-unimaginative.

Talbert, R. (2012). The trouble with Khan Academy. Retrieved April 9, 2014, from http://chronicle.com/blognetwork/castingoutnines/2012/07/03/the-trouble-with-khan-academy/.

Turkle, S. (2011). *Alone together: Why we expect more from technology and less from each other.* New York: Basic Books.

CHAPTER 15

LOADED QUESTIONS FOR AN EMERGING WORLD OF MUSIC EDUCATION TECHNOLOGY

JOHN-MORGAN BUSH

We are exposed to new technological devices/software in the marketplace, at home, in the media, and increasingly in the workplace. Over the past decade, we've seen how developments in information and communication technology (ICT) devices and software, the proliferation of social media platforms, and digital sharing have created an infinite number of combinations for creating, producing, sharing, and critiquing music. Now, more than ever, it is a wonderful time to be a student or teacher of music. In their Core Perspectives, Leong and Chrysostomou paint a landscape of uncharted terrain where we have only begun to tap into the extreme potential for the application of ICT in the music classroom. However, as part of the dialogue nature of this Further Perspective, I strive to set aside my optimistic partisanship toward the advancement of the application of technology in music education so as to respectfully investigate and offer alternate considerations. As we continue to negotiate the role, definition, and various applications of technology in the music classroom, we must also seek to find balance between traditionalist approaches to music education, which are still in widespread use, and the use of technology to achieve more successful learning outcomes.[1] This Further Perspective makes a beginning inquiry into three specific areas of this balancing act:

1. How do pressure points motivate or deter the use of ICT in the music classroom?
2. How can technology break down the long-standing model of apprenticeship to mastery in instrumental music?
3. How does unchecked use of ICT applications destabilize the social structures within musical learning communities?

The ever-steady stream of scintillating innovations from technology hubs around the world has created a demand and necessity for widespread technological competency in all facets of life. These innovations cannot be ignored. Like a swelling river, they hug the banks of the educational establishment with a gentle but ever-increasing power. Music educators are faced with the task of keeping pace with the brisk current while at the same time avoiding being swept away. The exponential growth, diaspora, and innovation in ICT impose the (sometimes perceived) need for a radical shift in classroom instruction, not just in music but in many other academic disciplines as well. Like the powerful river that can erode the riverbanks that cradle it, ICT has the potential to destabilize the education establishment in ways that may not benefit students in the long term.

Pressure Points

Through a variety of external pressure points, music educators can be made to feel that the implementation of technology in the classroom is imperative and nonnegotiable. Pressure from lawmakers, administrators, and school or district-based curriculum committees often all contribute to imposing the use of ICT as part of regular music instruction. In addition to these mandates, teachers may feel symptoms of anxiety that they must demonstrate competency with the latest and greatest forms of ICT in order to remain equal with their colleagues.

This can be problematic for a variety of reasons, but perhaps the greatest challenge comes when the traditionalist music educator is asked (or mandated) to switch to a contemporary practice that includes the use of ICT in the classroom. This conflict between expectation and personal experience is stacked in a way that inhibits the music educator in question from approaching the potential for application of technology in the classroom with an open mind. All too often, this results in what Chrysostomou cites as the use of ICT as an auxiliary component to enhance the administration or management of the music classroom, but not in such a way that implementation authentically impacts student learning outcomes.

When the teacher is exposed to a great many resources, is pressured to implement them without ample time, and is not given the professional development to establish competency in their use, while also experiencing fear of inadequacy within their professional sphere—the result is a rapid uptake of the least invasive technologies being introduced to the classroom. A few ubiquitous examples include:

1. Using the Internet during class to look up key vocabulary in real time during a lecture.
2. Using programs like Audacity to record individual playing tests (common to instrumental music education) and submit them for evaluation/grading by the teacher.
3. Using MOOCs (massive open online courses) to disseminate information and resources, while the responsibility lies with the student to log in, access, print, complete, upload, and interpret the information posted on the MOOC.

Educational policy encourages and expects technological competency as a major learning outcome on both macro and micro levels in the classroom. While the use of ICT in the music classroom demonstrates how instruction complies with and supports this pervasive belief in technology and better outcomes being linked, it may not be of enough substance to make an impact on the learning community. Put simply, teachers may believe that because they are using some form of ICT in the classroom, they are then rising to meet the pressures and expectations placed on them to do so. In many schools this is enough and does satisfy those who are mandating the implementation and use of ICT. However, the conversation all too often stops there when there are still other relevant questions that should be asked.

What is the litmus test that determines the overall effectiveness of using one element of ICT versus another? How does the ICT component being used assess student accessibility? When bombarded from all sides by newly emerging ICT platforms or devices, how do educators successfully assess whether their use of ICT is impacting their students' learning in an authentic way? As we press forward in innovation and in creating ICT applications for the classroom, we must also be cognizant of how these tools support our learning outcomes.

Breaking the Linear Model of Apprenticeship

When considering the long arc of music education and instrumental music instruction, we see a gradual evolution of the apprentice/journeyman/master system that dates back as early as the Middle Ages (Haar, 2010). Even today, the prevailing model is built in such a way that music students receive information about music from a teacher who achieved mastery of an instrument or of voice. In this system, the teacher serves as the repository of information that is made known to the student/apprentice at the moment the teacher/master decides it is relevant to share. For centuries this model was largely unchanged. With the dawn of public education, the model was slightly adapted to accommodate many apprentices to one master, which translates into the modern music classroom setting. Not only did the apprenticeship model endure for generations, it created a long-standing body of educational doctrine advocating for this traditional approach that persists in music classrooms today.

Reconciling the established doctrine with the Information Age has proven to be a difficult task. Traditionalists believe the key to the success of the apprenticeship model is in the power the master/teacher holds to disseminate information at appropriate junctures throughout the student's education. The proliferation of the Internet and the increasing presence of ICT in nonmusic classroom environments have begun to remove the barrier perception that one's teacher is the primary gatekeeper to information. It also opens students to myriad points of view, technical advice, and styles of music making (Olson,

2010). While it is undeniable that open access to relevant information creates new learning opportunities and enriches learning environments, educators may worry that the sequence of their curriculum is interrupted by it and that it will have a negative impact on their students' overall comprehension and progress.

Consider the following example: a middle school beginner trumpet player has only had her instrument for a few months. With enthusiasm, she ventures onto YouTube to learn more about trumpet playing. After watching instructional videos by various musicians, she returns to practice her instrument, convinced that she can mimic what she has seen and heard. This student is very dedicated to her pursuit. After many failed attempts, she believes that she will never be able to replicate the video. The student doesn't know that perhaps she hasn't yet developed the complex muscles of the embouchure. While admittedly trite, this example does illustrate a legitimate concern—that the consumption of developmentally inappropriate information by students does indeed have the potential to stifle long-term progress and participation in music learning. While we hope music educators can both identify this potential pitfall and provide safeguards against it, the fear of this scenario occurring at all may prevent many educators from encouraging their students to engage with ICT as a regular part of their music studies.

This reluctance to embrace new technologies and cling to the traditional apprenticeship model has another negative effect as well. When teachers shy away from the implementation of technology in their classroom, it deepens the generational gap between themselves and the students they teach (McAlister, 2009). The surge of options for sharing, cocreating, and manipulating music digitally created by ICT in recent years provides educators with some of the first truly viable options for establishing authentic relevancy in the eyes of their students.

We must work to increase understanding among music educators that there is a dual purpose for the implementation of technology in the music classroom. Not only does the appropriate application of technology-based pedagogical tools enhance and define students' learning environments while providing access to a multitude of resources, it also helps the student to bridge sociocultural differences that stereotype the study of music as an unrelated or foreign subject. Because today's students have grown up in a digital world, it is a challenge for them to thrive in a traditionalist music education environment because the lack of technology tools may feel foreign and irrelevant to the experiences of their everyday lives.

Social Structures and Music Technology

It is commonly asserted that the study of music encourages the development of social intelligence that is tangential to the core concepts of gaining proficiency on a musical instrument. Educators have built a strong case for the necessity of school-based music education programs, citing the development of nonmusical skills as an added benefit. Advocates have convincingly researched (Eerola & Eerola, 2014) and investigated the many ways in which participation in school music programs fosters nonmusical development in areas such as self-confidence, the ability to perform independently in front of others, teamwork, and goal-setting, to name but a few. For all of these areas to develop simultaneously with musical ability, students must learn in an environment that is rich not only in information resources but also in opportunities for student-to-student, in-person interaction.

In-person interactions are critical to the development of music students' relationships with their teachers and peers and to their individual identities (Hallam, 2010). The traditional model of instrumental music education in secondary schools in the United States is often structured around some of the same core tenets as these schools' athletic programs. Students in band, orchestra, or choirs typically train as a team (group rehearsals) and are taught to adopt a mentality that individual behaviors are to be dictated by the greater good of the ensemble. Participation in and association with the school band program form entire student social communities as well as shaping students' individual identities (Campbell, 2013). While many of these music students may not go on to pursue professional careers in music, the core values and identity they form while participating in music learning are essential life skills that they will depend upon in the future.

In his Core Perspective, Leong states that the use of ICT has the potential for effectively creating a highly personalized learning space. However, we must be judicious in ensuring that "personalized" does not become synonymous with "isolated." Leong is correct in his assertion that innovations in ICT have encouraged and enabled new forms of sharing and collaboration among students, and these new developments have the potential to make a positive impact on the individual learning space. Yet it is critical that these forms of group work/collaboration do not replace live, real-world, human interaction. If students place too much emphasis on the individuality of their own pursuits, they may fail to appreciate the experience of making music in group settings. Music education curricula that overemphasize ICT resources do have the potential to separate students from the network of peers, mentors, and often friends that can help to shape their emotional and social intelligences along with their musical competencies. In addition, overemphasis on the individual learning sphere can destabilize the music learning communities from which students derive at least a portion of their individual identities. When considering various pathways to reconcile traditionalist and progressive views toward the implementation of ICT as a tool for creating students' personalized learning environments, we must cautiously consider when it is appropriate to introduce any technology that limits person-to-person interaction.

After playing devil's advocate, I can once again voice my support for the increased implementation of ICT in the music classroom. Though, as educators, we remain optimistic about the future synergies between new technologies and their potential for use in music education curriculum, we must consider not only the impact a technology may have on a given learning outcome but also the ripples it causes throughout the learning community. In their Core Perspectives, Samuel Leong and Smargda Chrysostomou both offer well-articulated and detailed documentation of the increased use of ICT in classrooms across the globe. They present a worldview where music and technology share an assured, joined future. If this is the case, we must continue to ask the difficult questions of others and of ourselves to ensure that the way we move forward retains our students' learning outcomes as our primary concern.

Note

1. I shall use the term "traditionalist" to refer to music educators who generally prefer the repertoire of teaching practices built upon the traditions of instruction that were commonplace before the widespread use of the Internet.

References

Campbell, P. S., & Abril, C. (2013). Perspectives on the school band from hardcore American band kids. In Trevor Wiggins (Ed.), *The Oxford handbook of children's musical cultures* (pp. 434–448). Oxford, UK: Oxford University Press.

Eerola, P. S., & Eerola, T. (2014). Extended music education enhances the quality of school life. *Music Education Research*, 16(1), 1–17.

Haar, James. (2010). Some introductory remarks on musical pedagogy. In Russell Eugene Murray, Susan Forscher Weiss, & Cynthia J. Cyrus (Eds.), *Music Education in the Middle Ages and the Renaissance*, (pp. 3–10). Bloomington: Indiana University Press.

Hallam, Susan. (2010). The power of music: Its impact on the intellectual, social and personal development of children and young people. *International Journal of Music Education*, 28(3), 269–289.

McAlister, Andrea, N.C.T.M. (2009). Teaching the millennial generation. *American Music Teacher*, 59(1), 13–15.

Olson, C. A. (2010). Making the tech connection: are you ready to integrate technology into your classroom? Your students are. *Teaching Music*, (5). 30–35.

CHAPTER 16

MOBILE LEARNING IN MUSIC EDUCATION

JASON CHEN

The Core Perspectives by Samuel Leong and Smaragda Chrysostomou have provided an overview of technology in music and music education as global trends in both policy and practice. To follow up the trends from an Asian perspective in globalization and technology provided by Leong, this chapter further discusses the recent development of mobile learning and information communication technology (ICT) in music education in Hong Kong.

MOBILE LEARNING COMMUNITIES

Technologies can be used to support the formal education of children in mobile learning communities. Interactions with technologies have both enabled and constrained learning in mobile communities. Pachler (2010) explained that mobile learning is slowly establishing itself as a field in its own right. However, it is still unclear what is best understood by the term and if there is a need for a separate field of enquiry on mobile learning. One defining feature of mobile learning is "the need for individuals to go beyond the acquisition of knowledge relevant to issues encountered in the world but also to shape their knowledge out of their own sense of their world" (Kress & Pachler, 2007, p. 22).

Leong (2011) navigated through several key predictions and recent developments relevant to music education and technology, and called for a new music education future that is strongly linked with the global knowledge economy in the digital and conceptual age. The Horizon Report (2011) has identified six emerging technologies or practices that are likely to enter mainstream use (i.e., with widespread adoption) in the coming years:

- Within a year: electronic books and mobiles
- Within 2–3 years: augmented reality and game-based learning
- Within 4–5 years: gesture-based computing and learning analytics

Researching Mobile Learning

Mobile learning differs from learning in the classroom or on a desktop computer in its support for education across contexts and life transitions. Sharples (2009) stated that mobile learning poses substantial problems for evaluation, if the context is not fixed and if the activity can span formal and informal settings. There may be no fixed point to locate an observer. The learning may spread across locations and times. There may be no prescribed curriculum and lesson plan.

Most mobile learning research to date is limited and is situated in the music classroom or studio setting. Most of the research has concentrated either on the pedagogical use of various technologies and how to normalize them into specific pedagogy or to study independent and outcome-driven learning variables. Sole (2009) argued that the advent of mobile learning is a strong reminder that learning does not happen in neutral, aseptic places that just act as the background of learning. Technology opens up new possibilities for learners' agency and the creation of new social practices in learning.

In order to open up new practices in learning, an innovative methodology requires that the roles of researcher and research be democratized by inviting the learner to participate in the research process. In a recent study conducted to evaluate the effectiveness of using the application Auralbook to learn aural skills (Chen, 2015), the collection of the data for the study was done in a way that overcame the limitations of time and space and represents a new research methodology to evaluate the effectiveness and viability of mobile learning of aural skills.

In figure 16.1, the conceptual framework of the 3M model of mobile learning, motivation, and musicianship aligns with game-based learning such as learning aural skills in a mobile way through an aural e-book (Chen, 2015). Children learn aural skills in the digital age as they play aural games. Examples include singing with real-time recordings, clapping with musical excerpts, listening and identifying chords, intervals, and progressions and genres. Immediate feedback is provided, and evaluation can be sent to teachers and parents through cloud computing as records of assessment.

The Recent Development of Information and Communication Technology in Music Education in Hong Kong

In the digital era, the use of mobile learning resources for students to learn at their own pace in an interactive way has become a major trend in global education. To promote mobile learning, the Hong Kong government launched the e-Textbook Market Development Scheme in June 2012. Through the scheme, the government provides

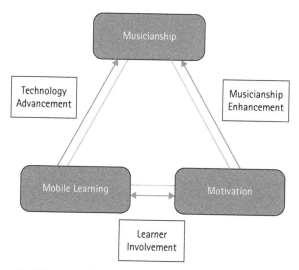

FIGURE 16.1 Theoretical framework of 3M – mobile learning, motivation and musicianship

grants to some schools to procure more than 100 iPad units for teachers to use as interactive teaching and learning tools in certain subjects, including music, in primary and junior secondary schools (Legislative Council, 2013).

According to the Education and Manpower Bureau report on the scheme (2013), it has approved a total of 30 applications for the development of e-textbooks for Chinese, English, mathematics, and other subjects, including general studies, computer literacy, *putonghua* (Mandarin), geography, life and society, and physical education. Twenty-one applications are for primary education, and nine are for junior secondary. At present, an application for music has not yet been developed. A detailed study of 120 teachers, including 60 in-service teachers and 60 preservice teachers, focusing on concerns and expectations regarding mobile learning in the music curriculum was conducted in 2014 and 2015, respectively in Hong Kong.

Figure 16.2 presents teachers' concerns about mobile learning in music education ranked according to the mean score. The most important concerns were ranked as 1 and the least important as 6. Their concerns were ranked as (1) equipment setup, (2) technical support, (3) financial burden, (4) teacher training, (5) classroom management, and (6) maintenance.

The teachers' expectations of mobile learning in music education were ranked according to the mean score (fig. 16.3). Their expectations were ranked as (1) e-learning resources, (2) interactive functions, (3) self-directed learning, (4) e-textbooks, (5) learning at the student's own pace, and (6) blended learning.

In this study, 53.5% of the in-service and preservice teachers agreed or strongly agreed that using a tablet or iPad could motivate students to learn music in a classroom lesson. In addition, 93.9% of the participants claimed that "listening" was the area where they mainly applied e-learning in classroom lessons. However, 65.9% said that "composing" was the most suitable area for applying mobile learning in music lessons. This implies that the Education and Manpower Bureau and the teacher training institution need to provide training to teachers in how to apply mobile learning in "composing."

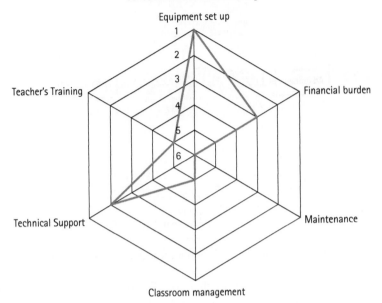

FIGURE 16.2 Concerns about mobile learning in music education

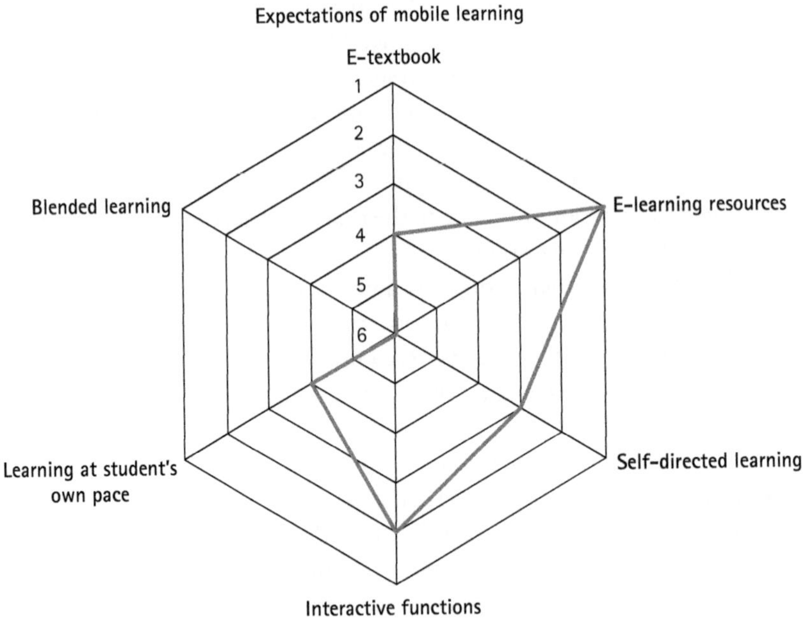

FIGURE 16.3 Teachers' expectations of mobile learning in music education

Implications of Using Information and Communication Technology in Music Education in Hong Kong

As displayed in figure 16.2, the top three concerns among teachers were equipment setup, technical support, and financial burden. In the focus group interview, teachers said that *preservice teacher training was important, as in-service teachers may not have time to attend training courses.* Furthermore, *technical support was a concern, as it strongly depended on the age of the teaching team and the school's IT department.* Moreover, *classroom management was a concern, as mobile devices could distract students.* These findings imply that course design for mobile learning in music for both in-service and preservice teachers is crucial for the successful implementation of the Hong Kong government's policy. Wi-Fi broadband width is another important issue that requires support from agencies such as schools and the Education and Manpower Bureau. An effective way to avoid distraction is to engage students in group learning, such as jamming using the application GarageBand with Bluetooth technology, to allow cooperative rather than individual learning. This can avoid some classroom management issues, especially for preservice teachers during their teaching practice.

As displayed in figure 16.3, the top three teachers' expectations were e-learning resources, interactive functions, and self-directed learning. In the focus group interview, teachers claimed that a mobile device *could help students explore possibilities and achieve higher standards of work.* Furthermore, *creative and group activities could be enhanced,* and *the mobile device could be more convenient than a laptop.* These findings imply that the interactive design of music applications will be critical during implementation, and that the device could actually reduce teachers' workloads, for example if teachers use the application NotateMe to compose music such that students can draw notes directly on the iPad. The idea of self-directed learning that moves from inside to outside the school can be interpreted as "game-based learning" (Leong, 2011), which allows the learning of music to extend into a student's daily life through mobile devices.

Creating an Ecology of Information Communication Technology in Music Education In Hong Kong

In figure 16.4, based on the theoretical framework (Pachler, 2010) of a sociocultural ecological approach that relates agency, cultural practices, and structures to mobile

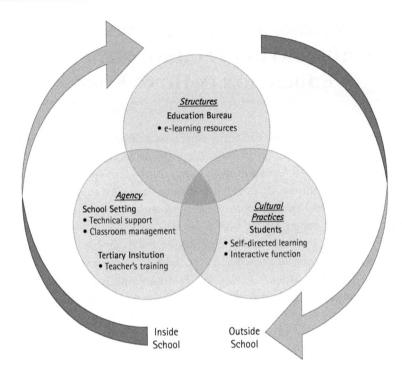

FIGURE 16.4 Ecology of ICT in music education in Hong Kong

learning, an ecology of ICT in music education as an "outside in–inside out" relationship is proposed, such that cultural practices involving mobile learning can be brought into the school, enhanced at school, and then fed back into the digital world at large. In the coming years, ICT in music education can be regarded as *cultural resources* that enable informal learning contexts in the world *outside the school* to be brought together with processes and contexts that are valued *inside the school*.

Agency: Both primary and secondary school students can increasingly display a new way of learning in which they see their lives and worlds framed as a challenge and as an environment with potential resources for learning. The primary concern of preservice teachers regarding mobile learning is teacher training. Courses provided by tertiary institutions could equip preservice teachers to start the process. The next stage is the school setting. The technical support must be sufficient, and in-service teachers must think about instructional design so as to avoid classroom management issues such as the distraction of students.

Structures: Student learning is strongly governed by the curricular frames of educational institutions and their specific approaches to the use of new cultural resources for learning. Sufficient e-learning resources and teaching packages have to be provided by the Education and Manpower Bureau to drive the initiative forward through seminars and workshops for in-service teachers on ICT in music education.

Cultural practices: Mobile devices are increasingly used for social interaction, communication, and sharing. Learning is viewed as culturally situated meaning making inside and outside educational institutions, and media used in everyday life have achieved cultural significance. Self-directed learning with interactive functions is recommended in mobile music learning to extend classroom teaching into the students' daily lives so as to facilitate their learning outside school such disciplines as composing, listening to different repertoires, practising aural skills, and recording one's own performances using mobile devices outside the classroom. This situated meaning making inside and outside school can be connected to the media that students use in everyday life. Finally, mobile music learning can be transformed into a *cultural practice*.

Digital Pedagogy for Creating, Performing, and Listening in the Music Curriculum in Hong Kong

According to the Digital 21 Strategy report (2014), the world competitiveness yearbook conducted by the International Institute for Management Development ranked Hong Kong first in technological infrastructure in both 2012 and 2013. Hong Kong's Internet connection speeds, and broadband and mobile penetration rates, at 85%, are among the highest in the world. These data revealed that Hong Kong has a good ICT infrastructure to support ICT in education.

However, being able to use technology effectively requires an understanding not only of technology itself but also of effective pedagogical approaches for utilizing it in a particular content area. The affordances and constraints of a technology for use in a specific instructional context need to be considered. Teachers must contemplate the dynamic relationship that exists among (1) content knowledge, (2) pedagogical knowledge, and (3) technological knowledge (Bauer, 2014).

Based on the ecology of ICT in music education in Hong Kong (fig.16.4), the potential of digital pedagogy in creating, performing, and listening will be explored in the music curriculum in Hong Kong. The critical issue in the ICT in music education in Hong Kong is the *teaching context*. This is the instructional environment, which includes the physical setup of the classroom, the quantity and quality of technology that is available to students and teachers, students' learning styles, and the atmosphere of the school. All these factors are crucial in launching digital pedagogy in the music curriculum. With massive development of technology, the Education and Manpower Bureau and the teacher's training institutions would gradually change both the preservice and in-service music teachers from *routine experts* to *adaptive experts* in the digital age to cope with the need and imagination of our students—the app generation (Gardner & Davis, 2013).

Funding Acknowledgment

The study of ICT in music education in Hong Kong that this chapter describes received funding from the Dean's Research Fund (ECR-006) of the Faculty of Liberal Arts and Social Science of The Education University of Hong Kong and from an Internal Research Grant from the same institution.

References

Bauer, W. (2014). *Music Learning Today*. New York: Oxford University Press.

Chen, C. W. (2015). Mobile learning: Using application Auralbook to learn aural skills. *International Journal of Music Education, 33*, 244–259.

Digital 21 Strategy. (2014). 2014 Digital 21 Strategy. Retrieved December 15, 2014, from http://www.digital21.gov.hk/eng/relatedDoc/download/2014D21S-booklet.pdf.

Education and Manpower Bureau. (2013). Textbook Selection Criteria. Retrieved April 16, 2013, from http://www.edb.gov.hk/atachment/en/curriculum-development/resourcesupport/textbook-info/emads/edbcm12087e.pdf.

Gardner, H., & Davis, K. (2013). Acts (and apps) of imagination among today's youth. In *The App Generation* (pp. 93–120). New Haven, CT: Yale University Press.

Johnson, L., Smith, R., Willis, H., Levine, A., & Haywood, K. (2011). *The 2011 Horizon Report*. Austin, TX: New Media Consortium.

Kress, G., & Pachler, N. (2007). Thinking about the 'm' in m-learning. In N. Pachler (Ed.), *Mobile learning: Towards a research agenda* (pp. 7–32). London: Institute of Education, London.

Legislative Council of Hong Kong (2013). LCQ20: E-textbook Market Development Scheme. Retreived January 5, 2017, from http://www.info.gov.hk/gia/general/201310/30/P201310300378.htm.

Leong. S. (2011). Navigating the emerging futures in music education. *Journal of Music, Technology and Education, 4*(2 & 3), 233–244.

Pachler, N. (2010). Charting the conceptual space. In N. Pachler, B. Bachmair, & J. Cook (Eds.), *Mobile learning: Structures, agency, practices.* (pp. 3–26). New York: Springer.

Sharples, M. (2009). Methods for evaluating mobile learning. In P. Lang (Ed.), *Researching mobile learning: Frameworks, tools and research designs* (pp. 17–40). Bern: International Academic Publishers.

Sole, C. (2009). The fleeting, the situated and the Mundane: Ethnographic approaches to mobile language learning (MALL). In P. Lang (Ed.), *Researching mobile learning: Frameworks, tools and research designs* (pp. 137–150). Bern: International Academic Publishers.

PART 2

LOCATIONS AND CONTEXTS: SOCIAL AND CULTURAL ISSUES

2 A

Core Perspectives

CHAPTER 17

CRITICAL PERSPECTIVES FROM AFRICA

BENON KIGOZI

THE application of technology in music education is currently popular everywhere, including Africa. There are many studies on the application and integration of technology in music education to meet specific curriculum targets, most of which suggest that prosperity for the skilled and educated workforce will occur only with technological competence and literacy. Being an educator and researcher, I have participated in music education research in the area of information and communication technology (ICT). Teaching at various institutions in different countries over the last 15 years has made me more aware of the unequal distribution of resources across schools, and of how the application and integration of technology in education is a cause for the poor status of music education across various regions in Africa (Kigozi, 2013). It is because school management teams do not usually understand fully the benefits in music teaching and learning of having technological gadgets like computers to record and play back.

WHAT IS MUSIC EDUCATION AND WHAT IS TECHNOLOGY?

Music education from an African point of view takes the form of socialization and learners, and of inducting them into the accumulated music, dance, and drama heritage of their predecessors (Mbabi Katana, 1971). This is done in a variety of contexts, including informal music education, which is the lifelong process by which every individual acquires and accumulates musical knowledge, skills, attitudes, and insights from daily experiences and exposure to the environment. Music education, therefore, is achieved through the nonformal context: systematic education activities carried on outside the framework of the formal context (Kigozi, 2014). Nonformal music education is mostly delivered orally by indigenous village musicians, primarily in rural areas. Formal music education does take place in many urban cities, as prescribed in school curricula. Music education is therefore perceived and acquired differently depending on the context.

As a member of the Pan African Society for Musical Arts Education (PASMAE), a society that concerns itself with the promotion of musical arts education on the African continent, I am involved with workshops and conferences organized by the Society. At the PASMAE workshop in Kisumu 10 years ago, I attended the technology sessions facilitated by Robert Kwami, who was then charged with the responsibility of developing music technology resource materials for African scholars within the PASMAE fraternity. At this forum the entire PASMAE jurisdiction of countries from Eastern, Southern, Western, and Central Africa were represented. At the end of the workshop, PASMAE members decided to come up with a working definition of "technology" from the African perspective. This is the definition they reached: "audiovisual aids and tools such as books, systems of musical transmission, aural-oral, mental and other mnemonic aids, [that are] indigenous [to Africa], even stories, language and literature—and other aspects of science, the arts and culture" (Pan African Society for Musical Arts Education, 2003a). With this definition, which includes mental and mnemonic aids, stories, arts and culture, language and literature, PASMAE refers to the immediate environment and the body, mind, and spirit as being the source of all technology. The human body embodies the finest technology, in my opinion, from which trickles down other artistic forms of technology in terms of construction and application. As musical artists we need to apply technology from an artistic point of view and work with it as an art without too much scientific and technical procedure. When we are confronted with a musical instrument, we seek knowledge of how to operate the hardware, including our body, even if the instrument is not the human voice. Technologically, we are challenged to produce a clear, effective, and soulful sound with the instrument in question. So we command the body, mind, and soul according to the musical instrument we possess, and not until that is done do we communicate musical information.

Where Is Technology within Music Education?

With technology, issues regarding technical knowledge, affordability, and maintenance always top the agenda more than the application itself. The question of where technology is within music education is determined by the resource bases and budgets of the different institutions involved. Schools don't have enough computers, and music educators need more training to apply technology effectively.

It is important to note that, unlike in the West, most of the use of technology within music education in Africa takes place outside the formal music classroom context, as earlier noted. Music education technology in the rural context differs from the urban context for various reasons, ranging from the level of development, accessibility of resources, the nature and kind of resources, and the traditions, norms, and cultures of the people. While nonformal music education is orally delivered by indigenous community musicians in rural areas, both the informal and formal contexts exist in urban areas. Therefore, musical memory as part of the body, mind, and soul is extremely significant with regard to technology in music education in theory and practice. Because the majority of people live below the poverty line, there is insufficient modern technological equipment for teaching and learning, and education is mainly by rote. Performances are usually memorized, in order to express oneself freely as well as to allow for improvisation without worrying about the score. Kwami reiterates that "the traditional context is the best environment for a student of African music" (1989, p. 24). Even with insufficient resources, the importance of technology in music education in Africa cannot be emphasized too much, as it is aligned with the general function of education, which is mostly community-based and communal, both in formal and informal settings. Technology in music education therefore offers far more than a mere heightening of the general quality of life.

In the big cities of Africa, technology in music education happens in commercial communities and, mostly, in educational institutions. In schools it offers an aesthetic experience to both learners and educators through the Internet. However, because of the inadequacy of technological resources at most local schools, music educators do not use much technology for teaching music in class, even though they use electronic instruments like guitars and keyboards for school concerts and assembly performances. Having worked in various African countries, I have witnessed educators using computers and the Internet outside the classroom, mostly for completing school assessments, grading, and reporting. Beyond schools, I have seen Internet cafés mushrooming in many cities. These cafés are flocked to by everyone, including illiterate people, because there is always an attendant inside who assists the customers. Many young people use cafés to download music for their own consumption, but also for commercial reasons. There is no doubt that "technology is deeply embedded in the contemporary lexicon of many young people's musical lives . . . and the internet is their playground" (Burnard, 2007, p. 201). In Africa today, expensive technological gadgets seem to find their way into the country, and many young people are surrounded by cell phones, iPads, MP3s, digital music players, video games, iPods, laptop computers, sequencers, and the like. It seems as though the young people do not know a world with no technology, especially playback gadgets.

Technology in rural Africa does not always rely on electricity, hence the PASMAE definition that encompasses "systems of musical transmission, aural-oral, mental aids and other mnemonic aids, [that are] indigenous African" (Pan African Society for Musical Arts Education, 2003a). More than half of the schools in rural Africa are not connected to electricity and do not use modern technological devices.

Oral-aural tradition *is* the tradition, and the most common educational context is the informal context, which happens through community musical activities, including marriage ceremonies, twin ceremonies, circumcision ceremonies, and local religious ceremonies. Other ceremonies include funeral ceremonies and music festivals. These activities are married with mental and other mnemonic aids, indigenous African instruments, songs and stories, language, and literature. Education in "music technology" is achieved through nonformal contexts. I have come together with other educators and students and participated in a technology project, collecting information about indigenous African musical instruments in order to research them as well as use them for teaching and learning. Information collected on indigenous instruments includes photographs, drawings, and sound recordings. In addition, we have projects on indigenous knowledge systems where we facilitate students in the construction of classroom instruments. Projects like these as part of course units are cost-cutting ventures because instruments are not purchased but rather are homemade or locally constructed and used for teaching and learning. The instrument construction activity is considered a part of music technology education. These projects have a great impact on communities in cases where groups of students and teachers come all the way from other regions to access information about how indigenous African musical instruments are constructed and repaired.

According to Herbst, technology in music education happens in these kinds of contexts through listening, observation, and participation, which constitute the reciprocal dimensions in the development of musicianship, and these begin even before birth (2005). Therefore, music education technology is transmitted and perceived through listening, watching, and getting one's "hands dirty" with the actual construction and repairing of instruments.

Modern Technology in Music Education

Modern technological developments in analog and digital electronics in Africa have allowed musicians to have access to a wide range of technological equipment and sounds. With the birth of broadband Internet connections, iPods, iPads, and cell phones provide a wide range of music technology options with regard to digital recording, sound editing, and playback. One particular recording studio I know in downtown Kampala, called Kasiwukira, has served a number of local artists because it is very accessible, is reasonably cheap, and has all that is needed for local artists to finalize tracks. There are a multitude of such studios in cities like Kampala, Harare, Nairobi, and Dar es Salaam. In these urban cities, recording and manipulating sound is much easier now because of the easy access to music gadgets, including sequencers, synthesizers, and nonlinear recording and editing systems interfaced with computers. In selected schools, students and educators alike arrange and compose easily, and are able to play back without assembling an ensemble. I have had an opportunity to work with International General Certificate of Secondary Education and

International Baccalaureate students at Kampala International School in Uganda, and it is interesting to note that all these students now complete their composition coursework technologically with programs like GarageBand. The St. Andrews School in Kenya and the International School of Tanganyika have this privilege currently. Even though other schools might not have the capacity to use sophisticated music programs, they too use computers and the Internet to a certain extent to search for as well as listen to music. I encounter students and parents alike bringing along iPods, iPads, and cell phones to record their children's music recital performances, assemblies, and other concerts. University students and the wider community take advantage of online tutorials; they also watch a multitude of performances from all across the world via the Internet using cell phones, laptops, iPods, and iPads. In addition to being a vital mode of communication, technology has become an increasing necessity to Africans for the successful dispensation and development of music and musical arts education as academic fields.

Survey I - Educators

Even though PASMAE has an African working definition of "technology," over the years there have been numerous developments in this field and subsequently many interpretations of what scholars perceive as music education technology based on the definition. In order to synchronize various perceptions of music education technology across various regions, it was necessary for me to conduct a qualitative inquiry in 2013.

I decided to send out a questionnaire and get feedback on participants' views on music technology in relation to how they define "music technology." In addition to questionnaires, impromptu qualitative interviews were conducted in order to probe participants' perspectives. Interviews and questionnaires were sent to two groups of educators: rural music teachers and urban music teachers. Altogether in this survey I engaged 50 participants from rural and urban contexts in Zimbabwe (5), Tanzania (3), South Africa (2), Kenya (2), Uganda (32), Zambia (2), Nigeria (2), and Ghana (2). I received back 46 responses. The information I sought to gather was based on the following inquiries: what equipment the educators perceived as music education technology resources; which educators engaged in music technology education and what actual levels of music education participants engaged in; how exactly educators applied technology in music education; challenges educators faced with regard to the application of technology in music education; and finally, who was affected by and possible solutions to challenges encountered. When asked about what they considered music education technology to be, respondents gave answers under three main categories: traditional indigenous musical instruments (TIMI), Western acoustic musical instruments (WAI), and electrical music/media equipment (EME). Thirty music educators confirmed that they considered TIMI—including drums, xylophones, tube fiddles, sticks, lyres, and shakers—to be music education technology resources. Forty-six of the respondents agreed that WAI—including acoustic guitars, modern drums, tambourines, cymbals, triangles, metallophones, glockenspiels, brass instruments, woodwind instruments, pianos, and electric keyboards—were part of music education technology resources. The same number of participants affirmed that EME—including cassette players, CD players, radios, videos, tape recorders, microphones, computer software, computers, and overhead projectors—constituted music education technology resources.

From the survey it is evident that a number of educators' perceptions on music education resources resonated with the PASMAE definition of technology. Now the question was: where was technology in music education? In order to address this subquestion, we needed to find out which educators engaged in music technology education and at what levels of music education. The questionnaire responses indicated that 35 urban educators were involved with music education technology, and 11 rural educators were. Among these were primary education teachers, secondary education teachers, and tertiary education teachers. It is interesting to note that tertiary and primary school educators were the most involved with music education technology.

While several educators were using recordings and music software for instruction and performances, all the respondents fancied the use of CDs and cassette players for playback as a form of technological application during their lessons. In addition, a large number of respondents fancied the use of audiovisual aids and tools such as "books, aural-oral, mnemonic aids, and other systems of musical transmission," including "indigenous African stories, language and literature, arts, and culture" (to use the language of the PASMAE definition of "technology") as other conduits through which technology was being applied in music education. From the survey results, it appeared that relatively few educators were engaging fully in the type of modern music technology education that calls for the use of modern gadgets and of programs like sequencers and GarageBand for composition, recording, and performing.

Survey II - Students, Teachers, and Parents

A second survey I conducted in 2013, across four main regions of Uganda, centered on the integration and application of technology in music education. The research instruments consisted of a survey in which 77 students, 20 teachers, and 23 parents took part. Employing purposive maximum variation sampling, 16 districts in eastern, western, central, and northern Uganda were picked. The number and percentages of participants in each district are shown in table 17.1.

This investigation was mainly based upon teacher and student samples. It consisted of a student-centered classification of two contexts: the out-of-school environment and the school environment. Table 17.2 lays out the two contexts; four categories within them—home, students, teachers, and infrastructure; and subcategories covering a wide

Table 17.1 Regions, districts, and types and regions of participants

Region	District	Staff members	Parents	Students	%
Eastern	Pallisa, Jinja, Iganga, Kamuli	3	4	16	20.6
Central	Mukono, Luweero, Kampala, Masaka	11	10	30	42.4%
Northern	Kitgum, Gulu, Arua, Lira	3	4	11	14.8%
Western	Bushenyi, Hoima, Kabale, Mbarara	3	5	20	23.2%
Total	16	20	23	77	100%

Table 17.2 Contexts, categories, and subcategories

Context	Category	Subcategory
Out of school	Home	*Access to*: electricity, telephone, computer, radio, television
School	Students	*ICT*: lessons, resources, application in classrooms
School	Teachers	*ICT*: policy, integration in curriculum, financing, application in curriculum
School	Infrastructure	*Access to*: radio, television, electricity, telephone, computer, overhead projector

range of aspects, including access to ICT resources and support from school management and from government.

The Student Survey

The 77 students in the survey above were both female and male, were aged 12–18, and were at levels ranging from primary 1 to primary 6. The data collected were based on the questions shown in table 17.3.

Of these students, 61% cited inadequate technological resources as a hindrance to teaching and learning technology in music education. Where schools had some

Table 17.3 Context, categories, questions, and participants

Context	Category	Survey questions	Participants
School	Resources	Are there keyboards in your class? Do you have access to CDs and DVD players? Do you have a television set at your school?	Students
School	Electricity	Is there electricity connection in your classroom?	Students
School	Internet access	Is your school connected to the Internet?	Students
School	Computer access	Are there computers and computer software at your school?	Students

computers, Internet connection was yet to be made. In some schools, for example the Kampala Parents' School, the unreliability of technological equipment was highlighted by students, yet that was about the best you could get in Kampala.

The First Teacher Survey

This survey consisted of music teachers teaching theory, voice, analysis, performance, and composition in four districts across the country. Data were based on school context, two categories of government and school, and four subcategories ranging from infrastructure, resources, training, and Internet access. The questions asked are shown in table 17.4.

The results from this survey reveal the kinds of resource available to music teachers, and their own opinions about applying and utilizing ICT in music teaching and learning in schools (Kigozi, 2013). As in the student survey, 60% of the teachers confirmed that inadequacy of technological resources was a hindrance to doing so.

The Second Teacher Survey

The second teacher survey sought to establish how music teachers in Uganda mostly applied technology in music education. Of 120 participants, 30 applied technology mostly for composition (CMPSE), 100 for performance (PFM), 10 for research (RSCH), 110 for listening sessions (LSTN), and 60 for recording (RC'D) (Kigozi, 2013) (see fig. 17.1).

The evidence shows that the majority of music educators, across the northern, southern, eastern, and western regions of Uganda, mainly use technology for listening

Table 17.4 Context, categories, subcategories, survey questions, and participants

Context	Category	Subcategory	Survey questions	Participants
School	School management	Training	Do you have training in ICT?	Teachers
School	School management	Resources	With money available, what would your priorities be? • Purchase teaching aids • Prosvide electricity access • Buy ICT equipment	Teachers
School	School management	Internet access	Do you have Internet connection in your office?	Teachers
School	Government support	Infrastructure	Is the government committed to supporting ICT? Is there electricity connection in your classroom? Do you have music software in your school? How do you apply ICT in music lessons?	Teachers

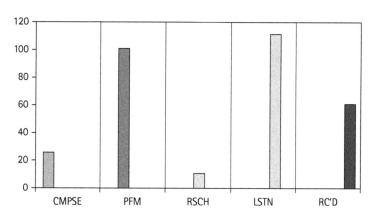

FIGURE 17.1 How music educators apply technology most often

sessions and performance sessions. I believe the reason is simply that in addition to the indigenous traditional instruments, the majority of schools have basic technological equipment, including cassette players, and CD players that mainly play back and record. A few schools have electric keyboards that use batteries and one or two acoustic guitars.

As I engaged more with music educators across the region, especially from the neighboring Kenya and Tanzania, I was able to ascertain that these data represent a wide spectrum of regions, representing various parts of East Africa, including Kenya and Tanzania.

Educators' Challenges in Applying Modern Technology

There are other factors that get in the way of educators' engaging in more sophisticated technological advances in music education. There are music educators all across Africa who are inclined to seek out and find ways to incorporate modern technology in their teaching; but among some music educators, moving away from the traditional methods to integrating and applying modern technology in their teaching has caused mixed feelings. A number of educators feel that they would rather spend their time teaching music, not technology. Music educators want to be sure that technology will improve their music teaching experience before embracing it. And while there are music educators who are genuinely interested in applying and integrating technology in music education, they are not convinced that they can do so with ease. Most of the music educators in this part of Africa need training before embarking on the application of technology in music education, and the cost of providing such training is enormous! In addition, it takes time for educators to become proficient at using software and hardware. They need to be as reliable and as easy to use as mbira, *engoma, hosho, endeere,* or a CD player, so that precious time is not lost due to technology glitches during class sessions.

Even with a technological world surrounding the young generation today, it is hard to foster enthusiasm for technological applications among music educators in many African countries like Uganda, where music budgets are cut every now and then, and local school authorities threaten to delete music as a subject. Uganda's Ministry of Education and Sports is obsessed with "academic subjects," not music. In many cases a school's location plays a part in determining its fate with regard to technological infrastructure. Whereas there are many schools and institutions with the proper infrastructure, the majority of schools and institutions in rural areas lack permanent classroom buildings, water access, furniture, electricity, and adequate road networks. The lack of electricity makes it hard to integrate modern technology in school curriculums, including music education.

Radio has instead emerged as one of the most attractive and cheap technological devices in rural schools. The surveys conducted in 2013 showed that 30% of the schools in Uganda have access to radio, whereas 5% of have both radio and television. Access to computers in schools is greater in towns where there is electricity than in isolated rural nonelectrified villages (Kigozi, 2013). Whereas it is still a primitive technological device, radio has the rare advantage of mobility, as it can be operated using alternate power sources, such as batteries. Radio is effective in reaching a wider spectrum of learners without extra infrastructural costs such as electricity. Some schools can afford a CD player and a pair of speakers as the standard technology unit, though most rural schools rely on two cassette players as the standard technological equipment for a long time. Usually one side of the cassette player has a record capability and an in-built microphone, and the other side is for playback.

Urban Areas

Urban schools identified as having computer initiatives usually have between two to five computers on the average for the whole school. Those that have bigger budgets for technology subsequently have more computers; however, it is important to note that school computer labs centrally managed within the school face an uncertain future and are not used solely for music education. Music educators are discouraged with regard to the integration and incorporate technology into music curriculums because ICT coordinators do not usually accommodate music teachers when it comes to installing and maintaining software that they need to use. Generally, however, technology and in particular computers are starting to permeate the general education environment.

A number of people have home studios where they easily record, mix, and edit music, and at schools like Kampala International School Uganda, the International School of Uganda, and Rainbow International School, students are able to create as well as perform their compositions via technological devices. A variety of students with whom I have worked in music classes feel that the computer is a gateway to various methods of music expression. They use computers to enhance their music creativity and for playback of prepared soundtracks for performance accompaniments in talent shows, school assembly presentations, and music festivals. In addition, music sequencer software, such as Logic Audio, and Pro Tools, are widely used in music production.

Despite the challenges surrounding technical knowledge, affordability, and maintenance, instructional technologies exist and have affected the ways educators deliver their music lessons. Despite the widespread use of technology and computers by educationists and other professionals outside the classroom, however, systematic, meaningful, and sustainable deployment of such technologies in our education system has yet to happen. School culture and instructional practices do not successfully incorporate computer-based technology at most schools, even at some of the few that can afford it.

POSITIVE IMPACTS OF TECHNOLOGY ON OUR COMMUNITIES

The advancement of technology in Africa has brought about, to some extent, not only the introduction of electronic equipment but also its integration and application in teaching and learning. Technology has created significant changes in music education through the provision of support as well as the offering of opportunities to spur recording and compositional development. Music software is used to enhance education and expand students' repertoires, thereby enhancing creativity. All this happens at urban schools and international schools, where students of and are required to submit coursework in composition and performance. There are newer ways of storing and transporting music that do not include CDs and cassette tapes. Using flash disks, iPods, iPads, cell phones, and the like is easier and less cumbersome because they are not as bulky and fragile as materials like CDs and cassette tapes. Cartridges and microprocessors within synthesizers and electronic keyboards have made it possible for instrumentalists to come up with various new sounds. There are multitudes of recording studios everywhere, and it is common to see software applications that support sequencing, editing, recording, and notation. Subsequently, the standard of recorded music has improved greatly, because a prepared studio recording of music usually sounds better than a live performance in terms of balance, sound quality, and timing. Regardless of the level of music education one has had, more people can now tell the difference between very good, average, and bad sound. Technology has allowed people who previously would not be considered musicians to handle, create, and communicate music via their computers (Wise, 2013). Many local people are engaged in business practices that have created jobs for music producers, suppliers, and vendors.

The growing availability on the market of music recorded by local musical artists and the advancements in studio technology for studio recording and stage performances have allowed competition, convenience, and better quality of the final products. Many people prefer the digital recording process because it is easier, less cumbersome, and shorter than analog recording. One fixes any small mishaps without having to repeat the whole recording. Digital recording has transformed the way music is produced and listened to. When I frequented Ebony Village Studios, recording was done analogically with a tape machine. When the tape machine was running during a recording session, everyone was on alert, with maximum concentration, because if any slight mistake occurred, the whole song would have to be redone, rather than cutting out the specific mistakes, unlike digital recording, where one only needs to cut out individual ones.

Negative Impacts of Technology on Our Communities

The introduction of smartphones among teachers and students has allowed constant access to the Internet, downloading, and listening to music, and teachers who have been the center of learning in the music classroom are now experiencing difficulty with distracted students struggling to stay focused. In addition, students, educators, and the wider community download and sample music without acknowledging its source. Music is then duplicated and sold on the local market without thinking about the intentions of the composers and producers who supposedly make their livelihood from it. Furthermore, technology has created unemployment for musicians. While it is cost-effective to use prepared soundtracks, it is disadvantageous to the instrumentalists who earn their livelihoods playing in bands accompanying local artists. As a result, instrumentalists no longer practice hard to attain the highest level possible, because they cannot count on the local artists' hiring them. This situation has greatly lowered the standard of instrumental music performance.

Mitigating the Negative Effects of Technology While Accentuating the Positive

There are rules prohibiting personal electronic gadgets in schools during school hours, but of course one or two students will use them when no one is observing. More stringent measures prohibiting unwanted gadgets from being brought to school are necessary. However, in order not to discourage students, exceptional permission for them to use their music electronic gadgets at school during music lessons under the guidance of the teacher should be allowed. Governments without copyright laws should develop them as a matter of urgency in order to curtail the escalating amount of unauthorized downloading and piracy of music. Music educators should adapt to the current trends and exercise innovative ideas while engaging old technologies, like radios and analog televisions, in poor areas. These old technologies still play an important role in advancing music education in rural regions across the continent (Kigozi, 2013).

Conclusion

The future of music education in developing countries will largely depend on the schools' and wider communities' level of aggressiveness in embracing ICT initiatives. Training of educators in the application of modern technology is important, as is the availability and accessibility of modern technological resources.

References

Agawu, K. (2003). Musical arts in Africa: Theory, practice and education. Pretoria: Unisa Press.
Bissell, P. M. (1998). Tune in to technology. *Music Educators Journal*, 85(2), 36–41.
Burns, A. M. (2006). Integrating technology into your elementary music classroom. *General Music Today*, 20(1), 6.

Burnard, P. (2007). Creativity and technology: Critical agents of change in the work and lives of music teachers. In Finney, J., Pamela B., Sue B., & Anthony A. (Eds.), *Music education with digital technology* (pp. 197–206). New York: Continuum.

Dictionary Unit for South African English (Ed.). (2005). *South African concise Oxford dictionary*. Cape Town: Oxford University Press Southern Africa.

Facts About Technology Users. (1994). *New ways in music education, spring, 1994*. Grand Rapids, MI: Yamaha Corporation of America.

Feldstein, S. (1988). Technology for teaching. *Music Educators Journal, 74*(7), 35–37.

Hanson, J. R., & Silver, H. (1991). Learning styles of at-risk students. *Music Educators Journal, 78*(3), 30.

MacDonald, R. A. R., Hargreaves, D. J., & Miell, D. (2002). *Musical identities*. New York: Oxford University Press.

Herbst, A. (2005). *Emerging solutions for musical arts education in Africa*. UNISA Press.

Kigozi, B. (2012). Technological survey in music education: The integration and application of ICT in music education—the case of Uganda. *Technologii Inforrmatice Si De Communicatie in Domeniul Muzical, 3*(2).

Kigozi, B. (2013). Critical impacts of music technology to music education: A restructuring of the teaching and learning of music at secondary schools in Uganda. *Technologii Inforrmatice Si De Communicatie in Domeniul Muzical, 4*(1).

Kigozi, B. (2014). *Music education in Uganda*. Scholar's Press.

Kwami, R. (1989). *African music education and the school curriculum*. Unpublished doctoral dissertation, University of London, Institute of Education.

Kwami, R. M. (1998). Non-Western music in education: Problems and possibilities. *British Journal of Music Education, 14*(2), 161–170.

Lamont, A. (2002). Musical identities and the school environment. In R. Macdonald, D. Hargreaves, and D. Miell (Eds.), *Musical Identities* (pp. 41–59). Oxford: Oxford University Press.

Mbabi Katana. (1971). *The role of indigenous music in modern African education*. Kampala: Uganda Publishing House.

November, A. (2010). *Empowering students with technology*. Thousand Oaks, CA: Corwin Press.

Pan African Society for Musical Arts Education. (2003a). *Musical arts in Africa: Theory, practice and education* (Anri Herbst, Meki Nzewi, & Kofi Agawu, Eds.). Unisa Press.

Pan African Society for Musical Arts Education. (2003b, July). Unpublished transcripts of Pan African Society for Musical Arts Education conference sessions. Kisumu, Kenya.

Pan African Society for Musical Arts Education. (2005). *Emerging solutions for musical arts education in Africa* (Anri Herbst, Ed.). Cape Town: African Minds. http://www.african-minds.co.za/wp-content/uploads/2012/05/EmergingSolutions.

Peters, G. D., & Miller, R. F. (1982). *Music teaching and learning*. New York: Longman.

Pitts, A., & Kwami, R. (2002). Raising students' performance in music composition through the use of information and communication technology: A survey of secondary schools in England. *British Journal of Music Education, 19*(1), 61–71.

Regelski, T. A. (1981). *Teaching general music: Action learning for middle and secondary schools*. New York: Shirmer Books.

Renee, C. (2008). Are resources solely to be blamed? The current situation on music education facilities, computer and music technology resources in Victoria. *Australian Journal of Music Education, 1*, 44–55.

Rudolph, T. (1991). Technology and music education. *ACM SIGCUE Outlook, 21*(2), 84–86.

Rudolph, T. (2004). *Teaching music with technology*. Chicago: GIA.

Scripp, L., & Meyaard, J. (1991). Encouraging musical risks for learning success. *Music Educators Journal, 78*(3), 36–41.

Shuler, S. (1991). Music, at-risk students and the missing piece. *Music Educators Journal, 8*(3), 21–30.

Start Wise. (2013). Students' perceptions of digital technology in music education. *Technologii Inforrmatice Si De Communicatie in Domeniul Muzical, 4*(1).

Tiberondwa, A. K. (1978). *Missionary teachers as agents of colonialism.* Lusaka: National Educational Company of Zambia.

Tomorrow, P. (2010a). Creating our future: Students speak up about their vision for 21st century

CHAPTER 18

INTEREST-DRIVEN MUSIC EDUCATION

Youth, Technology, and Music Making Today

KYLIE PEPPLER

An artist by training, I engage in research that focuses on the intersection of arts, new media, and informal, interest-driven learning. Over the course of my work, I have collaboratively investigated how new technologies are allowing youth to leverage their musical intuitions in the making, performing, and sharing of music. In this chapter, I share a range of work that examines how music education can benefit from the use of new tools and materials for music making that allow learners to combine their interests and prior understandings toward deepening their engagement in music.

NEW OPPORTUNITIES FOR MUSIC LEARNING

Where and how do youth learn music? For most, music education is a continuous, two-pronged affair. On the one hand is the formal education that begins for many youth in elementary school, with its acronyms for remembering the pitches on a staff and their first instrumental performances (whether on percussion instruments, recorder, or guitar). On the other hand we find the informal education that begins when youth overhear their first song and continues throughout their lifetime as they consume new artists and hear "new things" in their old standards. Both are relatively universal. For example, here in the United States, regardless of region or income level, the most current estimates state that 94% of all schools offer music education programs (Parsad & Spiegelman, 2012) and youth spend an average of two and a half hours a day listening to music and audio (Rideout, Foehr, & Roberts, 2010). Of the two, however, the former occupies a substantially shorter amount of time in the lives of most youth and often fails to activate youth interests in concert music, which has contributed to the declining numbers frequenting music concert halls and crippling today's concert music industry (Harlow, Alfieri, Dalton, & Field, 2011).

However, informal, interest-driven participation in music is increasingly dominating youths' out-of-school recreational activities (Peppler, 2014; Rideout, Foehr, & Roberts, 2010). One of the most common sights today is youth on the subway, at local coffee shops, in schools, and online listening to and engaging in music—most of which is afforded through the new digital music landscape. Moreover, notable pockets of youth are creating a diverse amount of media, including music videos (Knobel, Lankshear, & Lewis, 2010), original compositions, and visual animations or movies with original soundtracks (Luckman & Potanin, 2010; Thomas & Tufano, 2010). This type of music and media production, while not inherently rooted in the canonical arts, denotes a "creative turn" (Sefton-Green, Thompson, Jones, & Bresler, 2011) in how youth create with new technologies and it may very well provide a new informal pedagogy to support music learning today.

Taken together, all this suggests that today's youth have no shortage of access to and awareness of music in its various forms. However, the challenge is therefore to engage youths' preexisting interests in the music that is present in their everyday lives and move them to deepening this understanding and becoming original and high-quality music makers (Peppler, 2013, 2014). The promise of new technologies is that they seem to be lowering barriers to who can engage in music making and diversifying the methods of participation that can signify a twenty-first-century approach to music learning. This chapter takes a closer look at several examples of how new technologies are changing the traditional relationships between youth, music making, and performance. These examples highlight new opportunities for music educators, parents, and youth to question current misconceptions of how new technologies offer watered-down or inadequate versions of traditional music education and instead offer new pathways into a music education that is more aligned with youth culture and what we know about high-quality, interest-driven learning.

New Doorways into Music Learning and Performance

With the introduction of new technologies, the pathways into music education are changing. As a notable example, my research team at Creativity Labs and I found that "rhythmic video games"—virtual representations of rock music performance and practice—are changing the way young people learn authentic music concepts and notation commonly thought to be the sole domain of formalized music education (Peppler, Downton, Lindsay, & Hay, 2011). The majority of rhythmic video games (e.g., Rock Band and Guitar Hero) are characterized by players' use of a simplified instrument, such as a guitar or drum set, to execute, "in time," music that is simultaneously heard and notated onscreen. The in-game scrolling notational system—a combined form of guitar tablature and a modified MIDI notation that translates notes into vertical rectangular blocks that pass over a horizontal "finish line"—differs from modern staff notation but embodies several of its rich musical concepts, including models of metric hierarchy, subdivision, measurement, and pattern identification (Peppler, Downton, Lindsay, & Hay, 2011). And, by aligning intuitive notation symbols with prerecorded instrument tracks, rhythmic video games enable rare opportunities for real-time formative feedback ("notes" on the screen will fail to illuminate and your instrument temporarily drops out of the song when you hit the wrong keys) that lets players know if their "performance" corresponds with the notation onscreen, reinforcing the symbiotic relationship between a notation system and sound, and sensitizing players to the multiple parameters required to effectively represent music in a written form.

The tension between understanding music in both its aural and written forms is, of course, central to the nature of musical education—that learning about music is as much about becoming familiar with its notational and theoretical underpinnings as it is about its performance, the learning of each becoming increasingly specialized in advanced study. Research indicates that rhythmic video games, by disseminating notational concepts through the act of musical "performance," are best positioned to address the former. When an individuals purchase a guitar after spending an extended period playing rhythmic video games, the knowledge they appropriate from the game to the real world is less rooted in any instrumental facility they've developed than in the foundational concepts of music upon which they build their understandings of how music "works," a distinction voiced in the music education and popular musicology literature by pioneers in the exploration of youths' informal music practices (Green, 2002, 2005, 2008; Campbell, 1997; Clements, 2008). In this work, scholars cite examples of youth developing their aural, improvisatory, compositional, and/or theoretical intuitions, even haphazardly, through immersion in peer-led musical activities (in the absence of formal instruction) (Green, 2005; Abril, 2008; Lum, 2009). An element that Lucy Green (2002, 2005, 2008) identifies as the prime source of informal music learning is youths' emulation of recordings, emphasizing the connection between the listening experience and the act of performance. In traditional music education curricula, the listening experience and the act of performance are often kept separate by a focus on how music is notated, with the memorization of note names, key signatures, and tempo markings being a central focus—in effect designating the act of critical listening (e.g., to a master's performance) as a value-added activity to supplement, when possible, the primary goal of mechanical facility (e.g., pressing the right key at the right time, singing in tempo and at the correct pitch).

Yet the interlocking of aural and notational elements in rhythmic video game environments (i.e., by linking recordings of a master musician's performance to a scrolling notational system) could be said to present the "best of both worlds," with youths' musical understandings potentially augmented by connections they can make between sound and visual representations. The result is an immersion that players report experiencing during rhythmic video game playing, one that encourages them to follow and "read ahead" in a musical score as well as enabling them to listen to music differently; they pick apart different things in the song (e.g., bass, melody, rhythms, structure), thus elevating the gaming experience to a level steeped in music learning (Miller, 2009). It is the my contention that the music concepts that are central to the comprehension of traditionally notated music are represented in rhythmic video games' notation system, which serves as a novice-friendly method whose lessons can be applied to more traditional forms of notation, affording beginning learners a "doorway in" to more formal practices (Wiggins, 2009). This is particularly propitious for low-income and otherwise underserved youth who have greater access to video game consoles (Pew, 2008) than, one might assume, to private music lessons. Furthermore, video game consoles are becoming important music venues, with 71% of youth across all socioeconomic groups having played games like Rock Band and Guitar Hero (Rideout, Foehr, & Roberts, 2010), two of the most lucrative video games of the past decade (Quillen, 2008).

In our prior research, Michael Downton, Eric Lindsay, Ken Hay, and I sought to better understand the potential relationship between rhythmic video games and music learning among today's youth. Toward this end, we conducted a study in the after-school hours at a local Boys and Girls Club (Hirsch, 2005) located in a midsized midwestern city (Peppler, Downton, Lindsay, & Hay, 2011). At the time of the study, freely available music lessons were offered by volunteers at the Boys and Girls Club, most often on the violin or recorder. However, very few members took advantage of the opportunities for formal music instruction, and the few who did were primarily Caucasian and from middle- to upper-middle-class homes. This landscape quickly began to change when we introduced a Rock Band Club, as it quickly became popular with youth, with about 10 times the enrollment of the traditional music lessons.

As we investigated the relationship between the in-game notation system to other formal notational systems, we, aided by local music educators, developed a series of sight-reading, transcription, and echoing tasks, pulling measures from the local music textbooks based on the Kodály method (e.g., *Spotlight on Music*, 2005) that were age-appropriate and followed K–5 standards for music. This assessment was administered to all participants in the Rock Band Club as well as a group of nonparticipants and further analyzed, correlating the relationship between the number of Rock Band sessions and youths' scores on the traditional music assessment using a simple univariate regression. Collectively, our results showed that extended play in Rock Band positively and significantly correlated with the assessment results of youths' traditional music abilities (Peppler et al., 2011). The results provided evidence that the youth playing rhythmic video games saw a connection between the two notational systems and that extended Rock Band play was significantly correlated with how well youth were able to sight-read, transcribe traditional staff notation, and perform rhythmic echoing tasks. While we probably wouldn't go so far as to say that the game was teaching traditional notation, we believe that the game attuned youth to reading notational forms of music that made it easier and more relevant to them when they were presented with traditional notation back in the classroom.

As Green has observed, young musicians only interact effectively with music to the extent that they are enjoying themselves. In her observations, "cooperation, sensitivity to others, commitment and responsibility are explicitly highly valued by the young musicians" (2008, p. 8). Comparing the Rock Band Club to the private instrumental lessons at the Boys and Girls Club, the contexts for performance were quite different; the violin and recorder students never gave recitals or official performances for their Boys and Girls Club peers, whereas the Rock Band Club youth treated each session as a performance, with nearby members clapping and dancing to the music blasting from the television. With performance being one of the most rewarding aspects of being a musician, the Rock Band youth (through their displays for the audience and wide grins when their "band" played well together) appeared to get a sense of what that aspect of being a performer was all about much sooner than their private-lesson peers (who would also have to practice much longer on their instruments before having a piece "performance ready"). Without the opportunities for performance, the students in private lessons were engaged in an activity that seemed, at least to youth on the outside, lacking in context or function. Yet the members of the Rock Band Club eventually overcame their earlier apathy toward the idea of taking private lessons, possibly because the Rock Band Club provided them with a venue to forge their identities as musicians. In fact, almost all the youth enrolled in the Rock Band Club signed up for traditional music lessons for the first time following their Rock Band play, resulting in the first ever waitlist at the Boys and Girls Club for private instrument lessons. Asked why he had finally signed up for free violin lessons after playing in the Rock Band Club, an 11-year-old boy replied, "I want to learn guitar, and if I can do this (mimics the playing of a violin), then I can do this (mimics the playing of a guitar)" (Peppler et al., 2011, p. 1). In this response and several like it, we see that the young boy was able to see the relationship between what he liked and aspired to do with music and the kinds of traditional Western instruments and repertoire available to him. And the fact that several of the youth who signed up for private lessons after participating in the Rock Band Club verbally acknowledged that they had been sitting on latent desires to learn an instrument for some time further points to the connections that youth were forging between the Rock Band Club and their musical selves.

Perhaps this phenomenon can be explained by the fact that rhythmic video games differ from most video games in that they don't simulate as much as they represent actual experience, with a great deal of fidelity to the rhythmic aspects, especially, of music making. A significant reason why rhythmic video games could provide an advantageous introduction to music learning concerns the culture of video gaming and could provide its relationship to the performative aspects of being a musician and leverage what we know about effective video game design (Gee, 2003; Salen & Zimmerman, 2004). For example, youth were not sent into practice rooms to learn music fundamentals; their learning experiences were wrapped within the guise of game play and group activity. This proved important not only for youths' sustained engagement, but potentially for the learning itself. Furthermore, we forget the large chasm that exists between youths' music culture and that which we value in traditional music education, making Western music (like playing violin, practicing Bach, and reading modern staff notation) seem distant and unapproachable and leading them to opt out of even freely available opportunities.

When my colleagues and I recently followed up with our Rock Band Club cohort some years after this study, we found that most of them were still participating in the local youth orchestra and were deeply committed to playing the violin—years after their Rock Band Club experience.

One thing to note that emerged from our adult Rock Band expert-novice studies, however, was that when we invited professional rock musicians in to play the game (i.e., rock bands that had their own albums, played regular and professional gigs, and frequently covered the game's tunes as part of their covers) they adamantly disliked the game, performed poorly in-game, disliked the game's controllers, and felt that it was a pale imitation of actual onstage performance. They actually brought their own instruments to the study and, at the start of our observations, quit playing the game and picked up their instruments to play the songs in the game for us. One of the criteria for participation in our "rock musician" group that became important in our later interviews was that they learned to play "by ear" as a way of coming into music and had little to no ability to read traditional notation. This playing by ear meant that they had become accustomed to greater creativity and flexibility in the performance, had a greater reliance on listening to the music and to one another, and disliked the restrictions that the game's notation and restricted instrument had on their freedom to perform the music.

By contrast, we also invited groups of musicians with doctorate-level academic training in music, with prior experience with ear training, sight-reading, and writing traditional staff notation, to play Rock Band. In all instances, the academically trained musicians performed exceptionally well in the game. They went quickly through the various levels of game play, achieving expert in-game status easily, likely because of the alignment between traditional notation and the game's notation, as well as the game's privileging of the highly developed skill sets these musicians had in sight-reading, musical performance, and ear training. Moreover, they reported an overall enjoyment of the game, highlighted the immersive qualities (e.g., reporting a "performance high" after a successful performance), and a heightened aural perception of how rock music is put together that was not discernible in their everyday listening.

All of this highlights the fact that the ways in which new technologies are designed will privilege some ways of knowing and creating music while dismissing others that are valuable aspects of our music history and traditions. The music education community has long lamented that many youths fail to connect the repertoire, instruments, and skills embodied in informal music activities (e.g., rock music, garage bands, songwriting, and the cultural capital that comes with those activities) to formal music education. Music education that youths learn in garages and online environments is quite different from what their peers learn in high school bands and orchestras (Green, 2002) and could further inform designs for new technologies in interest-driven music education.

Reshaping the Music-Making Experience

New technologies are not just changing the nature of music performance, they have radical and far-reaching impacts on nearly all aspects of the music-making experience, including recording, distribution, sharing, composing, and even the genres of music itself (with new hybrids of previously distinct forms of music, new digital or computational forms of music and sound art, and so forth). For example, the widespread availability of inexpensive recording hardware and software (e.g., Audacity and Pro Tools), coupled with the expanding opportunities for amateurs to distribute and share their work online, has caused tremendous shifts in the music industry (Graham, 2009). For the first time in 2009, digital sales surpassed physical CD sales (Arango, 2008). Because musicians can now sell their work directly through online marketplaces such as iTunes, they do not need to wait for major record label deals and difficult-to-obtain album contracts. Some young artists have started their own DIY record labels and/or begun marketing themselves primarily using free services such as YouTube and other social media platforms. Instead of following the lead of professional artists, young artists are redefining how musicians create, produce, share, and distribute their work in the twenty-first-century digital marketplace.

For young musicians wishing to produce professional-sounding work, new technologies are further blurring the lines that once divided artist, record company, and distributor. Notably, Apple's Logic Pro and GarageBand, Sony's ACID Pro and Sound Forge, Cakewalk Sonar, ReCycle, FL Studio, Propellerhead's Reason and Ableton Live are all prominent examples of programs that enable musicians to create original compositions featuring realistic virtual instruments, radio-ready beats, and audio engineering effects—while using little more than an electronic keyboard and a laptop. Youths who have never learned to play an instrument can even create entire compositions by dragging and dropping various arrangements of prerecorded "loops"—royalty-free segments of drum patterns, musical gestures, and chord progressions. Although skeptics could argue that music built on others' loops is not wholly original, the use of prerecorded loops (or "samples" from other artists' work) is arguably an authentic practice of professional composing and producing, as is evidenced by its pervasive use in genres such as hip-hop, electronica, and rap. Many companies are dedicated to the sale of loops and sample music libraries. With the right tools, a music hobbyist can create something in 15 minutes that used to cost people $1,500 per hour to produce with live musicians in a studio. Further, a flood of new mobile apps, such as Beatwave, Sonorasaurus Rex, PatternMusic, and Looptastic, is expanding the opportunities for music creation.

Online web communities for digital music composition and performance have given young people a place to share, critique, and collaborate with others. New online communities for sharing music include MacJams, iCompositions, Circuitbenders Forum, General Guitar Gadgets, Create Digital Music, Facebook/MySpace, and SoundCloud.

Such communities represent a major shift in how music is consumed and created—from a solitary act of composing via paper and pencil to a worldwide collaborative and creative enterprise. Youth rock band rehearsals are even moving out of the garage and into cyberspace. One compelling example was TwitterBand: a group of 11 people who have never met in person but who share an interest in music and philanthropy have formed a band through the social media network Twitter and have released singles and videos that have helped raise money for charitable organizations.

The underlying story behind much of the market demand for these tools is that children (and adults) know more about music than they realize (Bamberger, 1991) but in prior generations have just lacked access to the right tools and materials for music making. Bamberger and colleagues have shown that both novice and expert musicians talk about music in much the same ways; that is, music is perceived in meaningful "chunks" rather than discrete properties (e.g., "notes") (Bamberger 1991; Downton, Peppler, & Bamberger, 2011). As new technologies enter this landscape, real or perceived barriers to music making are breaking down, opening new doors for young musicians and composers to create, perform, and share their music with others.

New technologies can make their greatest contribution to music education in the form of bridges between learners' intuitive and formal understandings of music when these technologies provide a welcoming environment in which users can explore the conceptual underpinnings of composition (i.e., by arranging and layering premade groups of melodies and rhythms, as a foundation) in the absence of prior musical training. Many programs are now available, including Bamberger's Impromptu, that allow youth to become, more than just consumers, creators of music—in effect moving creating music from the confines of professional recording studios to homes and classrooms throughout the world (Savage, 2005; Théberge, 1997).

Impromptu successfully engages youth in music composition and analysis by enabling users to construct, reconstruct, and remix tunes using "Tuneblocks"—virtual blocks that contain portions of melodies and/or rhythmic patterns—all while building an understanding of important musical concepts such as form, melody, pitch, rhythm, and structure (Bamberger, 2000). Bamberger's Impromptu is unique due to the high emphasis it places on learners' reflecting on the decisions they make in the construction process. In the process, these reflections can also reveal aspects of learners' cultural identities. For example, Bamberger's prior work demonstrates that when listeners are dealing with unfamiliar atonal music passages (i.e., music that does not have a recognizable tonal center) they will actively tinker with the compositions to establish a tonal center, even in the absence of a formal understanding of what sounded "wrong" about the atonal music or what sounded intuitively "right" about their tonal creations (Bamberger, 2003). Cases like these highlight how novice listener-composers will initially inject their own cultural preconceptions of music into their compositions in lieu of more formal intentionality. For example, when the Tuneblocks in the program feature patterns foreign to users' aural sensibilities (i.e., tunes from outside the users' own culture), the users then become more aware of their cultural biases through their reflections (e.g., some users might say that a pattern sounds "spooky" or "weird" when they are unfamiliar with it, whereas someone from that pattern's culture might find the music "comforting") (Downton, Peppler, Portowitz, Bamberger, & Lindsay, 2012).

As the arrangement of Tuneblocks makes some of the more imperceptible aspects of music apprehendable to the user—such as form and structure, as well as the construction of pitches and rhythms (Bamberger, 2000)—these apprehensions require users to listen critically and adjust their perceptions concerning what they are hearing. In this way, they are becoming aware of the commonalities and differences among differing cultural traditions (Bamberger, 1974). Furthermore, the task of creating music from the "building blocks" of other cultures' music is offered as a possible mechanism for restructuring thinking and adjusting perceptions in such a way that culturally diverse materials become recognized and accepted as different yet relevant to one's own culture (i.e., developing an awareness and respect for artifacts of a culture other than one's own). Technology may very well play an important part in maintaining the role of music in crosscultural understanding, as it can be a means through which to distribute music across traditional divides as well as a way to empower youth to write and share their music across cultural barriers.

How New Technologies Are Changing Music Education Today

In sum, we can see that new technologies are providing many opportunities to positively reenvision the performative and music creation aspects of music education today, including their capacity to provide (1) immediate feedback to players just learning to play a new instrument, (2) scaffolded ways of interacting with new and alternative forms of music notation (yet still deeply connected to traditional notation systems), (3) authentic contexts for performance, and (4) genres of music participation that are more connected to youth culture.

However, some notable limitations of today's music technologies are also illustrated in some recent trends. More recently, designers tried to respond to the common critiques of rhythmic video game play and offered new and more authentic peripheral guitar devices, complete with a full range of strings. As they did so, they were trying to make the video game experience simulate the "real thing." Despite these aspirations, these attempts were never as widely popular among the public. One explanation is that this drive to be more authentic in terms of traditional musicianship turned away from what was initially successful about the game—the focus on the notation, the immersive performance, and the easy entry to music performances—and replaced it with a focus on instrumental facility and the fine motor skills needed to command a traditional instrument.

Further, those new demands placed on the player/performer also run up against the limitations of the technology itself. The ways the games are designed (and even the authentic relationships to traditional staff notation) privilege particular types of music training and neglect other equally important traditions of learning about music. As noted, in the case of rhythmic video games, our expert-novice studies seemed to indicate that Rock Band resonated most with musicians who had traditional academic training in music (in our case doctoral-level musicians and composers). By contrast, musicians who had taught themselves to play "by ear" and had little to no ability to read traditional notation reported a dislike of games like Rock Band that embed a form of notation in the game play.

Furthermore, new technologies are providing easier entryways into pursuits like composing and novice music making, radically shifting the lines between performer, listener, and composer. The kinds of reflective meaning making that new technologies afford in the process can engage youth in very high-level conversations, similar to what might typically be expected at the postsecondary level. As music becomes more accessible, so does the opportunity for youth to listen to a wider array of music, becoming aware of traditions outside their cultural norms. Combined with new software for music composition, these opportunities allow for young people to remix and make existing music their own, forging new connections to other cultures in the process. All of which lays the foundations for new visions for a twenty-first-century music education that has the potential for engaging greater numbers of young learners in authentic music making and performance and for reconnecting today's youth with the rich traditions and repertoires of music that have accumulated over the years.

REFERENCES

Abril, C. (2008, March). *Culturally responsive teaching in secondary instrumental music: Mariachi as a case in point*. Paper presented at the Cultural Diversity in Music Education Nine Symposia 2008, Seattle.

Arango, T. (2008, November 25). Digital sales surpass CDs at Atlantic. *New York Times*. Retrieved July 6, 2009, from http://www.nytimes.com/2008/11/26/business/media/26music.html.

Bamberger, J. (1974). *The luxury of necessity* (Memo No. 312, Logo Memo 12). Cambridge, MA: Artificial Intelligence Laboratory, Massachusetts Institute of Technology.

Bamberger, J. (1991). *The mind behind the musical ear*. Cambridge, MA: Harvard University Press.

Bamberger, J. (2000). *Developing musical intuitions: A project-based introduction to making and understanding music*. New York: Oxford University Press.

Bamberger, J. (2003). The development of intuitive musical understanding: A natural experiment. *Psychology of Music, 31*, 7–36.

Campbell, D. (1997). *The Mozart Effect: Tapping the power of music to heal the body, strengthen the mind, and unlock the creative spirit*. New York: Morrow.

Catterall, J., Dumais, S., & Hampden-Thompson, G. (2012). *The arts and achievement in at-risk youth: Findings from four longitudinal studies* (Research Report 55). Washington, DC: National Endowment for the Arts.

Clements, Ann. (2008, April). *From the inside out: The case study of a community rock musician*. Paper and poster presented at the Music Educations National Conference Biannual Meeting, Milwaukee.

Downton, M., Peppler, K., & Bamberger, J. (2011). Talking like a composer: Negotiating shared musical compositions using Impromptu. In the Proceedings of the 9th International Conference on Computer Supported Collaborative Learning (CSCL) (pp. 92–93). July 4–8, 2011. Hong Kong.

Downton, M. P., Peppler, K. A., Portowitz, A., Bamberger, J., & Lindsay, E. (2012). Composing pieces for peace: Using Impromptu to build cross-cultural awareness. *Visions of Research in Music Education, 20*, 1–37. Retrieved January 5, 2017, from http://www-usr.rider.edu/~vrme/v20n1/index.htm.

Gee, J. P. (2003). *What video games have to teach us about learning and literacy*. New York: Palgrave Macmillan.

Graham, J. (2009, October 14). Musicians ditch studios for tech such as GiO for Macs. *USA Today*.

Green, L. (2002). *How popular musicians learn: A way ahead for music education*. Burlington, VT: Ashgate.

Green, L. (2005). The music curriculum as lived experience: Children's "natural" music-learning processes. *Music Educators Journal, 91*(4), 27–32.

Green, L. (2008). *Music, informal learning and the school: A new classroom pedagogy*. Hampshire, UK: Ashgate.

Harlow, B., Alfieri, T., Dalton, A., & Field, A. (2011). *Attracting an elusive audience: How the San Francisco Girls Chorus is breaking down stereotypes and generating interest among classical music patrons*. New York: Bob Harlow Research and Consulting.

Hirsch, B. (2005). *A place to call home: After-school programs for urban youth*. Washington, DC: American Psychological Association.

Knobel, M., Lankshear, C., & Lewis, M. (2010). AMV Remix: Do-it-yourself anime music videos. In M. Knobel & C. Lankshear (Eds.), *DIY media: Creating, sharing and learning with new technologies* (pp. 205–230). New York: Peter Lang.

Luckman, S., & Potanin, R. (2010). Machinima: Why think "games" when thinking "film"? In M. Knobel & C. Lankshear (Eds.), *DIY media: Creating, sharing, and learning with new technologies* (pp. 135–160). New York: Peter Lang.

Lum, C. (2009). Musical behaviours of primary school children in Singapore. *British Journal of Music Education, 26*, 27–42.

Miller, K. (2009). Schizophrenic performance: Guitar Hero, Rock Band, and virtual virtuosity. *Journal of the Society for American Music, 4*, 395–429.

Parsad, B., & Spiegelman, M. (2012). *Arts education in public elementary and secondary schools: 1999–2000 and 2009–10* (NCES 2012–014). Washington, DC: National Center for Education Statistics, Institute of Education Sciences, U.S. Department of Education.

Peppler, K. (2013). *New opportunities for interest-driven arts learning in a digital age*. New York: Wallace Foundation.

Peppler, K. (2014). *New creativity paradigms: Arts learning in the digital age*. New York: Peter Lang.

Peppler, K., Downton, M., Lindsay, E., & Hay, K. (2011). The nirvana effect: Tapping video games to mediate music learning and interest. *International Journal of Learning and Media, 3*(1), 41–59.

Pew Internet and American Life Project. (2008). *Teens, video games, and civics*. Retrieved May 1, 2009, from http://www.pewinternet.org.

Quillen, D. (2008). Rock Band and Guitar Hero turning gamers into musicians says Guitar Center survey. *1up News*. Retrieved March 19, 2009, from http://www.1up.com/do/newsStory?cId=3171507.

Rideout, V., Foehr, U., & Roberts, D. (2010). *Generation M2: Media in the lives of 8- to 18-year-olds*. Menlo Park, CA: Kaiser Family Foundation.

Salen, K., & Zimmerman, E. (2004). *The rules of play*. Cambridge, MA: MIT Press.

Savage, J. (2005). Working towards a theory for music technologies in the classroom: How pupils engage with and organize sounds with new technologies. *British Journal of Music Education*, 22, 167–180.

Sefton-Green, J., Thomson, P., Jones, K., & Bresler, L. (2011). *The international handbook of creative learning*. London: Routledge.

Spotlight on Music [curriculum series]. (2005). New York: MacMillan/McGraw Hill.

Théberge, P. (1997). *Any sound you can imagine: Making music/consuming technology*. London: Wesleyan University Press.

Thomas, A., & Tufano, N. (2010). Stop motion animation. In M. Knobel & C. Lankshear (Eds.), *DIY media: Creating, sharing, and learning with new technologies* (pp. 161–184). New York: Peter Lang.

Webster, P. R. (2006). Computer-based technology and music teaching and learning: 2000–2005. In L. Bresler (Ed.), *International handbook of research in arts education* (pp. 1311–1328). Dordrecht, The Netherlands: Springer.

Wiggins, J. (2009). *Teaching for musical understanding* (2nd Ed.). Rochester, MI: Center for Applied Research in Musical Understanding.

CHAPTER 19

SITUATING TECHNOLOGY WITHIN AND WITHOUT MUSIC EDUCATION

JOSEPH MICHAEL PIGNATO

THE editors have posed two overarching questions for consideration in this part of this book: "Where is technology within music education?" and "Who is affected and how?" In this chapter, I will consider each question individually in an attempt to situate technology within and without music education.

Before addressing these questions, I should offer a brief contextualization of my perspective. As a research scholar I have investigated improvisation, popular music, and alternative music education. I am active as a teacher and composer of experimental music, electronic music, and other technologically dependent multimedia works. Consequently, my views and understandings of technology in regard to music education as presented here reflect and inform those activities, interests, and identities.

In addition to contextualizing my perspective I think it necessary to clarify the definitions of two words that appear throughout the chapter, *without* and *musicians*. I have used "without" both in the archaic sense of "external to" but also in the literal sense, that is to say, technology without the paradigmatic assumptions of music education practice. When referring to "musicians," I do so in much the same way education scholars refer to "readers" or "speakers," with the assumption that doing music, in one way or another, is an innate ability in need of development through socially dynamic praxis. I do not refer to musicians as specialists, or an elite class of doers set apart from a majority spectator class.

Throughout the course of the chapter, I intend to describe and argue for the development of authentically *poietic* spaces (Nattiez, 1990) within music education. Further, I will recommend situating technology within those spaces in such ways as to empower young musicians to explore their own interests, idiosyncratic ways of learning, and the subjective-social worlds they inhabit. To do so, I have organized the chapter into three sections. The first section will offer consideration of the particularized nature of technology within music education contexts. The second section, substantially longer, will consider some examples of technology without music education. The final section will offer conclusions in an attempt to situate technology in and across those two broad domains.

Technology within Music Education

In order to consider the question "Where is technology within music education?" we might first ask closely related, constituent questions such as "What is technology in music education?" and "Where is music education?" Perhaps even "Where is music?" I do not pose such questions as a matter of semantics but instead to identify critical points of focus pertinent to considerations of technology within music education. Pondering such questions seems increasingly necessary for music education scholars, as the ways in which people experience music in their day-to-day lives and the practices of music education diverge (Kratus, 2007; Rodriguez, 2004). If we are to situate technology within music education, we must distinguish between technology in the institutionalized contexts where music education happens, technology in music learning experience, and technology in music experience—all closely related, albeit distinctly different, phenomena. To tease them out, I will start with the constituent question "What is technology within music education?"

Although technology in music education is, manifests as, and represents far too many things to detail in this chapter, I can offer one definitive answer. Technology in music education is particularized, bound to, and reflective of institutional learning environments, of music education practices, and of the prevailing value systems and political economies that have given birth to technology within music education. Technology within music education reflects a complex of circumstances, a "web" of scientific, cultural, economic, political, social, and pedagogic authorities (Nattiez, 1990). Technology within music education simultaneously represents, informs, and reproduces the contexts and circumstances from which it has emerged, the institutions that fund its existence, and the intents, interests, and agendas of those who invent it, manufacture it, distribute it, purchase it, and use it for music education.

Consequently, technology within music education serves many masters: legislative priorities, preferred methodologies and pedagogies, institutional interests, corporate interests, and a plethora of approaches to and value systems underpinning the doing of music, the place of music in general education curricula, and the purpose of and function of education. More often than not, young musicians have no role in determining those purposes, functions, or values.

None of this is to say that technology in music education does not serve students. It can, and to varying degrees has. Rather, I intend to highlight the relative vulnerability of young musicians, with little power, control, or recourse in regard to the types of, implementations of, or dispositions toward technology within their school music experiences.

Technology within music education represents what Michael Apple (2000) has deemed "official knowledge." It reflects sets of assumptions borne out of prevailing music education practices, the commercial marketplace that produces and sells music technology to schools and teachers, and the political, social, and cultural value systems that inform school music. Those underpinning points of influence, which we might view as an authoritative complex, set technology within music education apart from the subjective desires, intents, or dispositions of young musicians toward technology, toward music, or toward education in their everyday extrainstitutional lives.

That young musicians use technology, make, consume, and share music, and often engage in profound music learning without music education directs me back to my earlier questions "Where is music education?" and the closely related "Where is education?" and "Where is music?" I am not alone in asking such questions, in looking beyond institutional contexts to understand phenomena within music education. In fact, questions regarding "where" music learning happens seem to perennially percolate at the surface among music education researchers, scholars, and philosophers. That Estelle Jorgensen (1997) titled a book *In Search of Music Education* reveals at once the vexing nature of such questions and the degree of insight to be mined from their careful consideration. Jorgensen argued for an expanded conception of music education, for it to "be construed to encompass an array of societal institutions . . . [that] its conception as public or state-supported music education cannot suffice" (p. xiv).

A cursory review of works looking outside the parameters of traditional music education contexts reveals the ubiquity of and influence of technology on what some have called "informal" music learning (Green 2002; 2008). In 1980, H. Stith Bennett's prescient work *On Becoming a Rock Musician* devoted an entire chapter to technology, the roles it plays in the development of rock musicianship and in the autonomous, self-directed ways in which rock musicians often learn. Bennett (1980), writing nearly 35 years ago, described how rock musicians often engaged technologies in pursuit of highly personalized goals such as producing idiosyncratic sounds. Writing more than a decade later, Patricia Shehan Campbell's (1995) investigation of music learning in a "garage band" mentioned technology less explicitly. However, references to technological tools like "amps," "pickups," playback hardware ("sound systems"), and software ("tapes" and "CDs") peppered the narrative, indicating the degrees to which and myriad ways in which young musicians seek out and engage technology outside music education contexts.

More recently, John Finney (2007) posed the question "Where is Music Education?" in direct consideration of how young musicians learn with technology in and out of schools. In so doing, Finney delineated two discrete worlds, one associated with the institutions and practices of compulsory schooling and the other associated with the everyday musical lives of school-aged children.

Members of the College Music Society Advisory Committee for Education (2013) recently observed: "One of the most neglected questions in music education is 'Where is music?'" (n.p.). Pertinent to this discussion, the Committee framed their consideration in terms of increasing degrees of technological mediation in music listening, making, and sharing within and without institutional education: "It is easy to think of music education as happening mostly within the walls of our studios, practice rooms, rehearsal halls, and classrooms. However, once we consider all of the spaces where music making, learning and teaching occur (at home, school, online and the hybrid spaces in between) we encounter the reality that the majority of young musicians are learning music on their own, mediated by streaming music, video, and text-based forums on the Internet" (n.p.). Perhaps we might get to the crux of the editors' primary question then—Where is technology within music education?—by first considering music technology without music education: external to institutional contexts, integral within people's day-to-day musical lives, their subjective music experiences. By stepping outside, if you will, we might better understand that technology without music education has significance precisely because young musicians engage it from positions of choice, with greater power, control, autonomy, and agency than they do technology within music education.

Technology without Music Education

Technology without music education—that is to say technology accessed, learned, and used outside institutionalized contexts and without the paradigmatic constraints of traditional school music—permeates the musical lives of young musicians. Technological means for discovering, accessing, organizing, consuming, producing, learning, teaching, and sharing music have become increasingly ubiquitous, diffuse, personalized, and niche oriented in these, the early years of the twenty-first century.

Echoing philosopher Gilles Deleuze, Bridgette Jordan (2008) described ways in which people's experiences have been, to varying extents, "deterritorialized" (p. 30). The ways young musicians pursue knowledge, make social sense of their worlds, and connect with others are less bound by geography, culture, or localized identity than within music education. Deterritorialized experience has many implications, in general, but certainly for young musicians and, most pertinent to this chapter, for situating technology within and without music education. Deterritorialization is a byproduct of increasingly multilocated experiences. As described throughout this chapter, young musicians learn, share, and connect with others across great distances in locales that are both real and technologically enabled. In her article "Empowering Place" Margaret C. Rodman (1992) identified three essential characteristics of multilocated experiences: they are decentered and require "looking at places from the viewpoint of others"; they are comparative; and, they result in and are driven by reflexive relationships that transform those involved (p. 646).

My recent research interests have focused on music learning on the "outside," external to institutional contexts. As a primary investigator in the Music Learning Profiles Project (MLPP), I have sought to illuminate such locales and the musical experiences, music learning, and music making they engender. In the course of researching the MLPP (Cremata, Pignato, Powell, & Smith, n.d.), I have been repeatedly struck by the ways in which music learners rely on technology. Although the MLPP is an ongoing, ever developing project, some early themes regarding technology have emerged.

Participants in the MLPP have cited a range of software, hardware, and technologically mediated networks as integral to their music learning, often identifying specific technologies as points of inspiration for pursuing music. For example, one participant in the MLPP, Anthony "The Twilite Tone" Khan (2017), a BMI Award winner, multiple Grammy nominee, and influential figure in the worlds of hip-hop and Chicago House music, described the impact of seeing a DJ rig for the first time: "I just saw myself being that guy that was behind there, *that box*, you know, pressing those buttons or turning those knobs. I literally had visions of that" (emphasis added). For The Twilite Tone, the technology employed by DJs represented something more than its utilitarian purpose, spinning records. It represented an aspirational vision. The Twilite Tone's vision gave "that box" meaning, transforming it into something, maybe more precisely, *someone* he aspired to become.

Similarly, Amia "Tiny" Jackson (2017), a hip-hop artist, event producer, and photographer from New York City, described using streaming media and social networks to teach herself how to rap, work digital audio workstations (DAWs), software synthesizers, and beat generators. In addition to ascribing meaning to those various technologies, Tiny used technologically mediated platforms like streaming and social media to connect with other rappers and to organize *battles*. Battling, or competitive exchanges among rappers, plays an important role in the enculturation of developing rappers. (For historical background and further discussion see Chang [2005] and Katz [2012].)

For a period of time, Tiny produced online rap battles via Facebook. She coordinated the participant submissions, presided over the battling, and curated new talent for these virtual events. To find new talent as well as to learn, Tiny relied heavily on streaming media, YouTube in particular. Tiny's (2017) explanation of how she used YouTube revealed much about the ways in which many young musicians use technologically mediated social environments for learning:

> I go on YouTube a lot to figure things out. I might search something simple like how to become a better rapper and just listen to different people speak and how they rap, about how to build your flow. I do that all the time. I watch the same videos over and over and over again. . . . YouTube is my source. You can pause. You can advance. In class, you can't really pause the teacher, you can't really ask a lot of questions. You might sound like this is a stupid question but on YouTube, you can sit there; you can write notes. You can pause it. You could press play. It's up to you. (Cremata et al., 2017.)

Tiny's narrative reflects those of other musicians profiled in the MLPP who described navigating divergent technological planes in service of personalized learning objectives, in pursuit of like-minded musicians to share with and learn from, and in pursuit of new ideas, sounds, and possibilities for their own music making. A shortlist of technological tools, software, and networks referenced by participants in the MLPP includes streaming media outlets such as YouTube, Spotify, and iRadio; various social media networks, including Facebook, Instagram, and Myspace; self-publishing platforms such as SoundCloud, Bandcamp, and Reddit; microblogging platforms like Twitter and Tumblr; hardware, including turntables, digital DJ decks, digital audio interfaces, microphones, controllers, and digital production rigs like Akai's fabled Music Production Center series; software, including DAWs, video editing suites, software synthesizers, beat generators, and a plethora of iPad apps for accessing, organizing, creating, and sharing music, among many others.

These technologies offer an unprecedented panoply of possibilities for young musicians. In addition to access to information, such technologies offer access to highly particularized communities of response. Communities of response are groups of individuals that come together for a "gather round" experience, connecting around interests, content, cause, or activity. I began using the phrase in my teaching of popular music to describe the ways in which fans come together around an artist, a musical genre, or a subcultural movement. Communities of response represent a kind of precedent step to Jean Lave and Etienne Wenger's (1991) notion of "communities of practice." Participation in communities of response affords developing musicians connections with other musicians who share similar interests, have desired expertise, or represent potential collaborators or fans. In addition, such participation generates opportunities to participate in and help shape communities of practice. By using such technologies to engage multiple, distributed communities, young musicians forge new ways to learn about, produce, and share music.

Technology without music education, then, confounds traditional notions of music learning, of music learning environments, and of musicianship. Self-publishing platforms, such as SoundCloud or Bandcamp, allow musicians to share their work as well as to learn about the work of others, often working across continents. Voice over Internet Protocol (VoIP), social media, and collaborative communities such as Indaba Music, Hit Record, and SoundCloud allow young musicians to collaborate in real time, asynchronously, and, where they wish, anonymously. Such collaborations include so-called telematic performances, which rely on technologies that include digital networks like Skype, Internet2, or hardware devices like JamLink to join contributors separated by distance (Ascott, 2003).

Young musicians use such technologies in highly individualized ways. They curate their learning, seeking out particular types of music that are most meaningful for and reflective of their personal identities, desired social personae, and desired social memberships. Those who wish to learn about a particular aspect of music can purposefully search for precisely what they seek unfettered by institutional sloth, authoritative value systems, or institutional controls. Opportunities for exploratory learning based on a user's whim, interest, and personal agency abound. Further, with nearly all recorded music ever made available online, young musicians have access to recordings from every epoch and genre, as well as recordings from outside their cultural, musical, and social communities of response. New communities of response frequently form, sometimes resulting in small, largely anonymous groups sharing, collaborating, or acting in consort around niche genres, themes, techniques, or repertoires.

Other times, communities of response form around artistic or sociopolitical fronts. For example, I have coordinated a number of telematic performances, ones in which the performers, many of whom have never met, are in remote locations linked by VoIP connections. We are typically joined by a single theme, a common point of expression or concern. For example, in 2013, while planning one such collaboration with Israeli composers Kiki Keren-Huss and Lior Pinsky, hostilities between Gaza and Israel erupted. Our work shifted focus from purely artistic concerns to ruminations on the perpetuity of armed conflict. Our collaborative work exists in multiple places but also generates many digital artifacts as we share files, video, and audio assets. Those materials remain available for other musicians to use, manipulate, or otherwise repurpose.

Closely related to collaboration is the notion of recombinance, or the recombination of preexisting materials into new ordered forms (Cope, 2005). Musicians use technological tools to collect digital artifacts from recorded songs, sound effects, incidental music, theme songs, field recordings, and Foley sounds, among many other possibilities. Mixing, editing, and sound design software, samplers, DJ decks, and other production hardware allow musicians to manipulate and rework the original work of others, either as actual remixes, which are reengineered recordings of preexisting music, or through mashups—compositions based entirely on preexisting materials, often incongruously or ironically "mashed up" to create something new.

These ever expanding musical techniques, genres, distribution methods, and performance practices have generated new notions of virtuosity, facilitating the development of superstar DJs, internationally renowned breakbeat artists, master beatboxers, remix artists, mashup specialists, and countless singers, songwriters and rappers, and hybrid musicians.

Robert Ray Boyette, also known by his online name, roboyette, a play on the robot trope so pervasive in electronic music circles, provides a quintessential example of what I have described in the previous paragraphs. He is a master beatboxer, one who uses his voice, mouth, and breath to recreate percussive musical passages (Garfield, 2002). Beatboxing has long been a core element of hip-hop culture. In the past decade, however, beatboxing has, in large part due to technology, seeped far beyond the genre of its origins to influence other forms of music, such as the current a cappella singing vogue. Likewise, beatboxing has been influenced by an ever widening circle of musical practices, generating newly hybridized approaches. He has become quite celebrated for a particular hybrid approach, almost uniquely his own, that pairs beatboxing with the often callisthenic technical virtuosity of progressive rock drumming.

Boyette publishes his work, largely through YouTube, but also on sharing sites like SoundCloud and social media platforms like Facebook. In each of his video performances, roboyette plays multiple roles: producer, videographer, and performer. He even serves as a teacher for his fans, often other beatboxers, offering his followers brief explanations of the technology incorporated in the production. Those technologies include digital audio recording software for multitrack recording and playback, condenser microphones for performing live beatbox patterns above the playback, and video recording and editing software. His most popular video (2013) features a beatbox treatment of progressive rock group Dream Theater's original song "The Dance of Eternity." As a drummer for some 40 years and one active in online drumming communities, I can attest to the fact that the level of virtuosity exhibited in roboyette's work, and in this particular piece, is astounding.

The original song features the drumming of Mike Portnoy, who, according to an official bio (2013), won the "best progressive rock drummer" category in *Modern Drummer*'s Readers Poll for "a record 12 consecutive years." Boyette reproduced Portnoy's drumming by recording multiple tracks of beatboxed parts. For the final video presentation, roboyette beatboxed a lead part over the multitracks of his voice synchronized to playback of the original Dream Theater recording. The song features a dizzying 108 time signatures. He marked each time signature change on flash cards. During his performance, he raised each flash card for the camera on the downbeat of each corresponding measure with perfect accuracy. Within a few short weeks of publication, roboyette's mind-boggling treatment of the tune reached over 100,000 viewers and ended up on the Facebook pages of Mike Portnoy, Dream Theatre, and countless other progressive rock and drumming web outlets. Scores of progressive rock beatbox covers began to populate YouTube and other streaming and social media sites.

The popularity of new virtuosi like roboyette has elevated modes of music making heretofore inaccessible and, in the process, brought them to the fore of the culture. These virtuosi's work has inspired a torrent of other creative musicians, each of whom publishes sound recordings, videos, blogs, tweets, and countless other expressions of or about music.

Musicians like The Twilite Tone, Tiny Jackson, and roboyette represent but a few examples drawn from the MLPP's initial findings. They represent even fewer of the millions of young musicians who use technology without music education in poietic ways. That is to say, they use technology to give voice to highly personalized identities, idiosyncratic expressions, and socially constructed and socially negotiated musical meanings (Nattiez, 1990). Technology provides tools for "gathering the meaning of things" (Aristotle, 2014). In gathering and constructing meaning, young musicians create versions of themselves, of the worlds they live in, and of the worlds they would like to see exist.

Individual musicians use technology in ways that are initially subjective. Ultimately, however, they seem to engage technology to project the subjective in pursuit of social connection, social learning, and social meaning making; roboyette (2014) described a pivotal moment in his own learning that epitomizes the ways in which subjective, social, and technological experiences provoke young musicians:

> I really just experimented on my own for a long time, like adding accents to beats, kind of like a percussionist, you know, not a full drummer or drum set. Then, I got this recorder, like a digital recorder. At the same time I had this really good friend. We would just turn on the recorder and record. Play around. You know, record our shenanigans. I would say something or make a rhythm. He would add to it. That playing around helped me. Like that recorder and my friend helped me get to the

next level with my beatboxing, you know? From just by myself, like childhood messing around, to having an actual real rhythm with the purpose of being a rhythm, having development in the rhythm, musical development, to having purposeful musical ideas with beatboxing that I could share. (Cremata et al., 2017.)

Engaging technology in the manner described by roboyette differs from, and in some ways stands in opposition to, the authoritative complex of technology within music education and represents what we might describe as a subjective-social interaction complex. Lave and Wenger (1991) have described learning of precisely this type: "learning is not merely situated in practice—as if it were some independently reifiable process that just happened to be located somewhere; learning is an integral part of generative social practice in the lived-word" (p. 37).

Technology within and without Music Education

The multitude of possibilities engendered by technology enumerated in the preceding section should be viewed neither as a panacea nor as answers to where technology ought to reside within music education. Instead, those possibilities simply represent some manifestations of technology within music that are worthy of consideration among music educators, music education scholars, and policy-makers. It would be naïve to distinguish between inside and outside applications of technology in music learning and music experience in absolute or discrete terms. After all, many aspiring musicians who learn outside school music contexts still attend school. Peter Webster (2002) described the experiences of such young musicians: "Children entering formal education today are unaware of a world without computers, personal digital assistants, portable CD and MPG3 players, digital keyboards, and the Internet with its connection to vast amounts of information. Music is everywhere in these media, and music teachers are continually inspired to use these computer-based technologies in their work" (p. 38). David B. Williams (2011) called a subset of these aspiring musicians "nontraditional" in respect to music education, meaning that although such students do not participate in traditional school music programs, they have active musical lives that are at once dependent upon and mediated by technology. Williams explained, "more than 67 per cent of these students may play an instrument or sing, 28 per cent have an active music life outside of school, and many aspire to a career in music industry or performance" (p. 132). The experiences of such musicians within music education is constrained by the authoritative complex described earlier. This complex dictates implicitly and explicitly the purposes, accepted usages of, and representative value for technology within music education. In contrast, young musicians increasingly view technology as a means to enable or further musical identity, musical meaning, and social connections. For those musicians, the technology is secondary to and exists in service of a purpose rather than existing to dictate purpose. Their "peripheral participation" (Lave & Wenger, 1991) in music learning, their real activities as music consumers and makers without music education, often goes overlooked or gets undervalued in institutional music education contexts.

When compared to the ways in which young musicians use, rely on, and navigate technology, institutionalized implementations of technology can seem inherently limited, restrictive, and even repressive. Whereas young musicians use technology to socially construct idiosyncratic personal and social musical worlds, educational institutions often use technology to reproduce, affirm, or advance fundamental assumptions about music; for example, what music ought to be and how it is best learned. Dissonances between the authoritative complex of technology within music education and the subjective-social interaction complex of technology without music education seem to be a factor in such "peripheral participation." Such dissonances percolated throughout my discussions with participants in the MLPP. Some participants seemed particularly cognizant of the fact that their engagement with technology takes place outside of, and is unfettered by, the authoritative complex of technology within music education. One of roboyette's recollections draws a clear distinction between his experiences in school and his own musical pursuits:

> It was a little bit of the Internet, and then I had the recorder and between my brother and my friend, we just did that. It was outside of all of that [school music and school technology lab]. Maybe if I had a teacher who knew about it, I would be better, but really, maybe I wouldn't be as good at the specific things I do. It was very isolated, very individualized. (Cremata et al., 2017.)

Conclusions

Learning of all kinds is situational, contextualized, and multidynamic, shaped by the individual agency of learners but also by all that surrounds them and all with whom they interact. It would be impossible to replicate within music education the possibilities of technology without music education. One could hardly imagine the "two turn tables and a microphone," rapped about by Mantronix (1985) and later memorialized by Beck (1996), having quite the same impact in the context of institutionalized learning as they did on playgrounds and at block parties in the early days of hip-hop.

However, the social interactions of the playgrounds and block parties of early hip-hop, crucibles for purposeful and innovative usage of technology, represent a point of intersection between technology within and without music education. Technology in both contexts has meaning only inasmuch as and because of how people use it or interact with it in meaningful ways (Blumer, 1986). Two turntables and a microphone did not birth hip-hop. Thousands of New York City kids, disenfranchised by economic inequality, racism, and urban mismanagement did by interacting with their circumstances, with each other, and with the world at large to create meaning, to make things of value. The various technologies considered in this chapter ultimately serve and are subservient to human needs, interests, and interactions.

As alluded to earlier, technology within or without music education cannot be viewed through utopian eyes. As James Franklin (1983) noted over 30 years ago in a special issue of the *Music Educators Journal* on music technology, "education has frequently been guilty of a jumping-on-the-bandwagon, everyone-is-doing-it-attitude concerning educational trends" (p. 30). Fetishizing technology over music practice and experience, that is to say, elevating technology as an entity over the idiosyncratic musical uses of technology by young musicians, remains a real and pernicious tendency. The thoughts from participants in the MLPP shared throughout this chapter indicate the power of and value in honoring the ways in which young musicians use technology to navigate their worlds. Their expressions reflect values, meanings, and understandings of musical doing that differ from traditional modes of teaching and learning. Technology within music education often reflects the values of those more traditional modes: the values of institutions, of market interests, and of teacher experience and expertise. Christopher Small (1996) remarked that musicians have "become afraid of the encounter with new musical experience, where knowledge and expertise are no guide and only the subjective experience honestly felt can serve, and retreat in the safe past, where we know what to expect and connoisseurship is paramount" (p. 6).

Discrepancies between technology within music education—that is to say the known, the expected—and the experiences of young musicians engaged with technology outside, or without the knowledge or expectations, of music education will continue to generate an underlying level of discord, frustration for music educators, and alienation for music students. Those discrepancies and the corresponding dissonances they cause can be mitigated by situating technology within music education around identity expression, play, idiosyncratic purposeful creation, and socially dynamic sharing of the kinds described in this chapter. Working with students to establish practices reflective of their own musical lives, their own individualized explorations, and their own music learning tendencies might shift peripheral participants to the center of vibrant, democratically negotiated communities of response and communities of practice.

Efforts to create authentically poietic spaces within music education and to situate technology within those spaces face many challenges. First and foremost, young musicians will always be vulnerable within music education settings. Empowering young musicians requires recognizing that vulnerability by placing their interests, ways of learning, and the subjective-social complex they inhabit at the center of technology within music education. To paraphrase Christopher Small (1996), situating technology within music education requires enabling young musicians "to live in the world" (p. 4), to make sense of themselves, where they fit in, and where they might go. Young musicians use technology in their everyday lives to do just these things.

We might situate technology within music education by emphasizing multilocated spheres of experience and, as Rodman (1992) put it, "empowering place." Situating technology within music education for such a purpose would provide a foil to the vulnerability of young musicians described earlier. In such an environment, teachers would function like what Etienne Wenger, John Smith, and Nancy White (2009) have called "technology stewards," individuals who curate technology within particular communities in response to the needs of the communities and in response to emerging practices, social relationships, and individual needs of students. At the same time, teachers must empower students to share that stewardship in ways reflective of the subjective-social complex within which young musicians learn and as a counterbalance or a mitigating response to the authoritative complex that defines technology within music education.

References

Apple, M. W. (2000). *Official knowledge: Democratic education in a conservative age.* New York: Routledge.
Aristotle. (2014). The poetics. In *Aristotle Collection* (J. C. Douglas, Ed., and I. Bywater, Trans.), (location 3598). Amazon Digital Editions.
Ascott, R. (2003). *Telematic embrace: Visionary theories of art, technology, and consciousness* (E. A. Shanken, Ed.). Berkeley: University of California Press.
Beck. (1995). Where it's at. On *Odelay*, DGC. Los Angeles, CA.
Bennett, H. S. (1980). *On becoming a rock musician.* Amherst: University of Massachusetts Press.
Blumer, H. (1986). *Symbolic interactionism.* Berkeley: University of California Press.
Campbell, P. S. (1995). Of garage bands and song-getting: The musical development of young rock musicians. *Research Studies in Music Education*, 4(1), 12–20.
Chang, J. (2005). *Can't stop, won't stop: A history of the hip hop generation.* New York: MacMillan.
College Music Society Advisory Committee on Music Education. (2013). Thinking beyond curriculum and pedagogy. College Music Symposium, 53, n.p. Retrieved from http://symposium.music.org/index.php?option=com_k2&view=item&id=9699:thinking-beyond-curriculum-and-pedagogy&Itemid=126.
Cope, D. (2005). *Computer models of musical creativity.* Cambridge, MA: MIT Press.
Cremata, R., Pignato, J., Powell, B., & Smith, G. D. (2017.). *The Music Learning Profiles Project: Let's Take this Outside.* New York: Routledge.
Finney, J. (2007). Music education as identity project in a world of electronic desires. In J. Finney & P. Burnard (Eds.), *Music education with digital technology* (pp. 9–20). New York: Continuum International.
Franklin, J. L. (1983). What's a computer doing in my music room? Music Educators Journal, 93(5), 31.
Garfield, J. (2002). *Breath control: The history of the human beat box.* Documentary film. Brooklyn, NY: Ghost Robot Productions.
Green, L. (2008). *Music, informal learning, and the school: A new classroom pedagogy.* Burlington, VT: Ashgate.
Green, L. (2002). *How popular musicians learn: A way ahead for music education.* Burlington, VT: Ashgate.
Jackson, A. (2017). In R. Cremata, J. Pignato, B. Powell, & G. D. Smith, *The Music Learning Profiles Project.* Unpublished manuscript.

Jordan, B. (2008). Living a distributed life: Multilocality and working at a distance. *NAPA Bulletin, 30*, 28–55.

Jorgensen, E. (1997). *In search of music education.* Chicago: University of Illinois Press.

Katz, M. (2012). *Groove music: The art and culture of the hip hop DJ.* New York: Oxford University Press.

Khan, A. (2017). In Cremata, R., Pignato, J., Powell, B., & Smith, G. D., *The Music Learning Profiles Project.*

Kratus, J. (2007). Music education at the tipping point. *Music Educators Journal, 94*(3), 42–48.

Lave, J., & Wenger, E. (1991). *Situated learning: Legitimate peripheral participation.* New York: Cambridge University Press.

Mantronix. (1985). "Needle to the Groove." On *The Album*, Sleeping Bag Records.

Nattiez, J. (1990). *Music and discourse: Toward a semiology of music* (C. Abbate, Trans.). Princeton, NJ: Princeton University Press, 1990.

Portnoy, Mike. (n.d.). mikeportnoy.com. Website. Retrieved January 5, 2017, from http://www.mikeportnoy.com/default.aspx.

roboyette. (2017). *The Dance of Eternity Beatbox Cover* (YouTube video). Retrieved January 5, 2017, from http://www.youtube.com/watch?v=EkFNdi8gIMY.

Rodman, M. C. (1992). Empowering place: Multilocality and multivocality. *American Anthropologist, New Series, 94*(3), 640–656.

Rodriguez, C. X. (Ed.). (2004). *Bridging the gap: Popular music and music education.* Reston, VA: MENC.

Small, C. (1996). *Music, society, education.* Hanover, NH: Wesleyan University Press.

Webster, P. (2002). Historical perspectives on technology and music. *Music Educators Journal, 89*(1), 38–43, 54.

Wenger, E., Smith, J. D., & White, N. (2009). *Digital habitats: Stewarding technology for communities.* Portland, OR: CPSquare.

Williams, D. B. (2011). The non-traditional music student in secondary schools of the United States: Engaging non-participant students in creative music activities through technology. *Journal of Music, Technology and Education, 4*(2-3), 131–147.

2 A

Further Perspectives

CHAPTER 20

HUMAN POTENTIAL, TECHNOLOGY, AND MUSIC EDUCATION

SMARAGDA CHRYSOSTOMOU

THE three authors who deploy their views as Core Perspectives in this part of this book come from different countries and focus on different aspects and roles of the triptych of music, technology and education. However, certain key points emerge as common. Situating technology outside formal music education settings is the most prominent. Of course, Africa emerges as something unique, because, as Kigozi claims, "unlike in the West, most of the technology within music education in Africa takes place outside the formal music classroom context."

Nevertheless, the common theme relates to the boundaries and differences that exist between formal and informal, between school life and personal life, between school music learning and making and individual music learning and making. In the "era of hyperconnectivity" (Leong, 2011, p. 236), though, work and leisure are interlaced throughout the hours of the day. Students are making their own choices regarding their learning. This freedom to choose the source of information and knowledge they need, to select the context of learning, the setting and the location, is characteristic of the situations that the authors describe in their Perspectives.

Because technology within music education represents and serves the institutions and other authorities (political, economic, social, cultural, etc.) that have funded it, often young people choose to interact with music outside those formal settings, as Pignato affirms. They collaborate online with peers from around the world and participate in communities with similar interests, facilitated by digital technology's social networking tools. Moreover, new technology in music offers unprecedented possibilities, and young people use the available technology in highly individualized ways, learning while they are listening and making music, creating new genres, new hybrid forms, and expanding traditional techniques, "generating new notions of virtuosity" and musicality, as both Pignato and Peppler point out.

It is true (and has been for many years, regardless of the advancements of technology) that young people enjoy music much more outside than inside school. Music is a constant presence in all youth cultures (and microcultures) and an important factor in identity formation and self-perception. Nevertheless, technology developments that offer freedom in communication, connectivity 24/7, instant access to sounds, musics, and videos have as a result augmented music's presence and role in young people's lives.

As Green (2008) has observed in her research, young musicians in their informal interactions and settings where they make music, exhibit all those qualities that we set as goals in our formal music educational settings. They cooperate effectively, show commitment and responsibility to the task at hand, and show sensitivity to others. Above all, they enjoy themselves, and this is what sustains their interest and assists their perseverance (Green, 2008). Interest and enjoyment is what we aspire to in formal settings as well. These are the prerequisites for a "flow experience" (Csíkszentmihályi, 1990; Whitson & Consoli, 2009). As the boundaries between many areas of our lives have been dissolved, it may be the right time to bring formal and informal music settings closer, so that each can benefit from the other and at the same time take music education to another level, one that would be impossible without digital technology.

An additional aspect that emerges as a common concern is the role of technology in easing or hindering music and music education. Technology and, as follows, music technology are means to enable users, musicians, music educators. Young people take advantage of technology to further their musical identities and create new musical meaning; thus, technology should be a means to an end and exist "in service of a purpose," as Pignato asserts. This discussion is enhanced by Kigozi's worries that technology should be applied "from an artistic point of view," which, instead of hindering the process, will enable users to reach their final goal with less effort. From similar discussions we can discern two complementary areas of concern:

1. Is technology created to assist and enhance the human brain?
2. Can technology be applied from an artistic and musical intelligence perspective to assist and enhance such processes?

The use of technology as a tool instead of an end in itself has been a major issue for decades. Every time a new leap and technological advancement is achieved, the limelight tends to fall onto the tool as much as the end result. Particularly in education, teachers have expressed doubts about technology's effectiveness, remaining ambivalent in their intentions to adopt what appears as new technology every couple of years (Gall, 2013; Jimoyiannis & Komis, 2007; Kiridis, Drossos, & Tsakiridou, 2006). Not all teachers' concerns can be categorized though as fear of the new, uncertainty about the unfamiliar, and reluctance to change. Many writers have expressed their annoyance at the difficulties created by technology at the same time it claims to ease them. Software and hardware that are incompatible and in need of constant updates hinder teachers' efforts, causing more time lost instead of enhancing the efficiency of time spent (Greher, 2011). It appears that the "human user is often given less consideration than the technology that is supposed to serve him or her" (Leong, 2011, p. 238). Learning should be driving the technology, which should be "brain friendly," complementing the cognitive process (Leong, 2011). In much the same way, technology should enable and enhance music learning and music making by "serving and being subservient" to a young musician's learning needs, interests, and way of thinking, as Pignato affirms.

I would like to turn now to an area of formal education touched upon in the Core Perspectives that was revolutionized by the use of technology. With the recent advent of wireless broadband and Internet access, and the development of Web 2.0 tools, new blended learning models have emerged for twenty-first-century tertiary education. We have witnessed the materialization of what can be described as a matrix of synchronous/asynchronous and virtual/physical learning environments, with an "increased focus on the 'third space' that supports social forms of student interaction" (Fisher, 2010, p. 3).

Long-distance courses and e-learning have become part of traditional mainstream higher education courses. The majority of tertiary education institutions are fully equipped for online learning and teaching, and gradually more university teachers are offering lessons that are attuned to contemporary active learning environments, presenting information and assessing learning using multiple mediums and modes. However, new technology and tools integrate learning even deeper into everyday life. The walls and boundaries of classrooms, buildings, institutions, and countries are dissolved. Formal and informal settings are getting harder to differentiate as young people's personal and student lives are intertwined. Control of learning belongs to the student, whose choices are (and will be even more in the future) global, multidimensional, multidisciplinary and multimodal. Consequently, students are free to choose the institution that will offer them their education regardless of their location on the world map, and acquire certificates and degrees and experience programs of study that resemble very little of what I experienced as a student in Greece and a postgraduate student in the United Kingdom a few years ago. With all these possibilities and prospects, might this be a sincere opportunity for accomplishing education for all? Could this be a basis for a true democratic turn in education that has as a prerequisite a paideia that is accessed by all, regardless of social and economic status, cultural and ethnic background, or geographical location? What are the dangers and threats for this prospect?

Many of the contributors to this book acknowledge the fact that technology has lowered the barriers obstructing music learning and music making. Similarly, it has lowered the barriers obstructing participation in tertiary education. Web 2.0 tools (as well as the advancing Web 3.0) and the development of social media and the "third space" have enabled tertiary education institutions to reach a student base that would otherwise have difficulty in attending a university campus or would have to overcome severe barriers to access to education. Massive open online courses (MOOCs), which have expanded during recent years, are such an example. Hundreds of thousands of students from around the world have the opportunity to enroll and complete courses taught by top professors in top universities without leaving their own rooms.

No doubt this newest trend has numerous shortcomings as well as negative repercussions, some of which have already been mentioned in discussions and forums

(Friedman, 2013). Recent research has shown that only a very small percentage of the enrolled students ever complete MOOCs (an average of 4%); moreover, in contrast to the advocacy rhetoric of democratization, their most common users belong to the economic elite (80% of students come from the richest 6% of the world population) (Rengel & Fach, 2013).

The first threat, then, relates to low participation of students and loss of motivation. This could be mitigated by utilizing contemporary tools and applications that are closer to students' way of thinking and communicating in order to incentivize them. For example, research on the use of Twitter and other such tools for educational purposes showed elevated participation and motivation of students (Rengel & Fach, 2013). At the same time, we need to make sure that, even though the barriers to accessing higher education courses have been lowered, the quality of studies offered remains high. This is dependent mainly upon the teacher, the university professor, and that person's professionalism, qualities, and abilities. The majority of research and its interpretation points to the importance of the teacher's role and impact on any chance for quality education, regardless of the level of educational studies or the medium through which teaching and learning take place.

But, to go back to the study on MOOCs and their potential for a democratization in education and more particularly the results regarding participation by different population groups, evidence showed that the majority of MOOC participants had similar characteristics to the population around the world that benefits from traditional education: they are young, are male, and belong to the economic elite (Christensen & Alcorn, 2014; Fowler, 2013; Rengel & Fach, 2013). Surely this indicates that open access to knowledge and university studies is not enough if our goal is a true democratic world. Rather, the fact that technology has made all this available around the world does not really concern the world, as only those who already have a certain level of education can afford the necessary equipment and meet the expense of an Internet connection. Kigozi's description of African areas with no electricity, let alone Internet connection, sheds some light on the huge divide that exists between communities, students, and teachers around the world and even within the same country—a divide that is not only digital and technological but economic. In order for true revolution in education and true democratization to become a reality, broader changes are needed.

Technology has changed the definition of education, music, and music education, as is affirmed by many writers in the field and in this book, and it is time to be rigorous and adapt to new circumstances. But at the same time we need to discuss the ways we can use human potential and "tap into the well of its creativity" to enhance learning and, more particularly, music learning and music itself for humanity's sake. Discussing a tragic notion connected to human existence, author and philosopher Emmanouel Kriaras, a prominent Greek professor, recently deceased, commented on contemporary civilization: "with the predominance of technological civilization, humanistic education is forgotten. Today we all learn how to use the technology but can't elevate our souls. This is also tragic" (*To Vima*, November 1, 2009). We have to negotiate our tragic fate and find a way to use all our developments, inventions, and advancements in elevating our souls.

References

Christensen, G., & Alcorn, B. (2014, March 16). The revolution is not being MOOC-ized. *Slate*. http://www.slate.com/articles/health_and_science/new_scientist/2014/03/mooc_survey_students_of_free_online_courses_are_educated_employed_and_male.html.

Csíkszentmihályi, M. (1990). *Flow: The psychology of optimal experience*. New York: Harper and Row.

Fisher, K. (2010). *Technology-enabled active learning environments: An appraisal* (CELE exchange 2010/7). Retrieved August 9, 2014, from www.oecd.org/publishing/corrigenda.

Fowler, G. (2013, November 20). Survey: Mooc students are elite, young and male. http://blogs.wsj.com/digits/2013/11/20/survey-mooc-students-are-elite-young-and-male-2/.

Friedman, T. (2013, January 27). Revolution Hits the Universities. Retrieved January 5, 2017, from http://www.nytimes.com/2013/01/27/opinion/sunday/friedman-revolution-hits-the-universities.html?pagewanted=2&_r=0.

Gall, M. (2013). Trainee teachers' perceptions. Factors that constrain the use of music technology in teaching placements. *Journal of Music, Technology & Education, 6*(1), 5–27.

Green, L. (2008). *Music, informal learning and the school: A new classroom pedagogy*. Guildford, UK: Ashgate.

Greher, G. (2011). Music technology partnerships: A context for music teacher preparation. *Arts Education Policy Review, 112*, 130–136.

Jimoyiannis, A., & Komis, V. (2007). Examining teachers' beliefs about ICT in education: Implications of a teacher preparation program. *Teacher Development, 11*(2), 149–173.

Kiridis, A., Drossos, V., & Tsakiridou, H. (2006). Teachers facing information and communication technology: The case of Greece. *Journal of Technology and Teacher Education, 14*(1), 75–96.

Leong, S. (2011). Navigating the emerging futures in music education. *Journal of Music, Technology and Education, 4*(2-3), 233–243.

Rengel, A., & Fach Gomez, K. (2013). *The transformative potential of technology in higher education*. Retrieved September 2, 2014, from http://www.academia.edu/7888496/The_Transformative_Potential_of_Technology_in_Higher_Education_The_Shortcomings_of_MOOCs_the_Benefits_of_Face-to-Face_Learning_and_the_Hybrid_Model_as_a_Possible_Optimal_Solution_a_2013_Spanish_case_study_Rengel-Fach.

Whitson, C., & Consoli, J. (2009). Flow theory and student engagement. *Journal of Cross-disciplinary Perspectives in Education, 2*(1), 40–49.

CHAPTER 21

"PLACING" TECHNOLOGY WITHIN MUSIC EDUCATION COMMUNITIES

AILBHE KENNY

THE sociocultural focus of this part of this book on technology within music education immediately calls for a focus on context and community. The three Perspectives provided by Peppler, Kigozi, and Pignato from particular contexts reinforce the centrality of community in their studies, thereby reminding us of the need to attend to both musical and social interactions within music education. Technology within music education requires this same sensitivity, which provides the focus of the discussion here.

CONTEXT

It is all too easy to think of technology within music education as devoid of "place." The Core Perspectives presented by Peppler, Kigozi, and Pignato all call attention to the importance of context in various ways. First, all three authors outline site-specific features of the "musical worlds" (Finnegan, 2007) they are describing. This serves to ground the studies in the "real world" (Robson, 2002) but also ensures that contextual particularities are considered in both the analysis and the reading of findings. Both Peppler and Pignato describe research carried out in urban areas within the Unites States, while Kigozi writes from a rural African context. Immediately, a comparison is thus set up to consider technology within music education in countries frequently termed "developing" and "developed." Issues of economic resources become all-important where access to technologies that are taken for granted in the West is brought into sharp focus against the restricted access to technologies of those in Kigozi's study. As music educators, learners, researchers, and/or policy-makers, we are prompted to challenge our assumptions here about what technology within music education actually means to different contexts, places, and peoples.

Taking online spaces of musical participation as an example of the ever-expanding contexts of music education, researchers consistently point to the interrelatedness of music and "place," despite the absence of a physical, geographical environment (Karlsen, 2010; Kibby, 2000; Partti & Karlsen, 2010; Salavuo, 2006; Waldron, 2011, 2013; Waldron & Veblen, 2008). In a research study I carried out of an Irish traditional music teaching and learning web platform, it emerged that the online space projected an identity that was distinctly Irish, rooted in both musical and cultural traditions. This identity functioned as a means to create a sense of belonging among the web platform's members and in doing so fashioned its own sense of genre "authenticity" in this new online context (Kenny, 2013a, 2013b).

The music technology spaces described by Peppler and Pignato similarly project a certain social, cultural, and geographical context—in these cases U.S. youth culture. Within Peppler's study, video games and a Rock Band Club highlight the importance of self-chosen youth learning experiences. The experiences, which "were wrapped within the guise of game play and group activity," connected these youth to their preferred musical cultures (and genres), as well as preferred ways of interacting. These choices were reflective of the youths' multiple and overlapping communities as well as being "of their time." Pignato further emphasizes the importance of context for youth music by highlighting the need for control over music-making contexts, claiming that technology "has meaning only inasmuch as and because of how people use it." Pignato relates the many challenges formal educational institutions face in light of its authoritative structures. Is the often-claimed disconnect between formal and informal music education (Allsup, 2003; Burnard, Dillon, Rusinek, & Saether, 2008; Green, 2002; Jorgenson, 2003; Pitts, 2007; Wright & Kanellopoulos, 2010) then following suit in its use of technology?

Kigozi's account of the importance of local recording studios in Uganda to music technology developments reminded me of similar observations I made in Zambia in recent years. As part of the Zambia-Ireland Teacher Education Partnership program, I was afforded the opportunity to work with colleagues in teacher education there. Technology within formal music education courses was limited. However, within an informal context, the central importance of local radio stations and recording studios became very apparent. Local artists and festivals were promoted, with regular live broadcasts, recordings, and interviews. These were vibrant places of music making, capturing all manner of genres and musical activities with which people engaged in particular localities. Here, technology was clearly facilitating and expanding a sense of community and place, but as in Kigozi's study, this was at a remove from formal music education settings.

All of these three Core Perspective authors point to a disjuncture between formal and informal music education settings in the use of technology. They continually underline the power of informal youth culture and social group activity as directing technological musical participation. Peppler refers to the empowerment of new technologies, such as rhythmic video games, to provide avenues for "interest-driven participation in music," while Pignato writes: "whereas young musicians use technology to socially construct idiosyncratic personal and social worlds, educational institutions often use technology to reproduce, affirm, or advance fundamental assumptions about music." Pignato further describes technology engaged with outside institutional contexts as "technology without music education," thus probing our understandings of what music education is and is not. Perhaps a reclaiming and broadening of the term "music education" needs to occur within our discourses to encompass all manners of music teaching and learning, whether formal, informal or nonformal, online or offline?

The studies and views of informal musical learning expressed by these three authors build on seminal work within broader music education research (Green, 2002, 2008) to apply that work to technology in music education. Inherent in studying any one particular approach to technology within music education (such as informal music learning) highlights the need for caveats. Researchers have warned about a perceived dichotomy between formal and informal music learning (Espeland, 2010; Finney & Philpott, 2010; Folkestad, 2006; Veblen, 2012), claiming that both forms of learning often occur simultaneously within musical practices. The same is true of technology within music education, where the lines between formal and informal musical learning are often blurred (Burnard, 2007; Tobias, 2014; Webster, 2011). Rather than being in tension with each other, then, technology within music education can perhaps offer us innovative ways to act as a bridge between formal and informal musical learning.

Community

A growing body of research into "online music communities" (Salavuo, 2006) and "new forms of social media musicianship" (Ruthmann & Hebert, 2012), as exemplified in this handbook, reflects a significant growth area within music education that is very much rooted to the idea of teaching and learning within communities. Ian Cross has written critically of the ways in which music has been utilized in technology, claiming that it has largely retained "Western-folk conceptions of music," and calls for further exploration of music as "communicative interaction" within the digital domain (Cross, 2013, p. 415).

These three Core Perspectives all reflect a strong connection between musical learning and community participation. In order to participate, however, there needs to be a "way in" to these online music communities. Peppler points to the role of technology here as "lowering the barriers" to music making, particularly for those who have not had access to formal music instruction. It is true that on the face of it, online music communities are leveling the playing field for musical participation; however, there remains a need to be cautious of claiming a "victory narrative" in this regard. Kigozi reminds us that access to the Internet remains difficult for some communities, while Pignato refers to technology within formal settings as perpetuating an "authoritative complex." How can we as music educators and policy-makers further develop opportunities for sociomusical experiences within a technological frame? Should this be left entirely to informal settings to build in an organic manner or is there a role for more formal approaches here?

Research that has applied the concept of "communities of practice" to examine online musical communities (Kenny, 2016, 2013a, 2013b; Kibby, 2000; Partti & Karlsen, 2010; Salavuo, 2006; Waldron, 2009) highlights the interrelated nature of social and musical participation. "Communities of practice" (CoP), as coined by Lave and Wenger (Lave & Wenger, 1991; Wenger, 1998), represent groups of people who create a sociocultural learning framework around a mutual shared interest or common purpose. Collaborative learning, negotiated goals, social interaction, and shared repertoire help to define the dimensions of "mutual engagement," "shared repertoire," and "joint enterprise" (Wenger, 1998) that constitute a CoP. As with many other disciplines, the CoP model has been extended to understand learning and participation within technology and music education. Pignato promotes the use of technology within (and without) music education as a potential means to build "communities of response" where "groups of individuals that come together for a "gather round" experience, connecting around interests, content, cause, or activity." In doing so, Pignato highlights an opportunity for a more democratic form of musical learning within the area of technology and music education where teachers might act as curators of CoP or "technology stewards" (Wenger, White, & Smith, 2009). I have no doubt that many teachers already espouse pedagogical approaches that encourage collaboration, peer learning, varying levels of participation, and interest-driven content, in spite of the assessment-driven bureaucratic systems in which they may find themselves. While hearing the student voice is of the utmost importance, it is worth reminding ourselves that teacher voices also need to be heard within technology and music education discourses in order to ensure that challenges are met and understood in a holistic way.

One of the major feats within technology and music education is its ability to create global communities. Peppler points to music distribution, writing, and sharing of music as some of the ways technology can promote "crosscultural understanding" and forge "new connections to other cultures." Kigozi adds further examples to this function by acknowledging the role technology within music education has played in opening up music performances worldwide but also in opening up the academic pursuit of music and arts education. As well as broadening access, the formation of these new global communities often requires a reinterpretation of how music is made, received, taught, and learned. Pignato claims that online communities often go through a process of "recombinance, or the recombination of preexisting materials into new ordered forms." In this way, although the music may remain genre-specific, the online community negotiates and fashions distinct practices particular to their new web space.

Pignato refers to "communities of response" online where niche genre groups, although often anonymous, collaborate and share musical themes, repertoire, and techniques. This aspect of community brings the notion of musical identity into focus. Studies have repeatedly shown that regular dialogical and musical interactions online, coupled with a mutual shared musical interest where participation is paramount, foster a sense of music-related identities online that are both individual and collective (Kenny, 2016, 2013a, 2013b; Kibby, 2000; Partti & Karlsen, 2010; Salavuo, 2006; Waldron, 2013). Musical identities offline can be perceived as projecting "in-group" and "out-group status" (Turino, 2008, p. 106). A strong sense of belonging can often form within the "in-group" with long-lasting effects for musical participation (Parker, 2010; Pitts, 2005; Power, 2010). How established and emerging online musical communities self-fashion and project musical identities to promote participation will be of great interest to the music education field.

Conclusion

Technology within music education represents an ever-expanding form of teaching and learning. New cultures and pathways to engage with music have been opened up in a way that has the potential to have a significant impact both locally and globally. The real business of music education will be to continue looking at ways to harness these technologies to provide meaningful routes into musical participation and learning experiences. What is most interesting about the continuing technological developments in this field is that the quality of the musical experiences remains heavily dependent on a strong connection between community and context. Wenger, White, and Smith claim that where groups "experience this technology as a 'place' for community" (2009, p. 38), there is the potential to form a community of practice in a digital space. As music educators and policy-makers, we would be leaving too much to chance to assume that simply having access to technology is enough in itself for music education to occur. There are exciting opportunities, therefore, to explore how we might develop further ways to engage learners in musical experiences that espouse collaborative knowledge building and learning through participation. Fostering "communities of musical practice" where a sense of belonging is built up through musical and social practices (Kenny, 2016) seems like a good place to start.

Musical communities and contexts that involve technological mediation and practices require continued research to further problematize this contemporary form of music education. This book is a significant addition to this developing discourse and represents an important point of departure in placing technology firmly on the music education map. If, as Peppler claims, "new technologies are changing the traditional relationships between youth, music making, and performance," there is an onus on us in varying positions as music educators, learners, leaders, and policy-makers to seek to understand this phenomenon in more depth. Through the voices of experience on the ground (and in cyberspace!), evidenced though rigorous research, this relatively new form of music education has the potential to be informed and shaped by both teachers and learners. Thus, future directions for both research and practice in technology and music education can be guided from the grassroots.

References

Allsup, R. E. (2003). Mutual learning and democratic action in instrumental music education. *Journal of Research in Music Education, 51*(24), 24–37.

Burnard, P. (2007). Reframing creativity and technology: Promoting pedagogic change in music education. *Journal of Music, Technology & Education, 1*(1), 37–55.

Burnard, P., Dillon, S., Rusinek, G., & Saether, E. (2008). Inclusive pedagogies in music education: A comparative study of music teachers' perspectives from four countries. *International Journal of Music Education, 26*(2), 109–126.

Cross, I. (2013). "Does not compute"? Music as real-time communicative interaction. *AI & Society 28*, 415–430.

Espeland, M. (2010). Dichotomies in music education—real or unreal? *Music Education Research, 12*(2), 129–139.

Finnegan, R. (2007). *The hidden musicians: Music-making in an English town* (2nd ed.). Middletown, CT: Wesleyan University Press.

Finney, J., & Philpott, C. (2010). Informal learning and meta-pedagogy in initial teacher education in England. *British Journal of Music Education, 27*(1), 7–19.

Folkestad, G. (2006). Formal and informal learning situations or practices vs formal and informal ways of learning. *British Journal of Music Education, 23*(2), 135–145.

Green, L. (2002). *How popular musicians learn*. Hants, UK: Ashgate.

Green, L. (2008). *Music, informal learning and the school: A new classroom pedagogy*. Aldershot, UK: Ashgate.

Jorgenson, E. (2003) *Transforming music education*. Bloomington, IN: Indiana University Press.

Karlsen, S. (2010). Boomtown music education and the need for authenticity—informal learning put into practice in Swedish post-compulsory music education. *British Journal of Music Education, 27*(1), 35–46.

Kenny, A. (2013a). Between the jigs and the reels (in cyberspace): Investigating an Irish traditional music online community. In M. Waligórska (Ed.), *Music, longing and belonging: Articulations of the self and the other in the musical realm* (pp. 94–113). Cambridge: Cambridge Scholars.

Kenny, A. (2013b). "The next level": Investigating teaching and learning within an Irish traditional music online community. *Research Studies in Music Education, 35*(2), 234–248.

Kenny, A. (2016) *Communities of musical practice*. London: Routledge.

Kibby, M. D. (2000). Home on the page: A virtual place of music community. *Popular Music, 19*(1), 91–100.

Lave, J., & Wenger, E. (1991). *Situated learning: Legitimate peripheral participation*. Cambridge: Cambridge University Press.

Parker, E. C. (2010). Exploring student experiences of belonging within an urban high school choral ensemble: An action research study. *Music Education Research, 12*(4), 339–352.

Partti, H., & Karlsen, S. (2010). Reconceptualising musical learning: New media, identity and community in music education. *Music Education Research, 12*(4), 369–382.

Pitts, S. E. (2005). *Valuing musical participation*. Aldershot, UK: Ashgate.

Pitts, S. E. (2007). Music beyond school: Learning through participation. In L. Bresler (Ed.), *International handbook of research in arts education* (pp. 759–772). New York: Springer.

Power, A. (2010). Learning through participatory singing performance. *UNESCO Observatory, 2*(1).

Robson, C. (2002). *Real world research: A resource for social scientists and practitioner-researchers* (2nd ed.). Oxford: Blackwell.

Ruthmann, S. A., & Hebert, D. G. (2012). Music learning and new media in virtual and online environments. In G. McPherson & G. Welch (Eds.), *Oxford handbook of music education* (Vol. 2, pp. 567–583). Oxford: Oxford University Press.

Salavuo, M. (2006). Open and informal online communities as forums of collaborative musical activities and learning. *British Journal of Music Education, 23*(3), 253–271.

Tobias, E. S. (2015) Crossfading music education: Connections between secondary students' in- and out-of-school music experience. *International Journal of Music Education, 33*(1), 18–35.

Turino, T. (2008). *Music as social life: The politics of participation.* Chicago: University of Chicago Press.

Veblen, K. K. (2012). Adult music learning in formal, nonformal, and informal contexts. In G. E. McPherson & G. F. Welch (Eds.), *The Oxford handbook of music education* (Vol. 2, pp. 243–256). New York: Oxford University Press.

Waldron, J. (2009). Exploring a virtual music "community of practice": Informal music learning on the internet. *Journal of Music, Technology and Education, 2*(2-3), 97–112.

Waldron, J. (2011). Conceptual frameworks, theoretical models and the role of YouTube: Investigating informal music learning and teaching in online music community. *Journal of Music, Technology and Education, 4*(2-3), 189–200.

Waldron, J. (2013). YouTube, fanvids, forums, vlogs and blogs: Informal music learning in a convergent on- and offline music community. *International Journal of Music Education, 31*(1), 91–105.

Waldron, J., & Veblen, K. (2008). The medium is the message: Cyberspace, community, and music learning in the Irish traditional music virtual community. *Journal of Music, Technology and Education, 1*(2), 99–111.

Webster, P. R. (2011). Key research in music technology and music teaching and learning. *Journal of Music, Technology and Education, 4*, 115–130.

Wenger, E. (1998). *Communities of practice: Learning, meaning, and identity.* Cambridge: Cambridge University Press.

Wenger, E., White, N., & Smith, J. D. (2009). *Digital habitats: Stewarding technology for communities.* Portland, OR: CPsquare.

Wright, R., & Kanellopoulos, P. (2010). Informal music learning, improvisation and teacher education. *British Journal of Music Education, 27*(1), 71–87.

CHAPTER 22

THE PROMISE AND PITFALLS OF THE DIGITAL STUDIO

ETHAN HEIN

The digital studio and the Internet have the potential to unlock musical creativity for a much wider array of young people than the ones who are proficient instrumentalists or singers. While technology has turned the music industry upside down, to say nothing of the lives of individual musicians, I still believe that the impact of technology on music as an art form is an overwhelming net positive. I echo Kigozi, Pignato, and Peppler when they point to the benefits that the digital studio brings to musicians generally and to music students in particular. Kigozi and Howell also bring up the perils and pitfalls of the digital studio. I will discuss those perils and pitfalls, and how those might be avoided, or at least mitigated.

I agree with Howell that the impact of technology in music education (or in any field) will depend on social and economic factors. When a battery-powered radio is the most sophisticated piece of music technology in a rural African school, any discussion of the creative impact of the digital studio becomes irrelevant. Howell and Kigozi remind us, too, that the definition of "technology" is far broader than computers or other electronics. Instruments, books, and chalkboards are technologies as well. We in the United States take these tools for granted, and it makes us forget how heavily they have impacted our teaching practice already.

THE POSITIVE IMPACTS OF MUSIC TECHNOLOGY

Music without Music Education

The major positive impact of digital technology for education is the explosive growth of accessible entry points into meaningful musical participation. Pignato describes how most young people in highly industrialized countries use computers and smartphones to listen to and share music, and how a great many use their devices to make music of their own, and to engage in profound learning and teaching as well. The bulk of this activity happens outside school, which Pignato describes as music "without" music education. He uses the word "without" to mean both "outside" and "lacking." It is intriguing to consider the latter sense. If so much music production, learning, and enjoyment happens without formal school settings, where does that leave us as music teachers?

Pignato demonstrates that students' use of technology without music education is coming "from positions of choice, with greater power, control, autonomy, and agency than they do technology within music education." The digital studio and the Internet present opportunities for exploratory learning based on an individual's whim, interest, and personal agency. Pignato cites the young virtuoso beatboxer roboyette, who produces and performs in a series of widely viewed YouTube videos that both showcase his skills and act as tutorials for other beatboxers. Popular musicians are substantially self- and peer-educated, and roboyette is emblematic of the horizontal information transfer that the Internet affords.

How might we bring some of the organically emerging energy of bedroom producers and YouTube performers into the classroom? Can we seize the opportunity presented by digital technology to turn music class into a space for students to take ownership over their musical lives, to produce work that is authentic and meaningful to them? Can we bring supposed nonmusicians into musical practice through pathways they might not have realized were available to them? I believe, as Pignato does, that we can, but only if teachers act as "technology stewards," curators who are responsive to the needs of their communities—and to new and emerging practices.

Unexpected Uses of Technology

The most innovative musicians find ways to use equipment for expressive purposes not intended by its designers. Howell points to the ingenuity with which early hip-hop artists used the means at their disposal to find radical new musical forms. The rapper and producer Q-Tip describes the method he used to make his earliest hip-hop instrumentals with nothing more than a cassette deck and a record player. He would cue up the section of a record he wanted to sample, record the desired passage to tape, and quickly hit pause on the tape deck. He would then back up the record, unpause the tape at the right moment, record another pass, pause the tape again, back the record up again, and so on. Through laborious repetition of this process, Q-Tip was able to create "pause tapes," hip-hop instrumentals built from looped samples. Some of Q-Tip's pause tape productions are now regarded as hip-hop classics (Rapaport, 2011). This example makes me wonder what other new musical forms lie dormant in the unexplored possibilities of computers and smartphones.

A more contemporary music technology that has found unexpected creative use is Antares's Auto-Tune, which was designed to correct the pitch problems in a natural-sounding way. However, producers discovered that by setting the program's Retune Speed setting to zero, they could produce a strange and distinctive sound: voices snapping instantaneously to piano-key pitches with stair-step precision. This unearthly effect has become ubiquitous in Western pop and hip-hop and has found creative use in other cultures too. Clayton (2009) discusses the importance of Auto-Tune to Berber music, as in "Lkit Li Nebghih" by Cheba Djenet.

Melisma is equally if not more prevalent in Maghrebi music (compared to African-American vernacular music). This explains the plug-in's mind-boggling success across North Africa. Contemporary raï and Berber music embrace Auto-Tune so heartily precisely because glissandos are a central part of vocal performance (you can't be a good singer unless your voice can flutter around those notes): sliding pitches become startlingly strange when the effect is applied. A weird electronic warble embeds itself in rich, throaty glissandos. The struggle of human nuance versus digital correction is made audible, dramatized. Quite literally, this is the sound of voice and machine intermodulating (Clayton, 2009).

Trained singers complain that Auto-Tune and other pitch-correction tools remove the incentive for people to learn to sing properly, since any untalented singer can be made to sound decent. This may well be true. But as Clayton demonstrates, Auto-Tune opens up an entirely new vocal technique, one that requires a new kind of skill. Here in the United States, rappers like Lil Wayne and Kanye West, who might not be able to carry a tune unassisted, can create remarkably innovative vocal melodies by singing or speaking through Auto-Tune. Young people who may not consider themselves singers may well find Auto-Tune to be creatively emboldening as well.

Playing (with) Music

Peppler considers the phenomenon of the rhythm video games, which simulate the playing of musical instruments. These games do not enable musical creativity per se, but they nevertheless afford a genuine kind of musical experience. While they do not directly teach instrument mechanics or music theory, the experience of playing them is inseparable from the experience of music. This experience carries intriguing pedagogical possibilities. As Peppler observes, rhythm games can teach "foundational music concepts and notation commonly thought to be the sole domain of formalized music education." The games align intuitive notation symbols with real performances by master musicians, and they give real-time feedback on the player's performance as well.

Traditional music education focuses on instrument and vocal technique and the reading and writing of notation. Critical listening to master performances is usually supplemental to the main goals. However, informal music learning as described by Pignato consists almost entirely of the study of recordings of master performances. Furthermore, a large and growing segment of contemporary music consists of electronically produced recordings that never existed as instrumental performances to begin with. In this environment, critical listening ability is perhaps the most important skill a musician can learn.

Rhythm games are ideal tools for teaching analytical listening. I have a friend, a self-professed nonmusician, who loves the Beatles. He played through the Beatles edition of Rock Band as Paul McCartney, and told me afterward that it was the first time he had ever been conscious of the bass line in a song. He had heard all of the Beatles' songs hundreds of times but had never been attentive to anything below the surface level. Suddenly he gained access to a new dimension of these familiar recordings. Critical listening is crucial for learning to produce recordings, for creating your own parts in a group, and for any other kind of creative music making. While rhythm games may not foster musical creativity per se, they can help build a solid conceptual foundation. Indeed, would-be classical musicians may benefit more from the games than rock musicians. Rhythm games are score-following exercises that reward precise performance of prescribed actions. While video games based on rock music would seem to be as far from classical music practice as possible, Peppler discovered that classically trained musicians tended to enjoy the games more than rock musicians, who chafed against the lack of improvisation and spontaneity.

Speaking of music software more broadly, Peppler observes that "new technologies are providing easier entryways into things like composing and novice music making, radically shifting the lines between performer, listener, and composer." I would go further and say that the digital studio has not just shifted those lines but erased them. Technology has also collapsed the distinctions between recording, composing, and performing. Long before students attain any kind of instrumental or music theoretical proficiency, they can participate in high-level thinking about form, structure, timbre. This is a truly exciting development.

The Negative Impacts of Music Technology

The Costs

A major problem with contemporary music technologies lies not in any particular harm they do to learning or expression but in their expense. Kigozi points out that the majority of African people live below the poverty line, and more than half of the schools in rural Africa are not connected to electricity. Cost and lack of infrastructure put most cutting-edge technology far out of the reach of these schools. The costs are not just financial. Computers are complex and unreliable. It takes considerable skill to be able to use them effectively at all and even greater skill to be able to troubleshoot and fix them. All of those skills must be learned, and that learning takes time and effort. Kigozi points out, quite correctly, that software and hardware "need to be as reliable and as easy to use as a mbira, *engoma, hosho, endeere* or CD player, so that precious time is not lost due to technology glitches during class sessions."

Access to well-maintained technology is not just a problem in the developing world. In the United States, approximately 70% of elementary schools lack the broadband Internet connections that make possible audio and video streaming, the sharing of large files, and downloading and upgrading software (Dobo, 2014). Schools are also burdened with out-of-date and poorly maintained computers, inadequate wiring, and a shortage of tech-savvy teachers and staff. Computers and Internet access have steadily fallen in price and will continue to do so, but access to them is hardly universal.

Distraction and Disruption

Kigozi echoes a lament, widely heard from American teachers, that computers and smartphones divide students' already fragmented attention: "The introduction of smartphones among teachers and students has allowed constant access to the Internet, downloading, and listening to music, and teachers who have been the center of learning in the music classroom are now experiencing difficulty with distracted students struggling to stay focused." Popular culture and commercial technology change with lightning speed. Even schools without financial limitations will find it difficult to keep up due to intrinsic institutional conservatism. Should we simply invite students to bring in their phones and laptops? Should we embrace our students' own interests and priorities? Should we take their lack of engagement with their lessons as a sign that their lessons are less relevant to them than their self-directed online activities? Or is it our job to filter out the noise, direct our students' attention to the most worthwhile information and activities, and hold fast to our in-class phone bans? There is no easy answer to this question.

Kigozi is concerned that the culture of illegal downloading has thrown respect for intellectual property out the window. In Africa as in the West, "students, educators and the wider community download and sample music without acknowledging its source. Music is then duplicated and sold on the local market without thinking about the intentions of the composers and producers who supposedly make their livelihood from it." Furthermore, the ubiquity of recordings has eroded instrumentalists' livelihoods, as singers find it easier and cheaper to use canned backing tracks. American technologists use the word "disruption" as a term of praise. In their usage, to disrupt an industry or institution means to sweep away inefficiencies, entrenched powers, and outmoded ideas. Kigozi asks whether all change is for the better, whether we are sweeping away traditional practices and norms that still have value. I share his concerns, though I fear that at this point they are largely academic. The changes wrought by file sharing are not likely to be reversed, whether we are glad of them or not.

Technology Is Not Intrinsically Liberating

While the digital studio and the Internet have enormous potential to open wider access to musical expression, they also have potential to stifle it. Technology is a set of tools that can be used well or badly, and its use will depend heavily on its social and political context. Pignato reminds us that young musicians in school are vulnerable; they have little control over their school music experiences (or any other aspect of their schooling, for that matter.) Whether it is in the form of computer software or textbooks, technology in music education represents "official knowledge," a means of tacitly enforcing institutional values. Pignato is concerned that such values must necessarily be authoritative in nature, and that they will not support students' subjective desires and intents. Even if all of the logistical and financial obstacles to broader technology adoption in the classroom are removed, will students' learning and creative experiences be the better for it?

Pignato's concerns are valid, and worth deep consideration. Nevertheless, I retain my optimism. The one constant in the history of technology in music is surprise. Whether or not we make the best use of technology in the music classroom, young people will continue to find unexpected uses for it elsewhere. There is no historical precedent for the informal learning possibilities afforded by inexpensive and ubiquitous computers. My son Milo, who is not yet 2 years old at the time of writing, happily engages with a variety of music-making apps on my phone. Some of these apps, like Toca Band and Monkey Drum, are designed for young children. Others, like Animoog and Figure, are aimed at adults, but Milo enjoys them just the same. What will his musical experience be like in 10 or 15 years? Are young music learners like Milo best left to their own devices, literally and figuratively? Or can we structure a classroom around these devices, combining independent play with guided group activity?

Howell suggests that formal educational settings will always compromise or even negate young people's autonomy and independence. Is this true? Perhaps if we think of the music room as a maker space rather than a classroom, we can admit to it some of the imaginative play and authentic expressiveness that students find outside school. Students experience guided play in kindergarten; why should it end there? The MIT Lifelong Kindergarten group, as its name suggests, is committed to developing new technologies that expand the range of what people can design, create, and learn, in the spirit of blocks and finger paint. Kigozi observes that music education will happen wherever people gather together, using whatever materials are at hand. In the West, musical instruments (computers included) are mass-produced industrial artifacts. In the more traditional African societies, the building of instruments is a part of music technology education. We may hope that students feel a sense of ownership over their tools, whether they literally build them or use them to build their musical selves. A school is necessarily an ad hoc community; ideally, it can be a genuine artistic community as well.

References

Clayton, J. (2009, May). Pitch perfect. *Frieze*. Retrieved September 12, 2014 from http://www.frieze.com/issue/article/pitch_perfect/.

Dobo, N. (2014, September). When schools can't get online. *Atlantic Monthly*. Retrieved from http://www.theatlantic.com/education/archive/2014/09/when-schools-cant-get-online/379863/.

Rapaport, M. (2011). *Beats, rhymes and life: The travels of a tribe called Quest* [documentary film]. New York: Sony Pictures Classics.

CHAPTER 23

MUSICKING AND TECHNOLOGY

A Further Swedish Perspective

BO NILSSON

HOME and school are traditionally primary environments for young people's socialization. A "third environment" is found in leisure time, which is between home and school and away from parental control, although adults are still more or less present in these environments through organized activities such as sports and arts. This pattern has changed in recent decades, especially in Western society, mainly due to the rapid development of the media and the Internet. Sweden has the largest amount of Internet use on mobile phones in the world. One-third of children 3–5 years old and almost every Swedish teenager use the Internet daily, and nearly every 11-year-old Swedish child owns a smartphone. New digital media offer children and young people new types of freedom from adult interference. On the Internet, you can be someone else, play roles, surf, and obtain information. Popular culture is constantly available on the Internet via YouTube and a variety of social platforms. Social media, such as Facebook and Twitter, are just as real as the other meeting places where young people used to socialize 50 years ago. These platforms and smartphones provide important forums for socialization.

Children and young people with different types of disability are often in the presence of family members and other adults all day and do not experience the freedom from adult influence and control in the "third environment." The media and the Internet can become for them a free zone, offering music, film, and various types of interaction and information. Social media such as Facebook, Twitter, and various online communities can become a type of "fourth environment" to compensate for a weak third environment. Access to media and digital tools can provide a "window" to the world. Musicking and music creation can be vital for many young people in need of special support. In the world of music, these young people can develop their identities far beyond a focus on disabilities. In this Further Perspective, I will discuss some important issues that are partly inspired by children's creative music making with digital tools and music technology in relation to young people who require special support.

A Computer in My Brain—Music Technology and a Young Musician

The following episodes demonstrate how Gunborg, almost 8 years old, found her own way in the world of digital musical instruments. An encounter with a synthesizer keyboard triggered her imagination as she suggested that the keys c-c#-d-d#-e resembled a face (fig. 23.1): "It looks like a nose and cheeks," she said.

Gunborg was one of the participants in my PhD study of children's computer-based creative music making (Nilsson, 2002, 2003, 2007; Nilsson & Folkestad, 2005). The participating children created music using a MIDI keyboard and a professional computer sequencer program on a Mac. These digital tools provide great opportunities for young, untrained children to express their musical ideas through creative music making. However, for Gunborg, it did not matter if the tools she used were digital or not; when she wished to create music, she used any tool available, including her own imagination and fantasy. For Gunborg, it was normal to communicate musically with her surrounding environment, but she also enjoyed being alone with her music. One technique she used when she was alone and bored was to pretend that she had a computer in her brain:

> Sometimes when I have nothing to do, I put my hand just as if I had a mouse, and then I imagine that I have a computer in my brain. I go to "File," and there I click on "Music" in the menu. Then, I hear music in my ear. It's great fun! I can see it even with my eyes open, and I can also play computer games. I can make music like this [sings the opening phrase from Beethoven's Fifth] or just make up new music of my own [sings a melodic phrase]. Sometimes when I have a headache it usually gets better when I play some music.

Gena Greher reminds us in chapter 30 of Sherry Turkle's prophetic statement concerning the way computers change our ways of thinking about ourselves and other people (Turkle, 1995). This new way of thinking is clearly demonstrated in this

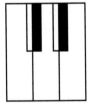

FIGURE 23.1 Gunborg recognized a face in these keys while improvising on a keyboard

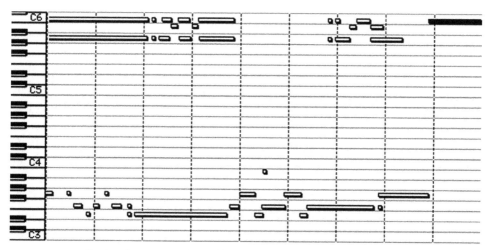

FIGURE 23.2 Gunborg's 1-minute-long final composition ends with "a long straight" note (highlighted)

episode: Gunborg found it natural to imagine a computer in her head and to manage it in the same way she would a "real" computer.

Let us talk a bit more about how Gunborg composed with her digital tools, keyboard, and computer. One of her pieces was prompted by an invitation from the researcher to compose music inspired by a painting by the Russian painter Wassily Wassilyevich Kandinsky (1866–1944). Gunborg's composition developed through four steps, each step a development and revision of the previous one. The first step was a revision of an earlier piece she brought to mind. Her final version (4) ended with "a long straight" note (fig. 23.2).

In the following, I will briefly discuss some important issues partly inspired by Gunborg's and other young children's creative music making in my study and partly based on projects regarding music and young people in need of special support.

Music Educators and Researchers Never Meet Musical Beginners

By taking part in the growing production of musical media, children learn a great deal about music by themselves, and they learn it fast. Margaret Barrett (1996, 1998) argued that children's creative activities always include aesthetic decision-making. I agree with Barrett that if we wish to access children's understanding of music, we must examine their musical responses rather than their verbal responses. When Kratus (1989) studied young children's musical creativity, he asked the children to compose a song that no one had heard before. However, as a researcher, his definition of a composition was "a unique sequence of pitches and durations that its composer can replicate." Kratus's view that a creative product should be able to be replicated if it is to be recognized as a composition is not shared by Webster (1992), who finds a piece to be compositional in its nature when the creator has the opportunity to *revise* it. As demonstrated above, in Gunborg's composition in relation to a Kandinsky painting, her creative process was facilitated and enhanced by the opportunity to repeatedly revise her music. When she listened to her composition, she was very pleased with it. She proudly drew my attention to the last "long straight" note (fig. 23.2), thus demonstrating her involvement in aesthetic decision-making.

Based on results from my own work (Nilsson, 2002, 2003, 2007; Nilsson & Folkestad, 2005) and the work of others (see Barrett, 1996; Folkestad, 1996), I propose this definition of "musical composition": "a piece of music that its creator experiences as meaningful" (Nilsson, 2003, p. 214). Research on musical creativity in childhood demonstrates how listening to children's music and what they have to say about it helps music educators and researchers to understand what music means to children. As we can see in the example of Gunborg's music, digital musical instruments have long been included in children's musical worlds.

Musicking and Musical Tools

Christopher Small developed the concept of "musicking" to address what he calls the *reification of music*. Music is "something that people do" and not a thing, he argues (Small, 1998, p. 2). Small defines "musicking" as "to take part, in any capacity, in a musical performance, whether by performing, by listening, by rehearsing or practicing, by providing material for performance (what is called composing) or by dancing" (Small 1998, p. 9).

The concept *affordances* (Gibson, 1979) describes suggestions of meaning perceived by the individual in a certain situation, including memories, expectations, knowledge, and meaning. In a musical situation, affordances interact with a physical instrument, digital setting, computer software, graphical interface, physical interface (e.g., head mouse), and other devices. From an ecocultural perspective (Nilsson, 2002, 2003, 2007), a musical instrument is a system in which the musician interacts with physical and psychological mediating tools (Vygotsky, 1978), forming a competent musical tool. Furthermore, objects other than obvious musical instruments can mediate music when that is the user's intention. A switch, for example, can be used for many purposes: to close a door, turn on a light, trigger a sound on a computer (Nilsson, 2013). Notably, even the clicking sound from the switch itself could be used as an acoustic musical instrument. The individual and the tool together constitute a system that is able to think, act, and learn. Again, artifacts can be used in countless different ways and have no predetermined uses.

When computers entered Western schools in the early 1990s, music educators and musicians immediately embraced the new digital tools and saw enormous possibilities for music education and music creativity. The media revolution produced new tools for creative musical activities, as many of the contributors to this book have noted. Computers and synthesizers represent important tools that young people can use to express themselves through music. For the young touchscreen generation, it is easy to create pictures, video clips, sounds, and loops and to communicate them almost immediately. However, it is not technology per se (in our case, many different digital musical tools) that is important; technology only provides new types of tools for creative thinking and doing. Rather, the core issue is the new possibilities that this new technology provides, the way we act on these affordances, and how digital media change our ways of thinking, which is demonstrated in the following two examples.

The traditional role of the musical instrument was challenged by Cappelen and Andersson (2011), who performed a study of families with disabled children using a digital cross-media interactive installation called ORFI. The researchers found that a representative family used ORFI in a number of different ways: as a musical instrument, comusician, communication partner, toy, meeting place, and surrounding musical landscape.

Students at our Pre-School Teacher Education program provided another example of changing expectations of a tool when I asked them to build an interactive music app using our most popular software for presentations. They were also asked to produce and record all sounds and music by themselves. This project pushed the students to think outside the box. The same software that the students recognized as the one used by their university professors in their long and boring lectures now became exciting multimedia software that was perhaps not very powerful but was nevertheless useful for creating interactive applications. All projects were demonstrated and discussed at a "digital vernissage" that was much appreciated by the students. Many students stated that they had changed their attitude toward technology and computers and had developed critical thinking about educational software on the market.

Today, many Western schools and preschools use touchscreen tablets. Most children can accomplish many things with tablets, but children with physical impairments may have problems using them. The touchscreen revolution may not be for everyone. The major smartphone manufacturers are developing new ways to control the screens of phones and tablets, such as motion controls using the front-facing camera of the device. These developments are promising because existing technology, such as eye-tracking technology, is still very expensive. A head mouse is cheaper, but this is still a type of assistive technology that is not compatible with touchscreen devices such as tablets or smartphones. Children and young people who need to use eye tracking with their computers are in many ways excluded, both by the expensive equipment and by the fact that they are not able to use the same software as their friends with tablets, laptops, and smartphones.

Music and Health

Our ability to recognize patterns of sounds is essential for us to be able to talk and to create music. We seem to share our ability to recognize sound patterns with squirrel monkeys. A research team recently found that squirrel monkeys were able to distinguish between tone sequences containing a dependency and those lacking it (Ravignani, Sonnweber, Stobbe, & Fitch, 2013). The researchers concluded that the ability to recognize this type of pattern is at least 30 million years old, when the last common ancestor of humans and squirrel monkeys lived.

Musicking is closely related to play and flow (Csíkszentmihályi, 1990, 1994). It has been shown that these types of activity also have measurable biological effects on people by activating the "feel-good" hormones oxytocin and dopamine. With support from Ruud (2008) and Bonde (2011), it could be argued that music making and music creation promote health even when this is not their specific aim. It appears that we have a quite an evolutionary heritage on which to build our musical activities.

Democracy is a significant part of health and public health. Active involvement in cultural events, such as musicking, is clearly a form of freedom of speech and expression and therefore vital in a democratic society. Music technology enhances new democratic opportunities for music making, musical learning, self-expression, and socializing, as noted by Heidi Partti in her Core Perspective (chapter 25). The aesthetic dimension itself is important in many ways. One of the key competences recommended by the European Union (2006) is *cultural awareness and expression*: "appreciation of the importance of the creative expression of ideas, experiences and emotions in a range of media, including music, performing arts, literature, and the visual arts" (European Union, 2006, p. 18).

The importance of self-expression and participation in cultural life is clearly stated in the International Classification of Functioning, Disability and Health, article d 9202 of which explicitly describes activities such as "engaging in, or appreciating, fine arts or cultural events, such as going to the theatre, cinema, museum or art gallery, or acting in a play, reading for enjoyment or playing a musical instrument" (World Health Organization, 2001, p. 169). This political angle makes the right to participate in cultural events a question of inclusion in a wide sense. It is not only a question of attending concerts or theatres; the right to participate in different types of cultural and musical event, both as a consumer and a performer, is a fundamental part of democracy. This is where technology can make a difference, as demonstrated in the following episode with Moha, an 18-year-old Swedish girl.

Moha creates music on her computer with professional software. She has cerebral palsy, and composing music has become an important way to express herself and to be "seen," as she puts it. Music means something important in her life, and she wants to become a music producer. She likes to chat online with new friends, who interact with her without being fully aware of her situation. Moha has difficulty controlling her voice and cannot sing or create voice tracks on her own. Fortunately, her personal assistant sings in a band and performs the song tracks in Moha's compositions. In this case, it is apparent that musicking also takes place when, for example, a personal assistant takes part in a musical activity by managing computer software, interacting with a user, or even creating a song track, thus acting almost as a musical tool for the user.

Finally

Play often takes place in the *flow channel* (Csíkszentmihályi, 1990, 1994), with its important balance between abilities and challenge, where the individual experiences a sense of control. We play for its own sake, and play always includes uncertainty, as noted by Huizinga (1955) and Caillois (1961). It seems reasonable to conclude that unpredictable events are important in creative processes such as musicking and improvisation. Therefore, good music technology should allow play and unpredictable events and should permit the users to find a balance between the challenge and their abilities. Digital instruments should also facilitate opportunities to revise and develop musical compositions. Most children practice interacting rhythmically during their play, for example by dancing, jumping, singing, or clapping. Children with physical impairments may not experience or practice in the same way through their bodies. Digital musical instruments can be especially helpful by promoting and facilitating musicking, with others or independently.

Many previous studies of creative music making with digital tools have examined music making on an individual basis. New digital music technology provides new possibilities for interaction on a collective level, as mentioned by Partti and others in this book. Musicking and technology is also about the plans, guidelines, and principles of society. In many cases, teachers, personal assistants, technicians, and music educators collectively contribute to the enhancement of participants' musical activities, affecting the accessibility of music for the individual.

References

Barrett, M. (1996). Children's aesthetic decision-making: An analysis of children's musical discourse as composers. *International Journal of Music Education*, 28, 37–62.
Barrett, M. (1998). Children composing: A view of aesthetic decision-making. In B. Sundin, G. E. McPherson, & G. Folkestad (Eds.), *Children composing*, 57–81. Malmö: Malmö Academy of Music.
Bonde, L. O. (2011). Health musicing—music therapy or music and health? A model: Empirical examples and personal reflections. *Music and arts in action*, 3(2), 120–140. Retrieved August 2, 2013, from http://musicandartsinaction.net/index.php/maia/article/viewArticle/healthmusicingmodel.
Caillois, R. (1961). *Man, play and games*. New York: Free Press.

Cappelen, B., & Andersson, A. P. (2011, May 30–June 1). *Expanding the role of the instrument*. Paper presented at the International Conference on New Interfaces for Musical Expression, Oslo. Retrieved February 3, 2017, from http://urn.kb.se/resolve?urn=urn:nbn:se:hkr:diva-7908. NOTE, THIS IS A PERMALINK

Csíkszentmihályi, M. (1990). *Flow: The psychology of optimal experience*. New York: Harper and Row.

Csíkszentmihályi, M. (1994). *The evolving self: A psychology for the third millennium*. New York: HarperPerennial.

European Union. (2006). Recommendation 2006/962/EC of the European Parliament and of the Council of 18 December 2006 on key competences for lifelong learning. *Official Journal L 394 of 30.12.2006*. Retrieved February 3, 2017, from http://eur-lex.europa.eu/legal-content/EN/TXT/?uri=celex%3A32006H0962

Folkestad, G. (1996). *Computer based creative music making: Young people's music in the digital age*. Göteborg: Acta Universitatis Gothoburgensis.

Gibson, J. J. (1979). *The ecological approach to visual perception*. Boston: Houghton Mifflin.

Huizinga, J. (1955). *Homo ludens*. Boston: Beacon Press.

Kratus, J. (1989). A time analysis of the compositional processes used by children ages 7–11. *Journal of Research in Music Education, 37*, 5–20.

Nilsson, B. (2002). *"Jag kan göra hundra låtar." Barns musikskapande med digitala verktyg* ["I can make a hundred songs": Children's creative music making with digital tools]. Malmö: Malmö Academy of Music.

Nilsson, B. (2003). "I can always make another one!" Young musicians creating music with digital tools. In S. Leong (Ed.), *Musicianship in the 21st century: Issues, trends and possibilities* (pp. 204–217). Sydney: Australian Music Centre.

Nilsson, B. (2007). Children's practice of computer-based composition. In G. Folkestad (Ed.), *10 years of research in music education* 135–154. Malmö: Malmö Academy of Music.

Nilsson, Bo. (2013). Performance and participation: A qualitative study of music education practices in digitally based musicking with young people with physical impairments. *Approaches: Music Therapy & Special Music Education 6*(1), 19–27.

Nilsson, B., & Folkestad, G. (2005). Children's practice of computer-based composition. *Music Education Research, 7*(1), 21–37.

Ravignani, A., Sonnweber, R.-S., Stobbe, N., & Fitch, W. T. (2013). Action at a distance: Dependency sensitivity in a New World primate. *Biology Letters 9*(6). doi: 10.1098/rsbl.2013.0852.

Ruud, E. (2008). Music in therapy: Increasing possibilities for action. *Music and Arts in Action, 1*(1), 46–60. Retrieved August 2, 2013, from http://www.musicandartsinaction.net/index.php/maia/article/view/musicintherapy.

Small, C. (1998). *Musicking. The meanings of performing and listening*. Hanover, NH: University Press of New England.

Turkle, S. (1995). *Life on the screen: Identity in the age of the Internet*. New York: Simon and Schuster.

Vygotsky, L. S. (1978). *Mind in society*. Cambridge, M.A.: Harvard University Press.

Webster, P. R. (1992). Research on creative thinking in music: The assessment literature. In R. Colwell (Ed.), *Handbook of research on music teaching and learning*, 266–280. New York: Schirmer Books.

World Health Organization. (2001). *International classification of functioning, disability and health*. Geneva: author.

CHAPTER 24

EXPLORING INTERSECTIONS OF TECHNOLOGY, PLAY, INFORMALITY, AND INNOVATION

GILLIAN HOWELL

IT is easy to be dazzled by new technologies and all that they make possible, but, as Joseph Pignato reminds us, humans are the ones who are making the music. The illustrations of practice in this section offer vivid examples of this: the authors Pignato, Peppler, and Kigozi describe an array of musical innovation and exploration with and through technology, but also shift our attention away from the technology and onto the individuals and their contexts. As such, the chapters work collectively to foreground three critical factors in the wider discussion of technology in music education: context, informality, and innovation.

The *context* for musical learning determines available technology, and can both afford and constrain. The individuals within their context may create in response to it, in spite of it, or in defiance of its limiting forces. *Informality* (in the form of unstructured and social learning environments that invite open-ended exploration of materials and ideas) opens up both learning possibilities and highly original outcomes. Interestingly, it may occur despite, rather than because of, the design of the learning program. Both context and informality are then shown to be in constant interplay with *innovation*, the third factor. Each of these illustrations of practice demonstrate the way that new innovations may be the result of deliberate, problem-solving explorations; they may also emerge as unanticipated outcomes in larger processes. This chapter offers a further perspective on the usefulness of attention to these three elements and attempts to map their interrelated, multidirectional forces.

Technology in Context

We can neither assess nor understand the impact of technology without first understanding the context—the social, cultural, political, and environmental characteristics and specificities of the site—in which it is used or has evolved. Context affects what technology is available, who can access it, how and why and when they access it, and who they are, within a wider community, through and beyond that process. Different settings afford access to different technologies. The three core perspectives in this section demonstrate this diversity: in the urban contexts in the United States that Pignato and Peppler describe, there is minimal display of particular wealth or privilege. Nevertheless, within the young people's social and cultural environments many of the commodities and gadgets of modern life are visible and available. Access to electricity is the norm. In contrast, Kigozi writes about both rural and urban settings, but in poor, developing countries where there are far fewer resources and where access to electricity can be undependable or even non-existent. We go from contexts where video games and complex compositional tools are the height of music technology in schools, to contexts where CD players and double cassette recorders are the big-purchase items.

Even the task of defining technology must be informed by context. Kigozi presents a working definition of technology that includes instruments, books, CD players, cassette recorders, and radios, as well as mnemonics and other memory aids, responding to a local reality of schools with few modern technological resources. In this definition, the body is the primary and central technology. Other aspects to technology—such as construction and application—come later. This example reminds us that while "technology" may seem synonymous with the most modern information and communication materials that enable making and sharing on a global and nano scale, it is more helpful and more inclusive to think of technologies as being anything that extends the capabilities of the body and mind, in any direction, at any time.

How do young people access new technologies, particular those living in precarious or disadvantaged socioeconomic circumstances? For some, it might be through home, school, or the wider peer group. For others, new technologies are accessed and explored in nonschool—although still somewhat structured or formal—settings, such as community centers, youth clubs, churches, and Internet cafes. Peppler describes a local youth center that offers free after-school activities for young people, including instrumental music tuition, and a Rock Band club, that enables the young people to take part in rhythmic video games.

In the modern world, culture and norms can change faster than large institutions. What is inaccessible today will be smaller, cheaper, faster, and more powerful in just a few months' time. Kigozi describes the world's new technologies flooding into Africa just as they are in other parts of the world. In such a context, it may be a long time before schools can match the technological capacities that young people can access through their own social and informal networks. This even more emphatically points to the relevance of informal, out-of-school learning for technology in music education.

The *social* context, in the form of hierarchies, exclusion, and marginalisation also influences the appeal and availability of particular technologies and how they are used to amplify or protest the social situation. Pignato cites the context of the early hip hop era as one where the sounds were a product of people responding to a particular context at a particular time with the technological tools available to them. However, it is hard to imagine today the raw energy and bold, challenging innovations of the early hip-hop movement emerging from the formality and restrictions of an institutional music learning environment, and this leads us to consider the next critical factor in technology in music education, that of informal and social learning environments.

Technology, Informal Learning, and Play

Informal learning has received increasing attention in music education in recent years since the publication of Green's (2002) research into how musicians in the rock and pop worlds learn their musical knowledge and develop their skills. Informal and non-formal learning (applying D'Amore's 2014, p. 9, definitions of these terms) can take many forms, and may be identified by pedagogy and learning approaches as well as location and personnel.

Informal learning—learning that takes place outside formal education institutions—looms large in many of the examples in the core perspectives of this section. In the informal realm, individuals make active choices about what, when, and how they want to learn. The technological tools at hand enable them to develop new skills through imitation and playful, engaged participation, and to explore, create, and share their own work with fellow members of supportive communities of interest. Many of these technological tools offer experiences that are complete within themselves—they are not dependent on the user being part of a larger group. They offer independence and portability, which makes them well suited to independent learners.

Pignato describes technology as having useful impact in music education inasmuch as it empowers young people to explore their particular interests and idiosyncratic ways of learning. He sees autonomy and independence as essential if young people are to be enabled to take charge of their own learning, but also notes that these qualities are invariably compromised, if not negated entirely, within the restrictions and rigidity of most formal music education institutions. Therefore, while access to new music technologies in formal learning is promising and positive, it is only a partial solution if innovation, autonomy, and originality are the hoped-for outcomes; the flexibility of the environment matters too.

Contemporary technologies have provided platforms through which young people can develop skills following a personally curated path rather than a set curriculum. The skills that are developed and shared are substantial and authentic: Pignato offers multiple potent examples of young people becoming highly skilled, versatile contemporary musicians and music consumers through following their own learning pathways. This knowledge has strong validity in contemporary culture.

Informal music learning is not only mediated through contemporary technologies. The vast majority of the world's musical traditions are transmitted through social interactional environments, where being present and participating are the critical elements for music learning. Kigozi describes the informal realm as an important part of the African music education landscape. Local community musicians—cultural knowledge bearers—teach and share the indigenous musical heritage with young people through various informal and nonformal (i.e., nonschool) settings.

Play—open-ended, exploratory, imaginative play—is an unsung part of the informal music learning landscape, and it plays an important role in the feedback loops between technology, context, informal learning, and innovation. Self-directed learning is at times exploratory and therefore open-ended, as players problem-solve and investigate solutions to challenges on the learning journey. Informal learning is also, like play, often a social experience through which self-identity and a sense of community can form. Imaginative play supports this development of identity, allowing players to "try on" and discard different versions of themselves in relation to what they are doing, as seen in the youths' engagement with the increased social status and "cool" factor described by Peppler among the Rock Band Club participants. And as with informal learning, play follows a nonlinear and fluid pathway, quick to adapt to changes in environment, interests, and energy.

This kind of free-form, open-ended, experimental type of play can be seen in Pignato's examples of the technology-filled learning pathways of the professional musicians he interviewed. Here, we see play as experimentation and curiosity, where access to tools (in this case hardware, software, and platforms for creating, producing, and sharing music) and communities of shared interest resulted in new musical ideas, unexpected hybrids, and brazen juxtaposition of hitherto uncommon bedmates.

More generally, musical play is an important way that foundational skills and knowledge about "how music works" can be established. Marsh (2009) offers many examples of multiple and complex forms of children's musical play in her comprehensive investigation spanning seven countries. Around the world, children's unstructured, free play has strong musical elements and the "players" reward high-level skills, such as focused ensemble playing, memory, and virtuosic technical ability. Other ethnomusicological studies also observe musical elements of rhythm, melody, and pattern repetition in the playful songs and games that children invent to enliven daily household chores in many agricultural and premodern societies, for example Timor-Leste (King, 1963).

The musical play described in Peppler's chapter certainly exemplifies this idea of foundational learning through play. Peppler's research found that the rhythmic video game experience may have provided two important foundations for acquiring further music knowledge. One of these was a deeper understanding of "how music works," and a familiarity with symbols denoting musical sounds in time; the other was a more developed self-identification as a musician and music learner. The young participants' subsequent participation in instrumental music lessons on violin and recorder and later membership of youth orchestras and other ensembles (which in the past had been notably undersubscribed) are suggestive of both these outcomes. Interestingly however, the rhythmic video games described by Peppler proved to be perhaps the more restrictive and rules-based technologies described in the core perspective chapters, despite being approached as games and therefore the most seemingly "play"-based. Success in these games required participants to adhere to particular rules of play, and rewarded in particular those players with a strong knowledge of Western music theory and notation. Professional musicians who had learned their music skills by ear did not excel in the game and did not enjoy it. While this is a welcome outcome for those that found success, it does indicate once again the way that a more restrictive environment (physical or virtual) can also constrain innovative and adaptive uses of the resources, and leads us to consider the ways that technology and innovation intersect with context and informality in a kind of self-perpetuating cycle of creative responses.

Innovation, Play, and Technology

The relationship between innovation and technology is important, but sometimes seems ubiquitous. It can imply that the technology itself is the essential ingredient for musical innovation; however, as this chapter shows, context and informality (or play) are also part of the picture. Technology merely facilitates, mediates, or provokes the exploration that leads to the innovation.; it does not assure innovation.

Innovation may result from the convergence of people, place, technology, and a creative urge or vision; it may just as often result from need, as a solution that emerges as a result of improvisation around the *absence* of something, or an imposed limitation. However, the restrictions, the predetermined, rigid structures, and the obligatory curriculums of most formal institutions are often anathema to this kind of improvisation, suggesting that the key to limitations enabling innovation is about having the space to *improvise* around the limitation or constraint.

History shows us that it is through exploring new materials in people's current context that new musical languages have been forged. Returning again to the example of the early days of hip-hop, created by disenfranchised young people living in a big city but excluded from the mainstream due to racism and social injustice, the frugality of "two turntables and a microphone," when coupled with the energy and social environment of the times, spawned a new musical language.

Reflecting on this, we are reminded that necessity is often the mother of invention and creation. It is through exploring the technologies of a particular context that distinctive musical languages, or musical identities, have been forged. Kigozi describes the limited resources available in many rural African music classrooms, where radios are an essential part of information communication technology, and a double cassette recorder represents modern playback and recording technology. With the early days of hip hop in mind, one wonders what musical innovations could be about to emerge from these settings, where young people may have limited access to digital technologies through institutions but in their leisure time are using functional, versatile widely-available gadgets like double cassette players and radios, as well as smartphones or MP3 recorders, in their musical explorations and expression.

In this way we would see these three recurrent themes intertwining and perpetuating themselves. Context determines the available technology but also the urge to express; informal, flexible, and social learning environments support playful exploration and creative solutions to environmental and resource limitations; and musical innovations (hopefully) emerge as a result. Technology plays an important part, but as Pignato reminds us, humans are the ones who are making the music.

REFERENCES

D'Amore, Abigail (Ed.). (2014). *Musical futures: An approach to teaching and learning* (2nd ed.). London: Paul Hamlyn Foundation.

Green, Lucy. (2002). *How popular musicians learn: A way ahead for music education.* London: Ashgate.

King, Margaret. (1963). *Eden to Paradise*. London: Hodder and Stoughton.

Marsh, Kathryn. (2009). *The musical playground: Global tradition and change in children's songs and games.* New York: Oxford University Press.

2 B

Core Perspectives

CHAPTER 25

PEDAGOGICAL FUNDAMENTALISM VERSUS RADICAL PEDAGOGY IN MUSIC

HEIDI PARTTI

THE use of technologies in education has been under debate most likely since the first technological tool was ever introduced. Regardless of innovations in hardware and software, the debate on the question of whether the ongoing introduction of new technologies and media will add to or rob from the quality of education continues. Along with the interest, doubt, and excitement toward the use of modern technological instruments and interfaces in music teaching, the discussion revolving around music technology refers increasingly, often specifically, to the advent of digital media. Indeed, the possibilities for music-related interaction and participation enabled by digital technology, particularly in informal music learning environments, have recently gained mounting attention in the music education research literature (e.g., Salavuo, 2006; Waldron & Veblen, 2008; Partti & Karlsen, 2010; Väkevä, 2010; Miller, 2012; Partti & Westerlund, 2012; Partti, 2014a; Michielse & Partti, 2015).

This chapter[1] is based on an understanding that "the ongoing transition of our culture into a media culture" (Hepp, 2013, p. 1) has brought about far-reaching and thorough changes in our daily lives and institutions of education. As pointed out by Hepp (2013), the transition should not be underrated, as "the media—or, more exactly, communication via media—have increasingly left their mark on our everyday life, our identity and the way in which we live together" (p. 1). Thus, there is, or should be, a robust connection between explorations into musicians' developments, identities, values, and out-of-school "digital habitat" experiences (see Wenger, White, & Smith, 2009) and the realities of the school.

In this chapter, I make a case for the importance of systematic development of media literacy in music education. I will first explore the social-cultural changes that technological innovations, particularly the so-called new media technologies, have brought about in the wider culture of music making and learning. In the second section I will focus on the role of the school in the face of these social-cultural changes, and introduce two pedagogical strategies—namely, "pedagogical fundamentalism" and "pedagogical populism"—that institutions of formal music education have adopted to respond to these changes. In the subsequent two sections I will reflect on some of the limitations of the strategies regarding music education, especially in terms of the development of music-related expertise and capabilities for equal, responsible, and sustainable participation. Finally, I will argue for the importance of *radical pedagogy* in music education by suggesting that institutions of formal music education, the school in particular, should aim to build balanced approaches to new media that will equip and empower students to act as moral agents across a range of communities in physical and cyber spaces.

This chapter builds on a theoretical framework, provided particularly by writers in media studies (e.g., Buckingham, 2003, 2010; Jenkins, Clinton, Purushotma, Robison, & Weigel, 2006; Davidson & Goldberg, 2010) and critical pedagogy (e.g., Giroux, 1999, 2003, 2011; Freire, 1970/1996), as well as by social theories of learning (e.g., Wenger, 1998, 2006; Paavola & Hakkarainen, 2005). Consequently, this chapter views music education as political education in which pedagogy is understood as a moral and political practice (e.g., Giroux, 1999, 2003, 2011) and learning is connected to students' growth as critical citizens with "the imperatives of social responsibility and political agency" (Giroux, 2003, p. 9). The development of such "radical pedagogy" (e.g., Suoranta, 2003; McGettigan, 2008) calls for an understanding of and commitment to the music teacher's role as a "critical companion" (Väkevä, 2013) supporting her students' growth into mature ways of thinking.

In addition to literature by other scholars, this chapter draws from my own research on music making and musical learning in the world of digital media. This chapter is motivated by an academic interest as a research scholar as well as by my personal experiences as a musician, music teacher, and music teacher educator. Rather than viewing myself as a "neutral" or "objective" observer on my mission to find and provide answers to questions or test theories, I consider myself a traveler engaged in a process of knowledge construction (see Kvale & Brinkmann, 2009). Consequently, the observations of and understandings about the relationship between (digital) technologies and formal music education presented in this chapter are my interpretations, shaped by my own experiences and background.

Mapping the Change: Learning Music in a Mediatizing World

Technological breakthroughs have generally influenced social, cultural, educational, and political conditions. Researchers have identified various stages in the process of *mediatization*, that is, the development of increasing media influence in society. This development ranges from the stage in which people adopt the role of spectator coming into contact with the world via representations offered by the media to the stage in which we adapt ourselves "to the working routines of mass media and the conditions that the media set up" (Asp & Esaiasson, 1996, p. 80; see also Lundby, 2009; Couldry & Hepp, 2013; Hepp, 2013). Currently, this adaptation to media logic is clearly seen in the ways people not only make media but also share it easily with others. This user-generated content creation can take place both by conceiving new works from scratch and by playing along with commercially produced content.

Due to the novel ways that music audiences and musicians from different backgrounds can now join in archiving, producing, appropriating, and recirculating media content in (online) affinity groups, this emerging cultural phenomenon is often referred to as *participatory culture* (Jenkins et al., 2006; see also Partti & Westerlund, 2012; Livingstone, 2013; Waldron, 2015). Although it has always been possible to learn music without entering into a music-learning institution, the opportunities for artistic expression, social participation, and musical learning, particularly those made accessible by the World Wide Web, are "unprecedented in both character and scale" (Gauntlett, 2011, p. 3). New media environments range from online composing projects (Partti & Westerlund, 2013; Partti, 2014b) to instrument or genre-specific virtual learning communities (Waldron & Veblen 2008; Waldron, 2009) and to online remix contests (Jansen, 2011; Michielse, 2013; Michielse & Partti, 2015), among many others.

With the arrival of new technologies, many of the old demarcations are being muddled. The clearly set limits between different musical cultures, styles, and genres, formally educated experts and informally trained amateurs, playing and working, and consuming and producing are getting more indistinct day by day with every new technological invention. However, as liberating as the boundary blurring may well be in terms of participation and creativity, it does not come without practical, political, and ethical ramifications. For example, opportunities to protect young people from seeing and hearing potentially damaging or unsuitable media content, such as music videos including violent or explicitly sexual imagery, seem to be disappearing. And access to the cultural and social worlds in which children participate seems to be slipping away from the reach of adults (Buckingham, 2003, p. 32). This is not simply related to adults' technical (dis)abilities—a deficiency of technological capital, as it were. Adult access is frequently withheld, particularly due to our failure to comprehend the relevance and, indeed, the meaning of the content and practices that seem to have such a strong impact on young people. In other words, the blurring of boundaries in one place appears to have resulted in the redrawing of them in another. Thus, this blurring challenges formal education institutions to keep pace and to find their own role in the change and standpoint toward it.

Choosing A Strategy for Facing the Change: Pedagogical Fundamentalism and Pedagogical Populism

The rise of new and wider opportunities for learning music and communicating and exchanging musical ideas through simultaneous participation in various global and local communities through digital music and information technology also presents new questions regarding the role of the school and other institutions of formal music education. While explorations into the natural territories of people sometimes referred to as "digital natives" (e.g., Prensky, 2001, 2010) have advanced our understanding about the new ways, places, and spaces in which a growing number of (young) people can carry out music-related tasks, these explorations have not always been considered particularly helpful by practitioners working in schools and other institutions of formal music education. The globe may well be open for "new, even radically new, ways of conceiving, manipulating, mediating, consuming, and recycling music" (Väkevä, 2009, p. 9), but what does that have to do with the reality of the music teacher facing her students on a Monday morning at a local school?

In the field of music education, teachers, researchers, and policy-makers across the world—not by any means limited to countries with advanced economies—are asking similar questions. Do children need to be saved from the corrupting influence of digital-technology-enabled mass media culture? Or are music teachers advised to stand back from the learning processes of their self-governing students? Are music teachers expected to protect their students from the ubiquitous influence of media or to zealously import models adopted from the ever-exciting virtual world, to jazz up the tediousness of the school classroom?

Next, I discuss two different pedagogical strategies that schools have adopted in the face of new, media-driven sociocultural changes. Following writings by the media education theorist David Buckingham (2003, 2010), I refer to these approaches as *pedagogical fundamentalism* and *pedagogical populism*.

Pedagogical Fundamentalism: A Great Escape from the Perils of Media to the Safety of Traditions

One strategy for addressing contemporary social change fueled by digital technologies has been to retreat from the risks and unpredictability of the mediatizing world to the (apparent) safety and clarity of the classroom's confines. Unlike constantly shifting new media environments, with their enormous amount of information, knowledge, and attitudes for which no one seems to be responsible, textbooks and clearly stated learning and assessment objectives—written by experts and published by verified institutes—may feel like welcome means of attaining stability. Consequently, schools have been accused of heading resolutely backward in the rhythm of a new educational fundamentalism.

Technology has irreversibly transformed the social and cultural experiences of students. At the same time, educational fundamentalism primarily views video games, tablets, and mobile phones as devices of distraction from instruction, and social networking sites as forums for the blossoming of narcissism and viciousness and the degradation of the quality of art and the diversity of knowledge (e.g., Keen, 2008; Aboujaoude, 2011; Koltay, 2011). For example, recently in music education, reservations about change have visibly been seen in regard to digital gaming. Rather than embracing the popularity of music video games and focusing on the benefits that gaming has been claimed to have in terms of deep learning, active participation, and acquainting oneself with various musical styles and skills (e.g., Gee, 2007; Missingham, 2007; Miller, 2012), many institutions of formal music education have exhibited caution and hesitancy in including music video games as part of classroom activities (Gower & McDowall, 2012).

Pedagogical fundamentalism can also manifest itself in more subtle ways. An alternative to explicit resistance against technology-related changes is a reluctant acceptance of students' relationship with digital technology, but seasoned with the conviction of its corruptive influence and the importance of the teacher in enforcing cultural, moral, and political authority over the media that students experience. Whether due to the scorn of "the media's apparent lack of cultural value, as compared with the 'classics' of great art," or the fear over "the undesirable attitudes or forms of behaviour which they are seen to promote" (Buckingham, 2003, p. 6), students' relationships with new media technologies, according to pedagogical fundamentalism, are understood to represent a problem that needs to be solved through education.

Pedagogical Populism: A Laissez-Faire Approach Based on Technological Determinism

At the other extreme of possible strategies for responding to technology-driven changes is what Buckingham (2010) calls *pedagogical populism*. If pedagogical fundamentalism is based largely on denial, fear, and contempt, pedagogical populism draws on "a kind of populist cyber-libertarianism which claims that ordinary people will somehow be empowered by technology, and that digital media are inherently liberating and countercultural" (p. 291). This approach is often justified by binary models of "old" versus "new," and posits that a more or less uniform generation now exists of "digital natives" who "think and process information fundamentally differently from their predecessors" (Prensky, 2001, p. 1) and should therefore be approached in completely new ways.

Based on "a form of technological determinism" (Buckingham, 2010, p. 292), according to which being exposed to technology will automatically result in social change and media literacy, pedagogical populism dismisses the need for teaching young people critical analysis, for instance. As Gauntlett (2007) states: "the patronizing belief that students should be taught how to 'read' the media is replaced by the recognition that media audiences in general are already extremely capable interpreters of media content ... thanks in large part to the large amount of coverage of this in popular media itself" (p. 3). As regards the music teacher, pedagogical populism results in a demand of continuous alertness. Unless teachers are fluent with downloading, editing, remixing, sharing, producing, deconstructing, reconstructing, and networking, they will "lose both the ability and even the right to teach" their students (Merrin, 2008). But even if the teacher would do her best to keep up with latest technological innovations, pedagogical populism ultimately leads to the question of the legitimacy of teaching. One wonders whether guidance in pedagogical situations is even necessary anymore. Buckingham (2010, p. 291) asks: "if ordinary people are already creating their own diverse meanings, participating and producing their own media, in [an] extremely capable and critical way ... then what do they need to learn, and what do we have to teach them?" With its focus on the importance of participation and creativity as goals in themselves and the significance of "the everyday meanings" (Gauntlett, 2007, p. 3) produced by people engaged in creative practices of media, pedagogical populism appears to suggest a haphazard music education where the teacher's role is to step back and leave students to their own devices, free from the patronizing efforts of helplessly dated "digital immigrants" (Prensky, 2001; 2010).

Although both pedagogical fundamentalism and pedagogical populism are common approaches in music education, the former is perhaps increasingly making room for the latter. However, and as I argue in the following section, neither strategy can be viewed as particularly helpful in terms of constructing educational practices that will support "the creation of a personal, life-long interest in any music" (Westerlund, 2008, p. 91) or guide students toward a stage of greater positive agency. Next, I will delve deeper into the implications that these approaches can be expected to have in music education, particularly in the school context.

Pedagogical Fundamentalism and the Problem of Ignoring Mediatization Cultures in School

In the previous section I discussed a pedagogical strategy based on a defensive attitude toward social-cultural changes driven by new media technologies, referred to as pedagogical fundamentalism. To an extent, this strategy, and educators' cautiousness toward new technologies, is understandable and even justified. Contrary to what is sometimes assumed, technology does not make life simpler for teachers—quite the opposite. Few music teachers are trained to use technology in creative and versatile ways; even fewer are trained in media education.[2] Introducing the latest gadgets in the classroom only for the sake of their novelty would hardly accelerate students' versatile learning. Furthermore, technological innovations undeniably entail phenomena with unattractive by-products related to, for instance, predatory capitalism (e.g., Suoranta, 2003; Schäfer, 2011) and individualism (e.g., Bakewell, 2011), as well as multiple legal issues of copyright and fair use (e.g., Demers, 2006; Dillon, 2006) that call for reflection and assessment.

However, as Buckingham (2003) reminds us, "if schools are to engage with children's media cultures, it is clearly vital that, in doing so, they do not simply attempt to re-inscribe traditional notions of what counts as valid knowledge" (p. 33). Approaching new media through teacherly attempts at protection that undermine students' tastes and pleasures will hardly be taken seriously, and may, at worst, be counterproductive. Davidson and Goldberg (2010) express their concern regarding the widening gap "between the excitement generated by informal learning and the routinization of learning common to many of our institutions of formal education" (p. 49). One could ask whether music education operating from a place of stability, fear, and defensiveness is likely to help students to make connections, for instance, between the information and skills obtained in the music classroom and those encountered in online music communities.

In order to help students to understand how what they learn in school is related to their life outside the school, it is crucial, as already pointed out by Dewey (1900) well before the arrival of digital technologies, that the school engages with and even reflects social reality. Digital music culture is not merely about technological tools and innovations; it is about new understandings regarding, in particular, learning and the development of musicianship. Thus, explorations into this culture provide music educators with important information about broader changes related to the values and experiences of their students' out-of-school musical environments.

The digital-technology-enabled opportunities to participate in musical practices by copying, manipulating, reconstructing, playing with, and redistributing various musical materials instantaneously, and on an extremely large scale, have opened new perspectives to understanding the nature of musical learning. Rather than viewing learning solely as an individual cognitive process with a beginning and an end, acquired as the result of teaching and administered by the expert teacher in the music classroom context, musical learning can now be seen as part of any musical practice (e.g., Green, 2001; Salavuo, 2006; Partti & Westerlund, 2012).

The view of musical learning as social participation draws on social theories of learning, with their emphasis on the significance of peers and communities for one's learning (Vygotsky, 1978; Lave & Wenger, 1991; Wenger, 1998, 2006) and their focus on mediated activity between an individual and the environment in knowledge construction (Paavola, Lipponen, & Hakkarainen, 2004; Paavola & Hakkarainen, 2005). Although the understanding of learning as "essentially a social process"

(Dewey, 1938/1998, p. 65) is not a result of the rise of digital technologies, new media environments provide an ever-increasing number of possibilities for processes during which the participants in (online) learning communities construct their practices, meanings, and identities with respect to those communities (e.g., Salavuo, 2006; Partti & Karlsen, 2010; Partti, 2013). Within this framework, learning can be compared to, for example, a trajectory (Wenger 1998, 2006; Barab, MaKinster, & Scheckler, 2004), social construction (Riel & Polin, 2004), and a social journey (Trayner, 2011).

If, once upon a time, there was a shared and well-established understanding about the development of music-related expertise as a linear process—guarded, initiated, and verified by the master teacher—that notion as the only option has by now been challenged by digital musicians' celebration of musical versatility, flexibility, and mobility in and between various musical communities (Partti, 2014a). Instead of understanding the teacher as the only expert in the classroom, with the authority to steer the novice toward increasing challenges and expanding musicianship, the music-related participatory culture, with its strong emphasis on the values of musical breadth as well as cultural and creative fluidity, highlights the nature of musical expertise as a constant and dynamic process—one that is inherently social, openly shared by nature, and based on appropriation and collaboration (Partti & Westerlund, 2012; see also Väkevä, 2010; Michielse & Partti, 2015). A young pupil stepping into the music classroom is thus increasingly likely to arrive from a musical environment that represents not only new kinds of musical sites, including the latest technological devices and related know-how, but also a new set of rules, understandings, and ideals about musicianship and the development of music-related expertise.

With its tendency of designing music education solely according to particular cultural traditions and norms, educational fundamentalism fails to take into account the student's earlier and outside-school (informal) learning experiences, as well as the whole social environment through which learning takes place. As stated by Dewey (2008; see also Väkevä & Westerlund, 2007; Westerlund, 2008), this kind of ignorance can have severe implications: "If we isolate the child's present inclinations, purposes, and experiences from the place they occupy and the part they have to perform in a developing experience, all stand upon the same level; all alike are equally good and equally bad. But in the movement of life different elements stand upon different planes of value" (Dewey, 2008, p. 280). One could ask whether neglecting or disapproving cultures of mediatization is likely to help the school to fulfill its role in supporting students to experience their global and local realities "as a holistic continuum" (Partti & Karlsen, 2010, p. 377). Or will it, on the contrary, drive the wedge deeper into the dichotomy between music learning environments outside and inside school?

It follows logically from pedagogical fundamentalism that the music teacher's task is understood, to paraphrase Freire (1970/1996, p. 57), as being to regulate the way media "enters into" the students. As I have argued elsewhere, this dichotomy could, in turn, lead to a situation where "students will regard the values and practices of school-based music education as increasingly alien and meaningless" (Partti & Karlsen, 2010, p. 377;

see also Greenhow & Robelia, 2009; Davidson & Goldberg, 2010). Therefore, with its tendency to disregard the school's role in helping to "bridge the widening gap between the school and the world of children's out-of-school experience" (Buckingham, 2003, p. 32), pedagogical fundamentalism offers too narrow an approach to issues related to the music-related mediatizing world.

Pedagogical Populism and the Issue of Music Education's Legitimacy of Teaching

If the school is required to be involved in and reflect social reality, as suggested above, one could ask whether formal music education should aim for a new kind of pedagogy altogether. The Pedagogical populism strategy questions not only the competence of digital immigrants to teach digital natives but the validity of what is taught.

The issue of the legitimacy of teaching is not a new one in music education, as it has been preceded by similar discussions in popular music pedagogy (see Green, 2001, 2008; Folkestad, 2006; Allsup, 2008; Väkevä, 2009, 2013; Karlsen & Väkevä, 2012). Over the last decade, in the search for alternatives for traditional classroom teaching, increasing number of investigations have been made into popular musicians' informal learning strategies. For some, understanding the importance of "garage rock band" practices has meant choosing a pedagogy in which the teacher's role is hardly more than that of a bystander observing students making their own decisions. Leaving the teacher on the sidelines of classroom practices has been justified by, for instance, the importance of providing students with learning experiences that are more authentic and meaningful to them.

According to this line of thinking—not unlike that in new-media-related pedagogical populism—young people develop, seemingly all by themselves, an ability to critically analyze music and performance quality, as well as global music industry manipulative practices, by learning music that they like (see Green, 2008). Such "informal pedagogy" is criticized by those who argue that the recognition of "the multi-faced and mutable ontological status for musical reality" should not result in leaving the student alone (Väkevä, 2009, p. 24; see also Allsup, 2008) but rather in facilitating a pedagogical environment in which the student's reality contributes to the educational purpose and direction (e.g., Väkevä & Westerlund, 2007; Westerlund, 2008; Väkevä, 2013).

While digital technologies have brought about great opportunities for artistic expression, sharing, and participation for many, they have hardly liquidated social inequality once and for all. There are still prominent social differences in access levels to new media and "the opportunities for participation they represent," both between and within societies (Jenkins et al., 2006, p. 12, see also Greenhow & Robelia, 2009; Santos, Azevado, & Pedro, 2013). Many young people seem to acknowledge the potential of the technology provided for sophisticated content creation and sharing but lack the social or personal motivation to actually generate original cultural content, even in countries with wide access to multimedia abilities and broadband. "To a large extent, the most active participants in the creative world of Web 2.0 are the 'usual suspects'" (Buckingham, 2010, p. 294)—those with enough social and cultural capital to participate and contribute in the society and culture already. The laissez-faire approach of pedagogical populism therefore appears to be based on rather naïve assumptions about inclusiveness and "the transformation of former audiences into active participants and agents of cultural production" (Schäfer, 2011, p. 10).

As a public institution, the school has the opportunity to provide young people with "social, intellectual and cultural experiences that they might not otherwise have" (Buckingham, 2010, p. 297). This, in turn, may well be one of the most important ways to prevent the emergence of "an educational 'underclass' that is effectively excluded from access, not merely to economic capital but to social and cultural capital as well" (Buckingham, 2003, p. 203)—especially as digital access often seems to correlate "with educational opportunity and wealth" (Davidson & Goldberg, 2010, p. 51). In the context of the still-existent "participation gap" of digital music culture, the importance of the school is thus highlighted, rather than diminished, in regard to equipping pupils with the requisite social-cultural competencies, skills, and knowledge to overcome digital divides.

As suggested above, the issue of access to media environments does not refer only to specific skill sets related to effectively using digital devices and information networks. Increasingly often, navigating rapidly changing digital habitats requires social skills that enable one to interact within larger communities. Although digital technologies have enabled new possibilities for creative activities and interactions that reach beyond geographical, cultural, and various other borders, merely bringing people together and exposing them to diversity will hardly generate tolerance and respect. In fact, as can be witnessed in the mounting hate speech in online communities, for example, encounters with varying perspectives tend to result in conflicting and competing values, norms, assumptions, and "claims about the meanings of shared artifacts and experiences" (Jenkins et al., 2006, p. 52). Instead of inspiring people to take action in response to suspicion and fearful stereotyping, facing diversity appears to frequently breed withdrawing from those who are different (Putnam, 2007; Sennett, 2012).

New media environments provide multiple opportunities to encounter different views and ways of life, but often in a form of anonymous opinions, comments, and appraisals. Learning to ethically and responsibly communicate, share, collaborate, and socialize in the world of "anonymous diversity"—the world where it may not be necessary, or even possible, to directly face the consequences of one's words and attitudes—calls for emotional maturity and critical and social agency (e.g., Dijck, 2009; Dijck & Nieborg, 2009).

The question arises whether or not the "hands-off" approach to teaching can successfully provide young people with opportunities for active reflection on their media experiences that would further help them to "master the skills they need as citizens and consumers" (Jenkins et al., 2006, p. 16). Will the music classroom based on the principles of pedagogical populism support students enough in developing "the ethical norms needed to cope with a complex and diverse social environment online" (p. 12; see also Schäfer, 2011)?

Buckingham and colleagues have shown (Buckingham, Grahame, & Sefton-Green, 1995) that the laissez-faire strategy according to which the teacher should deliberately refrain from direct instruction is likely to implicitly endorse values and divisions of labor that further, rather than prevent, the (re)production of stereotypes, discrimination, and inequalities. In addition, Greenhow and Robelia (2009) in their study on informal learning and identity formation in online social networks noticed that high school students participating in online communities demonstrated extremely limited understandings of the ethical and legal issues involved in their participation in semipublic online spaces. According to the researchers, the young peoples' disregard for social and ethical responsibility mirrored "the lack of understanding on the part of educators who are supposed to model digital citizenship behaviors" (p. 135; see also Greenhow, Walker, Donnelly, & Cohen, 2007).

In other words, although the widening opportunities for participation in digital-technology-enabled creative practices have provided important means for learning, participation and creativity are only one side of the coin. They should not be treated as end goals in themselves. Participation may or may not be democratic, empowering, countercultural, or liberating, and the experience of creative production is highly unlikely to automatically result in critical literacy (Buckingham, 2010; see also Luke, 2000). As such, it is important that one be able to "step back from immediate experience, in order to reflect and to analyze" (Buckingham, 2010, p. 299), to evaluate, and to discuss those experiences with others, thus gaining a deeper understanding of different media dimensions. It seems that pedagogical populism collapses from educational naivety; this naivety lacks the practical depth and breadth to prepare students for interacting with new media and approaching global music-making and learning environments on equal terms.

I have argued that it is just as unhelpful to adopt the strategy of pedagogical fundamentalism, according to which students are passive victims of harmful and damaging media, as that of pedagogical populism, which uncritically buys into utopian fantasies about social change autonomously produced by technology. Instead, I will suggest an approach that emphasizes media education in music education as a form of preparation. As stated by Buckingham (2003), this paradigm seeks to develop students' understanding of, and participation in, the media culture that surrounds them by highlighting "the importance of media education as part of a more general form of 'democratic citizenship'" (p. 13) while at the same time recognizing the significance of the enjoyment and pleasure people experience with media.

To conclude, I will discuss the relationship between formal music education and new media, particularly from the viewpoint of digital citizenship competency development. I will argue that the school, as a public institution, has a unique opportunity—indeed, the responsibility—to address "the unequal access to the opportunities, experiences, skills and knowledge that will prepare youth for full participation in the world of tomorrow" (Jenkins et al., 2006, p. 3). I will also envision the music classroom as a place of dialogue and reflection, contributing to the development of the media literacy and cultural competencies required for full involvement in the music-related mediatizing world.

Toward Radical Pedagogy in Music Education

As stated throughout this chapter, the phenomena currently emerging from a rapidly changing digital music culture are relevant also to the music classroom, as they inform the educator about new ways people are engaged in music and participate in music-related practices. However, these phenomena do not diminish the teacher's responsibility for organizing music learning and teaching contexts. In an age of digital technologies, the teacher has an even more importance, in recognizing musical growth opportunities and being a "critical companion" during the growth process (Väkevä, 2013, p. 102). The school classroom as "a little community," as envisioned by Dewey (1976, p. 20), should therefore be a place where students' musical creativity and aspirations are taken seriously, and where there is enough space for cooperative engagement between the students and teacher. Rather than adding a media literacy course to the music education curriculum, this continuous dialogue would enable different modes of learning, as well as critical discussions about values, ideas, norms, and subtle forms of propaganda behind various local and global environments and activities of music making. To paraphrase Dewey, such music education would provide fertile ground for the growth not only of the students' ability to "act with and for others" (2008, p. 98) but also of their capacity to independently consider the ways, places, and spaces of music-related activities and their possible moral consequences.

This kind of an approach to the relationship between the school and technology focuses beyond technology and media as teaching materials and curriculum content. Despite the importance of questions related to tools, technical skills, and teaching methods, we should ask more questions about *how* we teach, and what is *not* taught in school. Is the school dismissing or making the most of its chance for a radical—that is, far-reaching and profound—examination of the world? Are we considering "as broadly as possible the nature of education as it exists today—as well as how it might change as we move into the future" (McGettigan, 2008)?[3]

While the world keeps on changing through the impact of technological innovations and globalization, are we, as educators, aiming to create "the formative culture of beliefs, practices, and social relations that enable individuals to wield power, learn how to govern, and nurture a democratic society that takes equality, justice, shared values, and freedom seriously" (Giroux, 2011)? Echoing the voice of Dewey, Giroux states that this kind of formative culture arises from developing pedagogical practices that facilitate "conditions for producing citizens who are critical, self-reflective, knowledgeable, and willing to make moral judgments and act in a socially responsible way" (2011).

Digital citizenship, a notion referring to "an informed and participatory citizenry whose online behaviors simultaneously uphold standards for legal, ethical, safe, responsible, and respectful uses of technology" (Greenhow & Robelia, 2009, pp. 125–126), is advocated also by UNESCO, within its "media literacy" curriculum (Wilson et al., 2011), along with national curricula across the world. Indeed, these institutions support the development of digital citizenship competencies. Helping students to bridge the gap between the knowledge, skills, experiences, and ideals acquired through interacting with new media and those encountered in the music classroom calls for the recognition of the knowledge that students already possess and the significant roles that digital technologies play in their lives.

Hence, the school should neither be reduced to a mere irrelevance nor viewed as "a training ground for the new 'digital economy'" (Buckingham, 2010, p. 297). Instead, as suggested in this chapter, the school can become a seedbed for the growth of *educated hope*—hope understood as "a pedagogical and performative practice that provides the foundation for enabling human beings to learn about their potential as moral and civic agents" (Giroux, 2012). Music education in the mediatizing world thus becomes a collaborative inquiry within which the teacher can let go of the requirement of having all the answers. The teacher can now face the questions related to the mediatizing world as one who, along with her students, also asks questions, speculates, ponders, and wonders. As reminded by Suoranta (2003, p. 204), the teacher can feel at ease about experimenting with new things, and about expanding her field of teaching by allowing media culture meanings to flutter into the classroom with her students.

Epilogue: A Four-Track Recorder and Educated Hope

During a recent visit to get acquainted with the music education system of a postconflict society in Asia, I had the privilege to observe one of the most inspiring music classes I have witnessed in a long time. The music teacher, an elderly gentleman, introduced us to the technological equipment available in his music class, consisting of a hand-pumped organ, a four-track recorder, and a cassette player. In the context of a country where music teaching was not normally part of the school instruction, the tools marked a significant achievement. During the subsequent 45 minutes, the young pupils performed songs for us composed by the teacher, who, with the help of his four-track recorder, had created backing tracks to accompany the enthusiastic, sparkly-eyed singing. I could feel 30 pairs of passionate eyes (including those of the teacher) on us at the moment the teacher pressed "play" on his cassette player. We were equally impressed by the pupils' joyful excitement and the teacher's proficient use of what to us represented a rudimentary form of technology. I could not help wondering whether this was exactly what technology in music education looks like at its best: the teacher who utilizes the technology at hand, be it the most recent model of tablet or a secondhand track recorder, in order to manifest educated hope by caring for his pupils enough to invest all his humanity, compassion, creativity, pedagogical love, and people skills into helping them to grow into maturity and agency.

In the advent of digital technology and media cultures, Freire's (1970/1996) words about education as the practice of freedom are timelier than ever: "Education as the practice of freedom … denies that man is abstract, isolated, independent, and unattached to the world; it also denies that the world exists as a reality apart from people. Authentic reflection considers neither abstract man nor the world without people, but people in their relations with the world" (p. 62). Digital technology and media are not phenomena out there. In our swiftly mediatizing world, the music teacher, along with her students, is an indigenous species in digital habitats. Therefore, rather than observing the world of media from afar, the teacher is called to an explorative journey to gain understanding about a cultural phenomenon with which she is personally intertwined.

The role of the music teacher as a critical companion within cultures "'moulded' by the media" (Hepp, 2013, p. 2) is neither to impose her own views and meanings on her students' experiences nor to abandon children to approach global environments alone. Now more than ever it is time to make the classroom a place of dialogue and critical discourse, through which the students along with the teacher are constantly engaging in the world by conceiving and evaluating ideas and their moral consequences (see Dewey, 1980; Hansen, 2007). Regardless of how worthwhile investigations into new media and informal learning contexts are, it is crucial to open up formal spaces for critical discussions and active reflection (see Allsup, 2008). It is difficult to imagine any other place more suitable for engaging in such dialogue than the school classroom, which daily assembles the members of a whole generation from every possible background under its roof.

Notes

1. This research has been undertaken as part of the ArtsEqual project funded by the Academy of Finland's Strategic Research Council from its Equality in Society programme (project no. 293199).
2. For example, the results from a recent study on Finnish school music teachers' use of technology in their classroom activities suggest that, according to the teachers' own estimation, one of the key challenges preventing them from adopting digital technologies more actively and widely in their teaching is that they lack the skills to do so, not having learnt them during teacher training (Partti, 2015).
3. Importantly, this holistic approach to the development of media literacy in music education, woven through the entire educational system, highlights the significance of establishing teacher education that prepares teachers to analyze and understand different viewpoints and policy languages. Rather than preparing future teachers to merely become fluent in

their subject content, music teacher education should also, and in particular, provide for the development of competences and an outlook that enables teachers to cope with cultural pluralism in increasingly challenging and volatile settings (e.g., Darling-Hammond, 2005).

REFERENCES

Aboujaoude, E. (2011). *Virtually you: The dangerous powers of the e-personality*. New York: Norton.

Allsup, R. E. (2008). Creating an educational framework for popular music in public schools: Anticipating the second-wave. *Visions of Research in Music Education, 12*, 1–12.

Asp, K., & Esaiasson, P. (1996). The modernization of Swedish campaigns: Individualization, professionalization, and medialization. In D. L. Swanson, & P. Manicini (Eds.), *Politics, media, and modern democracy: An international study of innovations in electoral campaigning and their consequences* (pp. 73–90). Westport, CT: Greenwood.

Bakewell, S. (2011). *How to live: A life of Montaigne in one question and twenty attempts at an answer*. London: Vintage.

Barab, S. A., MaKinster, J. G., & Scheckler, R. (2004). Designing system dualities: Characterising an online professional development community. In S. A. Barab, R. Kling, & J. H. Gray (Eds.), *Designing for virtual communities in the service of learning* (pp. 53–90). New York: Cambridge University Press.

Buckingham, D. (2003). *Media education: Literacy, learning and contemporary culture*. Cambridge, UK: Polity Press.

Buckingham, D. (2010). Do we really need media education 2.0? Teaching in the age of participatory media. In K. Drotner & K. C. Schrøder (Eds.), *Digital content creation: Perceptions, practices and perspectives* (pp. 287–304). New York: Peter Lang.

Buckingham, D., Grahame, J., & Sefton-Green, J. (1995). *Making media: Practical production in media education*. London: English and Media Centre.

Collins, A., & Halverson, R. (2009). *Rethinking education in the age of technology. The digital revolution and schooling in America*. New York: Teachers College, Columbia University.

Couldry, N., & Hepp, A. (Eds.). (2013, August). Conceptualizing mediatization [Special issue]. *Communication Theory*.

Darling-Hammond, L. (2005). Teaching as a profession: Lessons in teacher preparation and professional development. *Phi Delta Kappan, 87*(3), 237–240.

Davidson, C. N., & Goldberg, D. T. (2010). *The future of thinking: Learning institutions in a digital age*. Cambridge, MA: MIT Press.

Demers, J. (2006). *Steal this music: How intellectual property law affects musical creativity*. Athens: University of Georgia Press.

Dewey, J. (1900). *The school and society*. Chicago: University of Chicago Press.

Dewey, J. (1998). *Experience and education* (60th anniversary ed.). Indianapolis: Kappa Delta Pi. (Original work published 1938)

Dewey, J. (1976). *The Middle Works, 1899–1924: Vol. 1. 1899–1901* (J. A. Boydston, Ed.). Carbondale: Southern Illinois University Press.

Dewey, J. (2008). *The Middle Works, 1899–1924: Vol. 2. 1902–1903* (J. A. Boydston, Ed.). Carbondale: Southern Illinois University Press.

Dewey, J. (1980). *The Middle Works, 1899–1924: Vol. 9. 1916* (J. A. Boydston, Ed.). Carbondale: Southern Illinois University Press.

Dewey, J. (2008). *The Later Works, 1925–1953: Vol. 6. 1931–1932* (J. A. Boydston, Ed.). Carbondale: Southern Illinois University Press.

Dijck, J. van. (2009). Users like you? Theorizing agency in user-generated content. *Media Culture Society*, 31(1), 41–58.

Dijck, J. van, & Nieborg, D. (2009). Wikinomics and its discontents: A critical analysis of Web 2.0 business manifestos. *New Media & Society*, 11(4), 855–874.

Dillon, T. (2006). Hail to the thief: Appropriation of music in the digital age. In K. O'Hara & B. Brown (Eds.), *Consuming music together: Social and collaborative aspects of music consumption technologies* (pp. 289–306). Dordrecht: Springer.

Folkestad, G. (2006). Formal and informal learning situations or practices vs. formal and informal ways of learning. *British Journal of Music Education*, 23(2), 135–145.

Freire, P. (1996). *Pedagogy of the oppressed*. London: Penguin Books. (Original work published 1970)

Gauntlett, D. (2007). Wide angle: Is it time for media studies 2.0? *Media Education Association Newsletter*, 5, 3–5.

Gauntlett, D. (2011). *Making is connecting: The social meaning of creativity, from DIY and knitting to YouTube and Web 2.0*. Cambridge, UK: Polity Press.

Gee, J. P. (2007). *Good video games and good learning: Collected essays on video games, learning and literacy*. New York: Peter Lang.

Giroux, H. (1999). Performing cultural studies as a pedagogical practice. In D. Slayden & R. K. Whillock (Eds.), *Soundbite culture: The death of discourse in a wired world* (pp. 191–202). Thousand Oaks, CA: Sage.

Giroux, H. (2003). Public pedagogy and the politics of resistance: Notes on a critical theory of educational struggle. *Educational Philosophy and Theory*, 35(1), 5–16.

Giroux, H. (2011). *On critical pedagogy* (Kindle version). New York: Continuum.

Giroux, H. (2012). *The Occupy movement and the politics of educated hope*. Retrieved January 31, 2014, from http://truth-out.org/news/item/9237-the-occupy-movement-and-the-politics-of-educated-hope.

Gower, L., & McDowall, J. (2012). Interactive music video games and children's musical development. *British Journal of Music Education*, 29(1), 91–105.

Green, L. (2001). *How popular musicians learn: A way ahead for music education*. Hampshire, UK: Ashgate.

Green, L. (2008). *Music, informal learning and the school: A new classroom pedagogy*. Hampshire, UK: Ashgate.

Greenhow, C., & Robelia, B. (2009). Informal learning and identity formation in online social networks. *Learning, Media and Technology*, 34(2), 119–140.

Greenhow, C., Walker, J., Donnelly, D., & Cohen, B. (2007). Fair use education for the twenty-first century: A comparative study of students' use of an interactive tool to guide decision making. *Innovate*, 4(2).

Hansen, D. (2007). Ideas, action, and ethical vision in education. In D. Hansen (Ed.), *Ethical visions of education: Philosophies in practice* (pp. 1–18). New York: Teachers College Press.

Hepp, A. (2013). *Cultures of mediatization*. Cambridge, UK: Polity Press.

Jansen, B. (2011), *Where credit is due: Cultural practices of recorded music*. Unpublished doctoral dissertation, Amsterdam University.

Jenkins, H., Clinton, K., Purushotma, R., Robinson, A. J., & Weigel, M. (2006). *Confronting the challenges of participatory culture: Media education for the 21st century*. Retrieved March 18, 2012, from http://www.digitallearning.macfound.org/atf/cf/%7B7E45C7E0-A3E0-4B89-AC9C-E807E1B0AE4E%7D/JENKINS_WHITE_PAPER.PDF.

Karlsen, S., & Väkevä, L. (Eds.). (2012). *Future prospects for music education: Corroborating informal learning pedagogy*. Newcastle upon Tyne, UK: Cambridge Scholars.

Keen, A. (2008). *The cult of the amateur: How blogs, MySpace, YouTube, and the rest of today's user-generated media are destroying our economy, our culture, and our values* (Rev. ed.) New York: Doubleday.

Koltay, T. (2011). New media and literacies: Amateurs vs. professionals. *First Monday, 16*, 1–3. Retrieved January 8, 2014, from http://firstmonday.org/htbin/cgiwrap/bin/ojs/index.php/fm/article/view/3206/2748.

Kupiainen, R., Sintonen, S., & Suoranta, J. (2008). *Decades of Finnish media education*. Finnish Society on Media Education. Retrieved February 8, 2014, from http://www.mediakasvatus.fi/publications/decadesoffinnishmediaeducation.pdf.

Kvale, S., & Brinkmann, S. (2009). *InterViews: Learning the craft of qualitative research interviewing*. Thousand Oaks, CA: Sage.

Lave, J., & Wenger, E. (1991). *Situated learning: Legitimate peripheral participation*. Cambridge, UK: Cambridge University Press.

Livingstone, S. (2013). The participation paradigm in audience research. *Communication Review, 16*(1-2), 21–30.

Luke, C. (2000). Cyber-schooling and technological change: Multiliteracies for new times. In B. Cope & M. Kalantzis (Eds.), *Multiliteracies: Literacy learning and the design of social futures* (pp. 69–91). London: Routledge.

Lundby, K. (Ed.). (2009) *Mediatization: Concept, changes, consequences*. New York: Peter Lang.

McGettigan, T. (2008, April 8). *What is radical pedagogy?* Retrieved from http://ezrawinton.com/2008/04/08/what-is-radical-pedagogy/.

Merrin, W. (2008). *Media studies 2.0*. Retrieved November 10, 2012, from http://twopointzeroforum.blogspot.com/.

Michielse, M. (2013). Musical chameleons: Fluency and flexibility in online remix contests. *M/C Journal, 16*(4).

Michielse, M., & Partti, H. (2015). Producing a meaningful difference: The significance of small creative acts in composing within online participatory remix practices. *International Journal of Community Music, 8*(1), 27–40.

Miller, K. (2012). *Playing along: Digital games, YouTube, and virtual performance*. Oxford: Oxford University Press.

Missingham, A. (2007). *Why console-games are bigger than rock 'n' roll: What the music sector needs to know and how it can get a piece of the action*. Retrieved March 1, 2012, from http://www.youthmusic.org.uk/assets/files/Console%20games%20and%20music_1207.pdf.

Paavola, S., & Hakkarainen, K. (2005). The knowledge creation metaphor—an emergent epistemological approach to learning. *Science & Education, 14*, 535–557.

Paavola, S., Lipponen, L., & Hakkarainen, K. (2004). Models of innovative knowledge communities and three metaphors of learning. *Review of Educational Research, 74*(4), 557–576.

Partti, H. (2013). Oopperasäveltäjäksi oppimassa: Opera by You—verkkoyhteisö musiikillisen asiantuntijuuden kasvualustana [Learning to be an opera composer: The Opera by You online community as a platform for the growth of music-related expertise]. *Musiikki, 1*, 33–50.

Partti, H. (2014a). Cosmopolitan musicianship under construction: Digital musicians illuminating emerging values in music education. *International Journal of Music Education, 32*(1), 3–18.

Partti, H. (2014b). Supporting collaboration in changing cultural landscapes: Operabyyou.com as an arena for creativity in "kaleidoscope music." In M. S. Barrett (Ed.), *Collaborative creative thought and practice in music* (pp. 207–220). Surrey, UK: Ashgate.

Partti, H. (2015, May 3–5). *The bliss and dread of creative music making: Finnish music teachers' approaches to teaching composing*. Presentation at the Nordic Network for Research in Music Education Conference, Helsinki.

Partti, H., & Karlsen, S. (2010). Reconceptualising musical learning: New media, identity and community in music education. *Music Education Research, 12*(4), 369–382.

Partti, H., & Westerlund, H. (2012). Democratic musical learning: How the participatory revolution in new media challenges the culture of music education. In A. R. Brown (Ed.), *Sound musicianship: Understanding the crafts of music* (pp. 300–312). Newcastle Upon Tyne, UK: Cambridge Scholars.

Partti, H., & Westerlund, H. (2013). Envisioning collaborative composing in music education: Learning and negotiation of meaning in operabyyou.com. *British Journal of Music Education, 30*(2), 207–222.

Prensky, M. (2001). Digital natives, digital immigrants. *On the Horizon, 9*(5), 1–6.

Prensky, M. (2010). *Teaching digital natives: Partnering for real learning*. Thousand Oaks, CA: Sage.

Putnam, R. D. (2007). E pluribus unum: Diversity and community in the twenty-first century (2006 Johan Skytte Prize Lecture). Scandinavian Political Studies, 30(2), 137–174.

Riel, M., & Polin, L. (2004). Online learning communities: Common ground and critical differences in designing technical environments. In S. A. Barab, R. Kling, & J. H. Gray (Eds.), *Designing for virtual communities in the service of learning* (pp. 16–50). New York: Cambridge University Press.

Salavuo, M. (2006). Open and informal online communities as forums of collaborative musical activities and learning. *British Journal of Music Education, 23*(3), 253–271.

Santos, R., Azevedo, J., & Pedro, L. (2013). Digital divide in higher education students' digital literacy. In S. Kurbanoğlu, E. Grassian, D. Mizrachi, R. Catts, & S. Špiranec (Eds.), *Worldwide commonalities and challenges in information literacy research and practice* (pp. 178–183). New York: Springer.

Schäfer, M. T. (2011). *Bastard culture! How user participation transforms cultural production*. Amsterdam: Amsterdam University Press.

Seel, J. (2001). Plugged in, spaced out, and turned on: Electronic entertainment and moral minefields. *Journal of Education, 179*(3), 17–32.

Sennett, R. (2012). *Together: The rituals, pleasures and politics of cooperation* (Kindle version). London: Penguin Books.

Suoranta, J. (2003). *Kasvatus mediakulttuurissa* [Education in media culture]. Tampere, Finland: Vastapaino.

Trayner, B. (2011, April 17). *Social learning—on a high horse*. Retrieved from http://www.bevtrayner.com/base/.

Väkevä, L. (2009). The world well lost, found: Reality and authenticity in Green's "New Classroom Pedagogy." *Action, Criticism, and Theory for Music Education, 8*(2), 7–34.

Väkevä, L. (2010). Garage band or GarageBand®? Remixing musical futures. *British Journal of Music Education, 27*(1), 59–70.

Väkevä, L. (2013). Informaali oppiminen, musiikin opetus ja populaarimusiikin pedagogiikka [Informal learning, music teaching and popular music pedagogy]. In M-L Juntunen, H. M. Nikkanen, & H. Westerlund (Eds.), *Musiikkikasvattaja: Kohti reflektiivistä käytäntöä* (pp. 93–104). Jyväskylä, Finland: PS-Kustannus.

Väkevä, L., & Westerlund, H. (2007). The "method" of democracy in music education. *Action, Criticism, and Theory for Music Education, 6*(4), 96–108.

Vygotsky, Lev. (1978). *Mind in society: The development of higher psychological processes.* Cambridge, MA: Harvard University Press.

Waldron, J. (2009). Exploring a virtual music "community of practice": Informal music learning on the Internet. *Journal of Music, Technology and Education, 2*(2-3), 97–112.

Waldron, J. (Ed.). (2015). Participatory culture. [Special issue]. *International Journal of Community Music, 8*(1), 3–6.

Waldron, J. L., & Veblen, K. K. (2008). The medium is the message: Cyberspace, community, and music learning in the Irish traditional music virtual community. *Journal of Music, Technology and Education, 1*(2–3), 99–111.

Wenger, E. (1998). *Communities of practice. Learning, meaning, and identity.* Cambridge, UK: Cambridge University Press.

Wenger, E. (2006). *Learning for a small planet—a research agenda.* Retrieved November 11, 2012, from http://www.ewenger.com.

Wenger, E., White, N., & Smith, J. D. (2009). *Digital habitats: Stewarding technology for communities.* Portland, OR: CPsquare.

Westerlund, H. (2008). Justifying music education: A view from here-and-now values experience. *Philosophy of Music Education Review, 16*(1), 79–95.

Wilson, C., Grizzle, A., Tuazon, R., Akyempong, K., & Cheung, C. (2011). *Media and information literacy curriculum for teachers.* Paris: UNESCO.

CHAPTER 26

THE IMPACT OF TECHNOLOGIES ON SOCIETY, SCHOOLS, AND MUSIC LEARNING

VALERIE PETERS

> We must expect great innovations to transform the entire technique of the arts, thereby affecting artistic invention itself and perhaps even bringing about an amazing change in our very notion of art.
> —Paul Valéry

> The digital revolution hasn't just made music-making easier or faster: It's made it possible.
> —Pete Thomas (2012)

Preface

I grew up without digital technologies. My music technology continuing education took place during my early teaching years (1989–2000) at the Récit des arts.[1] I learned about all the hardware and software for music classes and how to create pedagogical activities that integrated new digital music technologies. As one of four schools to receive funding for a computer-music lab in the province of Quebec, Canada, I was keen to innovate via a curriculum that focused on developing critical and creative thinking with students (see description of the curriculum, appendix). Composing with synthesizers and computer software allowed students to experience music in new ways and to think creatively and critically about how music is structured. I was surprised to learn that the school board had mandated a composer who worked for them to choose the equipment that would be installed in my lab, without consulting with me. The composer had envisioned a corner for splicing tape and rearranging sounds and equipping the lab with outdated synthesizers that were not MIDI compatible. I was opposed to many of the choices, but the school board was convinced that I, the music teacher, did not have the expertise necessary to make equipment decisions even though I would be the one teaching in the lab.

We received 16 synthesizers connected via a lab system and one computer for every two synthesizers. Students could work individually, in pairs, or in groups of four. The lab was used to teach students basic piano keyboard skills, to work with aural skills software, and above all to create and compose. In 2013, I visited the school and discovered that the lab had been closed and dismantled. It had never been updated!

During my graduate studies, I was quite involved with music technology. I even tried my hand at programming (Java) and discovered it was not for me. I also took a course in design of learning environments, which encouraged me to think more deeply about the educational environments that integrate technology and how they might facilitate new ways of learning. During my doctoral research, I used a particular knowledge forum technology as a complement to a conventional ethnographic approach to learning about traditional music cultures (Peters, 2007).

Technological tools have changed dramatically since my doctoral work in the 1990s when, for my qualifying exams, I wrote two essays about technology: one exposed technology as the key to transforming education, and the other discussed the nonneutrality of technology. In 2014, these two positions continue to be well represented in the educational discourse, as is evidenced by a recent review of literature that details the promotion of technology integration as well as the negative impacts of technology on society and classrooms (Després & Dubé, 2012).

In this chapter, I respond to several expansive, provocative questions. This is by no means an exhaustive discussion, but I hope to stimulate questioning in the field and provide some orientations for future pedagogical practices. In the next section I discuss the impact of technology on society and schools and our collective impact on music technologies. After that, I consider several social and cultural issues related to technology integration.

Technology

How is technology changing humans? According to Després and Dubé (2012), the debate about the impact of technology can be summarized by two sides: those who affirm positive impacts of technology on our world and those who point out the negative effects of technology on society at large. For those who promote technology integration, three themes regarding the impact of technology are common: (1) access to education, including rapid, instantaneous educational material and human resources from all over the world; (2) greater education quality because technology is motivating, facilitates acquisition of basic knowledge, and offers continuing education for teachers; and (3) an environment centered on the learner rather than the teacher, allowing for more open activities such as exploration and problem solving. On the other side are the people who insist that ICT (information communication technology) is not a panacea that will solve societal or school problems. They affirm that these technologies have had a negative impact on society, socially, physically, and psychologically. We can think of examples where people retreat from society, developing a type of addiction to "virtual" contact rather than real interaction with other people. Other negative impacts that have been cited are back problems, attention problems, cyberdependence, and the ecological impact of dated out of date technology in waste sites on the environment (Després & Dubé, 2012).

Technology is everywhere, and as educators we cannot ignore its existence in our evolving societies. We are living in an age of multimodal, participatory online learning environments that are changing dramatically how we engage with each other and the world. In a video created for the research group, Susan O'Neill describes how the globalized age has impacted artistic learning and created new opportunities for people to become involved in types of artistic meaning making, expression, and understanding that were never available before.[2] The incredible, fast-paced changes in digital technologies have created challenges for learners, educators, and policy-makers. O'Neill's research focused on how to best foster creative and collaborative learning opportunities in and through the arts in order to bring about transformation in thinking as well as in social and educational practice. According to O'Neill, nothing from the past can prepare us for the incredible changes that are going on currently in artistic learning in ways that are multifaceted, multidimensional, and multimodal. Art forms are no longer separate subjects that are learned in discrete places. Old and new media are being put together in innovative ways, creating what O'Neill describes as the "blending, blurring and braiding" that characterizes artistic learning for twenty-first-century learners. O'Neill's goals are to increase student engagement with the arts, to strengthen the arts cultures of schools and communities, and to create expansive learning opportunities in and through the arts using digital media technologies. Research on understanding artistic learning and youth arts engagement in a digital age is currently under way (O'Neill, Bosacki, Peters, & Senyshyn, 2012). This work should provide us with a better understanding of our students, their artistic worlds, and how we might engage with them more meaningfully in the music classroom.

It is difficult to state clearly how new technologies are changing us, but it is certain that we are being changed. It is important to examine how digital technologies are impacting our society, our human relationships. Carr (2010) states: "as our window onto the world, and onto ourselves, a popular medium molds what we see and how we see it—and eventually, if we use it enough, it changes who we are, as individuals and as a society" (p. 3). Carr goes on to describe a study of the effects of Internet use on young people, including the fundamental changes in the way people absorb information. How does technology mediate our understanding of the world? How should it? These are fundamental questions that we must answer in education as we propose innovative pedagogical practices in the twenty-first century, when learning can take place any time, any place, and with anyone in the contexts of participatory, online cultures.

Music Technology

How is music technology changing us? How are we changing music technology? There have been amazing transformations in music technologies in the last two decades, moving from the electronic keyboards and MIDI technologies I used in my music lab to tablets, smartphones, and MP3 players in more recent years (Després & Dubé, 2012). Technologies have changed the way we compose and share music.

The introduction of mobile apps has instigated a completely alternative way of becoming a musical creator. These technologies provide ease of entry and a space to advance skills and knowledge, thereby allowing any user to enter the world of music that was previously reserved for those with performing skills and knowledge of traditional music notation (O'Neill & Peluso, 2014, p. 119). In addition, young people have easy access to many musical styles. Music production, creation, and sharing have become less expensive, and traditional music skills and theory knowledge are no longer "necessary" in order to compose and perform music in particular music traditions. The "perfect" performance is now possible via recording technologies, and this is creating new challenges for "live" artists and their audiences. Students do not need to spend years learning to play an instrument in order to communicate musical ideas. They can learn from just about anyone on the web, and they themselves can share their accomplishments and teach others in participatory online communities. This has drastically changed the paradigm of teaching and learning music and has important implications for classroom pedagogy.

While no studies have examined technology integration in Quebec primary and secondary schools, a recent survey study has documented different aspects of technology-based music classes (TBMCs) in high schools in the United States (Dammers, 2012). The purpose of the study of administrators and music teachers was to determine the extent to which public high schools offer TBMCs and to describe their nature. Fourteen percent of schools indicated they offered these classes, with suburban high schools in the northeastern United States being the geographical location most likely to offer them. According to music technology teachers, classes were designed and taken by nontraditional music students. Given that 67% of the classes were created in the last 10 years, we do not yet have detailed descriptions of varied contexts and how music technologies are changing our students us in these situations.

How is society changing music technology? This is a more difficult question. I believe that we are demanding much easier access to different types of operations. For example, recording, notating, and other ways of inputting data to create music with computers has become more and more intuitive, easier to do, and more in line with computer operations that are used in other well-known software packages. There was a time when only the "programming jocks" (no offense to those readers who identify as such) were able to access higher-level operations of some software. Technology has become much more usable by the everyday human being (my mother, for example). In addition, we can now interact with music technology via gestures and our voices, opening up a whole new world of possibilities for artistic creation and performance. The tool is becoming more "human" in one sense. I remember improvising in the sound lab at the Experience Music Project Museum in Seattle in 2013 by gesturing "intelligently" with a machine. I was feeling my improvisation, and the technology allowed me to express myself without the necessity of mastering an acoustic instrument. I felt a sense of competency about this "performance," as the machine was able to effectively translate my musical ideas into sounds. I believe that society will continue to demand more seamless and highly intuitive technologies, which will take us in and out of different artificial worlds and allow us to express ourselves effortlessly. In addition, people with special needs are helping to transform music technology for everyone. They are demanding more kinesthetic, interactive, intuitive, and innovative technologies, and this is dramatically changing how we conceive of music making in the twenty-first century.

How are Music Educators Responding to Social, Cultural, and Economic Issues? How Should They?

According to Wellington (2005), recurring debates about the role of technologies in education are articulated around three principal areas: vocational (ICT and the workplace), pedagogical (added value of technology in learning), and societal (home learning v. school learning; equality of opportunities). In terms of the societal area, learning needs to be socially relevant for students, giving them some ownership of what, how, and when they are learning. Burnard (2009) discusses the importance of closing the technology gap between home and school as it relates to underachievement, dropout, and negative attitudes to secondary school music: "the challenge of technology is to find ways of developing the knowledge about digital music consumption and production brought from the home to school; moving technology from being an 'add-on' to being in the centre, embedded rather than integrated in the secondary music curriculum; employing technology to do more than merely 'serve' tradition; and enabling technology to bring 'real world' experience into the classroom" (p. 197). This is a very difficult shift for future music teachers, who continue to be influenced primarily by their private studio teachers, their high school music teachers (and especially their band directors), and their ensemble directors at the university (Woodford, 2002). With few exceptions, these contexts continue to employ a quite conservative and traditional pedagogy (skills transmission and knowledge telling) rather than one that seeks to build on the learner's interests and background.

Should music teachers continue to preserve and transmit traditional music knowledge or exercise leadership in developing technology skills through authentic, "real world" tasks? Is it possible to embrace the past, live in the present, and prepare for the future? If so, how? In the following section, I will discuss five different issues concerning the location and context of music education technology: curricular, sociocultural, ecological and economic, access, and gender issues.

Curricular Issues

There is certainly a need to discuss foundational issues related to technology integration in the curriculum. Dammers (2012) suggests that "a vigorous and challenging discussion of learning objectives and pedagogical practice is necessary during the formative stages to ensure that the establishment of a solid foundation of pedagogy will serve us well through the century" (p. 82). What are our pedagogical bases for TBMCs and how do we prepare our future music teachers for these contexts? Personally, I use multimedia creativity case studies for classroom music teachers (Peters, 2012, 2014) as a way for students to explore the pedagogical and organizational aspects of teaching creativities. In three of the case studies, music technologies are used extensively as part of the pedagogical approach to creating. Many of the undergraduate music education students I work with do not have the opportunity to work in technologically rich music learning environments during their student teaching, and this allows them to reflect on how "real" teachers integrate technology into the curriculum.

I do not think it is a question of whether we integrate ICT into the curriculum but how and why. In addition, my belief in the development of the "whole person" or the "whole musician" leads me to a very eclectic stance, conceiving of curricula as broad and varied. Students need to experience music in many ways, from participatory activities that engage the body through movement, dancing, singing, and playing to activities that incorporate ICT in composing, performing, listening, and other activities. I am especially drawn to pedagogical situations that combine these different "ways of doing music," blurring lines between the traditional and the nontraditional approaches. I am not a fan of choosing up—I want my students to "have it all"!

Sociocultural Issues

O'Neill and Peluso (2014) describe the importance of online, participatory cultures for young students. As mentioned, young people view artistic learning as multimodal (combining different art forms). The understanding of the contexts and ecologies of young people's music making is central to adapting instruction for them. Therefore, sociocultural issues are directly connected to curricular issues. We do not do a great job of this in music education, and I think attrition rates in our elective classes attest to this. We (I) are (am) not in touch with the sociocultural contexts of our (my) students. It is difficult to stay in touch with the latest trends, as technology innovations and life itself seem to move at a breakneck tempo. I think the idea of sharing with our students is a good place to start. In Quebec we use the expression *passeur culturel* (cultural mediator) to describe the "go between" role of teachers, moving between the "student culture" and "general culture." In many instances, moving toward our students will allow them to move toward us and create an open space for sharing the many and varied music cultures and practices of the world.

Ecological and Economic Issues

Absent from much of the discussion about technology integration are issues of ecology and economics. These issues are important because of the investment of substantial, nonrecurrent financial resources (Webster, 2007) and the important impact on the environment when we dispose of equipment. In addition, although technology has become considerably less expensive, society still needs to invest monies wisely (equipment and teacher education) in order to ensure that this investment contributes to students' authentic musical experiences. During a trip to Ghana in 2012 to study drumming and dancing, I was surprised and saddened to witness the environmental impact resulting from our disposal of dated technology. As our bus passed through the capital city, Accra, we noticed huge piles of used technology from across the globe that had been dumped in the river. This brings up a whole host of other ethnical issues, including the use and misuse of technology, the transport of technology waste to other countries, and the development of sustainable solutions for technology use and reuse. Some of the participants in our group were moved to tears as they contemplated the polluted river, choked full of technology waste.

Access Issues

Technology and music technology do not exist in a vacuum. They reinforce sociocultural norms, such as socioeconomic status and gender, to name two. However, according to Dammers (2012), technology classes offered in schools may provide a more accessible entry point for nontraditional music students with different backgrounds and motivations.

It is possible that technology may provide a focal point for new pedagogical practices that will attract more students to study music in middle school and high school. The efforts by teachers of TBMCs have clearly been successful in attracting new students in the United States since, on average, 69% of the students in these TBMCs are nontraditional music students (Dammers, 2012, p. 81). Music technology classes seem to have a broader focus and fewer restrictions than traditional ensemble groups. "This flexibility allows students with widely divergent backgrounds to be successfully served in the same class, but the teacher must apply very different pedagogical skills from those used in a performance ensemble" (p. 82).

O'Neill and Peluso (2014) describe technology as the equalizer in terms of accommodating students with diverse musical backgrounds. They describe technology integration with tablets for creative and collaborative composition with a group of university students with diverse musical backgrounds (not all the students in the class could play an instrument or read notation). It is important, however, not to assume that all students are digital natives and to differentiate instruction. I have found that among my own undergraduate music students quite a disparity still exists between the technologically savvy students and those who feel very uncomfortable with music technologies. While students do have extensive experiences with certain technologies, music teacher educators must not assume that they understand or have experience with the wide range of technologies available for music teaching.

The most exciting aspect of the access issue is how music technologies are transforming the lives of people who are unable to play conventional musical instruments, whether they are students with special needs or musicians who have had debilitating accidents. "Assistive music technology" is a term coined by Doug Briggs, who makes music accessible to people with special needs. This visionary initiative responds to both therapeutic and creative needs of professionals and music students. Ian Gibson of the Adaptive Music Technology Research Group describes this important work:

> "It looks at adapting technology in novel ways," says Gibson. "It could come from a disabled person battling against having a limited ability in a certain area. We can make it easier for those people to make music, a wonderful way of communication. The other side is that we are all in a digital world of music-making where the possibilities are infinite, and therefore it becomes harder and harder to control the way we make music. It works for disabled people to be able to make music, and for non-disabled people to do it more easily. If you have technology from the industry that makes it easier for a non-disabled person, you should be able to adapt that for people with special needs. It doesn't always work to design technology solely for disabled people." (Thomas, 2012)

I love how this process of designing for special needs is pushing the envelope of how we define music technology and music making and allowing us to open up new creative possibilities for all our students. Mark Hildred, who also works for the research group, continues: "it's an interesting side-effect that if you do design something for a specific disability, often you do end up with a musical instrument suitable for a whole range of people. It might introduce any very young children to music making, or you might design something for mainstream music production as an iPad app, for example, which you discover all of a sudden might be playable by somebody in a wheelchair using their nose" (2012). Adaptive music technology will have an important impact on music education contexts and will allow us to adapt instruction for all our students, as well as to imagine innovative ways to be involved in music making.

Gender Issues

Gender is one example of how accessibility to music technologies is connected to social and cultural contexts. While the Dammers (2012) study does not provide information about the gender makeup of music technology classes, 77% of music technology teachers in the study were males. In her book *Technology and the Gendering of Music Education*, Victoria Armstrong (2011) describes how boys and girls work with music technology during the various stages of the music composition process. This empirical study was carried out in four secondary schools in England from January to June 2003.

Armstrong describes the differences between how girls and boys interact with technology, specifically music technology. Girls often rely more on schools, manuals, and structured introductions to technology. In general, girls have less access to technologies than boys, and this leads to a lack of confidence. Boys on the other hand seem to be more informed and self-taught and often portray themselves as competent learners. They like being in control and controlling technologies. During an evaluation of music in schools, the UK Office for Standards in Education found that five times more boys than girls opt for music technology courses (Office for Standards in Education, 2009).

Armstrong argues for a deeper understanding of how social practices with music technologies can be highly gendered and calls for a critical examination of discourses that "reflect a disturbing technological determinism that uncritically embraces technology's supposed transformative possibilities" (2011, p. 19). For example, discourses of technological determinism embrace visions of technology that democratize, transform, and emancipate societies. However, particular music cultures with a strong technological focus have been traditionally much harder for women to break into. Have things changed dramatically in the last 10 years? Does there continue to be a digital divide between boys and girls?

We need to be mindful of difference and plurality in music technology classrooms. Gender is only one example of how music technology contexts mirror how society interacts with technology. If we are mindful that girls and boys do not approach music technology or music composition with technology in the same way, we can, at the very least, integrate strategies that might result in more inclusive pedagogical practices. "Uncritically celebrating claims made of digital technology as 'liberating,' 'empowering,' and 'democratizing,' offering unparalleled possibilities for children's creativity merely perpetuates myths about the so-called 'digital generation' because it does not acknowledge disparities regarding use and accessibility of technologies or the different social, economic and political contexts of their use" (Armstrong, 2011, p. 24). Despite arguments for technology's democratizing potential, it appears that there are gender differences, not in any innate, essentializing way, but differences produced through the reproduction of gendered understandings of technology within society. These differences are produced through discourses that posit boys and male teachers as the technological experts, where boys are given greater compositional autonomy in contexts in which boys' musical deviance will not only be tolerated but will contribute to teachers' perceptions of male pupils as confident and competent technologists and composers (p. 136).

Conclusion: The Potentialities of Music Technologies

I believe that technologies and music technologies have forever changed the face of our world and the way our students view music and music learning. I embrace the amazing potentialities of technology and what it can bring to music education. Students can express their musical ideas with more ease, and music can become participatory for everyone rather than remaining a spectator sport for some. At the same time, we must acknowledge that technology is part of the social and cultural fabric of a society, and therefore we cannot ignore the issues that are embedded in every context of technology integration. Music technologies open multiple possibilities for enabling students with special needs, allowing them to create sounds and to express themselves. Music technologies allow people to create music with little or no formal knowledge. Music technologies have endless possibilities. As a profession, we need to establish a strong philosophical and curricular foundation for our pedagogical practices that integrate music technologies in the classroom and identify the unique learning opportunities that technological advances have made possible.

Appendix

Creative and Critical Thinking in the Music Curriculum

Creative and critical thinking are important components of the music concentration program at Rosewood High School 1. We believe that student learning must go beyond regurgitation or replication, that students must think for themselves in critical and creative ways. Creative thinking is encouraged through exploration, improvisation, and composition. Critical thinking is incorporated into performance activities as well as proposing special projects for the students. While we encourage a traditional program of music instruction, we also make efforts to innovate in the curriculum, finding new ways of enriching student learning in order to provide a more holistic pedagogical experience for students. The foundations of the curriculum activities are current research, best practices as experienced by other professionals in the field, and teacher intuition. At this point in time, it is difficult to make definitive statements about the "most effective" methods to be used. The research base is not exhaustive, and the focus of the research itself is complex. However, some conclusions based on research in combination with adolescent development are plausible, and we use these as indicators in our structuring of the curriculum. However, the interaction with "real students" in "real situations" provides us with constant feedback and helps us to constantly modify things to reflect the needs and the reality of the students.

Secondary I (Grade 7)

Students beginning their studies at the secondary level bring openness to new things and a spirit of exploration. They are, at this point in their development, willing to experiment and try new activities. They have a naïve excitement about their new endeavors. Several activities seem well suited to this developmental stage in the young adolescent's life.

Music Technology Lab

All students are introduced to basic keyboarding skills at this level. They work at right- and left-hand technique as well as chording in the left hand near the end of the year. In addition, students work through several theory and ear training activities in a software program. They focus primarily on pitch reading, rhythm matching, and pitch matching. These skills reinforce student learning in the instrumental setting and help them to read more accurately, hence removing barriers to becoming better instrumentalists. Students also create soundscapes, using a visual aid as their inspiration and the sounds of the synthesizer as their palette. This type of an open-ended creative activity gives the students freedom to explore the many timbres that are available to them and combine them together to form a creative work. Students also create a "story-like" composition based on a cartoon of their choice. These two types of compositions provide an opportunity to discuss the difference between a linear type of compositional process and one that is more evocative and atmospheric, reflecting a particular feeling, idea, or emotion.

Composition

Special projects at this level include collaborative composition projects using percussion instruments. Students work in groups, make decisions, and notate their final products for other students to perform using nontraditional notation. This provides the students with a different type of experience from the individual creative process and introduces them to "collaborative creation."

Secondary II (Grade 8)

Music Technology Lab

Students continue to develop their keyboarding skills. They use their knowledge of chords and right-hand technique to write a pop song in collaborative groups of four. Each student is responsible for a different aspect of the song: lyrics, melody, chords, rhythm. The students record their creations in a software program. The students also continue their ear training activities, focusing on more complex pitch matching with intervals and rhythm matching using the library of musical excerpts available to them. When students finish their required segments, they may work on melody writing or compose rock or rap music in a CD-ROM.

Improvisation

Students are introduced to simple jazz improvisation. We use jazz method based on the imitation, assimilation, and innovation of short melodies. Students listen, sing, and then improvise, a sequence that is pedagogically logical and allows the students to experience immediate feedback and success. While the students have experienced many exploratory activities during secondary I (grade 7), we try to introduce them to creativity within an established tradition in secondary II (grade 8). Students also listen to improvisations by jazz masters and study improvisation transcriptions as a way of learning how an expert musician creates in the moment.

SECONDARY III (GRADE 9)

Music Technology Lab

Students work more closely with MIDI software during this year. They will learn to use notation software, sequencer software, and accompaniment software. MIDI will be explored in more detail, and students will have opportunities to use these tools to create a story composition or their own creative work. As the students become older and feel more comfortable with the tools, they are given more ownership of their project choices, and many different types of projects may be happening simultaneously. We begin here to encourage "distributed learning," where students become experts in an area and share their expertise with others. Special projects have included composing melodies along with other schools, sharing these melodies, and composing accompaniments. The final products were showcased as a virtual concert on the Internet.

SECONDARY IV, V (GRADES 10, 11)

In general, students have quite a demanding performance and academic schedule during these grades. Therefore, the curriculum is made up of special projects, including a major transcription for the band, research into different multicultural traditions, and collecting ideas and stories from a local music culture. At this level, we are trying to encourage higher-order thinking skills in our students and the ability to think outside the box. We want to encourage both their creative thinking and their analytic skills and offer them projects that will stimulate them to think differently about music and the world.

Notes

1. The Récit des arts is part of a Quebec network that provides professional development and teacher training regarding the use of ICT in theatre, visual arts, dance, and music classrooms. For more information, see Leading English Education and Resource Network, http://www.learnquebec.ca/en/content/recit/ or http://www.recitarts.ca/.
2. See http://www.modalresearch.com/video-artistic-learning-dialogues/.

References

Armstrong, V. (2011). *Technology and the gendering of music education.* Burlington, VT: Ashgate.

Burnard, P. (2009). Creativity and technology: Critical agents of change in the work and lives of music teachers. In J. Finney & P. Burnard (Eds.), *Music education with digital technology* (pp. 196–206). London: Continuum.

Carr, N. (2010). *The shallows: What the Internet is doing to our brains.* New York: Norton.

Dammers, R. J. (2012). Technology-based music classes in high schools in the United States. *Bulletin of the Council for Research in Music Education, 194,* 73–90.

Després, J. P., & Dubé, F. (2012). Une synthèse de la litterature portant sur les enjeux philosophiques liés à l'intégration des TIC en education musicale. *La Revue musicale OICRM, 1*(1). Retrieved January 5, 2017, from http://revuemusicaleoicrm.org/une-synthese-de-la-litterature-portant-sur-les-enjeux-philosophiques-lies-a-lintegration-des-tic-en-education-musicale/.

Elpus, K., & Abril, C. (2011). High school music students in the United States: A demographic profile. *Journal of Research in Music Education, 59*(2), 128–145.

Office for Standards in Education. (2009). *Making more of music: An evaluation of music in schools 2005/8.*: Office for Standards in Education. Retrieved from http://www.ofsted.gov.uk/resources/making-more-of-music-evaluation-of-music-schools-2005-08. Ottawa, Canada.

O'Neill, S. A., Bosacki, S., Peters, V., & Senyshyn, Y. (2012). *Understanding artistic learning and youth arts engagement in a digital age* (Insight Grant from the Social Sciences and Humanities Research Council of Canada).

O'Neill, S. A., & Peluso, D. C. C. (2014). Using dialogue and digital media composing to enhance and develop artistic creativity, creative collaborations and multimodal practices. In P. Burnard (Ed.), *Developing creativities in higher music education: International perspectives and practices* (pp. 115–126). New York: Routledge.

Peters, V. R. (2007). Collaborative knowledge building of ethnic musical communities in an urban high school: An ethnographic case study. *Dissertation Abstracts International, 68*(09), 3778. (University Microfilm No. 3278066)

Peters, V. (2012). L'approche par études de cas multimédia: Engager les étudiants dans une professionnalité en milieu scolaire. In M. Giglio & S. Boéchat-Heer (Eds.), *Entre innovations et réformes dans la formation des enseignants, Actes de la recherche, 9,* 185–198. Bienne, Suisse: Éditions HEP-BEJUNE.

Peters, V. (2014). Teaching future music teachers to incorporate creativity in their teaching: The challenge for university teacher practice. In P. Burnard (Ed.), *Developing creativities in higher music education: International perspectives and practices* (pp. 162–173). New York: Routledge.

Thomas, P. (2012). Music technology and special needs: Part 1. *Sound on Sound*. Retrieved January 5, 2017, from http://www.soundonsound.com/sos/dec12/articles/assistive-tech-1.htm.

Webster, P. R. (2007). Computer-based technology and music teaching and learning: 2000–2005. In L. Bresler (Ed.), *International handbook of research in arts education* (pp. 1311–1330). Dordrecht: Springer.

Wellington, J. (2005). Has ICT come of age? Recurring debates on the role of ICT in education, 1982–2004. *Research in Science & Technological Education*, 23(1), 25–39.

Woodford, P. G. (2002). The social construction of music teacher identity in undergraduate music education majors. In R. Colwell & C. Richardson (Eds.), *The new handbook of research on music teaching and learning* (pp. 675–694). New York: Oxford University Press.

CHAPTER 27

RE-SITUATING TECHNOLOGY IN MUSIC EDUCATION

EVAN S. TOBIAS

It was with a bit of nerve, naiveté, or both that I walked into the office of the school district's assistant superintendent of technology at the end of my first year of teaching to discuss my five-year plan for integrating technology into my middle school instrumental and general music program. After listening to me describe how technology could benefit students, Dr. M. let me know he would see what he could do. For several years I placed technology throughout my room for students to engage in projects ranging from creating original music with a digital audio workstation to sampling music from a CD featuring Creative Commons (2004) licensed tracks distributed by *Wired* magazine.

I can still hear strains of Zap Mama and Gilberto Gil emanating from inexpensive computer speakers while teenagers excitedly discussed what they wanted to sample. Throughout the class, students listened intently to the music, selected short segments of the audio recording, set markers to keep track of the parts they wished to sample, highlighted parts of the waveform, and then copied and pasted these samples into new audio tracks. A key interest of mine, then and now, is how technology might support a broad range of ways of being musical, including but extending beyond performing. Reflecting back on my ongoing interest in social justice and equity, experiences teaching and learning with students of diverse backgrounds in my early career, or how musicological inquiry sparks interesting ways of thinking about or engaging in music, I find myself excited by the potential for technology to mediate overlapping inquiry and musical engagement that addresses social and cultural issues or contexts.

My experience as a musician, educator, and researcher is primarily in the context of North America and specifically the United States. With this in mind, throughout this chapter I discuss technology in contexts with which I am familiar: where a significant segment of the population has access to technology, infrastructure exists to support societal functions mediated by technology, and technology impacts the ways many people interact with one another and the world. Throughout this chapter I explore how technology might be situated to address social and cultural issues or contexts and expand opportunities for people to engage with music.

Technology's place in music education is largely related to how it is socially constituted. Despite how technology enables intersections of and blurred boundaries between ways of being musical (Cain, 2004; Partti, 2014; Tobias, 2012a) it is often situated in terms of hard boundaries and compartmentalized notions of musical engagement. Music education's tendency toward cultural reproduction (Schubert, 1986) of existing paradigms of music and musical engagement often limits the potential of technology to support and mediate a broad range of ways people can be musical.

Wajcman (2002) suggests that "technology is a socio-technical product, patterned by the condition of its creation and use" (p. 351). Technological change, according to Wajcman, is "shaped by the social circumstances within which it takes place" (p. 351). As Wajcman argues, "a technological system is never merely technical: its real world functioning has technical, economic, organizational, political, and even cultural elements" (p. 352). This means considering technology in relation to people and to the places and contexts in which it is engaged.

Music education often situates technology as tools without necessarily considering related social, cultural, or musical contexts. The failure to apply critical frameworks can lead to technicist approaches to doing music without addressing the types of inquiry that arts and humanities offer. This can be observed in contexts ranging from professional development to practitioner-oriented publications that focus on how to use technology as tools and techniques without addressing musical, social, or historical contexts, let alone critical theories and frameworks. In this chapter, I suggest that we broaden our focus as music educators beyond technology itself to encompass related social and cultural issues.

I first discuss how philosophical, pedagogical, and curricular perspectives play a key role in the types and degree of change that occur in relation to technology and music education. I then suggest how music educators can enact a more reflective and critical engagement with technology that accounts for sociocultural issues and frameworks. To forward related praxis, I propose that music educators reconceptualize curriculum and resituate technology to address social and cultural issues explicitly. To this end, I invite music educators to consider the potential of digitally mediated musical engagement within the contexts of curriculum as experience and as social reconstruction (Schubert, 1986). To illustrate key principles, I include several vignettes based on composite accounts of my own and other music educators' experiences and discussions of the topics at hand. The content and organization of this chapter were catalyzed by and seek to address the Provocation Questions posed by the editors of this book.

Considering Change in Relation to Technology and Music Education

For the purposes of this chapter, I situate change in terms of practice, as in the ways that music educators engage people in music, musical engagement, and learning. Thus,

I sidestep questions about whether humans are changing physically, cognitively, or socially in relation to or as a result of technology. Discussions of change and technology risk lapsing into technological determinism, a stance asserting that technology determines or causes change to occur (Grint & Woolgar, 1997). Examples of music educators engaging in technological determinism might include their lauding technology for improving music teaching and learning or blaming technology for causing a decline in the quality of music or for hindering students from being musical. Claiming that technology is changing music educators or students can promote a deterministic outlook. Thus, music educators might problematize the notion that technology is causing change to occur.

As an alternative to technological deterministic perspectives on change connected to technology in music education, we might consider how people and technology are coevolving and impacting one another. Hayles (2012) refers to the coevolution of people and technology as a process of *technogenesis*, which recognizes that "both sides of the engagement (humans and technologies) are undergoing coordinated transformations" (p. 81). Hayles complicates notions of people being changed by technology by highlighting the "ongoing dynamic adaption of technics and humans that multiple points of intervention open up" (p. 81). From this perspective, changes that occur in music education can be attributed to the discourses, possibilities, challenges, sociocultural or musical contexts, and ways that people conceptualize, interact with, and integrate technology in their lives rather than to technology itself.

While in some cases change might be occurring in music programs, the presence of technology does not necessarily indicate or catalyze change. Music education can resist evolution by rejecting technology, sequestering it in particular aspects of music programs or applying it in ways that maintain the status quo. Determining how and the extent to which technology plays a role in change as discussed in the opening of this chapter, necessitates understanding how it intersects with music teaching and learning.

Strategies of Assimilation and Distinction

In discussing the role of technology in the digital humanities, Hayles (2012) describes strategies of *assimilation* and *distinction*, which may be helpful frameworks for similar discussions in music education. Strategies of assimilation involve using technology to extend what already exists into the digital realm (2012). Music educators who adopt such strategies can integrate technology in ways that maintain the status quo and can modify curriculum and practice without expanding or transforming music education. For instance, music educators might integrate technology to help students play better in tune or share and distribute recordings on social media or to allow students to create music using notation-based programs. These applications of technology replicate existing practices.

Music educators enacting strategies of assimilation might ignore or limit the potential of technology to change practice or support new ways of being musical. For instance, just as a mobile device can facilitate the performing of synthesized sound through gestural control, it can mimic acoustic instruments or be used by students to identify intervals or note names through computer-assisted instruction. Other music educators might include technology that fits within their paradigms of what constitutes musical engagement and reject that which they cannot assimilate. Thus, technology can be socially constituted in ways that resist significant change.

A strategy of distinction offers alternatives to integrating technology in music programs through assimilation. In this context, distinction emphasizes unique aspects of or new approaches afforded by technology (Hayles, 2012). This approach highlights what is distinct about technology in relation to music, musical engagement, and learning. Educators who support students' interacting with technology through strategies of distinction may enact related changes to curriculum, pedagogy, and types of musical engagement included in the program.

Many of the ways that musicians leverage technology to create, interact with, perform, and analyze music are atypical or nonexistent in elementary or secondary music programs. This can range from manipulating and processing live sound to live coding (Hugill, 2012) as means of creating and performing music. A strategy of distinction might help music educators integrate or adapt a broader spectrum of practices for music education contexts. For many, this would constitute a significant change from what typically occurs in their programs. While music education scholarship is beginning to address some of these issues in relation to musicians' practices (Burnard, 2011; Savage, 2005) and classroom contexts (Finney & Burnard, 2007), little evidence supports that such ways of engaging with music through technology are typical in music programs or have affected the types of curriculum or practices present in school music programs.

While change can be positive, technology is not inherently progressive or beneficial to students (Swanwick, 2001). Music educators sometimes focus on distinctive features of technology without considering curricular, pedagogical, philosophical, or sociocultural issues. Some may also perceive their integration of technology as transformative even while maintaining a particular paradigm of musical engagement. Analyzing how technology is socially constituted and integrated in music programs can help educators understand how and the degree to which it and related change contribute to students' musicianship, growth, and learning.

Degrees of Change: Three Images of Assimilation and Distinction in Practice

When making decisions regarding technology, music educators may find themselves negotiating a balance between fostering change and maintaining the status quo. This negotiation may impact how one approaches technology, through strategies of assimilation or distinction. To make informed decisions, music educators might consider the philosophical, curricular, and pedagogical foundations upon which they draw when determining why, when, and how they might change or maintain what currently occurs in their programs.

The following three vignettes portray how music educators might address technology through strategies of assimilation or distinction in varying degrees of change. I will then discuss corresponding images of transformation in music education (Jorgensen, 2003), in order to help situate technology in relation to change.

Looking in on Marty Christensen's Music Program

Marty Christensen's colleagues and students consider him an exemplary music teacher. Marty is dedicated to maintaining the excellence of his ensembles so that his students reach their maximum potential performing varied music. He recently began including mobile technologies in his ensembles to help students improve as performers and to increase the efficiency of his rehearsals. Students use apps for purposes such as referring to digital fingering charts or checking their intonation.

Marty recently used technology to automate the process of assessing students' pitch and rhythmic accuracy on select literature. Occasionally, students use the Internet to research composers or historical information about the music they perform in ensembles. In Marty's music theory class, students often work with computers or mobile devices to identify intervals and chord qualities. In place of practicing part-writing exercises using paper and pencil, students use notation-based software. Marty recently incorporated cloud-based resources in his program, such as a modular music history curriculum. His students seem to enjoy interacting with multimedia more than the textbooks they used in the past. Marty considers himself successful at staying current without changing much in his classes and ensembles.

Looking in on Paola Acevedo's Music Program

Paola Acevedo considers herself open-minded and willing to try new teaching approaches. Paola taught music for several years before feeling comfortable moving away from a curriculum consisting of existing literature and books that outlined what should occur on a given day in her class or ensemble or activities she learned during professional development sessions. Paola includes some student inquiry and collaborative problem solving, but her curriculum and pedagogy are mostly informed by her Orff and Kodály training. Her classrooms are an interesting and sometimes disjunct mix of pedagogies and curricula. This is paralleled in the way she applies technology.

Paola often adjusts both the types of technology she includes in her teaching and her approach to its application. In the past, Paola integrated technology that mirrored what already occurred in her classroom; students composed music using notation software, recorded musical excerpts for her to assess, and identified intervals or key signatures with computer-assisted instruction software. Students often played musical games on their iPads to learn symbols that Paola considered important, such as dynamic markings or notes on a staff.

While these forms of engagement continue, Paola also integrates technology in a more diversified manner.

Students now perform music with iPad apps in ways that were not possible with their use of Orff instruments. Instead of spending several weeks teaching students how to play existing music on recorders, Paola allows students to explore how they could amplify and process the sound of their recorders through analog and digital means. Along with including arrangements of music on Orff instruments she has purchased from catalogs, Paola allows her students to create their own Orff instrument arrangements using music creation applications. Since integrating music applications that display sound through MIDI notation or wave forms, Paola focuses less on having her students work on reading and writing staff notation and provides opportunities for them to express themselves through sound while exploring multiple ways it can be visualized. While Paola tries to keep up to date on contemporary technology by reading blogs and participating in conversations via social media, her curriculum is still focused on teaching students standard elements of music and helping them develop as musicians.

After realizing that some technologies such as advanced digital audio workstations or DJ applications addressed concepts that did not fit within her curriculum, Paola created a new music technology course. It focused more on creating and recording music and drew less upon her background in Orff and Kodály approaches. Somewhat new to incorporating technology in ways that did not correspond to her Orff and Kodály training, Paola often leveraged social media to inquire about and find activities she could implement in her music technology class. She felt comfortable with the mix of traditional and newer approaches to incorporating technology throughout her program and was open to changing what occurred in her classes as she developed a deeper understanding of what was available and how it related to her philosophical outlook on music teaching and learning.

Looking in on Leanna Diamante's Music Program

Leanna Diamante sometimes displays a peculiar facial gesture in response to the mixed messages her colleagues send her. Some praise her for innovative approaches to teaching and facilitating musical engagement and learning. Others critique her for ruining music education, labeling her irresponsible for encouraging other music educators to adjust their programs in ways that differ from the exemplary models of past decades. She occasionally receives passionate emails from colleagues or alumni of her school lambasting her for a controversial curricular change that moved a large ensemble to an after-school extracurricular program and modified others to allow space for a diverse range of ensembles and courses that were fluid and hybrid in nature.

Leanna and her students are less concerned with issues of standard instrumentation and more interested in musical possibilities. This stance was catalyzed when she began introducing mobile technologies and electronic instruments into the existing ensembles and asked her students to imagine what types of music they could create and perform with such a mix. Students in Leanna's program are often engaged in a wide variety of projects related to their interests, in driving questions that form the core of the curriculum, and in broad themes proposed by Leanna, her students, or both in collaboration with each another.

For instance, some students are designing instruments and interfaces by using computers, circuit boards, sensors, and a range of open source and commercial applications. In addition to exploring the ways science, technology, engineering, math, and music intersect, they relate their engagement to that of other instrument makers across cultures and history. Another group of students perform public domain music, preparing to record each part separately for others to use as they wish. Some experiment with improvising music to accompany live video game play, investigating relationships between music and media, while others create music for interactive media, trying to better understand issues such as nonlinear and dynamic music.

Leanna encourages students to take on multiple musical roles, such as performer, composer, musicologist, critic, facilitator, producer, and programmer, among others. Her students collaborate to reflect on and record what they learn through web-based interfaces, including standard and multimedia texts. She and her students often analyze the positive and negative potential of new technologies collaboratively and then share their analyses with others. While Leanna embraces many aspects of technology and sometimes changes what occurs in her program, she stays true to key principles of practice (ranging from student-centeredness to a desire that learning be situated in musical contexts) and a philosophical foundation that guides her pedagogy and curriculum.

Considering Technology in Relation to Change

These images of potential praxis in relation to technology highlight different degrees of change, goals, or outlooks on music education. These images also demonstrate that it is the ways that educators and students situate or leverage technology, rather than technology itself, that impacts what does or does not occur in music education. Whereas some educators seek to incorporate technology in ways that maintain the status quo through strategies of assimilation, others may adopt strategies of distinction to alter what music teaching and learning look or sound like by leveraging possibilities afforded by technologies.

Those who approach technology through strategies of assimilation in ways similar to that of Marty Christensen might provide resources and opportunities for students to engage with technology without altering their teaching or the notion of what constitutes an ensemble or music class. This is similar to an image of transformation as modification consisting of "the reorganization of some elements or properties short of changing a thing's central condition or function" (Jorgensen, 2003, p. 48). Jorgensen elaborates that modification "offers a way of negotiating change to accommodate to particular situations without fundamentally altering the tradition itself" (p. 49). Similarly, educators might enact transformation as accommodation "in which one thing conforms to another" (p. 49). This is apparent in the way educators similar to Marty Christensen might incorporate technology in ensembles but limit its application to a framework of rehearsing music for eventual performance. When operating from such mindsets, educators may situate technology through assimilation strategies that conform to preexisting structures and ignore or reject technologies that do not fit within these paradigms.

Educators whose praxis is more akin to that of Paola Acevedo may integrate technology in ways that provide new types of experiences while still maintaining what already exists. Such an approach is indicative of an image of transformation as integration, in which one includes technology in ways that allow new and standard forms of musical engagement to occur "in a mix that is sufficiently accommodating to enable them to coexist, but where one does not threaten the existence of another" (Jorgensen, 2003, p. 50). Educators who work within this paradigm might provide a place for newer or emerging ways of engaging with music through technology, but in a compartmentalized fashion. This may result in creating technology-focused courses apart from other curricular structures (Dorfman, 2013). This is similar to the "add and stir" approach that has historically been adopted to address gender inequities in curricula (Morton, 1994) by adding content to the curriculum, such as women musicians or music composed by women, without altering the underlying principles that inform one's teaching.

Music educators most open to strategies of distinction might approach technology in ways similar to Leanna Diamante. While Leanna's decisions regarding technology are grounded in principles of music teaching and learning, she considers the possibilities of music technology beyond replicating the status quo. In this way, one might not only transform traditional curricular structures such as general music or ensembles but envision new structures altogether. Music educators who approach technology through strategies of distinction in this way might be inspired by new possibilities that technologies offer, determine their appropriateness for music teaching and learning, and then adjust curriculum and praxis accordingly. Thus, music educators' adjustments to musical engagement, teaching, and learning in relation to technology, and vice versa, can be seen as processes of technogenesis (Hayles, 2012). Such transformational change occurs in contexts with varied degrees of teachers' autonomy and ability to make curricular changes. Whereas educators such as Leanna may create entirely new curricular structures, others may need to work creatively within mandates, policies, or existing structures.

Through these three images, I suggest that music educators operate from particular ontologies of musical engagement, teaching, and learning that impact the ways they address technology. Addressing technology without acknowledging the role that music educators' philosophies, principles of practice, or understanding of pedagogy, curriculum, and technology play in their integration of technology may limit possibilities for positive change. This might be seen in the ways that curricular and musical paradigms often remain the same in spite of the technology potentially available to music educators.

United States Advanced Placement music theory exams, International Baccalaureate curricula, many general music workshops or ensemble festivals, among other examples of music education, reveal notions of knowing and doing music that are fixed in Western classical music traditions focusing on developing literacy in staff notation.

Expanding music education to address a broader spectrum of musical engagement and learning through technology calls for educators to interrogate the social and cultural contexts upon which they draw for musicianship, musical understanding, pedagogy, curriculum, and inspiration. It is possible that an ethic of assimilation rather than distinction keeps change at bay in music programs and marginalizes new and emerging ways of being musical, regardless of technological development. On the other hand, while a strategy of distinction opens wider possibilities for change within music education, it can be problematic. When focusing on what makes technology distinct, music educators might conceptualize it as separate from other aspects of music programs. This can result in music educators' compartmentalizing and situating new technology in specific curricular structures while excluding it from those that currently exist. Furthermore, enacting a strategy of distinction without a strong philosophical, pedagogical, and curricular foundation can result in music educators' making ongoing changes based on the latest technology regardless of its implications for students' musical engagement or learning.

I contend that discussions of technology in music education are ideally situated in terms of larger conversations regarding philosophies, pedagogies, curricular inquiry, musical engagement, and sociocultural contexts. Research is needed to better understand the relationship between technology and change across varied music education contexts. This might include investigating the decisions that music educators make in relation to technology along with their philosophical, pedagogical, and curricular standpoints.

How Are Music Educators Responding Social and Cultural Issues?

Music educators are addressing technology, digital media, and related musical practices from a widening set of social and cultural contexts—such as aspects of DJ culture (Challis, 2007), audio production (King, 2008, 2012; Tobias, 2013), and digital media performance (Brown & Dillon, 2012)—and a widening set of interactive media, such as video games (Tobias, 2012b). Some educators have involved their students in leveraging technology to create music in relation to place and history (Savage & Challis, 2001). Attention to the ways that people leverage technology in musical engagement outside formal learning contexts (Partti & Karlsen, 2010; Waldron, 2013) can inform future directions of music teaching and learning. However, while broadening the social and cultural contexts addressed in which technology plays a role can be positive, additional work is needed to address social and cultural issues in music education.

Music education could benefit from more reflexive and critical engagement with technology while accounting for social, cultural, and economic issues. Explicit application and analysis of technology through critical and sociocultural frameworks currently resides largely in researchers' scholarship. This might be expected when the majority of professional development related to technology focuses on issues of awareness and techné, that is, "the material matrix and practical activity in and which we carry out our craft" and "instrumental, how-to thinking" (Henderson & Kesson, 2004, pp. 48–49). While techné is a critical aspect of technical and pedagogical knowledge (Koehler & Mishra, 2008), approaching technology comprehensively calls for music educators to address social and cultural issues in their practice.

Much of the scholarship that frames technology in music education within larger contextual and critical frameworks focuses on issues of gender. Caputo (1993–94) and Armstrong (2008, 2011) argue that the integration of technology should avoid reproducing social and cultural gendered biases. Music educators may not understand how their instruction can favor particular students' learning preferences or how students might frame peer expertise through gendered norms (Armstrong, 2008). Similarly, music educators might be unaware of how gender can be a factor in students' interactions with technology in relation to a classroom's physical environment (Pegley, 2000). While research addressing issues of gender, students' diverse abilities, social dynamics, and technology offers important considerations that could inform practice, these issues are often missing from discourse at the practitioner level.

Just as issues of gender are often neglected, music education often suffers from overly positive or negative assertions regarding technology and its impact on teaching and learning. Whereas discourse promoting technology can lack the nuance provided by critical and sociocultural frameworks, dialogue related to negative aspects of technology may lack awareness or understanding of contemporary technology and its potential affordances. Emerging scholarship on issues of technological determinism and its potential impact on music education can add nuance to positive and negative discourse on technology in music education (Ruthmann, Tobias, Randles, & Thibeault, 2015; Savage, 2012). This work addresses the ways technology is socially constituted and considers the ways potential affordances and constraints might play a role in people's engagement with technology. The following vignette highlights how music educators might analyze and resituate technology in response to such issues.

Looking in on André Newton's Program

Years ago, André Newton would never have devoted as much time to discussion in his music program as he has during the past few weeks. Students have been debating issues that never crossed André's mind while earning his degrees. The debate began when a music student proposed that the program host a virtual ensemble. André and students were now in the midst of varied hybrid musical engagement connecting people across physical and virtual places. Some students had formed ensembles that collaborated asynchronously, while others performed with people across the world synchronously in an ensemble connected via the Internet. While the discussion first centered on Eric Whitacre's virtual choirs (http://ericwhitacre.com/the-virtual-choir) and similar projects that students found online, it soon evolved into ongoing inquiry and engagement relating to deep philosophical questions generated by the students in collaboration with André. While some asserted that virtual ensembles were not real and were a waste of time, others were intrigued with how technology could afford ways of connecting with others through music.

What had begun as a simple suggestion had spun out into a full-scale project, with students addressing such questions as What is a virtual performance? What does it mean to perform together? What constitutes live performing? Who can and cannot perform together? What are the differences between performing with others in the same physical space and across physical spaces via online technologies? What skills, knowledge, and understandings are needed to collaborate with people across the world on a musical endeavor? How does performing for people compare and contrast with performing with people?

Students moved between working in varied groupings, as a class, and across classes performing music, discussing issues, and engaging in research on related projects. One student, troubled by the project, concentrated on performing music that could be shared with others online while also generating a position paper warning against the ways technology could ruin music and ensembles. André sometimes wondered if the students should spend more time rehearsing music, working on sight singing, practicing solfège, and engaging in other exercises he considered important. However, he found value in his students' thinking deeply about technology and how it impacted their musical experience and engagement. Along with leveraging technology and digital media to engage with music in new ways, they thought critically about the role of technology in their musical lives. "I'm probably learning just as much as they are," André thought to himself while checking on a group's progress and scaffolding their learning.

In this vignette, students engage musically in varied ways while enacting artistic and humanistic inquiry. Scholarship that synthesizes aspects of artistic engagement and inquiry with disciplines of humanities, musicology, and philosophy offers a promising avenue for developing the ways that music educators address technology. For instance, Thibeault (2014) unpacks and interrogates the role of algorithms in people's engagement with and experience of music. He offers compelling analyses of how algorithms can impact students' musical lives, for example their musical tastes in relation to music discovery applications, or their musical development when using technology such as SmartMusic. Thibeault (2013, 2014) also challenges music educators to resist the myth of cyberculture that promotes technological determinism and focuses exclusively on technology as the source of artistic creativity today. Upitis (2001) argues that "music educators can integrate technology into music classes in a humanistic way only if the technology enhances the music making and helps students become more artistic and more human" (p. 53). Upitis suggests that aspects of musical heritage ought to be preserved in relation to technology when they contribute to "fostering of a peaceful, tolerant, just, literate (in the broadest sense of the word), and joyful citizenry" (p. 50).

While notions of preserving heritage raise issues of whose heritage ought to be preserved and how music evolves in relation to society, these and similar issues are critical for music educators to unpack and consider as they make decisions related to integrating technology in their programs. Addressing technology with musicological and humanistic perspectives may be critical for addressing sociocultural issues in music education and developing more nuanced ways of considering relationships between people and technology.

Addressing Social and Cultural Issues Explicitly through Praxis

Throughout this chapter I have suggested that music educators might expand beyond a focus on the techné of technology to address social, cultural, and musical issues and contexts. In this section I propose that music educators foreground social and cultural issues by applying critical frameworks explicitly to their curricula and practice.

Mobilizing Critical and Sociocultural Frameworks

While scholarship addressing technology through critical and sociocultural theories or conceptual frameworks exists (see the above section), there is little evidence that music educators conceptualize or situate technology in their music programs through such lenses. Music teacher education and professional organizations could play a more prominent role in helping music educators develop critical lenses through which to view their integration of technology and inform pedagogical or curricular decisions. Such work, however, begins in one's own teaching and learning contexts.

Along with observing the social and cultural dynamics that occur among students in relation to their engagement with technology, music educators might ask students questions about how they experience technology and listen to the ways they describe their engagement. Educators might invite students to investigate what particular technologies can and cannot do or to identify assumptions made by those who design and create varied technologies. Identifying and negotiating the values, assumptions, and biases connected to music education and technology can inform and provide nuance to the ways technology is situated within curricula.

Informing Praxis for Integrating Technology with Social and Cultural Issues

Leveraging technology to address social and cultural issues and contexts calls for reconceptualizing music curricula. Barrett (2007) explains that "reconceptualizing the music curriculum necessitates that teachers and researchers examine and reflect upon taken-for-granted assumptions of musicianship and musical understanding. Meaningful change begins with the acknowledgment that traditional conceptions of the music curriculum have privileged the skillful performance of music, repertoire drawn primarily from the classical Western tradition, and academic study of common elements and structural properties of music" (p. 149). Along with thinking deeply about assumptions at play in curricula, music educators might consider the types of curricula most conducive for addressing social and cultural issues. Curricula that situate technology in terms of planned activities or discrete content and tasks (Schubert, 1986) might omit social and cultural issues or treat them as tangential to what students are doing. Other types of curricula that stress cultural reproduction (1986) might address social and cultural issues but in a manner that reinscribes particular aspects of music and musical experience, excluding those that do not fit within particular cultural frameworks. How, then, might music educators enact praxis that situates music technology in relation to cultural and social issues?

Digitally Mediated Musical Experience

In contrast to the aforementioned curricular frameworks, an approach to curriculum as experience (Schubert, 1986) emphasizes students' exploring possibilities rather than focusing on preset outcomes such as particular behavioral objectives or facts about music or technology. In this paradigm, "curriculum is the process of experiencing the sense of meaning and direction that ensues from teacher and student dialogue" (p. 30). Such experience and dialogue can involve in-depth engagement with social and cultural issues in relation to music.

Through curriculum as experience, students might leverage technology to support their inquiry and exploration of social and cultural issues. This contrasts with settings where technology serves as a mechanism for delivering facts about music or a tool for completing particular tasks. A curriculum-as-experience framework works well with inquiry-based or project-based learning, where students collaborate with educators to pose and investigate interesting and fruitful questions (Barron & Darling-Hammond, 2008). One potential avenue for leveraging technology within this type of curricular framework is a form of musicological inquiry and engagement that Marshall (2012) calls *technomusicology*.

Marshall (2011) demonstrates how musicological method can be a form of musical engagement and can take varied musical forms that fold "musical analysis into musical experience" (p. 307). Through leveraging technology to create "pedagogical mashups," Marshall employs rigorous analysis and rich musical context, enabling others to hear music in new ways and consider related social and cultural issues. Marshall's (n.d.) intersections of analysis, musicological thinking, media, and music demonstrate rich possibilities for digitally mediated musicology, or technomusicology. This work ranges from tracing the dembow rhythm pattern across time, place, genres, and musical cultures to discussing issues of gender, musical appropriation or transformation, and economics in relation to controversy over the company GoldieBlox's parody of the Beastie Boys' song "Girls" in a commercial (Marshall, 2013).

Music educators and their students might engage in technomusicology to study texts in new and emerging ways that embed musical, social, cultural, and historical issues and analysis. This might include interacting with multimedia musicological texts. For instance, providing students with opportunities to create montages of existing performances into metaperformances and to provide contextual information, as Marshall (2014) did with Mozart's "Queen of the Night" aria, leverages technology in making creative decisions, developing and applying aural skills, and addressing relationships among musical, social, cultural, and historical issues. This type of inquiry-based learning resituates technology to address sociocultural issues through musical engagement.

Technology, Music, and Social Engagement

Some music educators and students might be interested in extending beyond deepening their understanding of social and cultural issues by engaging with society more directly. Curricula designed for social reconstruction provide "an agenda of knowledge and values that guides students to improve society and the cultural institutions, beliefs, and activities that support it" (Schubert, 1986, p. 32). This type of curricular framework encourages students to inquire about what types of change ought to occur along with why and how they might act on such change "to build a better society" (p. 32). In music programs that foster curricula with an agenda for social reconstruction, technology might enable students to engage with social and cultural issues to benefit their communities or society in general. Such curricula can be considered controversial, given diverse perspectives of what constitutes building a better society and the role of education in students' lives.

To imagine possibilities, educators might observe artists who leverage technology to foreground social and cultural issues and engage society through their art. Paul D. Miller (a.k.a. DJ Spooky) has addressed issues ranging from Antarctica's transformations over time in his hybrid acoustic and multimedia work "Terra Nova: Sinfonia Antarctica" (Miller, n.d.), to what he describes as "colonial and postcolonial issues facing the digital economy of the 21st century translated into a string quartet" (DJ Spooky, 2014) in conjunction with multimedia, based on his exploration of the island of Nauru in "Nauru Elegies" (2014).

The intermedia work "SuperEverything," which involves documentary film footage, live music, recorded music, and a range of digital media, is described as follows:

> This live audio-visual performance explores the relationship between identity, ritual and place in relation to Malaysia's past, present and future. Revealing juxtapositions between tradition and modernity, it seeks to discover who we are and how our complex identities are connected.
> Engaging in social and cultural issues are at the core of "SuperEverything" as it asks "a timeless question about who we are and what we might become as both individuals and as a society" (Light Surgeons, n.d.)

Similarly, the performance piece "Beware the Dandelions" is described as "an immersive environment built on the aesthetics of hiphop designed to embody the communal lessons found within complex sciences. Through interactive hip-hop performance, handcrafted songs, video projection mapping, creative technologies, and performative installations explore the relationship between art, music, science and social justice movements" (Complex Movements, n.d.). Each of these projects forefronts social and cultural issues and situates technology as an aspect of humanistic and artistic inquiry and practice in which music plays a vital part. In this type of work, technology plays a critical role in enabling artistic practice, sociocultural inquiry, and action toward bettering the world. Music educators might incorporate and adapt such engagement for music teaching and learning contexts. This necessitates resituating technology to address social, cultural, and economic issues and expanding the ways students engage with music and society in music programs.

Resituating Technology in Music Education

As music education, society, and technology coevolve, music educators ought to consider broadening beyond techné and resituating technology in ways that address social, cultural, musical, curricular, and pedagogical issues and implications. Along with technological, pedagogical, and content knowledge (TPACK; Koehler & Mishra, 2008), this requires deep understanding of social, cultural, and economic issues connected to musical and societal contexts, along with critical frameworks. This is complex and hard work. Music teacher education programs might play an important role in providing preservice and in-service music educators with opportunities to engage with and think through aspects of technology addressed throughout this chapter.

While integrating technology in music education uncritically is problematic, so too is rejecting technology when it can forward students musical development and learning. Music educators reluctant to change aspects of their curriculum and practice in relation to technology risk reifying taken-for-granted ways of knowing or doing music and excluding others. As Upitis (2001) argues, "our challenge is not to deny new technology by refusing to use it, but to make use of technology in ways that advance our work with children in helping them become better musicians, better thinkers and most important of all—better human beings" (p. 49). It is up to thoughtful, creative, and critically minded educators to ensure that music education draws upon, fosters, and propels artistic and humanistic thinking so as to situate technology in the ways that are most beneficial to students' growth as musicians and people.

References

Armstrong, V. (2008). Hard bargaining on the hard drive: Gender bias in the music technology classroom. *Gender and Education*, 20(4), 375–386.

Armstrong, V. (2011). *Technology and the gendering of music education*. Burlington, VT: Ashgate.

Barrett, J. R. (2007). Currents of change in the music curriculum. In L. Bresler (Ed.), *International handbook of research in arts education* (pp. 147–161). Dordrecht: Springer.

Barron, B., & Darling-Hammond, L. (2008). How can we teach for meaningful learning? In L. Darling-Hammond, B. Barron, P. David Pearson, P, T. D. Zimmerman, G. N. Cervetti, & J. L. Tilson (Eds.), *Powerful learning: What we know about teaching for understanding* (pp. 11–70). San Francisco: Jossey-Bass.

Brown, A. R., & Dillon, S. C. (2012). Collaborative digital media performance with generative music systems. In G. E. McPherson & G. F. Welch (Eds.), *The Oxford handbook of music education* (Vol. 2, pp. 549–566). New York: Oxford University Press.

Burnard, P. (2011). Educational leadership, musical creativities and digital technology in education. *Journal of Music, Technology, and Education*, 4(2-3), 157–171.

Cain, T. (2004). Theory, technology and the music curriculum. *British Journal of Music Education*, 21(2), 215–221.

Caputo, V. (1993–94). Add technology and stir: Music, gender, and technology in today's music classrooms. *Quarterly Journal of Music Teaching and Learning*, 4-5(5-1), 85–90.

Challis, M. (2007). The DJ factor: Teaching performance and composition from back to front. In J. Finney & P. Burnard (Eds.), *Music education with digital technology* (pp. 65–75). London: Continuum.

Complex Movements. (n.d.). *Beware of the dandelions*. Retrieved January 5, 2017, from http://complexmovements.com/.

Creative Commons. (2004). *The wired CD: Rip. Sample. Mash. Share*. Retrieved January 5, 2017, from http://creativecommons.org/wired.

DJ Spooky. (2014). *The Nauru elegies*. Retrieved January 5, 2017, from http://www.djspooky.com/nauruelegies/.

Dorfman, J. (2013). *Theory and practice of technology-based music instruction*. New York: Oxford University Press.

Finney, J., & Burnard, P. (Eds.). (2007). *Music education with digital technology*. New York: Continuum.

Grint, K., & Woolgar, S. (1997). *The machine at work: Technology, work, and organization*. Malden, MA: Blackwell.

Hayles, N. K. (2012). *How we think: Digital media and contemporary technogenesis*. Chicago: University of Chicago Press.

Henderson, J. G., & Kesson, K. R. (2004). *Curriculum wisdom: Educational decisions in democratic societies.* Upper Saddle River, NJ: Pearson.

Hugill, A. (2012). *The digital musician* (2nd ed.). New York: Routledge.

Jorgensen, E. R. (2003). *Transforming music education.* Bloomington: Indiana University Press.

King, A. (2008). Collaborative learning in the music studio. *Music Education Research, 10*(3), 423–438.

King, A. (2012). The student prince: Music-making with technology. In G. E. McPherson & G. F. Welch (Eds.), *The Oxford handbook of music education* (Vol. 2, pp. 476–491). New York: Oxford University Press.

Koehler, M. J., & Mishra, P. (2008). Introducing TPCK. In AACTE Committee on Innovation and Technology (Ed.), *Handbook of technological pedagogical content knowledge (TPCK) for educators* (pp. 3–29). New York: Routledge.

Marshall, W. (2011). Mashup poetics as pedagogical practice. In N. Biamonte (Ed.), *Pop-culture pedagogy in the music classroom* (pp. 307–315). Lanham, MD: Scarecrow Press.

Marshall, W. (2012). *Technomusicologically speaking.* Retrieved January 5, 2017, from http://wayneandwax.com/?p=66141.

Marshall, W. (2013). *Boys v. girls.* Retrieved from http://wayneandwax.com/?p=7774.

Marshall, W. (2014). *Megamontage is the method: Mozart to K-pop.* Retrieved January 5, 2017, from http://wayneandwax.com/?p=7884.

Marshall, W. (n.d.). *Wayne&Wax.* Retrieved January 5, 2017, from http://wayneandwax.com/.

Miller, P. D. (n.d.). *Terra Nova: Sinfonia Antarctica.* Retrieved January 5, 2017, from http://www.djspooky.com/art/terra_nova.php.

Morton, C. (1994). Feminist theory and the displaced music curriculum: Beyond the "add and stir" projects. *Philosophy of Music Education Review, 2*(2), 106–121.

Partti, H. (2014). Cosmopolitan musicianship under construction: Digital musicians illuminating emerging values in music education. *International Journal of Music Education, 32*(1), 3–18.

Partti, H., & Karlsen, S. (2010). Reconceptualising musical learning: New media, identity and community in music education. *Music Education Research, 12*(4), 369–382.

Pegley, K. (2000). Gender, voice, and place: Issues of negotiation in a "technology in music program." In P. Moisala & B. Diamond (Eds.), *Music and gender* (pp. 306–316). Urbana: University of Illinois Press.

Ruthmann, S. A., Tobias, E. S., Randles, C., & Thibeault, M. (2015). Is it the technology? Challenging technological determinism in music education. In C. Randles (Ed.), *Music education: Navigating the future* (pp. 122–138). New York: Routledge.

Savage, J. (2005). Information communication technologies as a tool for re-imagining music education in the 21st century. *International Journal of Education and the Arts, 6*(2).

Savage, J. (2012). Driving forward technology's imprint on music education. In G. E. McPherson & G. F. Welch (Eds.), *The Oxford handbook of music education* (Vol. 2, pp. 492–511). New York: Oxford University Press.

Savage, J., & Challis, M. (2001). Dunwich revisited: Collaborative composition and performance with new technologies. *British Journal of Music Education, 18*(2), 139–149.

Schubert, W. H. (1986). *Curriculum: Perspective, paradigm, and possibility.* New York: Macmillan.

Swanwick, K. (2001). Musical technology and the interpretation of heritage. *International Journal of Music Education, 37*, 32–43.

The Light Surgeons. (n.d.). *What is supereverything?* Retrieved January 5, 2017, from http://supereverything.net/what-is-superevrything/.

Thibeault, M. D. (2013). The shifting locus of musical experience from performance to recording to data: Some implications for music education. *Music Education Research International, 6*, 38–54.

Thibeault, M. D. (2014). Algorithms and the future of music education: A response to Shuler. *Arts Education Policy Review, 115*(1), 19–25.

Tobias, E. S. (2012a). Hybrid spaces and hyphenated musicians: Secondary students' musical engagement in a songwriting and technology course. *Music Education Research, 14*(3), 329–346.

Tobias, E. S. (2012b). Let's play! Learning music through video games and virtual worlds. In G. McPherson & G. Welch (Eds.), *Oxford handbook of music education* (Vol. 2, pp. 531–548). New York: Oxford University Press.

Tobias, E. S. (2013). Composing, songwriting, and producing: Diversifying popular music pedagogy. *Research Studies in Music Education, 35*(2), 213–237.

Upitis, R. (2001). Spheres of influence: The interplay between music research, technology, heritage, and music education. *International Journal of Music Education, 37*, 44–58.

Wajcman, J. (2002). Addressing technological change: The challenge to social theory. *Current Sociology, 50*(3), 347–363.

Waldron, J. (2013). User-generated content, YouTube, and participatory culture on the web: Music learning and teaching in two contrasting online communities. *Music Education Research, 15*(3), 257–274.

2 B

Further Perspectives

CHAPTER 28

TECHNOLOGY IN PERSPECTIVE

Who is in Control?

PATRICIA A. GONZÁLEZ-MORENO

IN this part of this book, important philosophical and ontological questions have been addressed in relation to the role of technology within music education. Undoubtedly, technological advances have been closely linked to innovation in music education, as well as to music-making practices generally, and will continue with the evolution of society. This coevolution, or technogenesis (as described by Hayles, 2012, and Tobias, chapter 27 in this book), brings with it a large number of assumptions about, and implications for, how music educators are to face the affordances and constraints of a rapidly changing technological context. In this Further Perspective, I intend to expand the discussion of some of the points addressed by the authors of previous chapters, including technological determinism, and affordances and constraints in technology use. At the end, I reflect on the particular implications for teacher education and professional development that aims to better situate technology within music education.

TECHNOLOGICAL DETERMINISM VERSUS HUMAN DETERMINISM?

A recurrent debate in general education, and in music education in particular, relates to the influence of technology in fostering educational changes in schools. To what extent technology has determined changes in music teaching and learning (technological determinism; see Tobias, chapter 27) and to what extent technology has evolved as a result of the individual's search for innovative ways of music teaching and learning (human determinism) is in the midst of the debate. Peters (chapter 26 in this book) argues that technology does not provoke change by itself. In this sense, human decisions, usually made by those in higher posts within the educational hierarchy, have positioned technology in a predominant place in teachers' and students' lives, and as a result institutional expectations of the technologically enhanced provision of music education have risen. While this argument may disregard the notion of pure technological determinism, no doubt the specific educational contexts and institutional expectations for the implementation of technology to enhance teaching and learning processes affect the way technology is perceived and adopted by music educators and students. In this sense, the existence of technology influences change in teachers' use, as it is expected to be included within the curriculum. Unfortunately, a technological assimilation that is vertically mandated without consideration of the final user presents risks, such as inadequacy (see Peters, chapter 26) when technology has not been critically selected and implemented. The decisions on the type of technology needed and ways for its inclusion must be a task of music educators according to their particular needs and respective educational contexts (e.g., general music education, one-to-one tuition, music composition, etc.).

Without a strong technological background or training, music teachers are expected to intuitively respond and enhance their practice through diverse means, such as digital media, recording technologies, and virtual learning environments, among many others. It is not surprising to find that, when the technology is available, music educators' adoption of these tools is limited to only facilitating aspects of their traditional teaching practices, without consciously and deliberately changing previous pedagogical paradigms. Within this technicist approach, digital devices and media are usually considered technical tools, and their potential for transformative change in music teaching and learning is overlooked. However, as Rudolph (2004) suggests, "the place and purpose of technology in music education must be found before beginning to properly apply the technology. Simply making technology available is not enough" (p. 4).

Affordances and Constraints on the Use of Technology

New technologies provide tools that enhance teaching, cooperative learning, independent study, and hands-on learning. In chapter 26, Peters suggests that the most important outcomes of technology—ones that have positioned it among the top educational priorities around the world—include access to education (e.g., human resources, materials, updated information, new pedagogical strategies), greater education quality by introducing new innovative forms of knowledge transference (i.e., opportunities to acquire and apply new information), and a student-centered approach based on exploration and problem solving. For instance, the rapid technological progress of information and communication technology (ICT) has permitted open access and the development of new didactic approaches through web-based interactive environments, introducing students to self-directed learning. This and many other examples within general education suggest that technology is effectively transforming the way people learn.

In music education, the use of technology has been characterized distinctively across educational contexts. Even standards in particular music activities have evolved along with technological advances (see Tobias, chapter 27). While a higher inclusion has been observed and documented in areas such as music composition and the use of digital and electronic means to develop distinctive and innovative musical expressions (e.g., music composition laboratories, online composing projects, online remix activities), other areas, such as performance (e.g., voice and instrumental performance, music ensembles), still follow the tradition of cultural reproduction (Shubert, 1986; Tobias, chapter 27). Common paradigms of music performance limit "the potential of technology to support and mediate a broad range of ways people can be musical" (Tobias, chapter 27).

Differences in the degrees of assimilation are determined not only by the particular music education context and institutional expectations but also by teachers' individual responses, values, and attitudes toward the challenges of technology use. As Tobias describes in relation to students, frustration with new technologies can also be found among teachers, even when others could find them exciting. It is not a matter of implementing "technological recipes" but of critically and openly responding to new environments and changing preconceived paradigms. This implies an additional responsibility and workload that is not always fully supported by schools. Even when technological devices (e.g., computers, SMARTboards, projectors, Internet access, etc.) are provided in schools, institutions have not necessarily invested in teacher training to stimulate positive and constructive attitudes on the part of music teachers. As is the case with students who are not digital natives, many teachers have not acquired the requisite expertise for taking full advantage of innovative teaching and learning technologies. Under these circumstances, music teachers resist changes due to their own technological deficiencies (Núñez & Gómez, 2005).

Several authors also point to the constraints, risks, and dangers of an unreflective use of technology. For example, Clay (1999) lists some of the mistakes that new distance instructors commonly make (p. 6), including the use of cutting-edge technologies when simple measures would suffice, putting copyrighted materials online, failing to develop adequate structure, failing to maintain communication and interaction with students, and, particularly, not taking time to learn the technology in its fullest potential. In addition, ease in using technologically enhanced processes might also affect the development of teachers' problem-solving abilities. Without adequate training, music teachers may fall into repetitive and uncreative practices that could fail to engage students' attention, motivation, and creativity.

In order to provide a more differentiated music education provision fostered by the use of technology, it is necessary that music educators respond distinctively and effectively, taking into consideration the sociocultural and educational context. If this condition is not met, technology might create what Argersinger described as an "illusion of sophistication without nuance" (as cited in Rudolph, 2004). The possible approaches of music teachers may range from (1) relative resistance, by rejecting technology when it is perceived as an inadequate tool for curricular purposes; (2) maintaining the status quo; (3) assimilating technology in ways that do not radically change current pedagogical practices; to (4) accepting it as a necessary path for educational transformation. The decision depends on the teachers and the curricular needs, even when these are highly influenced by institutional expectations. It can be counterproductive to convince music teachers to transition from strategies of basic assimilation to strategies of distinction (see Tobias, chapter 27) when philosophical, pedagogical, and curricular considerations are lacking. Accordingly, music teacher education and professional development programs must focus on developing teachers' critical thinking about the place of technology in music education and the types of technology that better enhances their practice.

Implications for Music Teacher Education and Professional Development

In order to be knowledgeable about current trends in technology, music teachers may follow several paths to develop competencies that help them take full control as to the type of technology that more adequately fits into their curricular needs. It is not a matter of technology integration for its own sake but for the sake of a systematic development of the profession. Music teacher education and professional development initiatives are two means for empowering music teachers to critically decide the degree of assimilation or distinction of technology that is appropriate to their specific music education contexts in order to meet the standards of quality education. In this sense, teacher development programs must aim to enhance music teachers' technological competencies, and to foster philosophical inquiry and critical pedagogy.

The rapid spread of technology creates opportunities but also demands for collective reflection and decision-making. Cooperation among teachers from diverse disciplines can multiply the positive effects of technology (Camblin & Steger, 2000). In addition, in collaboration with ICT and digital media specialists, music teachers can provide their own expertise that can help to develop technological tools that are more appropriate to the wide variety of music education environments. Baldwin (1998) has argued that to succeed in a technologically advanced era, teachers need to "become more interdependent and mutually supportive" (p. 17). In this way, music teachers will collaboratively and collectively determine the specific roles and uses of technology within music education in a "human deterministic" approach rather than passively reacting to it.

Other types of collaboration and communication allow teachers to become proficient in technology use, reducing resistance to change, barriers, and feelings of security threat. Forming partnerships and participating in with local and global online communities (Sherer, Shea, & Kristensen, 2003) provide opportunities for music teachers to know about current trends in music and music education technology. In addition, professional conferences and international forums provide a space where music teachers can critically examine new pedagogical approaches for technology integration, learning from other music educators' successful approaches.

Conclusions

Undoubtedly, new technologies have changed the way people learn, but also have influenced the way music educators teach, with the aim of quality education. While technologies have facilitated many processes, such as connectivity, communication, integration, socialization, collaboration, and self-directed learning, they have also brought particular challenges and needs for professional development, not only to foster technological competencies but to enhance critical thinking and informed decision-making on technology use.

The literature suggests that great attention has been given to the impact of technology in changing teachers (technological determinism), but it is also important to emphasize the power, control, and free will that educators have with regard to technology use. As Ertmer, Gopalakrishnan, and Ross (2000) argue, "high levels of access and support, although desirable, are neither necessary nor sufficient for exemplary technology use to occur" (p. 26). In this sense, music educators must follow a more proactive approach, acknowledging that, despite the abundant literature on the topic, today's teachers and researchers have different views about what exemplary classroom technology use looks like in the wide range of music education scenarios.

The teacher must assume the role of instructional designer as a basic principle for educational technology integration (Dexter, 2002), planning the uses of technology that better fit curricular goals. Ideally, the school supports teachers by providing adequate technology resources and opportunities for continuous training. This implies that school authorities need to make changes that extend far beyond the mere installation of computers (Ertmer, Gopalakrishnan, & Ross, 2000). Change in pedagogical paradigms and practices requires time, commitment, hard work, and a wide variety of strategies. Therefore, teacher education and professional development programs must foster the teaching competencies that are required if teachers are to fully embrace innovative teaching and learning technologies that more effectively meet the expected quality of music education.

References

Baldwin, R. (1998). Technology's impact on faculty life and work. *New Directions for Teaching and Learning, 76,* 7–21.

Camblin, L., & Steger, J. A. (2000). Rethinking faculty development. *Higher Education, 39*(1), 1–18.

Clay, M. (1999). Development of training and support programs for distance education instructors. *Online Journal of Distance Education Administration, 3*(3). Retrieved January 27, 2014, from http://www.westga.edu/~distance/ojdla/fall23/clay23.html.

Dexter, S. (2002). eTIPS—Educational technology integration and implementation principles. In P. L. Rogers (Ed.), *Designing instruction for technology-enhanced learning* (pp. 56–70). doi: 10.4018/978-1-930708-28-0.ch003.

Ertmer, P. A., Gopalakrishnan, S., & Ross, E. M. (2000). Technology-using teachers: Comparing perceptions of exemplary technology use to best practice. *Journal of Research on Technology in Education, 33*(5), 1–26.

Hayles, N. K. (2012). *How we think: Digital media and contemporary technogenesis.* Chicago: University of Chicago Press.

Núñez, M., & Gómez, O. (2005). El factor humano: Resistencia a la innovación tecnológica. *ORBIS Revista Científica Ciencias Humanas, 1*(1), 23–34. Retrieved November 5, 2013, from http://www.redalyc.org/redalyc/pdf/709/70910104.pdf.

Rudolph, T. E. (2004). *Teaching music with technology* (2nd ed.). Chicago: GIA.

Schubert, W. H. (1986). *Curriculum: Perspective, paradigm, and possibility.* New York: Macmillan.

Sherer, P. D., Shea, T. P., & Kristensen, E. (2003). Online communities of practice: A catalyst for faculty development. *Innovative Higher Education, 27*(3), 183–194.

CHAPTER 29

THE CURIOUS MUSICIAN

LEAH KARDOS

Music technologies can lead us to a transformation of perceptions and the reinvention and refinement of our processes—from the way we see, interact with, and understand the materials of sound and music to the way we learn new skills, communicate, and share with each other, the way we represent ourselves to the world as music creators and professionals, and especially, the way we teach. Technology has and is transforming our language around music content and consumption ("I streamed a podcast of glitchcore mashups, reblogged it and gave it a 'like'"). It is creating musical and sonic possibilities that transcend the facilities of traditional music notation and analysis. It sometimes requires interdisciplinary and collaborative approaches to bring projects, artworks, and products to fruition. (Music technology resides not in the field of music only but also in the fields of media; science, technology, and society [STS]; electronics and computer science.) Finally, it grants music creators agency and control of their works (Taylor, 2014).

As a composer, I am completely enchanted and continually inspired by the way new music technology applications so readily challenge my own understanding of what music is and can be. The proliferation of digital applications, computerization, and online connectedness has given rise to a diverse and evolving collection of practices or "literacies" that are advantageous skills for creative musicians working in commercial and contemporary new music scenes to possess (Durant, 1990; Hugill, 2012). At the time of writing, these literacies can include skills associated with multitrack recording and production using digital audio workstations, MIDI sequencing, audio editing, sound design, synthesis, sampling, looping, triggering, live sequencing, coding, controlling music and sound with interfaces and apps, instrument and effect building, app development, hacking and circuit bending, mixing, remixing, and mashing up, score typesetting, publishing, broadcasting, and contributing knowledge and expertise to online communities of practice. To the composer in me these technologies represent an opportunity to expand my creative vocabulary with pure magic: to capture any sound and turn it into music that is meaningful; to conjure up ghosts of the past; to bend space and time; to hold the air. Speaking from my perspective as a teacher, they represent a new promise of freedom: never before have the materials of music been so pliable, touchable, easy to understand and access.

A lot has been said about the potential for Internet-capable handheld devices used in learning scenarios to raise students' achievement (Freidel, Bos, & Lee, 2013; Norris, Hossain, & Soloway, 2011). More than a technology of convenience, computerization and our connectedness to the Internet (and to each other by way of social networking) has changed how we see the world, how we form and maintain relationships and networks, how we talk to each other, how we share knowledge, how we create and consume media and art, and how we teach (Draper, 2008; McCormack & d'Inverno, 2012); yet there are times when it feels like the presence of such devices, with unfettered access to the Internet and social networking sites, is too enticing a distraction for any student to resist. In Castilla-La Mancha, Spain, access to smartphones in the classroom, linked to interruptions, cheating in tests, and cyberbullying, was banned in 2014 (Kassam, 2014), and a recent story in the United Kingdom about Nigel Mills (MP) being caught playing Candy Crush Saga while attending a committee meeting shows us that the problem of device distraction is more widespread than the classroom. "Know-how" and "know-what" is being supplemented with know-where: the understanding of where to find knowledge (Siemens, 2005), giving rise to new kinds of learning that are not always adequately catered for in our assessment methods, which can also be a source of tension in student-teacher relationships.

In a world that promises for every problem "there's an app for that," technology can seduce us into relying on quick fixes, which almost always seem a more attractive option than actually engaging in critical analysis, problem solving, or coming up with an original creative solution from scratch. Recently I was working through some aural analysis exercises with recording students in a session designed to develop critical and diagnostic listening skills and specialist vocabulary. During the session, a student concluded that a noise reduction third party plug-in would eliminate the "problem" in the audio being examined; while he did not engage in the deep-level critical listening I was trying to promote, he was right: there was an app for that. I had to then "sell" the idea back to the group that deep understanding and mastery of such skills would set them apart from the other button clickers and quick fix users in the worlds of work and creative practice research.

This issue speaks to an anxiety I've observed among older generations of teachers and practitioners: the fear that the technology is simplifying and streamlining processes, making hard-won skills redundant, promoting a shallow engagement with important concepts. Our little corner of very specific and complicated technology is getting increasingly easier to operate and understand, and it seems that our tech native students can easily teach themselves how to operate the software that is industry standard, by watching YouTube tutorials or even just by clicking around the interface and figuring it out. How many staff room conversations have I taken part in that begin with the words "Back in my day it wasn't so easy . . . [insert any tale of hardworking woe from a bygone era here: *I had only eight tracks to work with / my outboard compressor was made of mud and straw / I had to enroll in a postgraduate research degree to figure out how to program the MIDI messages on my Roland AX-7]."*

But if it's true that our students have access to digital literacies and possess the requisite fluency in them to operate the technology already, then why do they enroll in our music tech programs? I believe that what they seek is the secret of execution: a highly developed ability to listen critically; a flexible, future-proof skill set that allows for maximum control of musical and sonic materials in a range of contexts; the ability to engage in fluidity of approach; to develop personally situated mastery and an awareness of the field—agency to navigate routes to employment and future artistic engagement. It is no longer efficient to dedicate whole swathes of curriculum to teaching the finer details of operating a specific version of software (Pro Tools, Logic Pro, Ableton Live, Max, etc.). By doing so we build obsolescence into our programs and our students' skill set. A better aim would be to somehow enable our students to develop the skills necessary to make technology bend to their ideas and not the other way around.

Meanwhile, we have generations coming up the ranks who might challenge the abilities of their teachers—with iPad integration in classrooms from an early age, with coding and app building being taught to primary school children using MIT's Scratch programming language, in addition to other apps and tools such as Hopscotch, Kodable, and Objective-C (Dredge, 2014). If the arrival of the last generation of so-called digital natives (sometimes referred to as millennials) sent shockwaves though pedagogic practice and research with their intuitive fluency in technological applications and adapted learning styles, I wonder what the impact will be when the new generation of code-literate software-designing interdisciplinary creative practitioners begin enrolling on our music courses? It seems clear to me that we need to find a way for students and teachers to grow and learn together.

"Nothing is a mistake. There is no win and no fail, there's only make."
—Sister Corita Kent, *Rules and Hints for Students and Teachers*

In creative practice research, learning occurs through the processes of making, touching, doing, working, editing, experimenting, collaborating (insert almost any verb here, including *failing*), and reflecting. The outcome is the creation and sharing of "personally situated knowledge" (Barrett & Bolt, 2007, p. 2), generating experience and expertise and, hopefully illuminating the path to a deeper understanding of self. Music technology, as a collection of creative and interdisciplinary practices, represents a wonderful location for such learning to occur. In the foreword to the second edition of *Re-thinking Pedagogy for a Digital Age*, Diana Laurillard observes that "our digital native students may be able to use technologies, but that does not mean that they can learn from them . . . learners need teachers . . . to guide the learner's journey to a particular and productive end" (Beetham & Sharpe, 2013). Our role as teachers is to design the learning experience, not to necessarily be or embody it. Once I realized this, I felt such a weight lift from my mind: I don't need to know everything, I just need to design an experience that we can all learn from—making stuff while being mindful of context, culture, and process.

And we need not fret about old skills dying out—why not preserve and analyze them (for the furtherance of musicological and cultural perspectives) and at the same time jump in the sandbox and play with the new toys too? Through making together, students and teachers (and others connected via social networks and online communities of practice) can grow together through the ongoing exploration of new digital creativities and the potential for innovation offered by technology tools. Can new forms of expression be found? In what ways can original ideas be communicated using the cultural sonic languages and digital literacies unique to our age?

Because of music technology we see in contemporary and commercial music practices a blurring of the lines that once marked the boundaries of musical disciplines (as already mentioned in this book by Heidi Partti). It is not uncommon to see composers also identifying as producers and sound designers—as if the materials of music (melody, harmony, rhythm), sound (texture, timbre), and space (position of elements in the listening environment) coexist on the same painter's palette, which of course in the age of self-production they absolutely do. Performers of commercial, contemporary art and experimental music might enjoy designing their own digital effects or engaging in live coding using applications such as '(fluxus)' or Max/MSP. From what we see in industry and research, the successful career musician would ideally have a well-rounded knowledge of music and music technology—and it could be argued that the line separating those fields is blurring also.

Creativity challenges technology, and technology inspires creativity. Because practitioners and performers are pushing the envelope, we are all rewarded by the emergence and development of brilliant technology such as Imogen Heap's "gestural music ware" glove controllers, Celemony's Melodyne plug-in with "DNA" (direct note access) technology, allowing us to reach inside an audio wave form and reconfigure its harmonies and overtones, and Bjork's bespoke iPad-controlled acoustic instruments (specially created and developed for her Biophilia project). Since the emergence of dub reggae in the late 1960s and hip-hop in the late 1970s, music technology has been a unifying and driving force in the development of styles, scenes, and cultures. Whether they'd been to music lessons or not, these practitioners found new sounds by pushing the limits of the technology in their hands, shaping aesthetics, and claiming new sonic territories. Computer-aided music sequencing and sound design are continually shaping the aesthetic landscape of dance and electronic music styles. Whole music scenes are built around the sound world of "mistakes" and technological errors and accidents, as seen in the cases of "glitch," "lo-fi," and "noise" music.

As we explore the potentials of music and sound using technology, we shape the development of technology at the same time that we encourage the development and evolution of music cultures that we are part of. We can achieve it in the classroom. When learning (the coming together of the experience, modality, and environment) integrates technology linked to real-world practices and scenarios, a student can sense the "ownership of his or her learning" (Salavuo, 2008) and authenticity in his or her developing arts practices.

At the start of this think piece I separated my perspectives into "composer" and "teacher," implying that there are two versions of me, each with a separate vested interest in music technology. That is not really true; I teach and compose (among many things), and those outcomes spring from the fact that I am simply a curious musician. When I'm making music, I enjoy exploring the ways technology helps me to communicate effectively and uniquely. Of course my personal perspectives, discoveries, and experiences filter into my teaching; my enthusiasms for the subject flavor the resources I devise and the delivery of my lessons. Similarly, the knowledge that my students discover through the outcomes of their music making and experimentation comes back to influence, enrich, and enliven me: I am open to being exposed to new ideas, fresh, original perspectives. The classroom is not the place where I broadcast the knowledge, it is the place where we discover it and grow together.

References

Barrett, E., & Bolt, B. (Eds.). (2007). *Practice as research: Context, method, knowledge.* London: IB Tauris.

Beetham, H., & Sharpe, R. (Eds). (2013). *Rethinking pedagogy for a digital age: Designing for 21st century learning.* London: Routledge.

Draper, P. (2008), Music two-point-zero: Music, technology and digital independence. *Journal of Music, Technology & Education, 1*(2-3), 137–152.

Dredge, S. (2014, January 27). Tablets in schools: Coding, creativity and the importance of teachers. *Guardian.* Retrieved February 10, 2014, from http://www.theguardian.com/technology/2014/jan/27/tablets-schools-coding-kids-education-ipad.

Durant, A. (1990). A new day for music? Digital technology in contemporary music-making. In P. Hayward (Ed.), *Culture, technology and creativity in the late twentieth century* (pp. 175–196). London: Arts Council and Libbey Press.

Friedel, H., Bos, B., & Lee, K. (2013). Smartphones—smart students: A review of the literature. *Society for Information Technology & Teacher Education International Conference, 1,* 1862–1868.

Hugill, A. (2012). *The digital musician.* New York: Routledge.

Kassam, A. (2014, November 24). Spanish schools clamp down on smartphones in classrooms. *Guardian.* Retrieved December 10, 2014, from http://www.theguardian.com/world/2014/nov/24/spanish-schools-clamp-down-smartphones-in-classroom.

Kent, C., & Stewart, J. (1992). *Learning by heart: Teachings to free the creative spirit.* London: Bantam.

McCormack, J., & d'Inverno, M. (2012). *Computers and creativity.* Berlin: Springer.

Norris, C., Hossain, A., & Soloway, E. (2011). Using smartphones as essential tools for learning. *Educational Technology, 51*(3), 18–25.

Salavuo, M. (2008). Social media as an opportunity for pedagogical change in music education. *Journal of Music, Technology and Education, 1*(2-3), 2–3.

Siemens, G. (2005). Connectivism: A learning theory for the digital age. *International Journal of Instructional Technology and Distance Learning, 2*(1), 3–10.

Taylor, T. (2014). *Strange sounds: Music, technology and culture.* Hoboken: Taylor and Francis.

CHAPTER 30

ON BECOMING MUSICAL

Technology, Possibilities, and Transformation

GENA R. GREHER

> Computers don't just do things for us they do things to us, including our ways of thinking about ourselves and other people.
>
> —Sherry Turkle (1995)

How Are Technology and Music Technology Changing Us?

THE epigraph from Sherry Turkle (1995, p. 26) is particularly meaningful to me with regard to the work I do with students using technology, in both high-tech and low-tech ways. In the program notes for a fairly recent concert our youth orchestra presented in our local community, I wrote: "What does it mean to be musical and how might that musicality manifest itself in someone who is physically incapable of either singing or playing an instrument? This is but one of the many challenges that our featured guest artist, Dan Ellsey, has surmounted in his young life. . . . The fusion of music and technology is fueling the growth of digital musicianship, creating spaces for all students to engage in music making." As discussed by the contributors to this part of this book, we risk not connecting with and possibly turning off a great many of our students by strictly employing methods of instruction developed before the advent of digital media. In particular we risk marginalizing even further a segment of the population who are often underserved by the music education community. The young composer featured in the concert I mentioned is but one example of how technology can transform a person from someone who previously could only listen to music into an active music maker. Not only did technology transform Dan Ellsey's perception of who he was and what he was capable of, the other performers and the audience members were equally changed regarding their assumptions about who is and isn't a musician.

Ellsey became an active music maker who discovered his inner musician as a composer and a conductor thanks to Hyperscore, a music creation tool created by Tod Machover and his team at MIT's Media Lab, which was used in partnership with Tewksbury State Hospital, where Machover and his team were tasked with employing this new technology as part of a music therapy program (Moss, 2011). Ellsey's transformation was the fortunate and perhaps unintended consequence of a software program originally created to engage young children in composing music being adapted for use by the patients in this program. The software was further tweaked for Dan Ellsey's particular sets of abilities and to accommodate his physical limitations so that he ultimately could harness the technology to perform his compositions live.

Though this example represents the extreme end of how technology can change us, and is most likely beyond the scope of most music teachers, it does highlight the positive role technology can play in music education for the many students whose language, behavioral, cognitive, and physical challenges would preclude them from being able to participate in many of the more traditional modes of music making. Music technology, along with a variety of low-cost digital multimedia tools, has great potential to provide multiple simple entry points for nontraditional learners to connect with their inner musical selves.

Several authors in this part of this book note that at-risk learners, including those with special needs, are among the many underserved populations who find technology particularly advantageous. In this Further Perspective I will home in on the challenges that confront the music education field in serving this population, as well as the benefits for incorporating technology. As Darrow (2007) suggests, music educators will come into contact with a great many students with special needs over the course of their teaching. In the music education program at my university, as in a great number of music education programs in the United States (Colwell & Thompson, 2000), the focus on students with special needs is covered as part of the required coursework rather than being offered as a stand-alone course or courses. Music teachers' attitudes toward special needs students can negatively or positively impact the outcomes for these students when they are mainstreamed into music classes (Colwell & Thompson, 2000; Hammel, 2001; VanWeelden & Whipple, 2007). Yet there are often few context-specific experiences that enable music education students to work with special needs populations so as to gain an understanding of effective teaching strategies for working with them. This often causes tension in the field when music teachers' expectations conflict with the reality of working with this population (Colwell & Thompson, 2000; Hammel 2001; Devito 2006; VanWeelden & Whipple, 2007; Blair, 2009; Hourigan, 2009).

Adding to the challenge of scant preparation and scarce clinical experiences for working with special needs and at-risk populations, the amount of technology there is to learn can be daunting. Boehm's (2007) research indicates how varied and fragmented the field of music technology is with regard to areas of specialization and concentration, making it difficult to even define what exactly the study of music technology is. As Webster and Williams (2013) have highlighted in their latest progress report on undergraduate music technology competencies, they began with a set of 51 competencies and through a series of national surveys have whittled it down to a core of 8. As someone who teaches the one and only required technology course for music education students at my institution, there's a great deal to cover in one semester in just learning to work with technology, let alone learning to teach with it. And, as many of the contributors to this book point out, just learning how to use the technology without the pedagogical application piece will leave both teacher and students fairly frustrated. When working with technology and special needs populations, context is everything. One needs to be equally aware of the constraints and affordances of a particular technology in order to create effective lessons, as well as knowing the limitations and possibilities of the population one is working with.

Context-Specific Experiences for Teacher and Student Transformations

How I went from no experience working with special needs students to commissioning an original composition from Dan Ellsey for my university's String Project and Youth Orchestra students serves as a foundation for the various context-specific field experiences and outreach projects developed for music education students at our university. In a similar vein to teaching special needs students, if music education students aren't given enough field experiences to apply this vast body of knowledge and competencies with actual students, chances are they will be less willing to employ technology in their teaching. Through my own immersion in working with students with behavioral, cognitive and physical challenges, I found it didn't take a great deal of technology to make a tremendous difference with these students.

Reading Heidi Partti's description (chapter 25 in this book) of the teacher with the four-track tape recorder brought back memories of one of my first forays into teaching with very low-tech equipment: a portable cassette recorder. I was a teaching artist with the Creative Arts Laboratory (CAL) at Teachers College, Columbia University, when a cassette recorder moment represented a turning point in the behaviors of a class of behaviorally challenged second graders. What I discovered was how valuable a tape recorder is as a feedback, analysis, and assessment tool. Lee Pogonowski, the founder of the Creative Arts Laboratory and its director, drummed into us teaching artists the need to record student performances. It is one thing to read about how much students intuitively know and understand, but witnessing first-hand the results of this process in action made everything come together for me. I was the nice lady who came in with the fun instruments, but a typical day for the classroom teacher would involve a student or several students throwing things at her or using language that was far from school appropriate. When she enlisted my aid in teaching the students a song for a performance at a holiday assembly, we both had a very dim outlook of what could actually be accomplished.

The classroom teacher had had little luck getting them to pay enough attention to learn the song, let alone sing together. When she and the students sang what they had been working on, I recorded them on my portable cassette recorder, which sounds very "old school" now in a culture of smart phones. Before the classroom teacher could comment on their performance, I hit the play button and then asked the students how they thought their performance was. In complete sync with each other, their response was a unanimous "EEEEWWWWWW!!!!!." The fact that technology made it possible for people to hear professional performances through all sorts of media meant that these students had enough exposure to what music is supposed to sound like; their performance was nothing like anything they had heard before. Much as Bamberger's (1995) experiences show, students really do have an intuitive understanding of music, and these students were even able to articulate what the problem was. When asked about their disappointment, they informed us that they weren't singing together. When asked what it would take for them to sing together, they told us that they would need to listen to each other. That was their assessment, not mine or the classroom teacher's. For the next several weeks the teacher borrowed the tape recorder so that she and her students could assess their progress. Wishing to sound like the music they heard on the radio and not wanting to sound bad in front of their parents and the whole school was enough of a motivating factor for them to get their act together. At their performance, no one was more surprised than the classroom teacher and the principal when they received a standing ovation.

I experienced similar discoveries at a hospital school with a class of students in an inpatient ward consisting of many high-functioning autistic students, as well as students who had severe behavioral issues. As part of a unit where we were integrating music to support literacy development, we videotaped their weekly work in which they were creating a story, with music, about outer space. Ironically, the video camera turned out to be the unifying and motivating force in this classroom, though for the protection of the students it was technically against hospital regulations to photograph or videotape them. It was agreed from the outset that I would be allowed to tape their work, for documentation purposes, providing that the students could not be identified visually or aurally.

At a midsemester point the classroom teacher and I decided to screen the video for the class as a way of validating for them what a great job they were doing. On the one hand they were so excited that we must have watched the video at least a half a dozen times. Yet the looks on their faces told another story. When asked what was wrong, they mentioned that they were disappointed that they were not able to see themselves in the video. One student suggested that they could do it again with paper bags on their heads. When probed further about what we could do with these paper bags, they came up with the idea of making masks. The following week, they surprised me with the masks they had created. We augmented this with construction paper spaceships and drawings of flying saucers. We spent the next several weeks expanding our ideas, retaping the sequence with their masks, adding alien sounds, and putting some movement into it. The students had generated all these ideas after viewing the video that first time.

By turning the video camera on them in the form of video feedback, we developed a style of working that was both appealing and engaging for them while also serving as a tool for reflective practice in the classroom. We gradually moved from a working environment of teacher-directed comments and criticisms to one where the students could monitor their own behaviors, actions, and music-making activities, all through the use of video playback. What began as a simple exercise in analytical listening eventually grew into an 8-week project involving deep learning. The idea of seeing themselves on "TV" was a powerful incentive, but the real pride they felt was in knowing this was something they had created. The video feedback was the turning point in this project that put them at the center of their learning. As suggested by Veenema and Gardner (1996), they were given the means to demonstrate their understanding through media that were relevant to them. They were able to evaluate their own work without the implication of right and wrong that pervades the atmosphere when a teacher gives an evaluation. They were engaged in a learning process that, while relevant to their own experiences, was helping to change their behavior as well. Here was tangible proof that they were learning to work together, negotiate with each other, and make compromises. These were concepts that seemed unthinkable when we first began working together.

As a result of these early experiences, I have been proactive in creating partnerships and outreach projects so that my students can get the field experiences needed to boost their comfort level with working with special needs populations and teaching with and through technology. These context-specific experiences under the guidance of classroom mentor teachers help the music education students to understand how to design lessons so as to capitalize on each student's potential rather than focusing on his or her limitations.

When Dan Ellsey's expressive therapist initially contacted me, I had no idea what would come of our meeting. Upon meeting Ellsey and seeing the absolute joy on his face when the discussion of music and composing came up, I was affected in much the same way that Machover's team was when meeting and working with him (Moss, 2011). Over the next year we had multiple meetings and conversations with his expressive therapist and his music therapist. The concert preparation involved our ensemble director and one of our student arrangers working with Ellsey and his music therapist in order to adapt and transcribe the Hyperscore file for the live student performers. In order to allow the audience to fully see the way Ellsey was conducting the group with his head gestures, we arranged for video record and playback on a giant screen facing the audience. What better way to demonstrate to my students the role technology can have on nontraditional students than to witness firsthand a world premier performance of a composition for our String Project and Youth Orchestra students? This could not have taken place were it not for the researchers who took the time to understand what Dan Ellsey could accomplish with a little boost from some technology.

References

Bamberger, J. (1995). *The mind behind the musical ear.* Cambridge, MA: Harvard University Press.

Blair, D. V. (2009). Fostering wakefulness: Narrative as a curricular tool in teacher education. *International Journal of Education & the Arts, 10*(19), 2–22.

Boehm, C. (2007). The discipline that never was: Current developments in music technology in higher education in Britain. *Journal of Music, Technology and Education, 1*(1), 7–21.

Colwell, C. M., & Thompson, L. K. (2000). "Inclusion" of information on mainstreaming in undergraduate music education curricula. *Journal of Music Therapy, 37*(3), 205–221.

Darrow, A. (2007). Looking to the past: Thirty years of history worth remembering. *Music Therapy Perspectives, 25*(2), 94–99.

Devito, D. (2006). The communicative function of behavioral responses to music by public school students with autism spectrum disorder. Ph.D. dissertation, University of Florida.

Hammel, A. M. (2001). Preparation for teaching special learners: Twenty years of practice. *Journal of Music Teacher Education, 11*(1), 5–11.

Hourigan, R. M. (2009). Preservice music teachers' perceptions of fieldwork experiences in a special needs classroom. *Journal of Research in Music Education, 57*(2), 152–168.

Moss, F. (2011). *The sorcerer's and their apprentices: How the digital magicians at the MIT Media Lab are creating the innovative technologies that will transform our lives.* New York: Crown Business.

Turkle, S. (1995). *Life on the screen: Identity in the age of the Internet.* New York: Simon and Schuster.

VanWeelden, K., & Whipple, J. (2007). An exploratory study of the impact of field experiences on music education majors' attitudes and perceptions of music for secondary students with special needs. *Journal of Music Teacher Education, 16*(2), 34–44.

Veenema, S., & Gardner, H. (1996). Mulitmedia and multiple intelligences. *The American Prospect, 29*, 69–75

Webster, P. R., & Williams, D. B. (2013). *Defining undergraduate music technology competencies and strategies for learning: A third-year progress report.* Paper presented at the fifty-sixth national conference of the College Music Society/Association for Technology in Music Instruction, Oct. 30–Nov. 2, 2013 Cambridge, MA.

CHAPTER 31

EBOLA SONGS

Exploring the Role of Music in Public Health Education

CARLOS CHIRINOS-ESPIN

> The curious beauty of African music is that it uplifts even as it tells a sad tale. You may be poor, you may have only a ramshackle house, you may have lost your job, but that song gives you hope. African music is often about the aspirations of the African.
>
> —Nelson Mandela

From January 2014 to March 2016 there were 28,616 cases of the Ebola Virus Disease (EVD) and 11,310 deaths in Guinea, Liberia, and Sierra Leone, with some cases in other countries in West Africa, Europe and the US (World Health Organization, 2015). The health systems in West Africa were not prepared to contain a disease outbreak of this scale, lacking the infrastructure and human capacity to treat the sick and educate the public on how to prevent human-to-human transmission of the disease. Radio and TV played a key role in communicating the risk of Ebola to the public and many West African and international artists produced songs to raise awareness about the disease, receiving recognition from public health officials, agencies, and the international press for their contribution to public health communication during the crisis.

Ebola is a viral infectious disease that is transmitted from human to human through direct contact with the bodily fluids of sick and deceased Ebola-infected people. Initial reports and testimonials indicated that general ignorance about the symptoms and modes of transmission—plus poor hygiene and sanitation, and mistrust in the public health systems—were risk factors contributing to the spread of the disease. The 2014 outbreak of Ebola, a relatively new and unknown zoonotic disease only discovered by Western science in 1976, had an unprecedented geographical and demographic impact, expanding significantly over international borders and affecting urban areas for the first time (Alexander, Sanderson, Marathe, Lewis, Rivers, Shaman, Drake, Lofgren, Dato, Eisenberg and Eubank, 2015). Without a known cure or vaccine, and lack of public familiarity with the new disease, some communities in West Africa reacted with confusion and fear, and in some cases health workers were physically attacked, with some fatalities, when people suspected that they were bringing Ebola to their villages (Hamilton, 2014). Acknowledging that education and information were a big part of the problem, UN secretary general Ban Ki-moon declared that "we must also fight the virus of fear and misinformation" in his address to the General Assembly on the Ebola crisis in September 2014 (Ban Ki-moon, 2014).

Evidence from the Ebola outbreak in Uganda in 2000–2001 suggests that the most effective way to educate the public about Ebola was the use of a media awareness campaign led by artists and celebrities appearing on radio and TV (Lamunu, Lutwamab, Kamugishaa, Opioa, Namboozec, Ndayimirijed, & Okwarea, 2004). But the broad impact of Ebola in West Africa in 2014 raised questions about the transnational capacity and preparedness of public health institutions in developing countries to respond to a crisis of this scale, and quickly educate and engage communities in Ebola prevention to bring the disease under control. Can music contribute to public health education and community engagement during acute health emergencies? In the following pages, I explore the role of music and celebrity advocacy in raising awareness about Ebola during the outbreak in West Africa in 2014.

Public Health and Music

The role of music and celebrities in advocating for social change is not new. From John Lennon's opposition to the war in Vietnam to Bono's stance against global poverty, artists have been actively engaging audiences in humanitarian causes, defining celebrity activism and fundraising for charitable causes (Easterly, 2010). Songs themselves—when performed by influential artists—can be powerful means of communicating social issues as lyrics in song can relay information in plain language and artists can persuade fans to follow their advice. For example, in Nigeria, the artists King Sunny Adé and Onyeka Onwenu recorded two songs –*Choices* and *Wait for me*- that addressed family planning and sexual health. The evaluation of the campaign suggested that the songs helped increase the use of contraceptives from 16% to 26% and helped change public attitudes toward family planning (Singhal, Cody, Rogers, & Sabido, 2003, p. 98).

To raise awareness and educate citizens about a public health threat like Ebola, health services use formal written warnings and descriptions of symptoms in leaflets, posters, and public announcements in the media. Although this information is essential to informing government officials, literate audiences, and opinion leaders, this form of communication fails to reach disenfranchised and illiterate communities living in rural areas. "Entertainment-education," on the other hand, is also commonly employed during health emergencies to reach poor and rural communities to influence citizens' attitudes and health behaviors, using popular communication strategies that can include interactive radio programs, soap operas, TV programs, folk theatre, music videos, and songs (Singhal & Rogers, 1999; McPhail, 2009).

In health emergencies that require individuals to act rapidly to prevent a communicable disease, the speed of information distribution is critical to raising early awareness and mobilize communities to adopt prevention behaviors. Community radio stations in particular are able to produce cost-effective messages that reach large audiences much more quickly than print, TV, or the Internet. In West Africa, radio broadcasting has an average audience penetration of 90%, compared to the average 35% penetration of TV, and the limited availability of printed newspapers and the Internet. In rural and urban areas, where electricity supply is intermittent and costly, radio is accessible via cheap battery-operated radio transmitters and mobile phones with FM tuning capability, as well as in taxis, buses, shops, and loudspeakers in public spaces (Myers, 2011). Mobile phones are also playing an important role in expanding the reach of telecommunications and facilitating interpersonal communication, but their role in increasing risk awareness and knowledge about health has been tested with mixed results. Limitations include the reliance on text-based information, the use of official languages, and the cost of using mobile phone networks. (Jamison, Karlan and Raffler, 2013).

Marketing research confirm that celebrities appearing in media campaigns can help create consumer awareness and favorable attitudes about a product or cause, as celebrities have the public's trust, which is essential in communicating effectively about risk (Reynolds & Seeger, 2005). Through social learning and role modeling, celebrities are able to influence social beliefs and promote changes in health attitudes and behaviors (Bandura, 1977; Singhal et al., 2003). The degree of *identification* between performers and fans—or how individuals see performers as role models—can determine the artists' ability to persuade the public to adopt healthy behaviors (Brown & Fraser, 2003).

Up until this outbreak, the communication capacity of the health services in the affected countries had not been tested at this scale, and this emergency necessitated using all the communication channels available to help disseminate information as quickly as possible. But information alone is not always understood or trusted and this creates a need for more culturally relevant communication methods to effectively engage communities in disease prevention.

Ebola Songs

Music is an important communication tool in West Africa, where culture and traditional knowledge are customarily transferred from generation to generation by Griots, who—like shamans in indigenous cultures—are in charge of transmitting oral traditions and advice in their communities, using music and storytelling (Hale, 1997, p. 251).

During the Ebola outbreak, many West African and international artists produced Ebola-themed songs, some of which had an unusual popular response due to their broad dissemination in social media and a large input of commentary generated in the international press. Thirty six Ebola-themed songs and music videos produced since the outbreak began in 2014 have generated over 66 million views and listens on social media and music websites, and many of these songs also aired on local radio and TV in West Africa. The songs produced during this period can be broadly classified into two categories: (1) songs to raise awareness about Ebola, produced by West African artists based in Africa, and (2) songs to raise funds for Ebola relief efforts, produced by both Western and West African artists.

In the first category, two songs produced in Liberia were reported to have been very popular and helped raise public awareness about Ebola. The songs, *Ebola in Town* by Shaddy, D12, and Kuzzy, and *Ebola Is Real* produced by HOTT FM and UNICEF Liberia, have lyrics that reflect the realities confronted by artists and communities living in areas affected by Ebola. *Ebola in Town* deliver warnings about risks in phrases like "Don't touch your friend" or "Ebola is worse than HIV AIDS." The song *Ebola Is Real* reinforces the idea that Ebola is not a myth but a real threat, and includes a readout of an official public health message describing the symptoms and modes of transmission of Ebola. These songs were created in popular genres that have wide local appeal such as hipco, hip-hop, ballad, and reggae, and their lyrics were relevant to the local population. Some reports indicated that the songs were very popular on local radio and TV, leading the press to say that "everyone in Liberia was talking about the songs" (Beaubien, 2014) (see the appendix for details about Ebola songs).

"Do They Know It's Christmas?"

In the fundraising category, the 2014 recording of the song *Do They Know It's Christmas?* by Bob Geldof and Midge Ure, featured Bono (U2), Clean Bandit, Paloma Faith, Guy Garvey (Elbow), Ellie Goulding, Niall Horan (One Direction), Angélique Kidjo, Zayn Malik (One Direction), Chris Martin (Coldplay), Olly Murs, Sinéad O'Connor, Rita Ora, Liam Payne (One Direction), Emeli Sandé and Seal. The song was originally written and recorded to raise funds for the Ethiopian famine of 1984 and kick-started Band Aid, a series of charitable concerts and records that claims to have raised over $200 million dollars for humanitarian relief (Johnson, 2009).

The rerecording of the song in 2014 was done specifically to raise funds for Ebola, and was the subject of extensive criticism in the press. Journalists and advocates pointed out that the lyrics of the song perpetuated and reinforced the image of the African continent as a place of poverty, disease and death, in phrases that suggest that Africa "is a World of dread and fear", "Where there's death in every tear" and "the Christmas bells that ring there are the clanging chimes of doom". Other commentators questioned the role Western artists play in portraying a negative image of the developing world, in particular Africa, and the impact such portrayal has in the economic and social development of the continent (Adewunmi, 2014). Others pointed to the lack of African artists participating in the recording, with only one, Angélique Kidjo, participating in the recording of the song. Fuse ODG, a Ghanaian artist based in London who was invited to perform in the song, pulled out of the recording at the last minute in disagreement with the lyrics and the way the project portrayed Africa. The artist stated that he was worried about how the song "would play into the constant negative portrayal of the continent of Africa in the West" (Fuse ODG, 2014).

International music celebrities have contributed to defining global advocacy and have helped shape the culture of charitable giving in the West, turning it into something "cool" that appeals to young audiences who were not traditionally concerned with global suffering. (Franks, 2013, explores this effect in detail.) The Band Aid Charitable Trust reported that they distributed over 1.8 million pounds sterling to 18 relief organizations in January 2015 from funds collected through sales of downloads and special edition CDs of "Do They Know It's Christmas?" which aligns with their stated aim of raising funds for relief efforts (Band Aid 30, n.d.). Yet, the controversy over the narrative of the song and video clip raised questions about the perceived benefits of fundraising campaigns by Western artists, and how they contribute to a negative portrayal of the developing world.

AFRICA STOP EBOLA

In October 2014, I coproduced a song called *Africa Stop Ebola* with some of West Africa's most famous artists that included Grammy award winner Oumou Sangaré; Mory Kanté and Salif Keita -two widely recognized griots from Guinea and Mali-, Amadou & Mariam, Tiken Jah Fakoly -a reggae artist from Ivory Coast who has a strong reputation as a musician committed to social development in Africa- Hip Hop artists Mokobé, Didier Awadí, Markus, and the singers Sia Tolno, Barbara Kanam and Kandia Kora. The idea of making a song was not new—as other songs preceded this one—but this was the first one produced with a group of well-known West African music celebrities. I worked with the artists to write the song lyrics to convey specific health prevention messages approved by the World Health Organization, and intended to send a clear message about Ebola prevention to raise hope, and increase public trust in the local health services. The song was recorded in French, English, and five vernacular languages widely spoken in the region: Malinké, Kissi, Bambara, Sousou, and Lingala.

Arts and development journalists quickly acknowledged and recognized the quality of the song because it featured well-established African artists performing together for the first time, and reported the contribution of the song in relaying accurate health information to raise Ebola awareness (Kozinn, 2014; Jones, 2014). The song became caught up in the controversy over the song *Do They Know It's Christmas?* as the press needed an alternative story to demonstrate that African artists were also producing quality songs for awareness and fundraising.

The controversy helped promote the song in Europe and America as well as in West Africa, where the song was reported to be on rotation in radio stations in Guinea and Mali. The popularity of the artists who took part in the recording of the song, and the use of approved lyrics led to a strategic association with Médecins Sans Frontières-MSF France (Doctors Without Borders) to use the song in their fundraising campaign for Ebola relief efforts in West Africa. The song was released commercially through Universal Music France and was used in TV commercials, radio ads, and online content, with all profits donated to MSF-France. The involvement of recognized artists also helped secure a donation to Médecins Sans Frontières of over US$60,000 from Roskilde Festival, a Danish not-for-profit music organization.

In early 2015, Africa Stop Ebola received USAID's special award "Fighting Ebola: A Grand Challenge for Development" (USAID, 2015) to produce a song contest in Guinea to engage local artists in the creation of new songs about Ebola, after attending workshops with health promotion workers from Médecins Sans Frontières. The song contest concluded with a live reality TV show hosted in Conakry on July 29, 2015, that was attended by over six hundred people and broadcasted on TV and Radio nationwide (Africa Stop Ebola, 2015). An evaluation of the campaign found that lay people and health workers considered that the campaign was effective in increasing public trust in health workers, in raising awareness about Ebola, and in effectively engaging youth in Ebola prevention. (Chirinos, 2015)

Conclusions

The Ebola songs of 2014 helped raise awareness about the risks associated with the disease and, according to testimonial evidence, played a vital role in creating talking points for communities, helping to overcome the limitations of the official scientific language that was being used to communicate the risk of the disease to the public (Friedman, 2014). Monies raised through fundraising campaigns using songs contributed significant funds to the relief efforts, and the controversy over the song *Do they know it's Christmas* drew international public attention that generated an alternative narrative for talking about the disease in the Western media.

The role music played in the Ebola crisis in West Africa would not have been possible without the wide availability of music production and distribution technology. Whether through traditional FM radio, social media, or files shared through mobile phones, music technology is helping define new models for cultural creation and consumption in the developing world. Demand for locally produced music is increasing as music distribution models proliferate on the continent through partnerships between record labels and mobile phone operators in promising emerging markets for music like Nigeria (Akwagyiram, 2017). Open source technology and low-cost hardware is democratizing, unlike anything before, people's ability to make music in Africa, thanks to the availability of cheap laptops loaded with free software like FL studio (previously known as Fruity Loops), Audacity, Logic and other software. Affordable production technology, local radio, and music distribution through mobile phones is sparking a wave of electronic dance music with musical roots on the continent that includes Afrobeats in Nigeria, Kwaito in South Africa, Bongo Flava in Kenya and Tanzania, Coupé Dècalé in Ivory Coast and Hipco in Liberia –just to name a few- that is reshaping local music consumption at a time when the global music industry is being restructured and reformatted. It is this new music production scenario, enabled by accessible technology, that can help communities use songs and local celebrities as an effective communication method to improve public health education.

From "Bangla-desh," the song recorded by George Harrison in 1971 to raise funds for the famine in East Pakistan, to "We Are the World," the song written by Michael Jackson and Lionel Ritchie in 1984, and to countless other initiatives that have included charity singles, charity supergroups, festivals, and advocacy campaigns, music has played a critical role in the foundations of global advocacy for the developing world. Music celebrities' contribution to advocacy campaigns for international development is unquestionable and highlights the important role of music and media advocacy in shaping public opinion about humanitarian causes and global social change.

Understanding the close relationship between music and human behaviors could help artists and music entrepreneurs to become instrumental in designing effective public health information campaigns. Public policies mainly recognize the value of music as an art form and as part of basic education, and this determines how resources are allocated to music in education budgets and unfortunately, music education is usually the first item to be reduced when education cuts are needed. In this regard, the contribution of music and entertainment to public health communication should be acknowledged so as to take into account the aggregated value that music can add to health education and community engagement. Music education policies should consider this unaccounted value and promote music making as an interdisciplinary area of work that can convene music actors (composers, educators, performers, the music industry), health promotion experts and social change advocates to develop more effective public health communication strategies.

With this in mind, music actors and the public health sector should work together to develop innovative approaches in public health communication that embrace the power of music and the promotional tactics employed by the music industry to mobilize communities for social action. Such approaches could open new opportunities for musicians, music educators, industry leaders, and entrepreneurs to create socially responsible business models that use songs and music celebrities to improve public health education.

APPENDIX

Selection of songs by African artists to raise Ebola awareness. Listen to a podcast and interview with Carlos Chirinos-Espin on Ebola songs: https://www.mixcloud.com/SOASradio/rita-ray-presents-ebola-music-special/:

"Ebola in Town," by Shadow, D-12, and Kuzzy of 2kings, Liberia

"Ebola Is Real," by F.A., Soul Fresh, and DenG; hipco track produced by Hott FM and UNICEF; Liberia: https://soundcloud.com/unicef-liberia/hott-fm-ebola-song

"Ebola," by Black Diamond: criticizes the government and blames the United States and Europe for Ebola; Liberia

"Un Geste Pour La Vie contre Ebola," by Degg J Force 3, Sekouba Bambino, BANLIEUZART, Didier Awadi, William Balde, Soul Bangs, Petit Kandja, Takana Zion, FADJIDIH, G Force Sotigui, Khady Diop - Aminata Kamissoko, DJESSOU MAMA, Instinct Killers, Fakri, ONE TIME, PRINCE, Zawagui, Maitre Barry, Kabakoudou Alhassane, TAMSIR.

"Ebola awareness song Sweet Africa," by Yung Muse, Liberia

"Ebola Rap," by Dr. Ground Zero, Nigeria: https://www.youtube.com/watch?v=utqGiUiOPVc

"Stop Ebola," by Mali Rap All Star, Mali: https://www.youtube.com/watch?v=ezCFtqz7DBs

"Stop Ebola," by Iba One, Mali: https://www.youtube.com/watch?v=izKIotQbQOw

"Ebola Stay Away from Africa," by Just King, Liberia: https://www.youtube.com/watch?v=m4viKtJa8ug

Songs produced by international and African artists to raise funds:

"Do They Know It's Christmas?," produced by Bob Geldof's charitable organization Band Aid, featuring: Bono (U2), Clean Bandit, Paloma Faith, Guy Garvey (Elbow), Ellie Goulding, Niall Horan (One Direction), Angélique Kidjo, Zayn Malik (One Direction), Chris Martin (Coldplay), Olly Murs, Sinéad O'Connor, Rita Ora, Liam Payne (One Direction), Emeli Sandé and Seal. https://www.youtube.com/watch?v=-w7jyVHocTk

"Africa Stop Ebola," produced by a collective of West African artists including Salif Keita, Amadou & Mariam, Oumou Sangare, Mory Kanté, Tiken Jah Fakoly, Kandia Kora, Marcus, Mokobe, Didier Awadi, Barbara Kanam, Sia Tolno and Kandia Kora. Co-produced by Leon Brichard and Carlos Chirinos. https://www.youtube.com/watch?v=ruYQY6z3mV8

References

Adewunmi, Bim. (2014, November 11). Band Aid 30: Clumsy, patronising and wrong in so many ways. *The Guardian*. Retrieved April 28, 2016, from http://www.theguardian.com/world/2014/nov/11/band-aid-30-patronising-bob-geldof-ebola-do-they-know-its-christmas.

Africa Stop Ebola (2015, October 13) Africa Stop Ebola Song Contest in Guinea, July 2015. Retrieved February 4, 2017, from https://www.youtube.com/watch?v=1bzmUqkRabg

Akwagyiram, Alexis (January 20, 2017) Sony Music seeks Nigeria streaming growth to build on ringback market. *Reuters*. Retrieved February 4, 2017, from http://www.reuters.com/article/us-nigeria-music-sony-idUSKBN1541XK

Alexander, Kathleen E., Sanderson Claire E., Marathe, Madav, Lewis Bryan L., Rivers Caitlin M., Shaman, Jeffrey, Drake, John M., Lofgren, Eric, Dato, Virginia M., Eisenberg, Maria C., Eubank, Stephen (2015) What Factors Might Have Led to the Emergence of Ebola in West Africa? *PLoS Neglected Tropical Diseases* 9(6)

Band Aid 30. (n.d.). Where the money went? Phase 1. Retrieved March 1, 2015, from http://www.bandaid30.com/money-went-phase-1/.

Bandura, Alexander. (1977). Self-efficacy: Toward a unifying theory of behavioral change. *Psychological Review*, 84(2), 191–215

Ban Ki-moon. (2014, September 24). Address to the UN General Assembly. Retrieved December 12, 2014, from http://www.un.org/sg/statements/index.asp?nid=8037.

Beaubien, Jason. (2014, October 12). Liberian singers use the power of music to raise Ebola awareness. NPR. Retrieved on March 4, 2016, from http://www.npr.org/2014/10/12/355427316/liberian-singers-use-the-power-of-music-to-raise-ebola-awareness.

Brown, W., & Fraser, P. (2003). Celebrity Identification in entertainment-education. In A. Singhal, M. Cody, E. Rogers, & M. Sabido, *Entertainment-education and social change: History, research, and practice*. Mahwah: LEA.

Chirinos, C. (2015). *Africa Stop Ebola: Music and Media Campaign to Raise Public Health Awareness in Guinea, West Africa*. Evaluation Report. Washington DC: USAID

Easterly, William. (2010, December 9). John Lennon vs. Bono: The death of the celebrity activist. *Washington Post*. Retrieved March 20, 2016, from http://www.washingtonpost.com/wp-dyn/content/article/2010/12/09/AR2010120904262.html.

Franks, Suzanne. (2014). A revolution in giving: Live Aid. In *Reporting disasters: Famine, aid politics and the media* (pp. 71–86). London: Hurst Publishers.

Friedman, Uri. (2014, August 25). How to make a hit song about Ebola. *Atlantic*. Retrieved on February 4, 2017, from http://www.theatlantic.com/international/archive/2014/08/how-to-make-a-hit-ebola-song/378980/.

Fuse ODG. (2014, November 19). Why I had to turn down Band Aid. *The Guardian*. Retrieved March 17, 2016, from http://www.theguardian.com/commentisfree/2014/nov/19/turn-down-band-aid-bob-geldof-africa-fuse-odg.

Hale, T. (1997). From the griot of roots to the roots of griot: A new look at the origins of a controversial African term for bard. *Oral Tradition*, 12(2), 249–278.

Hamilton, Jon. (2014, October 28). An Ebola strategy brings good news to one Liberian town. (Blog entry). NPR. Retrieved March 19, 2016, from http://www.npr.org/blogs/goatsandsoda/2014/10/28/359355410/an-ebola-strategy-brings-good-news-to-one-liberian-town.

Jamison J., Karlan, D., & Raffler, P. (2013). Mixed method evaluation of a passive health sexual information texting service in Uganda (NBER Working Paper No. 19107). New Haven, CT: NBER.

Johnson, Andrew (2009, November 21). Feed the world? Band Aid 25 years on. *The Independent*. Retrieved on February 4, 2017, from http://www.independent.co.uk/news/world/africa/feed-the-world-band-aid-25-years-on-1825385.html

Jones, Sam (2014, October 29) African musicians band together to raise Ebola awareness. *The Guardian*. Retrieved on February 4, 2017, from https://www.theguardian.com/global-development/2014/oct/29/african-musicians-record-song-ebola-awareness

Kozinn, Allan. (2014, October 29). How to protect yourself from Ebola, in song. *New York Times*. Retrieved on February 4, 2017, from http://artsbeat.blogs.nytimes.com/2014/10/29/how-to-protect-yourself-from-ebola-in-song/?_r=0.

Lamunua, M., Lutwamab, J. J., Kamugishaa, J., Opioa, A., Namboozec, J., Ndayimirijed, N., & Okwarea, S. (2004). Containing a haemorrhagic fever epidemic: The Ebola experience in Uganda (October 2000–January 2001). *International Journal of Infectious Diseases*, 8, 27–37.

McPhail, TL. (2009) *Development Communication: Reframing the Role of the Media*. Oxford: Wiley-Blackwell.

Myers, M. (2011). *Voices from villages: Community radio in the developing world.* Washington, D.C.: Center for International Media Assistance.

Reynolds, B., & Seeger, M. (2005). Crisis and emergency risk communication as an integrative model. *Journal of Health Communication, 10,* 43–55.

Singhal, A. & Rogers, E. (1999). *Entertainment Education: A Communication Strategy for Social Change.* Mahwah, NJ: L. Erlbaum Associates.

Singhal, A., Cody, M. J., Rogers, E. M., & Sabido, M. (2003). *Entertainment-education and social change: History, research, and practice.* LEA.

USAID. (2015, February 11). *United States announces additional results in Grand Challenge to fight Ebola* (Press release). Retrieved on February 16, 2015, from http://www.usaid.gov/news-information/press-releases/feb-11-2015-united-states-announces-additional-results-grand-challenge-fight-ebola. Washington, DC.

World Health Organization. (2016, March 30th). *Ebola situation report.* Retrieved February 3, 2017, from http://apps.who.int/ebola/current-situation/ebola-situation-report-30-march-2016.

CHAPTER 32

THINKING AND TALKING ABOUT CHANGE IN MUSIC EDUCATION

ROGER MANTIE

In the way that constructionism has become de rigueur in learning theory, a standard theoretical paradigm in the study of language is "linguistic relativity." This paradigm, sometimes referred to as the Sapir-Whorf hypothesis after the linguistic anthropologists Edward Sapir and Benjamin Lee Whorf, who theorized it, postulates that various cultures understand the world differently according to the specific concepts developed in their language, illustrated in part by the problem of translatability, where words in one language do not always have an adequate equivalent in another language. In brief, the theory emphasizes the role that language plays in structuring and delimiting the ways in which we understand the world. There are stronger and weaker versions, but at heart linguistic relativity stands in opposition to universalist paradigms, such as structuralism, that posit stable versions of meaning and cognitivist development (e.g., Jerome Bruner's "structures of the discipline"). In a vein similar to linguistic relativity, George Lakoff and Mark Johnson argue in *Metaphors We Live By* that the meanings we make of the world are conceptually driven: "the concepts that govern our thought are not just matters of the intellect. They also govern our everyday functioning, down to the most mundane details. Our concepts structure what we perceive, how we get around in the world, and how we relate to other people" (1980, p. 3). This governing, they suggest, depends not just on semiotic-like correspondence of signifier and signified—"c-a-t" representing feline (but even then the poststructuralist skeptic asks: house cat? lion? stuffed? cartoon?) but also on the metaphoric connections we make: argument is war (attack, defend, win/lose), time is money (spend, cost, invest, spare, surplus), and so on. Like the conceptual differences between languages, then, metaphors reveal how we understand the world.

Specifics of the Sapir-Whorf hypothesis aside, acknowledging linguistic relativity, or the metaphoric nature of understanding, raises the prospect that the words we choose to discuss and make sense of the world help to structure and delimit how we understand experience. My point is that in order to answer a question about how technology may or may not be changing us, one needs to pay attention to how we talk about technology and/in music (education). A more finely grained understanding of the discourse of technology and music education provides insights into how we understand the experience of music learning and teaching as affected by technology, and may provide clues for imagining new vocabulary and ways of thinking. For example, while some new terms have undeniably entered our lexicon over the past couple of decades—Internet, MP3, iPod, iPad, DAW (digital audio workstation), smartphone, and so on—the ways in which we discuss music learning and teaching today are not appreciably different from the way they were discussed 50 (if not 100) years ago. Thus, while we can point to and label new devices (or "gadgets"), our fundamental understandings are likely to remain unaltered if we do not generate new conceptual and metaphorical ways of thinking. The question of change—the topic of this part of this book—in other words, cannot be divorced from an examination of the discourses that surround music and technology. It is not just a matter of what we do but of how we talk about what we do.

Arguably, some new conceptual vocabulary is beginning to enter music education discourses. In this book, for example, one finds many instances of words, phrases, and acronyms mostly absent from conversations 10 years ago: "user-generated content," "convergence culture," "participatory culture," "TPCK (TPACK)," "technology-based music classes," "social media," and so on. At the same time, however, the basic framework we use to discuss these new concepts and ideas remains largely the same. We still speak of a "role" for technology and of "teaching," "embedding," or "incorporating" technology, as if it were a tool or a thing discrete from the norms of music learning and teaching. We still speak of culture in terms of "ours" and "theirs"; we still speak of participation in terms of access and exclusion—the haves and the have-nots.

Partti, Peters, and Tobias have all offered perspectives with the potential to help us think differently about our musical and pedagogical practices. Tobias, for example, encourages us to think beyond deterministic conceptions of technology by invoking Hayles's (2012) term "technogenesis" (the coevolution of people and technology) and her associated concepts of *assimilation* and *distinction*. Rather than thinking in positive/negative, agential/deterministic ways about technology, coevolution suggests a more reciprocal relationship. Rather than thinking of technology as something one accepts or rejects, the word "technogenesis," if more widely embraced, could advance a more "living with" conception of technology and music education. At the same time, however, I would argue that the potential of Hayles's concept of technogenesis is weakened by its modernist underpinnings of knower and the known—technology as something outside the "autonomous individual," as if technologies are inhuman or other to the purity of human experience, a rather familiar trope in the history of technology and music education.

Partti likewise offers vocabulary that might help us to talk about music education differently. To think in terms of "pedagogical fundamentalism" or "pedagogical populism," for example, is to consider more deeply how knowledge has been historically conceptualized as a commodity and how teachers have thought of themselves as curators or stewards (see Pignato, chapter 19 of this volume). Similar to the "banking model" of education that Paulo Freire presents, where teachers deposit knowledge in the heads of students fortunate enough to receive the "gift," pedagogical fundamentalism reminds us of the cultural reproduction tendencies in music education to which Tobias alludes. Teachers curate and hold all the precious cards (i.e., the Western musical canon) that students, if they work hard enough, might inherit. Rather than providing an alternative to the banking model, pedagogical populism—predicated on the idea of liberation—further entrenches the conceptualization of knowledge as a commodity. Technology, to this way of thinking, provides the egalitarian solution whereby the walls protecting the sanctity of knowledge are torn down.

Partti underscores the prevalence of this particular way of thinking about learning and teaching through the phrase "the participatory revolution in music making and learning." To speak of a participatory revolution or a "participation gap" is to imply that participation in music has been, and continues to be, predicated on a metaphor of exclusion for which technology might provide the cure or antidote. Quite rightly, I think, Partti points out that participation "may or may not be democratic, empowering, counter-cultural or liberating" and that the simple capacity or ability to create does not necessarily result in "critical literacy" with respect to mediatization. While technology may (or may not) result in more people creating music, this way of thinking about participation reduces or confines the issue of music making to a quantitative matter, as if participation is a one-dimensional, yes/no construct. Along slightly different lines, cultural stewards in the Matthew Arnold tradition often continue to think of participation in hierarchical terms, with some kinds (most often Western classical) superior to other kinds (most often "popular"), as if participation were a valid/worthy-invalid/unworthy construct. Our music education vocabulary seems particularly impoverished with respect to concepts that might better capture and explain participation qualitatively—not as the word "quality" is understood in evaluative terms (i.e., better/worse) but as various colors or moods describe facets of our experience (i.e., as qualities, not as "quality"). Rather than reducing participation to a yes/no or a pejorative valid/invalid, employing a richer vocabulary with respect to the nature and meaning of participation might help us to think differently about music education.

Perhaps the most intriguing term emerging in the Core Perspectives in this section is *passeur culturel*. As Peters explains, this term refers to teachers acting as cultural mediators—as go-betweens for the culture of students and the culture at large. Although similar in some respects to Partti's discussion of cultural literacy and digital citizenship, *passeur culturel*—on the surface at least (to my limited understanding of it)—might engender a less paternalistic orientation to the issue of cultural literacy. Conceiving of teachers as mediators, for example, is slightly different from conceiving of teachers as those who save students from being uncritical cultural dupes at the mercy of a predatory mediatized world. As mediators, teachers might still be tempted to impose a strong hand, of course, but, as an ideal, *passeur culturel* may differ from the idea of digital citizenship in that teachers, as mediators, might be less inclined to see their role as leading students from ignorance to enlightenment (pedagogy in its literal sense: *to lead a child*, especially toward a predetermined goal) and more as someone who facilitates and negotiates understandings.

In *The Structure of Scientific Revolutions*, Thomas Kuhn advances an argument that, in the scientific world, new paradigms of thought reduce older ways of thinking to myth or error. Where at one time it was commonly accepted that the sun revolved around the earth, new discoveries falsified previous beliefs. The social world doesn't provide a perfect analogue to this, but it nevertheless provides an interesting thought experiment to consider how our thinking about technology and music education might come to regard previous ways of thinking as myth or error if only we could break out of our existing conceptual paradigms. What if a new vocabulary made a phrase like "technology-based music classes" nonsensical? What if we eschewed the word "integration" and started talking only about use-spaces or user-spaces instead of classrooms? What if we avoided the terms "teacher" and "student" and started talking exclusively about "learning communities" or "participatory communities" or "contribution communities," where people were simply expected to contribute and share? What if we twisted a phrase like "replication with extension" and used it as a learning metaphor?

Clearly I am being somewhat utopian in my commentary. As with many endeavors in life, it is relatively easy here to be an "armchair quarterback." I am offering analysis and critique without offering solutions. This is quite deliberate on my part, for, as Kuhn has demonstrated, change must be authorized by a community. While it is true that new concepts and neologisms are often the product of a lone individual (Shakespeare, for example, contributed countless words to our English vocabulary), paradigms only shift when new ways of thinking are adopted by a legitimized community of practice—in this case, the field of music education, especially its scholarly arm. My lone voice is inadequate to the task of creating new concepts and metaphors for music learning and teaching. It takes a village, so to speak.

Technology in music education is accepted and promoted so often because technology writ large represents for many people a form of twenty-first-century literacy, not because the profession (outside most readers of this book) truly believes in its value. I submit that without sensitivity, imagination, and inventiveness, our profession will continue to deploy the same tired vocabulary, and "new" technologies will simply be assimilated into existing thinking structures in music education, based in large part—especially in music—on human exceptionalism: real music making is that done acoustically; to digitize is to bastardize in some way, and thus legitimate music making can never be done outside the realm of acoustic instruments (and the voice). So long as terms such as "literacy" (students learning to read staff notation) or "musicianship" (students learning to play or sing in large ensembles) and metaphors of virtuosity, talent, and expressiveness continue to drive our thinking, technology will continue to exist outside the human endeavor of music learning. Change—real change—will only occur when enough of us generate new ways of thinking about, and talking about, what we do.

References

Hayles, N. K. (2012). *How we think: Digital media and contemporary technogenesis.* Chicago: University of Chicago Press.

Kuhn, T. S. (1970). *The structure of scientific revolutions* (2nd ed.). Chicago: University of Chicago Press.

Lakoff, G., & Johnson, M. (1980). *Metaphors we live by.* Chicago: University of Chicago Press.

CHAPTER 33

A SOCIOLOGICAL PERSPECTIVE ON TECHNOLOGY AND MUSIC EDUCATION

RUTH WRIGHT

THIS writing will consider *Where is technology within music education?* and *Who is affected and how?* from a perspective drawn from sociology and in particular the sociology of music education. I will discuss the emergence of a totally technologized society, akin to the totally pedagogized society identified by Bernstein (2000), and consider its implications for music technology and music education. I will suggest that many students exist in an educational technotopia that offers great potential within models such as informal music learning for more equitable models of music education. In this connection I will discuss the potential for technology to provide what Rancière (1991, p. 6) calls "a thing in common," a pedagogical tool that bridges the musical worlds of student and teacher, the informal and formal in music education. Such tools allow students to learn in naturalistic ways without the need for extensive explication and bring students and teachers into more equal learning relationships. I then turn to questions of equality in the global and local distribution of music technology resources and pose some questions to be considered by music educators as we move forward into the twenty-first century.

In some of his last writings, British sociologist of education Basil Bernstein (2000) identified a new stage of development in late twentieth-century society. He described this as the emergence of a *totally pedagogized society*. By this he meant "a society that [has] introduced pedagogy in all possible areas of human agency" (Wright, 2010, p. 22). In a similar manner, I would suggest that early twenty-first-century society has seen the rise of the *totally technologized society*. Just as the totally pedagogized society introduced pedagogy in all possible areas of human agency, I would suggest that the totally technologized society has inserted technology into all areas of human existence. This can be seen clearly in education and, of key interest here, in music education.

Technology now plays a central role in education in many countries, although, as the contributors to this book who represent countries from the so-called developing world remind us, this is not universally so, and we would do well to be constantly mindful of this. In many schools, however, technology enters the students' pedagogic lives, from the moment they arrive in the morning and electronic attendance registers are taken, throughout their pedagogic experiences, mediated by interactive whiteboards, networked wireless computer software, tablet devices, and smartphones (to name just a few), and to their homework experiences, which are accessed, shared, and assessed though websites or cloud sharing applications. In music classes specifically, students may be presented with opportunities to perform, compose, and listen to music using music technology. The use of electrified instruments, amplification, sound manipulation, recording, and editing is standard in many curricula, as is the use of interactive whiteboards and composition and scoring programs, used more or less creatively according to the level of comfort, experience, and ingenuity of the teacher. Many students might be said to be living in a veritable technotopia. Some countries' national examination systems offer specific qualifications in music technology, and many permit its use for the presentation of performances and compositions. The use of technology for listening to music has of course become mainstream since the invention of the gramophone and is now increasingly diverse and available on many budget levels.

It is arguable that the totally technologized society carries within it the seeds of both the destruction of humanity's musicking and its salvation. On the one hand the ubiquitous presence of and easy access to recorded music appears to be posing severe problems to the art music sphere. Orchestras and opera companies are facing increasingly difficult economic situations, and audiences for these performances are declining. Recordings of art music are subsidized by massive sales of popular music in its multitude of genres. On the other hand audiences who could never previously have afforded to attend concerts and performances such as those of the Metropolitan Opera now have access to high-quality art music performances on digital radio, iTunes, and similar platforms and through broadcast performances in cinemas; there has never been such easy access for many to professional quality recorded popular music and self-instructional videos. The universal access to online music instruction afforded by music technology through sites such as YouTube offers potential for radical new modes of music learning and teaching. It enables access to music instruction for those who cannot afford years of private instrumental tuition and for those to whom the rituals and regimes of traditional Western art music instruction do not speak. Models predicated on such an understanding of the potential of music technology, like the informal learning model pioneered by Lucy Green (2001, 2008), place the learners at the center of the music education experience and allow them to use technologies, such as the Internet and smartphone audio/video recordings with their infinite stop, start, rewind, and play features, as pedagogic tools. Such technologies become new pedagogic authorities in the music classroom, as students use recordings to learn instrumental and vocal skills and to analyze musical material and reproduce it in ways that are socially and culturally relevant to them right there and then. This requires of teachers a very new approach to music education, however. Rather than disappearing from the pedagogic encounter or fading away to the edges of the picture, such teachers become key members of the musical pedagogic experience. Moreover, they do so in new relationships to their students that can be as affirming of the teacher's musicianship as they are of the students', as all become equal members of the musicking encounter. Teachers then function as comusicians alongside their students and as equal learners within the learning experience, just as the members of an informal music-making ensemble do. It requires of teachers that they have the humility to acknowledge that they are no longer the fount of all knowledge concerning music in the classroom, that some of their students may know more than they themselves do about some genres, and that some students may be more skilled than they are in certain types of instrumental or vocal performance. Moreover, it requires that teachers embrace music technologies and utilize them not just as pedagogic tools but as pedagogues in and of themselves.

In this way, creative teachers embrace the potential of music technologies and allow them to teach students in a reciprocal pedagogical arrangement. This reminds me of some of Jacques Rancière's pedagogical discussion in *The Ignorant Schoolmaster: Five Lessons in Intellectual Emancipation* (1991).

In this work, Rancière presents and reflects upon an unusual pedagogic experiment conducted in 1818 by Joseph Jacotot, a lecturer in French literature at the University of Louvain, Belgium. After a long teaching career in university in France, Jacotot found himself exiled to Belgium as a consequence of political circumstances after the return of the Bourbons to power in France. He was appointed to the University of Louvain, where he expected to quietly complete his days. However, his lectures became so popular that many students wished to attend them, despite the fact that they did not speak French and Jacotot did not speak Flemish. They had no pedagogic language in common. To teach them, Jacotot realized the need for the link of a "thing in common" to be established between himself and his students. This presented itself in the form of a bilingual edition of the work *Télémaque*, a didactic French novel written in simple vernacular prose by the author François Fenélon. Through an interpreter, Jacotot asked the students to read the book, to learn the French text with the help of the Belgian translation, and to arrive at the point where they could recite the French text by heart. At the end he asked the students to write, in French, their thoughts about the work. To Jacotot's complete surprise, the Flemish students managed this task as well as many French natives could have done. Somehow, through the link of a thing in common, these students had managed to teach themselves French.

From this story, Rancière (1991) draws some powerful conclusions about the perceived need of explanation within education: "according to the unequal returns of various intellectual apprenticeships, what all human children learn best is what no master can explain: the mother tongue. We speak to them and we speak around them. They hear and retain, imitate and repeat, make mistakes and correct themselves, succeed by chance and begin again methodically, and, at too young an age for explicators to begin instructing them, they are almost all—regardless of gender, social condition, and skin color—able to understand and speak the language of their parents" (p. 5). From this point forward, however, Rancière explains, "everything happens as though he could no longer learn with the aid of the same intelligence he has used up until now, as though the autonomous relationship between apprenticeship and verification were, from this point on, alien to him" (p. 6). Rancière suggests that this division is due to the word "understanding": "understanding is what the child cannot do without the explanations of a master." Rancière then explains that a revelation came to Jacotot, that it was this erroneous "logic of the explicative system" that had to be undone. The perceived requirement for explanation divided the world into those who know and those who do not know, into an inferior and a superior intelligence: "explication is the myth of pedagogy, the parable of a world divided into knowing minds and ignorant ones, ripe minds and immature ones, the capable and the incapable, the intelligent and the stupid"(p. 6). Yet Jacotot's experiment had shown that we can learn without explication if we have a link between the known and the unknown provided by a thing in common—in his case the French original and the Belgian translation of *Télémaque*. I would suggest that music technology acts in informal learning situations in the manner of the translation of *Télémaque* used by Jacotot to teach French to his Belgian students when he himself could speak no Flemish. The technology provides a thing in common between the musical worlds of student and teacher. Students are able to access the sound and visual worlds of music through technology and thus learn to play music alongside their often formally trained, notation-reading music teachers. The technological representations of music and instrumental skill therefore provide the thing in common that can be used to build bridges between the informal and formal musical worlds. In this way, technology can be an ever-present member of the pedagogic team, assisting but not replacing the teacher, just as *Télémaque* assisted but did not replace Jacotot.

I have discussed above the presence and effects of technology within specific classroom music learning situations, but there is also a macro dimension to the questions guiding this part of this book: *Where is technology within music education? Who is affected and how?* We are living in an era of globalization. As global governments adopt increasingly similar policies and ideologies, particularly with respect to education and its relationship to the economy, it is perhaps no longer relevant to think in terms of national education policies and ideologies; rather, perhaps we should consider such matters in terms of global education policy and ideology (Ball, 2014). In terms of music technology, globalization offers much potential for online sharing and collaboration. For example, applications such as SoundCloud allow users to upload their music and receive feedback from others or to collaborate in cyberspace with other composers. It is possible to set up safe pedagogic collaborative spaces within SoundCloud where students can "meet" and make music with students in other countries and from other cultures. This process offers powerful pedagogic opportunities for intercultural and multicultural music education and for creative collaborative musical work. NuMu has been offering a similar opportunity for online music sharing of students' compositions for some time now. NuMu also has a download chart where the popularity of different musical offerings can be seen. This may prove motivating for students.

Alongside such positive affordances of music technology, however, there are also more negative considerations. The power and speed of working with music technology contains within it possibilities of unequal distribution of resources on a grand scale. In chapter 17 Benon Kigozi indicates this. While many children and young people live in the musical technotopia I described, their counterparts in other places may lack even the prerequisite for use of much technology—access to electric power. How can music education function as a global community with such inequalities of access as this? It behooves us as a music education community to consider and act upon such distributive injustices together with our students. As a profession, we also do not need to look so far afield to encounter variable access to technology in education. Such injustices exist much closer to home as well, and distribution of technology often varies, alongside other more general educational resourcing, by postal code. Bernstein's (2000) observation that educational resources tend to be provided in inverse proportion to need is as true of much of music technology as of other technological resources. In a globalizing world, it is no longer possible to plead ignorance of such injustices. At some point we have to decide as a music education community to take positive action to redress the balance. There is great potential, through online collaboration and sharing, for more and less advantaged students living geographically close to each other to be brought together in virtual learning environments and to experience the realities of each other's daily musical lives. Such projects require great sensitivity and careful management on the part of teachers, but they can be the springboard for important social and philanthropic projects joining socially disparate schools and their communities.

I do not believe we have even begun to scratch the surface of what music technology is capable of doing from a social justice perspective. Important questions for music educators to consider will be:

- How do we fully embrace the potential for music technology to make music education socially inclusive?
- How could we use it to further this end?
- How can we use this technology to promote intercultural understanding and respect?
- What economic issues does music technology raise for twenty-first-century society?
- What does technology mean for the music industry?
- How, by considering these issues with students, can we best prepare them for living as musical beings in a totally technologized society?

References

Ball, S. (2014, November 21–22). *Global education policy: Is research looking in the right place?* Keynote address to Learning in a Changing World Conference, Institute of Education, London.

Bernstein, B. (2000). *Pedagogy, symbolic control and identity: Theory, research.* (Rev. ed.). Lanham, MD: Rowman and Littlefield.

Green, L. (2001). *How popular musicians learn: A way ahead for music education.* Aldershot, UK: Ashgate.

Green, L. (2008). *Music, informal learning and the school: A new classroom pedagogy.* Farnham, UK: Ashgate.

Rancière, J. (1991). *The ignorant schoolmaster: Five lessons in intellectual emancipation.* Palo Alto: Stanford University Press.

Wright, R. (Ed.) (2010). *Sociology and music education.* Farnham, UK: Ashgate.

PART 3

EXPERIENCING, EXPRESSING, LEARNING, AND TEACHING

3 A

Core Perspectives

CHAPTER 34

POWER AND CHOICE IN THE TEACHING AND LEARNING OF MUSIC

CHEE-HOO LUM

My earliest recollection of listening to music through a box was the Rediffusion (Singapore's first and only cable-transmitted radio station) in the 1970s. The Rediffusion was always an accompanying background at home, a white wash of sound that I never really paid much attention to but would notice when someone in the family switched it on in the morning and turned it off at night. I vaguely remember the Cantonese storytelling and all kinds of Chinese dialect popular songs being broadcast and would only pay more attention to it aurally when my grandmother or parents would turn the volume up when they heard a familiar song or wanted to listen to Lee Dai Sor. I never had the choice or power to access the Rediffusion radio set because it was literally out of my reach—it was placed on a high shelf in the kitchen. In addition, there were limited program channels (only two, in fact) to choose from even if I had access to it.

I must have been 9 or so when I chanced upon a small electronic keyboard while with my parents at a shopping mall. After listening to a live demonstration by the salesperson (it must have been something familiar like "Jingle Bells"), I was motivated and wanted to learn to play it. I urged my parents to buy it for me, and they happily agreed, since it was not an expensive gadget. Within a week or so, my father noticed that I had picked up all 12 songs in the repertoire booklet that came with the keyboard. I was thrilled by the various accompaniment possibilities that came with it, like "swing," "rock & roll," and "waltz," which I could easily access by just the push of a button. I was much entertained and would play the melodies of the songs in the booklet to the family with these refreshing and new (to me) "groovy" accompaniment patterns. As I got bored with the 12 songs, I started to explore and play around to create new melodies with the sound palette provided (all Western-styled instrumental sounds, like the piano, violin, organ, harpsichord, etc.).

Advice from a music teacher convinced my father that my electronic keyboard was just a plaything and if he really wanted me to learn music properly, I should pick up a real instrument like the piano. My father valued the advice from the music teacher and bought a piano (a huge expense in our humble household), and that was how I started my journey as a piano student, playing only Euro-American classical music, listening exclusively to CDs of similar repertoire thereafter, and coiffured into a particular musical trajectory that was deemed worthy and sanctioned by my piano teachers and my parents.

As an undergraduate music student, I was exposed to electronic music during our composition modules and spent hours exploring and recording sounds, fiddling with MIDI, and working on Pro Tools in the computer music lab to create my compositions. Despite my amateur skills in these tools and my attempts at composing using such tools, I had fun. I was motivated and inspired, particularly because the university had invited visits from internationally renowned electronic music composers who showed us examples of their works and processes and provided valuable feedback on our works-in-progress. We students also had the chance to work collaboratively with one another across different digital platforms, sharing and working together on compositions and performances.

Technological advances have been relentless and become even more sophisticated ever since. My initial encounters from the 1970s through the 1990s were vastly different from those of the digital native learner in the classroom of the present. However, there are commonalities in our experiences as learners in our encounters with technology, which I will flesh out in this chapter. Technology, at least in my experiences as a music learner, has (1) provided access to the musical world as sound through digital forms and formats; (2) provided greater opportunities for the most basic beginner in music to explore and experiment with sound; (3) provided wide/r choice and variety in listening, performing, and creating for both the "trained" and "untrained" musician; (4) compressed time and space in musical experiences and environments; and (5) opened up discussion about what constitutes musicality, a musician, and skill acquisition. The introductory narrative above has considerable significance for the music teacher, especially in sanctioning (and censoring by consequence) what is "worthy" in musical learning through "turning up the volume" in guiding, giving particular critical views, showing, and drawing attention to music learners as they encounter the musical world through technology. Technology, in this instance, can be read as access to the Web, Web 2.0, and technical gadgetry (soft- and hardware) that allows for skill-based acquisition and the personal ability to enact and exact change through material and immaterial means and media.

> ... I chanced upon ... listening to a live demonstration ...wanted to learn to play ... not an expensive gadget ... thrilled by the various ... possibilities ... easily access ... entertained ... refreshing and new ... explore and play around to create ... different sound palette provided

The digital native learner who is a present reality in music classrooms is often engulfed, immersed, submersed in a sound surround (Lum & Campbell, 2007) within a multimedia/ted and multisensorial space. The advancement of technology has brought about for the individual a whole new sound palette to play with. Any sound idea can easily be captured on a recording device and manipulated using a growing array of software applications, creating all kinds of musical possibilities for the individual. As Lamb (2010) points out, individuals use such devices to "tell stories of their lives, to actively craft their identities by engaging creatively and productively with . . . technologies. They use these devices as tools within their everyday lives to enhance social networking and express their individuality" (p. 35).

Just like children creating artworks from the get-go, with their pencils, crayons, and watercolors, technology has now afforded learners with a similar option in music; notation and sequencing software can be used as "sketch pads" for a student's musical ideas—a place where "wrong answers" can create opportunities for learning. Why does my music sound the way it does? . . . What if a different instrument played this melody? . . . These questions and the lessons that can be learned from them are powerful creative experiences—experiences that can be greatly enhanced by music technology (Frankel, 2010). The advancement of technology has thus provided learners with the ability to create and compose music that they cannot physically play (Odam & Paterson, 2000, as cited in Cain, 2004). Thus, digital native learners enters the music classroom with a far more complex sound base and connectivity than their teachers, who ironically might have become considerably myopic through their previous musical training.

The Fluidity of Technology

The availability and fluidity of technology has opened up a plethora of choices and distributed power to the learner. The active immediacy of choice and the wielding of power by the learner through technology directly impact the way teaching and learning is perceived. This power differential between the teacher and the learner given access by technology is of significance in transforming music classroom practices, as it allows both the teacher and learner to constantly reflect and question definitions of musicality, musicianship, and the expansion of musical creative work.

Technology has allowed the individual to gain access to a sound world far beyond anything imaginable just a few decades ago. With a click of a button, an individual can be transported from the throat singers of Siberia to the kulintang ensemble in the Philippines. The sound choices are limitless, as Leonhard (2008, p. 37) points out: "Music 2.0 allows us the sort of access that we have to water: 'music [is] like water,' available on tap" (as cited in Schroeder, 2012, p. 28). Technology has also dislocated time and space for the music user, shifting "listening from being tied to a specific environment to being less location-bound as it becomes accessible to the user from her (often) portable digital device" (Schroeder, 2012, p. 28). One possible outcome of the instantaneity of all these sound possibilities through the advance of technology is the eventuality of boredom: "the insubstantial, instantaneous time of the software world is also an inconsequential time. 'Instantaneity' means immediate, 'on-the-spot' fulfillment—but also immediate exhaustion and fading of interest" (Bauman, 2000, p. 118). Maintaining interest toward self-sustainment in musical learning becomes a huge challenge for the music teacher in this technological age.

An intention to learn comes from intrinsic motivation and interest in the topic at hand. Boredom can be a challenge for educators if individuals who passes through their music classroom is not engaged or has no interest in what is being presented because the experience or the repertoire introduced has no connection and relevance to their live/d experience. The music educator is competing for attention with the pervasiveness and fluidity of technology that vies for the attention of any individual, who, once entrapped, has the illusory power and control of consumption at any time in any space. The last bastion of power seems to lie with the conviction and overpowering of compliant behavior of institutionalized musicality, "often limited to formal works, or texts, composed by especially gifted artists and performed by skilled musicians trained to capture the essence of the composer's intentions" (Cavicchi, 2009, p. 99). The music teacher must hold onto and turn up the volume of institutionalized musicality in order to regain control and power from the learner.

Alternatively, the music teacher can attempt to ride on the fluidity of technology alongside the learner, although "keeping fluids in shape requires a lot of attention, constant vigilance and perpetual effort—and even then the success of the effort is anything but a foregone conclusion" (Bauman, 2000, p. 8). Taking the fluidity of technology into the classroom is a complicated matter for the music teacher. For one, buying the latest gadget and equipping the music classroom in the hope of motivating the learner takes the music teacher down a slippery road of budget impracticalities. As Webster (2007) points out, "large costs are incurred by equipment and software, to say nothing of the hours of teacher-time in learning the technology and planning for its use. Policymakers must be as sure as humanly possible that such costs are expended wisely" (p. 1312). The music teacher is then responsible for the justification of the acquired hardware and software, which eventually leads on to the learner being held hostage by the inevitable outdating of the technology in the classroom. What is also worrying is the fact that, while the power and availability of music technological hardware and software for music teaching and learning have grown over the last few decades, "in-service teachers lagged behind in their application of these resources." There is also little evidence to suggest "how committed music teachers are in the integration of technology into music instruction" (p. 1324), particularly if we consider Hooper & Rieber's (1999) suggestion that teachers process through different stages (familiarization, utilization, integration, reorientation, and evolution) in working with technology integration.

Power

> I never had the choice or power ... out of my reach ... limited program channels ... to choose from even if I had access to it

The fluidity of technology has perhaps turned Jeremy Bentham's Panopticon on its head. Music teachers who used to be able to control and dictate students' content learning and skill acquisition through their "surveillance towers" are losing their vantage point, as they are being questioned and challenged by students and other stakeholders about their roles and functions in this age of technology. The sound world of the child has exploded because of technology that makes the expanding and changing repertoire that the child has access to ever more difficult for the music educator to grapple with and pin down. Even conceptions of what constitutes music and who is or is not a musician are constantly being questioned by critically reflective students in the music classroom through the music encounters they have that are fueled by the advancement and fluidity of technology. Music teachers can no longer hold onto the power of institutionalized musicality without having to explain and defend themselves. Music programs are constantly needing to justify their continued existence, as they are at risk of alienation by students who have deeply felt musical experiences through technology or otherwise outside the school environment, something that challenges the legitimacy of a curriculum that is wavering in its affirmation of the true nature of "good music" (Cavicchi, 2009, p. 100).

> *advice from a music teacher convinced my father that my electronic keyboard was just a plaything ... to learn music properly ... pick up a real instrument ... my father valued the advice from the music teacher ... deemed worthy and sanctioned by my piano teachers and my parents*

The systemic structure of institutionalized musicality is no longer on solid ground, as it is constantly being shaken up by the availability of technology. Evoking Bauman's (2000) discussion on liquid modernity, the solids that used to bind these musicality structures, which have "come to be thrown into the melting pot and which are in the process of being melted at the present time, the time of fluid modernity, are the bonds which interlock individual choices in collective projects and actions—the patterns of communication and co-ordination between individually conducted life policies on the one hand and political actions of human collectivities on the other" (p. 6). The power in any contextualized music education community is thus reliant on the strength of the politics that surrounds the institutionalized musicality, a commitment to convince the learner of the value and worthiness of the musicality in question, preferably sanctioned by parents, school leaders, and other stakeholders. This political strength is fast weakening because of the proliferation of technology that is widening the bifurcation of everyday and institutionalized musicality. Again, as Bauman (2000) reminds us, "we are presently moving from the era of pre-allocated 'reference groups' into the epoch of 'universal comparison,' in which the destination of individual self-constructing labours is endemically and incurably underdetermined, is not given in advance, and tends to undergo numerous and profound changes before such labours reach their only genuine end: that is, the end of the individual's life" (p. 7). The manifestations of power in the contestation of musicality through technology from the teacher and learner interact in complex ways. Taking the general music classroom as a point of departure: a well-intentioned music teacher who wants her students to pick up an acoustic instrument (say the ukulele) so that they may enjoy practicing and playing on the instrument while learning musical concepts through the reading of particular repertoire and may eventually play and sing as an ensemble for a school performance might be met with resistance from the learners. Assume that there are readily available ukuleles in the school for every child to use within the music classroom and the first few lessons began with the teacher getting the students to hold the ukulele correctly, figuring out their fingering positions to strum a few basic chords. The technologically savvy child in the music classroom who had difficulty pressing her fingers on the strings, remembering the chords, and wasn't able to bring the ukulele home to practice felt unmotivated, lost interest, and decided on an alternative. She went home and easily located an iPad app for ukulele and strum-ready chords on her app. She figured out numerous songs and repertoire for the ukulele online and started to accompany her own singing with the newly found ukulele app within a couple of hours. Because of technology, the student had, in a way, transgressed the limits of time, space, and the physicality of her body and brought her renewed sense of interest and motivation for the ukulele (albeit in a different form) back into the classroom. The musical instrument (in this instance, technology) is "a means to actualize music as sound" (Echard, 2006, p. 11), so the learning of technique in any musical instrument "is not in order to allow a more accurate rendition of one work, but to create a flexibility and creative potential that can actualize many different musical ideas" (p. 15). Technology for this student just opens up possibilities for her to access a broader musical vocabulary more easily without necessarily having to go through time in mastery to arrive at technical proficiency.

To continue the story: the student shares what she found on her blog with her classmates, and many of them follow suit, finding other ukulele apps and picking up different repertoire. The student returns to the music class the following week, tells the music teacher what she has done, and asks if she and her classmates can use their iPad ukulele app (with small speakers in tow) instead of the ukulele during class time. She suggests a myriad of popular music repertoire for the school performance that she found more enticing to her and her classmates than the tunes picked by the music teacher. The music teacher is left with a conundrum for next steps and felt somewhat powerless.

It is difficult to still pretend that musical knowledge and skill in this technological age is held on a pedestal by music teachers, dishing out words of wisdom as they deem fit. Agency provided to the learner through technology has considerably weakened the power of music teachers in controlling mastery over time and space. Thus, music teachers need to be constantly vigilant in the programs they create, and critically reflect and remind themselves not to become agents of an institutionalized power and "be mindful ... of the social structures that stand in the way of free-play of energies, the wide-awake-ness, the authenticity, and the moral sensitivity" (Greene, 1995, p. 50) that they would like students to develop in their classrooms in the twenty-first century.

IDENTITY AND THE INDIVIDUAL

The technological age of the modern world is destined for the lonesome traveler. Technology has effectively cut away the "middlemen" and left learners to fend for themselves, creating "an individualized, privatized version of modernity, with the burden of pattern-weaving and the responsibility for failure falling primarily on the individual's shoulders" (Bauman, 2000, p. 8).

Another story: an aspiring general music student who wanted to be a YouTube sensation saved enough money to gather her cadre of technological gadgets by googling the latest and cheapest software and hardware, assembling a personalized recording studio in the comfort of her own home. She searched and watched endless YouTube videos of cover versions of the song she intends to record so that she might come up with something unique that others have not created. Her developing musical identity was colored by the constant bombardment of media and technology, with their intentions to "discourage enterprise and independent thinking and to make the individual distrust his own judgement, even in matters of taste. [Her] own untutored preferences, it appeared, might lag behind current fashion, they too needed to be periodically upgraded" (Bauman, 2000, p. 85). As she posted her cover song online, she awaited with bated breath comments from the globe of netizens who might chance upon her post. After getting through the emotional turmoil of some negative comments she received, she did eventually find some like-minded souls across the Internet space.

Individualization through technology has brought forth a wave of possibilities and freedom to experiment, but at the same time, individuals needs to also deal with the consequences of putting their experimentation onto an open stage where capacity is given for the right to self-assertion but not necessarily the capacity to control the social backlash that might happen. The aspiring student formed a virtual pop band with these like-minded souls from various parts of the world, and they worked across time and space to record a couple of original covers. New musical ideas came along, but parts of the collective were no longer interested in the genre they were exploring and moved on. The aspiring student swiftly moved on to pursue her new-found musical interests. Her musical identity continues to be shaped by these experiences through technology, and her journey moves on, but where it will lead is a huge unknown.

> Living in the modern world and being modern means suffering the possibilities of
> the inability to stop and pause. One needs to constantly get ahead of oneself—to

constantly transgress which forewarns the instability of identity. The search for musical identity in this age of technology is thus an ongoing struggle to arrest or slow down the flow, to solidify the fluid, to give form to the formless. We struggle to deny or at least to cover up the awesome fluidity just below the thin wrapping of the form. . . . Yet far from slowing the flow, let alone stopping it, identities are more like the spots of crust hardening time and again on the top of volcanic lava which melt and dissolve again before they have time to cool and set. (Bauman, 2000, p. 82)

What might the possible responses of the music teacher be in the general music classroom to this aspiring student who is actively engaged and immersed in a musical environment of her own and with others in virtual space outside the music classroom?

Music Teacher as Exemplar, Critic, Facilitator

The power shifts between the teacher and learner and the turn toward the responsibility of the individual in the twenty-first-century music classroom begs a close examination of the changing roles of the music teacher.

Exemplar

The music teacher, through his general music program, can demonstrate to students the diverse range of repertoire that he listens to, performs on, and creates with, either by modeling through the playing of various acoustic and electronic instruments, working on the latest music software applications, and/or showing varied audio/visual examples. The point is to show, by example, open-mindedness and receptiveness when approaching all kinds of musical genres within the music classroom, with or without technological means. In the particular instance of the musical genre(s) that the aspiring student is currently focused on—like Green (2002) and others who have connected formal musical learning with informal and nonformal ways of engaging students—the music teacher is striving to impress on students that their music lessons can be connected with their daily musical experiences. The iPads or whatever technological tools and applications students might come across are used to aid in the music-making process, providing alternative sound and performance choices for students.

Critic

The music teacher should always provide a critical listening ear for the student when the student is prepared and ready for it. The music teacher, as a musician, should be able to make judgments on the merits of the artistic performance and composition being presented, providing constructive feedback to allow students to move steadily forward. The caveat here is, of course, the attachment of the music teacher to the institutionalized musicality that defines her as a musician, which will imply varying subjective judgments of worth and value attached to what students put forth musically.

Compounded by the proliferation of music in technology, which creates changing and limitless sound examples and possibilities, "digital music environments bestow greater ambiguity to the material itself, as they allow for individuals' interpretations of creative processes . . . draws attention to a fluidity in music which ultimately, rather than being savoured like bottled wine, becomes poured into the performance space—more like water!" (Schroeder, 2012, p. 38). Critical reflection of the learners then becomes even more crucial to develop, as they are bombarded by and work with these growing sound worlds. The music teacher would perhaps be helpful to learners in constantly questioning their sound experiences and choices, using the Socratic method of dialogue to encourage critical thinking and to clarify the learners' underlying motivations and intentions.

Facilitator

Technology has opened up many self-directed learning opportunities for individuals to develop their musicianship, that were not possible even a decade ago (Lamb, 2010). Instead of the traditional music classroom setting, more and more, "pupils are working either in pairs or as individuals, each with a workstation and a set of headphones," which "demands a different approach to teaching, and also requires teachers to be technicians, performing 'regular maintenance and careful management of resources'" (Odam & Paterson, 2000, p. 35, as cited in Cain, 2004). The music teacher can no longer be adamant about the teaching of particular repertoire, technique, and skills in the changing landscape of the twenty-first-century general music classroom. The power shift between the teacher and learner prompted by technology calls for a learner-centered pedagogy in music education that ascribes much less significance to content delivery. The teaching paradigm has to be one that emphasizes process over product and the ability to deal with ambiguity (Davies, 2008), while at the same time developing students' voices in relation with others in terms of criticality and creativity amid a communicative and collaborative environment. The music teacher needs to facilitate a more creative and imaginative musical learning environment for students, developing "independent musicians who are responsible for the thinking and doing and musical decision making within a teacher-supported learning environment" (Blair, 2009, p. 45). As a facilitator, the music teacher needs to cultivate a learning environment where "the learner is free to explore, discover new things, express ideas, and create. The teacher is a collaborator in problem solving rather than someone who has all the answers. The teacher's role in the interaction is to guide the learners . . . [in] discussions while they are working, suggest ideas that might solve problems or challenges, allow learners several tries at solving problems. . . . ensure students do not feel they have 'failed' if their solutions do not work; and emphasize the problem-solving aspects of an activity" (Mans, 2009, p. 186).

Community, Communication, and Collaboration

> I was motivated and inspired ... showed us examples of their works and processes and provided valuable feedback on our works-in-progress. We students also had the chance to work collaboratively with one another across different digital platforms, sharing and working together ...

With the advance of technology and the media, one cannot ignore the complexities and multiple channels via which communication and collaboration can happen. Communication and collaboration allows for a dialogic to unfold. As Richard Sennett (2012) suggests, while the discussion within the dialogic does not resolve itself by finding common ground, and "no shared agreements may be reached through the process of exchange, people become more aware of their own views and expand their understanding of one another" (p. 19). This is a needful skill set in the changing world of the twenty-first century. When communicating and collaborating with others, one becomes more self-aware, and listening skills are enhanced. Sennett further suggests that the idea of "we," through means like communication and collaboration in the modern world, is "an act of self protection. The desire for community is defensive. . . . To be sure, it is almost a universal law that 'we' can be used as a defense against confusion and dislocation" (p. 138). In the growing individualized existence that is the twenty-first century, the community or the notion of "we" offers at least a momentary shelter for the individual.

Communication and collaboration should thus be made manifest through students' musical connections with their physical and virtual communities so that they begin to understand and situate themselves within the context of their own changing soundscapes in the global environment. The music teacher can also be helpful in bringing in the many connections, networks, and partnerships that music can offer to students, so as to deepen their artistic orientations and allow them to see and experience the relevance of the arts within their changing contexts. Examples include online communities encouraging self-directed musical skill development, and musical collaborations that have opened up avenues of participation across diverse communities that are outside the boundaries of a traditional music classroom.

Summary Thoughts

Returning to the original questions posed in this part of the book, *How does technology impact the learning and teaching of music? How should it (or shouldn't it)?* In a way, what has been articulated thus far in terms of technology's impact on the teaching and learning of music is in tandem with the framework for twenty-first-century learning suggested by the Partnership for 21st Century Skills (2011), a framework that includes the development of skill sets for the learner in terms of (1) creativity and innovation; (2) critical thinking and problem solving; and (3) communication and collaboration. The development of these skill sets suggests the expansion of the music learner's ability to listen, perform, and create music as an independent individual as well as in a variety of networked spaces, and to explore and experiment musically in a collaborative or on one's own terms. While technology has opened up access, exploration, and independence to the music learner, the music teacher's roles of exemplar, critic, and facilitator are still helpful in guiding the learner along and to encourage the development of these skill sets in a positive learning environment.

The acknowledgment that technology has created an unprecedented fluidity in the teaching and learning of music signals a constancy of negotiation and navigation between the teacher and learner. Technology has, in a way, "unhinged" and "unframed" the music learning environment, leaving a sense of vulnerability and uncertainty, particularly for the music teacher, who needs not just to be cognizant of the most current music technological fad and movement but, more important, to be a lot more aware of the learner's musical being and becoming in the context of the impact of technology. There is perhaps a greater need for the music teacher to draw closer the sides of the widening gap between school music and learners' daily musical experiences that are swiftly propelled by technology. The music teacher should also become more cognizant of the multimedia and multidisciplinary space in which the digital native learner is engulfed, so as to further notions of creative work within the music classroom that might include these experiences.

Taking on the metaphor of the rhizome, from Deleuze and Guattari (1987), consider the possibility of the learner of the twenty-first century being situated within a rhizomatic structure, accumulating sound experiences based on an assemblage of multiplicities connected on a nonstratified plane of consistency. Because of technology, an opportunity exists for the music teacher to rethink music programs so as to signify the importance of freeing our children to tell their stories, not only so that we can hear them but so that they can make meaningful the birth of their own rationality. The expanse of information accessible through technology may remind us, too, of the importance of affirming the validity of many kinds of experience, even those that seem incompatible with our own interpretations of the world (Greene, 1995, p. 54).

Thus, the power and control that technology has given dramatically changes the relationship between the teacher and the learner in the music classroom. As Bauman (2000) reminds us, "in this territory only such things or persons may fit as are fluid, ambiguous, in a state of perpetual becoming, in a constant state of self-transgression" (p. 209). There is no longer an easy compromise of the life power and political power between the teacher and the learner, and indeed, everything can be pulled asunder if the teacher and the learner do not take active steps to shift and adapt with the times.

References

Bauman, Z. (2000). *Liquid modernity*. Cambridge, UK: Polity Press.

Blair, D. (2009). Stepping aside: Teaching in a student-centered music classroom. *Music Educators Journal, 95*, 42–45.

Cain, T. (2004). Theory, technology and the music curriculum. *British Journal of Music Education, 21*(2), 215–221.

Cavicchi, D. (2009). My music, their music, and the irrelevance of music education. In T. A. Regelski & J. T. Gates (Eds.), *Music education for changing times: Guiding visions for practice* (pp. 97–107). Dordrecht: Springer.

Davies, J. H. (2008). *Why our schools need the arts*. New York: Teachers College Press.

Deleuze, G., & Guattari, F. (1987). *A thousand plateaus: Capitalism and schizophrenia*. Minneapolis: University of Minnesota Press.

Echard, W. (2006). Sensible virtual selves: Bodies, instruments and the becoming-concrete of music. *Contemporary Music Review, 25*(1-2), 7–16.

Frankel, J. (2010). Music education technology. In H. F. Abeles & L. A. Custodero (Eds.), *Critical issues in music education* (pp. 236–258). New York: Oxford University Press.

Green, L. (2002). *How popular musicians learn: A way ahead for music education*. Aldershot: Ashgate Publishing Limited.

Greene, M. (1995). *Releasing the imagination: Essays on education, the arts, and social change*. San Francisco: Jossey-Bass.

Hooper, S., & Rieber, L. P. (1999). Teaching, instruction, and teaching. In A. C. Ornstein, & L. S. Behar-Horenstein (Eds.), *Contemporary issues in curriculum* (pp. 252–264). Boston: Allyn and Bacon.

Lamb, R. (2010). Music as sociocultural phenomenon: Interactions with music education. In H. F. Abeles & L. A. Custodero (Eds.), *Critical issues in music education* (pp. 23–38). New York: Oxford University Press.

Leonhard, G. (2008). *Music 2.0: Essays by Gerd Leonhard*. Hameenlinna, Finland: Creative CommonsLicense.

Lum, C. H., & Campbell, P. S. (2007). The sonic surrounds of an elementary school. *Journal of Research in Music Education, 55*(1), 31–47.

Mans, M. (2009). *Living in worlds of music: A view of education and values*. New York: Springer.

Odam, G., & Paterson, A. (2000). *Composing in the classroom: The creative dream*. High Wycombe, U.K.: National Association of Music Educators.

Partnership for 21st Century Skills. (2011). *A framework for 21st century learning*. Retrieved February 28, 2013, from http://www.p21.org/.

Schroeder, F. (2012). Shifting listening identities—Towards a fluidity of form in digital music. In S. Broadhurst & J. Machon (Eds.), *Identity, performance and technology: Practices of empowerment, embodiment and technicity* (pp. 24–38). Hampshire, U.K.: Palgrave Macmillan.

Sennett, R. (2012). *Together: The rituals, pleasures and politics of co-operation*. New Haven, CT Yale University Press.

Webster, P. (2007). Computer-based technology and music teaching and learning: 2000–2005. In L. Bresler (Ed.), *International handbook of research in arts education, part 2* (pp. 1311–1328). Dordrecht: Springer.

CHAPTER 35

MUSIC FLUENCY

How Technology Refocuses Music Creation and Composition

BARBARA FREEDMAN

flu•en•cy
: the ability to speak easily and smoothly; *especially*: the ability to speak a foreign language easily and effectively
: the ability to do something in a way that seems very easy (Fluency, 2014)

lit•er•a•cy
: the ability to read and write
: knowledge that relates to a specified subject (Literacy, 2014)

THE ability to communicate verbally freely and clearly in any language is what we call *fluency*. The ability to read and write in a language is what we call *literacy*.

When I was first hired to teach music technology classes at Greenwich High School in 2001, I honestly had no idea where to start. It was my fifth year teaching music, having started my career in my thirties as a general music and chorus teacher in a public high school in New York City. The electronic music courses at Greenwich High School had been created by Ann Modugno in 1969 and may very well have been the first public school music technology classes in the country, if not the world. Three other teachers have taught electronic music at Greenwich High School since Ann's retirement, but because music technology was rapidly evolving, the curriculum was in constant flux.

At first, I tried teaching with a text on recording. In class we discussed cables and recording techniques, and we used what little recording equipment we had, including a couple of four-track tape decks. But soon enough I saw that the kids were not happy or engaged. One of the electronic music classrooms, converted from a long and narrow closet, had eight PCs (personal computers) with sound cards, a software program called FreeStyle by MOTU, and "cable spaghetti" running in and out of mixing boards, sound modules, and synthesizer keyboards. I had no idea how to use the equipment, let alone how to teach it, but the kids were eager to get in there. One day a student asked, "Can we just make our own music?" I said, "Sure," and they ran to the computers. What was I to do? Just watch them? Leave them alone to freely create? No formal lesson? No homework? No tests? How bizarre and unfamiliar. That's exactly what I did and hoped that an administrator would not walk into the room and ask what we were doing.

Soon enough, students started asking me how to do something or to listen to their music and give them feedback. If I didn't know the answer to a technical question about the computers, keyboards, or software, there was always a student in the room who did know the answer or knew where we could look it up. I was very fortunate to have a few real "geeks" in the room at all times! I was learning from my students. Through listening to their music, I soon discovered that most students had something in common: a lack of understanding of some very basic music concepts and skills. Most of them had not played an instrument or sung in a chorus. They didn't understand how rhythms related to one another. They didn't know what scales or arpeggios or chords were. They didn't understand how to compose a melody, how to add chords, or how to create a bass line. They didn't understand anything about basic forms of music and how one might structure music over time. They lacked a basic understanding of the "mathematics" or "mechanics" of music: music theory. In other words, they had no skill with the language of music. Instead they relied solely on their ears. Although I had no problem with them engaging in this kind of creativity, they were upset and frustrated. They didn't like their music. They were somewhat ashamed of their music. It didn't sound "good" to them. So I started to teach. I taught music. They took care of the technology. Soon enough, I taught both. We were on a happy path.

It was clear that many of my students hadn't studied music in a traditional performance setting of band, orchestra, or chorus. If they had come up through our district elementary schools, they would have had music classes that provided a fine education and musical experience. However, they didn't remember much about music theory and certainly not much about notation. If they weren't using it, they forgot it. Many students had "general music" in middle school, and others played guitar, bass, or drums in pop music settings, but most of them could not read standard music notation. I wanted to come up with a way to teach them what I thought they needed to know to make their compositions sound "good" and to deliver the instruction in a manner that was engaging. The "engaging" part can be the most difficult. What I discovered was that the students really needed very little knowledge of music notation. They needed to know the mechanics of Western music—music theory—the foundation of which is the piano keyboard, not the printed page.

Music Literacy

Twenty-five years ago, Bennett Reimer noted that a person's ability to experience music changed drastically with a "major technological advance of recent history" (1989a,

p. 28). Before the phonograph, patented in 1878, the only way people could experience music was by listening to others perform live or by performing it themselves. According to Reimer (1989a), "Once it was refined, the phonograph allowed all people full access to music without having to produce it themselves or be within earshot of those producing it" (p. 28). Technology changes society, society's needs influence education, and history repeats itself.

Reimer challenged us to expand our concept and understanding of what it means to be musically literate.

> We now have a task in general music that differs from that of the world before phonographs: to enhance the music literacy of all people in a completely different sense from what that term meant before. Musically literate people are now those who know a great deal about the art of music. They understand its history, its techniques, its many styles, and its major practitioners; they know where to go to hear good examples of it, how to make discerning judgments about it, and how to respond to it appropriately and sensitively in its many manifestations. Such people can be considered musically literate in the fullest sense of that term: educated, perceptive, knowledgeable, sophisticated, and discerning about music. (1989a, p. 28)

Unfortunately, music literacy has been relegated to its most simplistic definition: the ability to read and write music. We have become so focused on the mechanics of standard Western music notation as the sole means for teaching music that it has diluted our ability to allow students an opportunity to fully experience and create music. If children's experience with music becomes limited to only what they can read or notate, their involvement with and expression of music is then further limited by their formal education and developmental stage.

Patricia White, a music educator in upstate New York, puts it well:

> ask a class of second-graders to create a story and have them tell you the story. What wonderful stories they create! Now, tell your class of second-graders that they are to write down their story but only use words that they know how to write down. Imagine how limited they now become. Isn't that exactly what we do as music educators when we limit our students' creativity to only using standard music notation? (2009)

In my book (Freedman, 2013) I posit that "melody is melody, and harmony is harmony. The principles of music theory, structure, and form are the same whether you are using an electric guitar, synthesizer, trumpet, violin, or voice and applying it to music of living composers or the dead ones." (p. vi) Howard Goodall puts it well in the 2006 documentary *How Music Works*,

> Whatever type of music you are into, it may surprise you to learn that the things that sound very different to each other on the surface are, in fact, using the same basic musical tools and techniques. Looking at the mechanics of a beautiful tune, a sweet chord or a hot driving rhythm is a gratifyingly democratic process. When you

analyze the nuts and bolts of music you find that the apparent differences between musical cultures, between Eastern and Western, between folk and jazz, or between classical and pop, start to melt away. The underlying techniques and tricks of good music can be and are applied to virtually any and every style (2006).

Do we say that a student who creates music in a contemporary pop style is less fluent in the language of music than one who creates a string quartet? Might it be that these students lack access to ongoing music education in their formative years because our music programs do not offer enticing programs to retain students' interests? It might be true that much of the music by contemporary pop artists lacks the sophistication of music by composers who create so-called serious music. It is therefore imperative for us as music educators to do what we can to reach as many students as possible to teach them how to communicate their musical ideas clearly and easily by becoming fluent in the language of music, thereby enabling them to produce more sophisticated work.

Standard Music Notation

The purpose of standard music notation is to preserve musical sound in writing so it can be re-created. A universal system of written signs and symbols that represents musical pitch or sound allows anyone with the knowledge of that system the potential to recreate the sounds represented. This realization became important to the Catholic Church in the fifth century as the codification of a written system to represent musical sound allowed the transport of music throughout Christendom. Those that knew the system anywhere in the world could recreate music. Thus, the creation of modern music notation contributed to the uniformity of Christendom.

Like any other codified system of signs and symbols, music notation is a language and, as with any other language, it requires constant use if one is to retain its "grammar." As with any other language, if it is not used frequently, people forget it. Elementary age students may have learned music notation but if they do not play an instrument or sing in a chorus they may forget it when the reach middle school or high school. Continual exposure may be the key to language mastery, expansion, and retention. Just as language teachers experience students speaking a language more easily than they can read or write it, music teachers also know that sound comes before sight.

Given technological advances, students do not first need to master standard music notation in order to compose (Reimer, 1989b). Technology allows students an opportunity to explore sound before sight by much as they would first explore music on an instrument or by singing. Using technology, students can create, save, edit, and listen to music creations instantly. They can even produce standard music notation if needed. "Furthermore, and possibly more importantly, teaching music through composition might provide a new paradigm in which people experience music. This approach could potentially transform music education." (Freedman, 2013, p. xxi) Reimer puts it best:

> Electronic technologies now allow students to accomplish all the essentials of genuine composition: to produce and retain a musical idea by recording it directly; to review it and make whatever refinements they choose; to extend it, enrich it, and develop it while keeping it available for further refinements. When it is finished, it exists immediately and permanently for others to experience by listening to it, and an accurate notation of it can be produced by pushing a button. . . . The effects on young people's musical understandings through composing involvements may be so dramatic as to change forever our present notions of what quality of musical experiences are possible for the nonprofessional populace. And that, in turn, would change the standards of music education dramatically. (1989b, p. 28)

Reimer continued by offering that knowledge and use of standard music notation could only deepen a person's understanding of and, possibly more important, experience with music. I agree. Even if our students are not going to become professional musicians, understanding some of the functions of standard music notation can lead to a deeper understanding of music. The very purpose of notation software is to capture music so that others can re-create it by reading and performing the music notated on a printed page. Accordingly, it becomes obvious that students need to learn the fundamentals of standard music notation in order to use notation software and the notation features in sequencing and recording software. Knowledge of music notation is also beneficial to students who use software primarily for audio recording or MIDI sequencing. If nothing else, understanding the basics of standard music notation helps students learn how to use tools in these types of software to help them compose, record, and edit their music.

Children develop skills that allow them to comprehend language, incorporate symbols and new words into their personal use, and demonstrate mastery of specific skills at different ages. English language arts and foreign language teachers understand this implicitly. We can have certain expectations of complexity and sophistication in children's speaking and writing at different stages of their development. The ability to write in a language requires comprehension of the letters (alphabet), words, and basic grammar of that language. To become fluent, that is, to be able to freely and clearly communicate in that language, requires practice. This is why, for instance, the academic essay is taught from the earliest elementary years and is retaught, evaluated, and practiced throughout the entire K–12 curriculum and beyond. Deepening levels of sophistication in written language take time and practice to develop.

Over the last 15 years, I developed a curriculum for teaching the basics of the language of music that allows students to create music they like. (The curriculum is available in my book *Teaching Music through Composition: A Curriculum Using Technology*.) My teaching is predicated on the idea that the necessary music skills for composition, the mechanics of music and theory, should be acquired at the piano keyboard, the foundation of Western music. If students can understand rhythm, melody, harmony, and all the increasingly complex details of music theory and composition, their ability to create more and more sophisticated music will be enhanced, no matter the genre. My goal is to teach the language of music so that students can freely communicate with as much ease as possible, given their musical developmental stage. As students learn more details of the language and practice using the language, more sophisticated use of the language appears in their creations. They become more fluent. They make music that they think sounds "good." In the end, what the music sounds like is what really matters. Technology allows us instant access to those sounds.

Rethinking Symbols

Sound does not exist on a two-dimensional plane that we can visualize. However, many ways of visualizing sound have been adapted over time. We are accustomed to viewing a sound along a graph. A basic graph can tell us two things about the items placed on it: (1) defined quality or quantity, and (2) spacing over time (see fig. 35.1).

The staff in standard music notation is a graph. Low and high sounds are positioned on the graph vertically, and time passing is shown horizontally. The clef sign determines the range of pitches available on the music notation "graph" (see fig. 35.2).

A variety of clefs in standard music notation allow music to be printed clearly on the staff, saving space and ledger lines. Music software for creation has an edit window that simply changes what the graph looks like. In this case, there is little need for saving space, as the notes appear up and down the piano along the side of the edit window. It could be argued that this is a more accurate representation of pitch, especially for Western music creators, as it more precisely depicts where the pitch is along the piano keyboard than does the elusive notation of the staff and clefs.

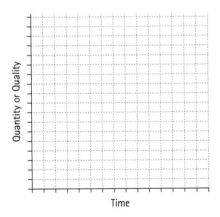

FIGURE 35.1 Graph showing quantity or quality of sound over time

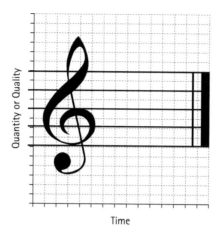

FIGURE 35.2 Staff with treble clef shown as a graph

Rhythm

Contemporary music technology can help students learn standard music notation and produce a more sophisticated music product through *quantizing*. Quantization is a scientific and mathematical term, defined by Media College (2016) as "the process of converting a continuous range of values into a finite range of discrete values"; in other words, to rounding something to an absolute value. Both audio and MIDI parameters can have quantization. Its application to standard music notation happens within the editing grid in software.

Rhythmic figures, notes and rests, tell us where the desired sound is to begin, on the beat or at a precise moment between beats. The subdivision of a beat into "slices of the pie" is determined by the basic rhythmic value given to the beat in the context of meter. In "common time" there are four beats in a measure, and the quarter note gets the beat. We can divide the beat into ever smaller components of eighths, sixteenths, thirty-seconds, and so on. Quantization is basically the same thing, breaking something down to its most basic, smallest part. Theoretically, we can continue to divide and subdivide further, but to be practical, we stop at sixty-fourth notes, and in general we even avoid that level. Most of our students will only need sixteenth notes or even thirty-second notes (double bass–drum patterns in rock and metal music and hi-hat patterns in electronic dance music and hip-hop). Our need for rhythmic quantization is not true to the scientific-mathematical definition of the word, but is limited to the desired accuracy along a designated subdivision of the beat. The more complex the rhythm, the more need for further subdivision of the beat. A student's age can also be a factor. Students may not be able to comprehend increasing subdivisions, smaller and smaller slices of the pie, until certain stages of development.

Rhythm is defined not only as the subdivision of a beat but also as duration of sound. A rhythm assigned to a note indicates how long that note is to be sustained. The sound may start on the beat or at any point in a subdivision of the beat, but the rhythm will determine just how long to sustain the sound. One can play any number of rhythmic figures directly on the beat. The value of the note, an eighth or sixteenth, for instance, will determine exactly how long that note is sustained. A rhythmic symbol in standard music notation, therefore, does two things: it informs us (1) when a sound should begin, and (2) how long it should be sustained along the "grid" of beats and subdivisions of beats. In many ways, using software for creating music can aid students in understanding this complex two-part concept.

Most music creation software edit windows are visually similar and have similar functions regardless of the brand of software. The grid is based on the subdivision of the beat. It would be safe to say that most of us do not perform and record music with absolute rhythmic perfection. Given that, it is fortunate that computer software can "correct" our rhythmic inaccuracies. Notes can be "snapped to a grid." The user can customize the grid by assigning a note value, such as quarter, eighth, or sixteenth. After selecting the notes to be "corrected," the user pushes a button, and the notes move to their precise location along the grid. If I play four notes in a measure and I want all four notes to begin exactly on each quarter note of that measure, I would choose to quantize to the quarter note. If I want to have each of the four notes to begin exactly on the second eighth of each beat, I quantize to an eighth note, making sure the notes "snap" to the correct eighth, the second eighth of each beat.

This is where understanding the relationship of eighths and quarters is important. It could be argued that students merely need to understand the relationship of quarters and eighths as they relate to the horizontal grid—in other words, to simply count boxes. My experience has proven this to be completely true. For students with little to no experience in music, the standard music notation symbol of a quarter note is as foreign to them as a letter in an alphabet they don't know. What's important for them to understand is that the symbols represent an actual duration of sound and that they have a mathematical relationship to one another. Students are better able to find the beat with an audible metronome click and when the lines on the grid, which mark the beats, are more prominent. The grid can show heavy thick lines where the beat is and thinner or dotted lines for the subdivisions of the beat. Students can "hear," or, "audiate," (Gordon, 2016) what they see along the grid. They see a long line or they see a short line placed at a specific spot in a measure. The graphed measure is a visual "snapshot" of beats and subdivisions of beats that repeats. In software for music creation, the grid in the edit window provides a more precise graphic representation of note placement over time than does standard music notation (see fig. 35.3).

Note starting points can be quantized. The user can edit the starting point of rhythmic figures by selecting the MIDI notes or audio transient spikes in the edit window and "snap" them to a specified grid. Rhythmic values of music, note lengths, can also be quantized. The note quantization grid durational values on the grid that represents note

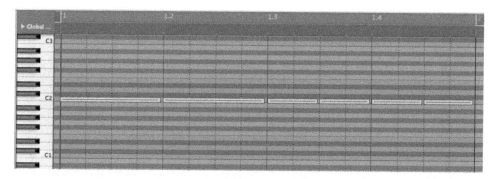

FIGURE 35.3 Typical edit window in music creation software (Logic Pro X 10.1.1)

lengths, can also be quantized. In other words, the note's rhythmic value or length on the grid, can be customized.

The edit window shown in figure 35.4 could be any software for demonstration purposes. Note values shown are eighth note, eighth rest, eighth note, eighth rest. Although MIDI notes are used here, the basic concepts of quantization apply to both MIDI and audio. The grid is set to 1/16 by default. Each quarter note displayed in the edit window occupies four sixteenth notes or four boxes. In some software, the grid can be customized and set to any note value the software allows; more sophisticated software allows more subdivisions. Less sophisticated software allows only for a subdivision of the beat up to 1/64, a variety of triplets, and other settings called "swing" parameters.

Teaching Music Theory: A Matter of Perspective

> I heard someone say, "Teach music. The technology will follow" [Barbara Freedman]. I say, "Teach technology and the music will follow."
> —Will Kuhn, Ohio Music Educators Conference (February 2012)

Will Kuhn has a degree in music technology. This major was just not available when I was in college and graduate school. Will's experience and background using technology as the instrument of musical creation and performance give him a distinctly different perspective from those of us who had no training with music technology in school. Teachers teach to their strengths and comfort levels. We bring our background and training into the classroom, and each teacher has a unique perspective on what should be taught in a classroom. It is not to say that one way is better than another, but that they are simply different; teachers' perspectives and approaches to education also differ.

FIGURE 35.4 Edit window in music creation software showing available subdivisions of the beat (GarageBand 10.1.2)

With more sophisticated concepts of music, such as arpeggiations, whether students use the software "tricks" to accomplish them becomes an issue of the learning outcomes that the teacher desires for her students. In many software applications, there is a button or simple process students can use to input a chord and then have it play in any number of iterations of arpeggiations. The software displays a block chord played harmonically, all notes at the same time, but plays it back melodically, one note at a time, in the order prescribed. It is not a "bad" thing to show students these pathways for creating arpeggiations. It is simply another tool. We can either teach arpeggios before we show students the arpeggiator button or use the button and then explain arpeggios. Again, it's really a matter of perspective.

When considering music fluency, I prefer to teach music theory at the keyboard. This is why I choose to use the piano keyboard as the MIDI input device rather than entering each note with the mouse, using the computer keyboard as an entry device, a device that uses just buttons or some other means of input. This approach helps students focus on using the piano keyboard to understand the relationships among pitches. I also teach basic piano skills as the primary means of teaching music theory, through playing and listening. The notation of these skills and patterns, standard notation or graphic, comes later. I am not teaching students to become pianists. I am teaching them to use the piano to discover the foundation of Western music creation. Touch-typing is a skill students learn to help them through school, business, and life. We don't teach them to touch-type so they will become professional typists. Western music theory is based completely on the structure of the piano. If most student's don't know the structure of the piano and the relationship among notes on the piano keyboard, Western music theory is more difficult to comprehend as a tool for analysis or creation.

I really like the idea of "sound before sight," and I wanted to incorporate that approach in training students at the piano. In this "YouTube age," with the advent of the "flipped classroom," I decided to create a series of digital videos that demonstrate basic piano skills, without any music theory explanation: five-finger position, one- and two-octave scales and arpeggios, and chord progressions in root position and inversions. These simple demonstration videos are each 30 seconds to 2 minutes long. Students can work independently at a workstation equipped with a computer or tablet and a piano keyboard. The student plays the video on her device and then rewinds to repeat, plays the keyboard with the video, and proceeds to learn at her own pace. Students basically imitate what they see on the video. These videos are not meant to train students to become pianists but rather to develop both physical and aural skills at the piano. Although the presentation is visual, students are, of course, listening to all these patterns and getting the sounds in their heads. They clearly know what all these patterns sound like. The explanation of how to build these patterns, music theory, is done in class after the students have practiced the concept at the keyboard. Practicing piano with the videos is part of the routine at the beginning of class.

Once students begin to "get" the sound, I incorporate lessons that teach music theory and demonstrate notation, graphic or standard. For instance, Figure 35.5 shows what I use to teach students an accompaniment pattern. Students will then practice playing and recording accompaniment patterns as an exercise.

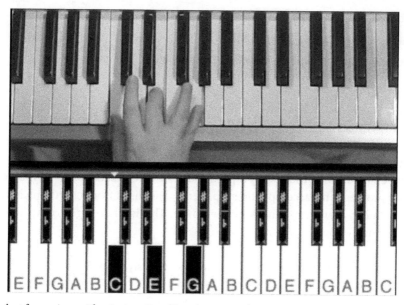

Screenshot from piano video instruction (Freedman, 2013)

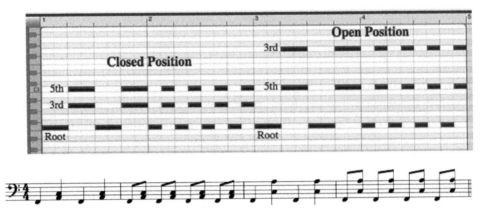

FIGURE 35.5 Image of music theory instructional materials demonstrating accompaniment patterns (Freedman, 2013)

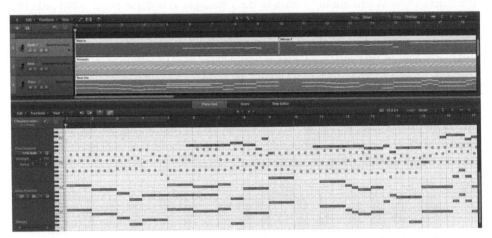

FIGURE 35.6 Image of Beethoven Moonlight Sonata in MIDI format (Logic Pro X 10.1.1)

Figure 35.6 shows a MIDI file of the first few measures of Beethoven's Moonlight Sonata in software displayed graphically. When the piece is played, students can see and hear its different parts and analyze its components. This is a great exercise for discussing elements of music: bass, harmony, melody, and arpeggiated accompaniment patterns.

Using this process, I teach many aspects of music theory at the piano keyboard, such as chord progressions, accompaniment patterns, bass lines from chords, melody writing, and others. Moreover, I sneak in Beethoven! This is where I get to pay homage to Bennet Reimer (1989a),

> They understand its history, its techniques, its many styles, and its major practitioners; they know where to go to hear good examples of it, how to make discerning judgments about it, and how to respond to it appropriately and sensitively in its many manifestations. (p.28)

Literacy includes a thorough understanding of music in its historical context. I don't just play Beethoven because I think students should be listening to Beethoven. Beethoven becomes an important tool for teaching music concepts, music theory, and composition skills. I use the first movement of Beethoven Symphony Number 5 to discuss motives and motivic treatment. Most students have at least heard portions of this piece. Now we get to use it as a tool for composition. I take a little freedom in the classroom to talk about Beethoven. When he lived, what was happening in Europe and world history at the time. We try to place Beethoven in context of history. I do this with any composer I "sneak" into my lesson. Copland for parallel fourth and fifths, massive contemporary chorales juxtaposed to Bach or Gabrielli, and intimate moments of solo instrumental pieces in comparison to the instrument used in a large orchestra. It depends on the topic or what the student is focused on. I had a student who loved writing percussion music and he used a lot of overlapping, repeated melodic fragments. We then listened to Steve Reich and Dan Levitan. I posed the question, "instead of just repeating your melodic fragments over and over (looping) and layering them, what can you do to effect change toward a particular goal in the piece?" Minimalism is a wonderful way to introduce the young listener melodic ostinati and then analyze how ostinati are treated in Hip Hop. My goal is to raise the bar. I want to increase the level of sophistication in how students use compositional techniques even if they are composing in contemporary popular music genres.

What I think is paramount is to be open to what students bring to the creative process. They only have their experiences to draw upon and for many their experiences and listening is limited to contemporary popular music, even if they call it "Classic Rock". Validating and giving honor to their music preferences keeps them open to what you have to offer as a mentor. If they don't trust you then they won't be inclined to try something new like listening to something very old!

I do not spend a lot of time going into historical details when referencing or playing examples of music. It's music appreciation on a different level. There is a great deal of value in the intense study of music history and it is very important for young people to know and understand European Western music history and the history of contemporary popular music. Much like the ensemble classroom where music performance is the goal, the music composition classroom has limited time and the focus is on composition skills to help develop more sophisticated music. Beethoven, Mozart, Stravinsky, Bernstein, Copeland, Arlen, Berlin, and so many others are included in the analysis of composition, orchestration, and the creative process.

Having been exposed to more advanced compositional and music theory skills or just the sounds and textures of great composers past and present, students who never played piano or studied another instrument suddenly create music, of their own volition, incorporating so many sophisticated compositional skills and music concepts as they learn them. It's like learning a new word or phrase; they start using it as part of their vocabulary for communicating. The more they learn, practice, and experiment in their own compositions, the more fluent in music they become.

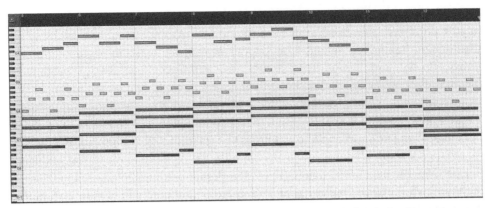

FIGURE 35.7 Image of a portion of the MIDI display of a student piece (Logic Pro X 10.1.1)

Figure 35.7 shows an excerpt of a piece written by student Tommy R., who didn't have much, if any, formal music training other than his elementary and middle school music classes. He did not study an instrument or sing in a chorus. Tommy is a very gifted musician and a great student. Here he demonstrates what he has learned about melody, harmony, accompaniment patterns, and bass lines. There were no assignment parameters that required Tommy to compose using these compositional tools or music elements. He created a piece that he wanted to create, and he intuitively utilized what he had learned. There is no better assessment or acknowledgment of students' learning than when they freely incorporate new concepts and tools that you are teaching in the classroom.

Music is a language that takes time for students to master. If we are teaching performers, instrumentalists, and singers to re-create music composed by others, it is critical that standard music notation be at the forefront of what we teach. However, if we are teaching students to create music using available technology, standard music notation can take a back seat. The use of technology in the music classroom forces us to reevaluate what skills are necessary for fostering creativity in our students. We no longer need our primary focus to be on teaching students to be readers and writers of standard music notation. We can begin to focus our teaching on whatever music "language skills" will allow students to freely communicate their ideas in music, to have them become *fluent in the language of music*. My experience has shown that the more fluent students become in their knowledge and use of the language of music, the more sophisticated music they produce. That's music to my ears.

References

Bell, A. P. (2013). *Oblivious trailblazers: Case studies of the role of recording technology in the music-making processes of amateur home studio users* (Unpublished doctoral dissertation). New York University.

Blake, A. (2010). Ethical and cultural issues in the digital era. In A. Bayley (Ed.), *Recorded music: Performance, culture, and technology* (pp. 52–67). New York: Cambridge University Press.

Escudé, L. (2011, April). Masterclass: Ableton Live. *Electronic Musician, 27*(4), 30–34.

Freedman, B. (2013). *Teaching music through composition.* New York: Oxford University Press.

Fluency. (2014). In *Merriam-Webster.com.* Retrieved January, 22, 2014, from http://www.merriam-webster.com/dictionary/fluency.

Goodall, H. (Writer), & Jeffcock, D. (Director). (2006). Melody [Television series episode]. In F. Hanly (Producer), *How Music Works.* London, UK: Tiger Aspect Productions.

Gordon, E. (2016). [Audiation]. Retrieved from The Gordon Institute for Music Learning website: http://giml.org/mlt/audiation.

Jeffcock, D. (Dlrector). (2006). Melody [Television series episode]. In D. Jeffcock (Producer), *How music works with Howard Goodall.* United Kingdom: Channel 4 Television Corporation.

Literacy. (2014). In *Merriam-Webster.com.* Retrieved January, 22, 2014, from http://www.merriam-webster.com/dictionary/literacy.

Prince. (1990). Joy in Repetition. On *Graffiti Bridge* [CD]. Minneapolis: Paisley Park/Warner Bros. (August 21, 1990).

Reimer, B. (1989a). Music education as aesthetic education: Past and present. *Music Educators Journal, 75*(6), 22–28.

Reimer, B. (1989b). Music education as aesthetic education: Toward the future. *Music Education Journal, 75*(7), 26–32.

Resnick, M., & Rosenbaum, E. (2013). Designing for tinkerability. In Honey, M., & Kanter, D. (Eds.), *Design, make, play: Growing the next generation of STEM innovators* (pp. 163–181). Routledge.

Seabrook, J. (2012). The song machine. *The New Yorker.* Retrieved from http://www.newyorker.com/magazine/2012/03/26/the-song-machine.

Wavelength Media. (1995–2012). *Quantization.* MediaCollege.com. Retrieved January 23, 2014, from http://www.mediacollege.com/glossary/q/quantization.htm.

White, P. (2009, August). *A dozen ways to use technology to improve music instruction.* Presentation at New York State School Music Association Conference, Albany, NY.

CHAPTER 36

PLAYING (IN) THE DIGITAL STUDIO

ETHAN HEIN

Technology has made it much easier to make music—not just to listen to music, or to perform other people's music, but to create music: to conceive new ideas and realize them in polished performances, recordings, and scores. The digital studio presents an opportunity to open up meaningful music making to a great many more young people.

Too frequently, computers in music classrooms are used as a more expensive platform for the same curricular materials that were formerly delivered on paper and the blackboard. Ruthmann (2012) criticizes bad technologically mediated classroom tasks as "information technology tasks applied in the context of learning about music, rather than engaging students directly in making, creating and responding to sound and music" (p. 80). Teachers should instead find ways to leverage technology in support of active, social music making—doing music, rather than simply learning about music.

To make the best use of technology in music teaching, we should embrace the new musical forms and practices that technology enables: sequencing MIDI, triggering samples, and manipulating preexisting recorded material. We should also embrace the musics that are digitally native. A MIDI file of a string quartet played back on sampled instruments will sound stilted and lifeless, but the same MIDI file played back on synthesizers over beats might sound delightful. While popular culture has enthusiastically embraced the language of digitally produced dance grooves, the music academy's reaction to electronic pop has ranged from indifference to open hostility. To fully realize the potential of music technology in the classroom, we need to open ourselves to the expressive language that is most compatible with that technology, and that is most authentic to the young people using it. My own formal music education experiences suffered from the extreme cultural disconnect between the classroom and the broader world.

Most of my music education has happened outside the classroom. It has come about intentionally, through lessons and disciplined practice, and unintentionally, through osmosis or accidental discovery. There has been no separation between my creative practice, my learning, and my teaching.

My formal music education has been a mixed bag. In elementary school, I did garden-variety general music, with recorders and diatonic xylophones. I don't remember enjoying or not enjoying it in particular. I engaged more deeply with the music my family listened to at home: classical and jazz on public radio, the Beatles, Paul Simon, and Motown. Like every member of my age cohort, I listened to a lot of Michael Jackson, and because I grew up in New York City, I absorbed some hip-hop as well.

In middle school we started on traditional classical music. I chose the cello, for no good reason except that I had braces and so was steered away from wind instruments. I liked the instrument, and still do, but the cello parts in basic-level baroque music are mostly sawing away at quarter notes, and I lost interest quickly. Singing show tunes in chorus didn't hold much appeal for me either, and I abandoned formal music as soon as I was able.

I found my way back into music through the back door: self-taught blues harmonica, rock and folk guitar, bluegrass mandolin, and unfocused poking around on the keyboard. My adolescent obsession with the Grateful Dead was the gateway into a broader musical universe: country, ragtime, free jazz, fusion, funk, reggae, and abstract electronica. As a senior in college, I started a sprint through formal jazz theory, which gave me a useful toolkit for all of the other music I was interested in.

I spent my twenties gigging with rock, jazz, and country bands, writing and performing music for theater both traditional and wildly experimental, and fumbling with recording and production. I tried Cubase and Rebirth before settling into an extended trial-and-error study of Reason and Pro Tools. I suddenly found myself in possession of the ability to produce the hip-hop and synth-oriented pop I had liked as a kid, but had long since written off as not being "real" music. I have always envied drummers, and dance music software gave me the ability to become one, and to visualize musical time in new ways.

I entered New York University's music technology masters program in my late thirties with a well-established personal style that blended improvisational jazz grooves with the polished rectangular surfaces of electronic dance music, along with a strongly felt belief that the job of a musician (and music teacher) is to help make life more bearable. Graduate school was my first brush with prescriptivist formal theory since eighth grade, and I was repulsed. My encounter with the modernist avant-garde also did not give me much enthusiasm. I much preferred studying with the film music composers, whose eclecticism, pragmatism, and "whatever works" attitude aligned more closely with my own. I also gravitated toward music education, and was delighted to discover the constructivist movement, as personified by Alex Ruthmann and his merry coterie of makers and tinkerers.

So now, here I am, continuing to teach and study privately and informally, but also teaching and learning in two very formal settings, New York University and Montclair State University. I feel that I've arrived home; as a music technologist I have more freedom to teach using creativity and so-called popular music than I would as a theorist or composer (even though I am both of those things as well.) My main goal is to help my students use technology as a tool for liberating their expressive potential.

Before asking what types of music we should teach and how we should teach them, it is worth asking a deeper question: why teach music at all? People enjoy music, but there are plenty of other activities we enjoy that are not taught in school. What makes music so special that it is worth spending finite educational resources on? Dillon (2007) argues, following Mihályi Csíkszentmihályi, that the primary purpose of music is to create deeply gratifying flow states, for both performers and listeners. He proposes that flow is a matter of public health, calling it "a powerful weapon against depression" (p. 48).

If we take a flow-centric view of music education, we are freed from the pressure of having to decide which kinds of music should be taught. The specific means by which the music creates flow is less important than the fact that it does it at all. Regardless of the kind of music students are making and the tools they are using to make it, if the music induces flow, then it is a worthwhile pursuit. The question then becomes, how do we create a flow-conducive environment in the classroom? How can we use technology to support meaningful musical thinking, knowing, and understanding? How can digital music in particular support an embodied and enactive combination of musicianship and listenership (Elliott & Silverman, 2014)? I believe that the answer lies in helping music students see themselves as producers: music creators with agency and ownership over their work.

In the digital studio, the distinction between composing, improvising, performing, recording, mixing, and editing has collapsed. Bell (2013) describes how musical actions like listening, performing, improvising, and composing have coalesced with technical actions like tracking, editing, mixing, and mastering into a single action that he refers to as "DAWing." I prefer the term "producing," which has become the de facto umbrella term for myriad forms of digital music making. While it was once used to describe a specific managerial job function within the recording industry, the word "producer" has expanded to refer to any musician who primarily creates music electronically (Pras, Cance, & Guastavino, 2013). Anyone with access to a computer and DAW (digital audio workstation) software is a potential producer.

The Digital Studio Has Expanded the Definition of Musicianship

In the first decades since the advent of digital audio editing, the approach of most recording artists changed little. Editing and mixing in the DAW was faster and easier than doing so with tape, but the workflow was essentially the same. It is only recently that new musical practices have emerged to take advantage of the DAW's unique capabilities. Chief among these practices is the idea of using the studio itself as the compositional medium, rather than as a tool for documenting or realizing preexisting compositions. While a few adventurous analog producers set a precedent for "playing the studio" (Eno, 1979), DAWs have taken the idea into the mainstream. Digital producers have unprecedented freedom to stitch together improvised and otherwise loose performances, decontextualized audio fragments, and a dizzying variety of timbres both natural and artificial into a seamless and coherent whole.

Thibeault (2011) urges us to think of the studio not just as a means of documenting "real" performances but as a musical instrument in its own right, carrying with it an entire philosophy of music making. Central to such a philosophy is the idea of the production process as a continuously iterative feedback loop. This has always been true for recording artists to an extent, but the digital studio dramatically reduces the cost of iterating in time, money, and frustration. The reflexive process of the digital musician invites a spirit of experimentation, improvisation, and self-guided learning. Music creation in the analog studio is an expensive, complex, and collaborative process akin to making a film. Producing in the digital studio is more like writing a novel or a play: inexpensive and conducive to working in small groups or alone. The digital producer requires a different skill set from traditional instrumentalists and composers. Below, I explore the nature of that skill set.

Marrington (2011) draws a contrast between the computer as a musical tool and the computer as a musical medium. We use computers as musical tools when they make predigital practices easier or faster. I use the computer as a tool when I make jazz lead sheets in Sibelius rather than with pencil and paper. By contrast, when we use the computer as a medium, it enables musical practices that would be impossible or inconceivable otherwise. I use the computer as a medium when I manipulate audio samples with Ableton Live.

Wise, Greenwood, and Davis (2011) describe the ways that students use Sibelius as a medium rather than a tool. Like all notation software, Sibelius is effectively a specialized MIDI sequencer. Younger students are likely to regard their scores as the end product rather than as an intermediate stage culminating in live performance. While the performance of students' scores by instrumentalists may be purely hypothetical, the playback from Sibelius itself is immediate and real. Furthermore, the software is not limited to the possibilities of human performers; users can and do create wildly complex patterns, played back on improbable instrument combinations.

It is not only the end result that is different when working with the computer as medium. A software visualization system can change one's entire conceptual imagining of music. Rather than solely experiencing it as events unfolding in time, one can also conceive music as a group of objects one manipulates in visual space. Notation serves this function to an extent, but it is limited in its granularity. Digital audio workstations have no such limitation; the user can zoom in to manipulate fragments less than a millisecond long, or zoom out to view an entire piece compressed to fit into a single screen. The full creative implications of this conceptual shift have yet to be borne out, but we can expect them to be profound.

Electronic musicians tend to begin their work by playfully experimenting with a piece of equipment or software, a period of open-ended "knob-twiddling." The discoveries made during this period are crucial raw materials for the more formal sequencing and editing that follow. A producer interviewed by Gelineck and Serafin (2009) described his tools as having a life of their own. We may consider the digital studio to be not just a passive set of tools but more, an active participant in the creative process.

A central difference between traditional musicianship and digital production is the role of recorded sound. Before the DAW, most musicians regarded recordings as a final product, a few avant-gardists notwithstanding. Digital producers can and do use recordings as a starting point, the way composers use melodies, harmonies, and rhythms. The digital studio makes it possible to actively engage the artifacts of our culture, to remix and recombine them, to personalize and mold them, and to use them as raw materials for entirely new work.

Formally trained musicians tend to take a dim view of dance music producers. There is a widespread sense that playing loops and samples created by other people is not legitimate musicianship, that it is "just pushing buttons." This attitude is unfortunate, because dance musicians have a great deal to teach other musicians and composers. Rather than listening to recordings as complete and inviolable, dance music producers listen for the ways that the recordings could be altered, customized, or combined with other recordings. I believe that reworking recordings is the most sensible artistic response to a world saturated with them. "Until 1877, when the first sound recording was made, sound was a thing predicated on its own immediate disappearance; today it is increasingly an object that will outlast its makers and consumers. It declines to disappear, causing a great weight of dead music to press upon the living" (Cutler, 2004, p. 138). Recorded music need not stay dead. Sampling and remixing can give old recordings unexpected new life.

While the DAW was conceived as a replacement for the tape recorder, it has in recent years become a tool for live performance as well. My DAW of choice, Ableton Live, was designed specifically with performance in mind, as its name indicates. When you first open it, you do not see a familiar DAW layout of horizontal tracks. Instead, you see its Session view, an austere grid. In most DAWs, music is represented on a linear timeline, with a horizontal row for each track. Session view uses a quite different representation: a spreadsheet, with each cell containing a *clip* of audio or MIDI. You perform by triggering the loops and samples, separately or in groups. Cells aligned in the same row can be triggered to play back simultaneously, in what Ableton calls a *scene*. Ableton Live can also play back sequences of loops automatically, in sequence or at random. Precise timing is not required; when triggered, clips automatically begin playing on the downbeat of the next measure, or at whatever time interval you designate.

Different clips can have different quantization settings and *Launch modes* (playback behaviors). Escudé (2011) calls these modes "game-like challenges" (p. 31). Ableton Live also includes automated clip-triggering modes called *Follow Actions*. These can be as simple as playing the next clip in the column, or as complex as choosing a clip at random that is not the clip that has been immediately played. The user can also override the Follow Actions at any time. The interplay between conscious clip choices by the user and the mindlessness or randomness of Follow Actions is fertile soil for new musical ideas.

A family of specialized controllers has emerged designed specifically for performative use of Ableton Live's Session view. These include hardware like Ableton's Push,

Novation's Launchpad, and Akai's APC series. There are also a variety of mobile applications offering similar functionality from the touchscreen. *Controllerism*, the performance style these tools were designed for, erases the distinction between the computer as recording playback device and musical instrument.

Before the digital studio can reach its fullest potential as a tool for student creativity, we must first overcome the disconnect between popular music and school music. While this disconnect is typically framed in terms of low versus high culture, the true conflict is between two different conceptions of what the most important and salient components of music are: melody and harmony, or rhythm and groove.

Western tradition treats melody as the fundamental basis of music. In the dance music derived from the African diaspora (including nearly all of American popular music), rhythm is the fundamental basis. McClary (2004) describes groove-oriented music as "converting our collective sense of time from tortured heroic narratives to cycles of kinetic pleasure" (p. 295). Prince (1990) drives home the point directly by singing:

> There's joy in repetition
> There's joy in repetition
> There's joy in repetition
> There's joy in repetition
> There's joy in repetition
> There's joy in repetition

Copy-and-paste is the defining gesture of digital editing tools, and infinitely looping playback is their signature sound. Loop-centrism is ubiquitous not just in pop but in contemporary art music as well, with African-American dance music as its major vector of cultural transmission. McClary argues that the music of Missy Elliott, Steve Reich, and John Adams are fundamentally more similar than different, united by their shared cyclic structures.

Monson (1999) proposes the *riff* (her term for a phrase-length loop) as the fundamental unit of the music of the African diaspora, the morpheme-level building block of much popular music. Since American popular and dance music dominates global musical culture, and African-Americans dominate American dance music, we need to consider the riff to be a foundational element of music, alongside scales, chords, and meters. This will require a major shift in thinking. The music academy mostly agrees with Theodor Adorno that the repetition in popular music is at best childish and worst a kind of fascism. The Western European classical music term for a continually repeated phrase is *ostinato*, an Italian word related to the English word "obstinate." This is not an attractive quality in a person, and the European classical world does not think too highly of it as a quality of music either.

What specifically is the value of the loop? Chernoff (1979) cites the "dynamic and open structure" (pp. 111–112) of riff-based Afrocentric music as a liberating force for self-expression and community building. And why should cyclical music have this effect? Perhaps it is due to its effect on our perception of time. When we listen to repetitive music, time is progressing in its usual linear way, but the cyclic sounds remove us from the stream of events, creating a feeling of timelessness and eternity. Malawey (2010) describes loops as having a feeling of alternating repose and tension. The first and third repetitions have a *call* feeling that is *answered* by repetitions two and four. The continuing reversal of call and response, of front and back halves of a phrase, can evoke other image schemas as well. Malawey lists swinging, fluttering, quivering, jiggling, hovering, and flickering as apt image schemas for repetitive music.

Nearly all world music uses repeating phrases grouped into longer phrases and groups those metaphrases into meta-metaphrases. Entire sections get repeated to form still higher-level structures. To my ears, the most satisfying music is the most modular and recursive. The joy of moving between different levels of recursive loops is one of the purest pleasures I know.

Most widely used DAWs come packaged with libraries of loops. In her discussion of GarageBand, Blake (2010) addresses the issue of whether assembling preexisting loops together constitutes fully creative music making: "GarageBand does not facilitate composition at the micro level: instead the user is invited to manipulate pre-existing loops of material, chaining them together to make new music from relatively large and undigested segments, to reform rather than to transform" (p. 55).

This might be true of the simplest and least expert use cases, but GarageBand enables extensive loop transformation through transposition, timestretching, and dozens of effects. Loop manipulation can entail significant microlevel musical thinking. This raises the question: where does loop manipulation end and composition begin? I do not believe there is a meaningful difference between the two acts. Loop manipulation is within reach of most students; therefore, composition is too.

The digital studio's loop-centrism has changed the mainstream pop songwriting process. Songs are increasingly assembled in the studio from improvised fragments. It is common practice in the digital studio to set a particular section to run on a loop, automatically recording each pass as a separate take. Performers can take as many tries as they want, with no need to stop and rewind tape in between each one. Because hard disk space is inexpensive, there is no barrier to running the loop endlessly, trying out different ideas until the right one pops out. Furthermore, edits are nondestructive, so there is no harm in a trial-and-error process.

Loop recording was originally conceived as a way to help performers nail especially tricky passages. As with so much music technology, though, it has taken on a broader creative function. The typical contemporary pop song is a collaboration between a producer (or producers) and a *top line* writer (or writers.) The top line writer creates the vocal melody and lyrics, and the producer creates the instrumental parts. The "writing" consists mainly of recorded improvisation that is then refined, edited, and looped into song form. Ester Dean, the top line writer cited by Seabrook (2012), prefers to hear instrumental tracks for the first time in the vocal booth with the proverbial tape running, so she can record her first reaction to it. The beauty of the digital studio is that one no longer has to choose between a spontaneous process and a smooth finish. In this process, editing skill is as important as the ability to perform in real time, if not more so.

In the digital studio it is not necessary to have instrumental proficiency in order to produce harmonically and rhythmically "correct" music; my own hunt-and-peck keyboard playing is proof of that. Just as word processors facilitate correct grammar and spelling, functionality like rhythmic quantization, pitch correction, and software keyboards lower the technical barriers to creative music making.

Rhythmic quantization is ubiquitous among DAWs. Notes and drumbeats can automatically be moved to the closest user-specified beat subdivision. It is also possible in many DAWs to do something more sophisticated: line up events to an imperfect human-performed rhythmic groove. For example, Ableton Live can analyze the placement of drum hits in a breakbeat and extract a stretchy, rubbery *grid*, to which you can then quantize audio and MIDI. There is a close analogy between groove and blue notes. Just as singers and instrumentalists can add melodic expression by stretching and bending the piano-key pitches into blue notes, digital producers can do so by stretching and bending strict metronomic time into grooves.

Pitch correction software and the customizable keyboards in mobile apps are the harmonic equivalent of quantization to a groove. Many such tools are set up in the manner of an Orff xylophone. Just as the xylophone's bars can be selectively removed, pitch correction software like Antares's Auto-tune and mobile DAWs like Apple's GarageBand enable me to perform melodies using only certain pitch sets. Garageband's graphical keyboard can be reconfigured to allow only the pitches from various of scales and modes. (The same applies to the graphical fretboards used to play guitar and bass and the fingerboards used for strings.) Moog's Animoog synthesizer app has similar keyboard functionality, with the added bonus of being able to customize scales one pitch at a time.

Antares's Auto-tune uses a similar scale selection and customization interface to Animoog. But instead of configuring a keyboard, users of Auto-Tune create an Orff xylophone for the voice. Setting Auto-Tune's Retune Speed setting to zero creates the ubiquitous "Cher effect," restricting the processed voice only to those pitches chosen by the user. The software gives singers real-time auditory feedback as well; the further the sung pitch is from an allowed pitch, the more it is processed and *artifacted*. Just as Animoog and GarageBand can embolden nonkeyboardists like me to explore melodic ideas, Auto-Tune can embolden nonsingers. This is another example of the ways technology invites a wider range of people to participate in musical creation.

Toward a Pedagogy of the Digital Studio

I have found that the best way to teach creative music making with the DAW is the method that I used to learn it: open-ended experimentation. Formal theory need not be a prerequisite to such exploration. Students receive immediate auditory feedback of their every move, and therefore quickly discover through experience what works and what does not.

In the absence of formal digital production training, the presets and default sounds of my audio software were a crucial educational resource. We bedroom producers may learn everything we know about or reverb simply by scrolling through the built-in settings in the software's plugins. As Bell (2013) puts it, "purchasers of computers are purchasers of an education" (p. 316). Software designers are curriculum designers as well, whether or not they realize it.

Resnick and Rosenbaum (2013) use the term *tinkering* to describe working without a clear goal or purpose, or without making noticeable progress. While classroom activities are usually highly planned and predictable, tinkering is a playful, exploratory, iterative style of engaging with a problem or project. How should we design pedagogical materials for tinkerability? Resnick and Rosenbaum list three qualities that such materials should offer: immediate feedback, fluid experimentation, and open exploration. The DAW is ideal for tinkering. You can hear the consequences of your actions instantaneously, and undo them at will. As teachers in the digital studio, our main job is to facilitate tinkering. This does not just entail opening up options to our students; we also need to help them select from the digital studio's vast musical possibilities.

Computers and synthesizers have given us unprecedented control over the minutest parameters of audio. However, our virtually unlimited sonic freedom does not always result in richer creative output. The most sophisticated audio production tools can just as easily stifle creativity under the weight of option paralysis. For this reason, music made with the most advanced tools too seldom makes it past the experimentation stage into fully realized works, and performances too often take the form of technical demos.

Simple, limited interfaces have two major virtues. First, a small feature set can be learned quickly. Second, the most obvious uses will quickly become tiresome, forcing the user to push the tool's limits. Magnusson (2010) speaks approvingly of interfaces that "proscribe complexity in favor of a clear, explicit space of gestural trajectories and musical scope" (p. 62). If presented with a finitely bounded feature set, users are more likely to move quickly past the knob-twiddling stage into a search for musical expressiveness. I have recently found more creative use for Propellerhead's ultraminimalist mobile app Figure than for Reason, their staggeringly complex desktop DAW. One elegant solution for option paralysis is to restrict not the software's functionality but the musical raw materials. Just as the blank page is daunting for writers, the empty session can be a difficult starting point for producers.

Many traditional music and nonmusic software programs (e.g., notation, sequencing, looping, audio editing, word processing) are based on the metaphor of a blank canvas or void. When the program is launched, the user is presented with a blank slate upon which to place notes, audio waveforms, images, or words. For many students, it can be intimidating starting from scratch. In my own teaching, I have seen students' reluctance to add their first notes to the page or sequencer, and heard them wondering whether they have anything of value to say. Of course they do have something valuable to say, but scratch is not always the best place for them to start (Ruthmann, 2012).

Ruthmann suggests making a game of a subtractive process. Students can be given a dense collage of loops, a "sound block," which they must transform into a new work by slicing and subtracting pieces only. More sophisticated students can be given a small collection of audio samples and told to create a piece of music using no other sounds. I am a member of the Disquiet Junto, an electronic music collective organized by Marc Weidenbaum. Shared-sample projects are a staple of the Junto, and the members consistently demonstrate the dazzling variety of musical works that can result from the most constrained source material (Nelson, 2012).

Beyond the Digital Studio

Electronic music production practices and techniques have a lot to offer students of voice or nonelectronic instruments. Music software is fairly easy to learn, so the major challenge for digital producers is learning how to listen. Active listening is a skill that generalizes to any kind of music making. Formal music training focuses on rhythm and harmony, but to consider recorded music on these terms alone is to miss an entire dimension of its creative content. We also need to consider timbre and space, those aspects of recorded music beyond the notes being played and words being sung. For many pop and dance styles, timbre and space are the most salient expressive component of the music. Shaping timbre and space with software is technically not difficult, but to do it well requires intense close listening. When I'm producing, I need to continually ask myself whether the sounds I hear are pleasing or not. If they aren't, why not? Is something missing? Or does something need to be removed? Is everything placed in the right position in perceptual space? Moylan (2007) poses a series of analytical listening challenges designed to help producers answer these questions, and it is not just producers who will benefit from them.

Like listening analytically to timbre and space, the practices of remixing and mashups have broad pedagogical potential for musicians far beyond the digital studio. Composers and improvisers have always reworked existing material; the language of the remix simply gives the practice a name. Remixing need not involve computers at all. Arranging and orchestration are forms of remixing, as is writing variations on a preexisting theme. There are three major pedagogical benefits of the remix, digital or otherwise: (1) remixing transforms consumers of music into producers, (2) remixing requires the same kind of deep analytical listening discussed above, and (3) remixing is a form of nonverbal critique, a way to provide musical commentary in the form of music rather than words. I will discuss each of these in turn.

Students are awash in popular culture as never before, but they need not just passively absorb it. The digital studio makes it possible to actively engage the sonic artifacts of our environment, to personalize and mold them. Remixes and mashups combine musical analysis with musical experience. The ability to claim creative ownership over pop culture is a tremendously empowering sensation, especially for young people who may not feel much empowerment otherwise.

Marshall (2010) argues that by deconstructing and recombining familiar pop texts, remixes and mashups open the door to broader critical thinking. Through mashups, "we discover correspondences, connotations, and critical readings of performances that we may not have given a second thought—or even a first listen" (p. 307). Even if they are unable to produce their own mashups, students can find considerable food for thought by listening to and discussing mashups made by others.

Marc Weidenbaum explains how remixes can focus attention on certain parts of a recording, making the internal experience of remembering the song into an external listening experience that can be shared with others:

> Remixes form a huge part of the way that I understand music. I remember when I was attending college, in the mid-1980s, buying an extended version of a song that I liked, by an Australian band called INXS, and I remember being astonished by how listening to the remix could kind of make you completely rethink the way that you relate to the original, and that moment was really important, realizing that altering something does not detract from the original, but can enrich your understanding of it. Part of the reason that particular remix registered with me was because it sounded the way the music sounded in my memory—the parts I liked, the parts my memory would often play on repeat when I wasn't actually listening to the original version of the song. (cited in Korth, 2014)

This kind of active, imaginative engagement with recorded music is fertile soil for new interpretations and new ideas.

If the source material is sufficiently transformed, what is the difference between a radical remix and a new work? The distinction is blurry at best. Four decades of hip-hop and electronic dance music have demonstrated that sampling existing recordings can result in startling musical innovation. Does this kind of production count as original work? The question is a contentious one. I believe that originality is a matter of degree, not kind.

The remix is a way for students and teachers to engage not just with the broader culture but with one another as well. It poses a new answer to the question of how we should evaluate creative musical work. Verbal discussion of music can be illuminating, but language is not adequate for all musicians in all situations (Elliott & Silverman 2014, p. 218). I find the best way to engage someone else's music is to remix it, so I can "discuss" the music in musical terms. Repeating and highlighting a passage implies praise. Excising a passage implies criticism. Altering or recontextualizing an idea is a way of suggesting different musical possibilities.

Remixing is time consuming, and one might reasonably object that most music teachers do not have time to remix dozens of student works. We can work around the time limitation by creating pedagogical remixes during class. Students can observe the process, ask questions, and offer their own suggestions. We can also assign students to remix the work of their peers.

Conclusion: Teaching Music, Not Technology

When schools address music technology, they tend to focus on the nuts and bolts of the technology itself, rather than its creative applications. This is ironic, since common-practice notation and instrument design are themselves forms of technology. Dillon (2007) observes:

> "The violin bow and the saxophone mouthpiece are perhaps the most expressive pieces of music technology in Western history yet composers and virtuoso performers did not undertake courses in these technologies. To understand them they actively explore what the expressive capabilities of these technologies enable, what they revealed and concealed to us as musicians" (p. 80).

So it should be with electronic music production tools. But to truly engage such tools for creative music making, we must address their most culturally significant context: electronic dance music and hip-hop. This music falls well outside the canon of what is widely considered suitable music for the classroom. Nevertheless, I believe that such music should be included, and not (just) because young people enjoy it. Rather than "dumbing down" music education, the inclusion of popular dance music would significantly enrich the curriculum, particularly in areas traditionally neglected: groove, timbre, and space.

The digital studio requires new skills and new vocabulary. But it also lowers barrier to entry for musical participation and creativity. Digital musicians at all skill levels have the advantage of being able to immediately hear flawless renderings of their ideas, with instant feedback for changes. As with good video games, the dynamic and interactive music software experience is its own motivator and its own psychic reward. I have experienced countless hours of flow while exploring the possibilities of the digital studio, and in so doing have enriched my musical knowledge and experience immeasurably. I would wish the same pleasures to be available to any music student.

References

Bell, A. P. (2013). *Oblivious Trailblazers: Case Studies of the Role of Recording Technology in the Music-Making Processes of Amateur Home Studio Users*. New York University.

Blake, A. (2010). Ethical and cultural issues in the digital era. In A. Bayley (Ed.), *Recorded music: Performance, culture, and technology* (pp. 52–67). New York, USA: Cambridge University Press.

Chernoff, J. M. (1979). *African rhythm and African sensibility: Aesthetics and social action in African musical idioms*. Chicago: University of Chicago Press.

Dillon, S. (2007). *Music, meaning and transformation: Meaningful music making for life*. Newcastle, UK: Cambridge Scholars. Retrieved September 20, 2015, from http://eprints.qut.edu.au/24153/.

Elliott, D. J., & Silverman, M. (2014). *Music matters: A philosophy of music education* (2nd ed.). Oxford: Oxford University Press.

Eno, B. (2004). The Studio As Compositional Tool. In C. Cox & D. Warner (Eds.), *Audio culture: Readings in modern music* (pp. 127–130). London: Continuum International Publishing Group.

Escudé, L. (2011, April). Masterclass: Ableton Live. *Electronic Musician, 27*(4), 30–34.

Gelineck, S., & Serafin, S. (2009). From idea to realization—understanding the compositional processes of electronic musicians. Paper presented at Audio Mostly, Glasgow, Scotland (pp. 1–5). New York: ACM International Conference Proceeding Series.

Green, L. (2002). *How popular musicians learn: A way ahead for music education.* Aldershot, UK: Ashgate.

Korth, N. (2014, March 19). *CC talks with Marc Weidenbaum* [Blog entry]. Retrieved April 11, 2014, from http://creativecommons.org/weblog/entry/42415.

Magnusson, T. (2010). Designing constraints: Composing and performing with digital musical systems. *Computer Music Journal, 34*(4). doi: 10.1162/COMJ_a_00026.

Malawey, V. (2010). Harmonic stasis and oscillation in Björk's Medúlla. *Music Theory Online, 16*(1). Retrieved September 20, 2015, from http://www.mtosmt.org/issues/mto.10.16.1/mto.10.16.1.malawey.html.

Marrington, M. (2011). Experiencing musical composition in the DAW: The software interface as mediator of the musical idea. *Journal on the Art of Record Production, 5.* Retrieved September 20, 2015, from http://arpjournal.com/845/experiencing-musical-composition-in-the-daw-the-software-interface-as-mediator-of-the-musical-idea-2/.

Marshall, W. (2010). Mashup poetics as pedagogical practice. In N. Biamonte (Ed.), *Pop-culture pedagogy in the music classroom* (pp. 307–315). Lanham, MD: Scarecrow Press.

Mcclary, S. (2004). Rap, minimalism, and structures of time in late twentieth-century culture. In D. Warner (Ed.), *Audio culture.* Continuum International. Retrieved September 20, 2015, from http://www.getcited.org/pub/100457949.

Monson, I. (1999). Riffs, repetition, and theories of globalization. *Ethnomusicology, 43*(1), 31–65. Retrieved September 20, 2015, from http://www.jstor.org/stable/10.2307/852693.

Moylan, W. (2007). *Understanding and crafting the mix: The art of recording.* Abingdon, UK: Taylor & Francis.

Nelson, P. (2012). *Kudos: Marc Weidenbaum.* Hilobrow. Retrieved August 7, 2014, from http://hilobrow.com/2012/03/02/marc-weidenbaum/.

Pras, A., Cance, C., & Guastavino, C. (2013). Record Producers' Best Practices For Artistic Direction—From Light Coaching To Deeper Collaboration With Musicians. *Journal of New Music Research, 42*(4), 381–395. http://doi.org/10.1080/09298215.2013.848903.

Prince (1990). Joy In Repetition. On *Graffiti Bridge* [CD]. New York: Warner Brothers. (August 21, 1990)

Resnick, M., & Rosenbaum, E. (2013). Designing for Tinkerability. In M. Honey & D. Kanter (Eds.), *Design, Make, Play: Growing the Next Generation of STEM Innovators* (pp. 163–181). Abingdon, UK: Routledge.

Ruthmann, A. (2012). Engaging adolescents with music and technology. In S. Burton (Ed.), *Engaging musical practices: A sourcebook for middle school general music* (p. 233). Lanham, MD: Rowman & Littlefield Education.

Seabrook, J. (2012). The Song Machine. *The New Yorker.* Retrieved September 20, 2015, from http://www.newyorker.com/magazine/2012/03/26/the-song-machine

Thibeault, M. D. (2011, October). Wisdom for music education from the recording studio. *General Music Today.* doi: 10.1177/1048371311425408.

Wise, S., Greenwood, J., & Davis, N. (2011). Teachers' use of digital technology in secondary music education: Illustrations of changing classrooms. *British Journal of Music Education, 28*(2), 117–134. doi: 10.1017/S0265051711000039.

3 A

Further Perspectives

CHAPTER 37

CONSIDERING MUSIC TECHNOLOGY AND LITERACY

JAY DORFMAN

THE Core Perspectives in this part of this book raise several interesting issues that I am glad to be able to explore through my own lens. In this short chapter I discuss my views on how music literacy might be viewed differently in the context of technology-based music instruction. I also discuss the ways in which teachers' roles can support that new view of literacy, and how technology can be seen as a support for creativity in music students.

My formal musical training is as a classical guitarist. Reared on the etudes and solo pieces of Heitor Villa-Lobos, Matteo Carcassi, Fernando Sor, and transcriptions of J. S. Bach (in D, of course), I most closely identify my musicianship with the guitar. While my dedication to practicing and performing may ebb and flow, my devotion to the infinite tonal, melodic, and harmonic possibilities of the guitar has never decreased, and I imagine it never will.

But the last 15 years or so have expanded my musical identity, and have broadened the ways I engage in making music. Now, when I wish to engage in music making, I am equally as inclined to launch Logic Pro or Sibelius as I am to pick up a guitar. Whether using standard notation, MIDI, or recording others or myself engaging in a variety of types of music making, these tools aid me in being musical. Through each of these "instruments," I am able to think in sound, practice and revise novel ideas and those influenced by others, and make audible that which might otherwise not be. Indeed, these activities represent different kinds of music making but all allow me to make music creatively.

As I consider the ways in which school-aged children engage in music making, it occurs to me that not all people go through the same evolution I experienced. Classical training is not the entry point into music making for everyone. A recent poll showed that technology has enhanced interest in learning to play an instrument in Americans (National Association of Music Merchants, 2012), providing evidence that entry points into music making have expanded beyond traditional avenues. Technological means of making music are perhaps a more universally accessible means for beginning to make music than the traditional music teaching and learning of the twentieth century, and educational practices should account for that possibility.

Similar to the realization Freedman expresses in chapter 35, my realization has been that it is quite possible that students who approach music from a technological standpoint might not understand—or even need, or have a desire to understand—musical fundamentals in the same way that traditionally trained students would. My perspective as a teacher and teacher educator is that whether an entry point into music making is traditional—as mine was—or is based on some alternate way of making music, students can gain valuable musical experience and learn to develop their musical literacy through either means of initiation, or through variously weighted combinations of both.

A New Music Literacy

The assumption that there are many entry points into music leads me to unpack the term "music literacy" a bit, and to wonder about the ways in which the concept is applied in current music education circles. "Literacy" itself is a term fraught with tremendous variability, so adding temporal and expressive layers to the mix can result in a half-baked comprehension of music literacy. "Music literacy" is a difficult term because many who examine it do so from the perspective that music learning can be used to develop literacy in other disciplines, such as reading or math. For example, Runfala, Etopia, Hamlen, and Rozendal (2012) showed that training teachers in pedagogical techniques can influence both musical and nonmusical outcomes. While music teachers need not ignore nonmusical outcomes of our teaching, our focus should be on musical development, with growth in other disciplines as a happy byproduct. Even among those definitions that focus appropriately on the idea that a separate, though perhaps analogous, musical literacy exists, many of them do not encapsulate the nuanced skills and understandings that musicians and music teachers associate with the term. Before we can even begin to agree about what music literacy means in the technology-enhanced music education world, we must agree that music literacy should not necessarily be entangled with development of skills in other areas.

The fundamental skill ascribed by most definitions of music literacy is the ability to read notated music (Gudmundsdottir, 2010). Among the most reasoned definitions of music literacy is Gordon's (2004), which suggests that it involves coding and decoding musical symbols, and presumes the ability to audiate notated music. This belief is reflected in the structure and content of Gordon's well-known and often-used Iowa Tests of Music Literacy (Boyle & Radocy, 1987; Schleuter, 1974). I suggest here, quite in agreement with Freedman's notion that reading and writing staff notation need not be the ultimate goals, that the idea of decoding musical symbols need not be restricted to staff notation; rather, we can develop literacy with other kinds of information that represent musical events, such as MIDI regions, piano scroll editors, and audio waveforms.

Regardless of the type of musical representation involved, however, as music teachers we somehow know what we want our students to know and be able to do in order to call them musically literate. Essentially, music literacy implies that our students are able to function in musical settings by performing, responding to, or creating music, and that they can understand the elements of those kinds of involvement in developmentally appropriate ways. These types of musical behaviors alone imply tremendously varied activities, so musical literacy can be demonstrated in many ways.

In the scenario of technology-based music instruction, where, as I have previously described, students are directly engaged with technology for learning, practicing, and demonstrating mastery of musical skills and concepts (Dorfman, 2013), educational goals are not necessarily the same as they are in traditional music performance. Perhaps, then, we should be using a term other than "music literacy" to capture the goals of these educative scenarios. The idea of being able to use technology as a sketch pad, get instant feedback, and revise one's ideas based on that feedback is similar across artistic areas (music, visual art, etc.). Given this similarity, perhaps the end-state for all of these domains should be referred to collectively as technology-enhanced creative literacy. Still, given the multiple paths toward engagement with music, I am forced to wonder about some fundamental and long-held components of musicianship and how they might be seen differently when music is being made (performed, composed, listened to) with digital technology rather than in other ways. Put succinctly, can we use the term "music technologist" in the same way that we use "vocalist" or "violinist"?

Simple web searches for "music technologist" turn up few uses of the term other than pages where people use it to describe themselves. What this should indicate is that, while "music technologist" may not be a widely used expression, individuals do indeed identify themselves in this way. To me, a music technologist is one who considers technological devices (such as computers, controllers, or handheld devices) to be the primary means by which they make music. Just as I consider myself a guitarist, I can understand how one could self-identify as a technologist. While the type of literacy that a technologist develops may be different from that which a violinist develops, it is musical literacy nonetheless. Technologists and violinists, for example, may be differently musically literate, but both may be sophisticated musicians with the ability to code and decode musical representations. We should feel comfortable applying this nomenclature to musicians, and we should do so without prejudice. On a personal level, I know I am being just as musical when I compose in a technology application as I am when I play guitar.

Changing Roles of Teachers

It is certainly far beyond the scope of a brief chapter such as this to consider all of the pedagogical implications associated with technology-based music instruction. Therefore, I will limit my pedagogical suggestions to the realm of the foundational "sound-before-sight" idea that Freedman mentions in chapter 35. While I believe fully in this principle, especially in the context of early instrumental and vocal music education, I often find it hard to reconcile in the technological context. Software is visual by its very nature, and a massive field of literature about multimedia learning has led to the conclusion that students who learn using multiple channels of perception learn better than those who use fewer channels simultaneously (see, e.g., Mayer, 2009, 2005). Because of the design of most media software, I would caution teachers against being too rigidly bound to the sound-before-sight principle when teaching in a technology-based context. Certainly students can develop their aural skills in traditional ways, even when they engage with music through technology, but denying the careful design of software and its ability to support learning through multiple channels of perception may not set up students for their greatest chances of success.

Lum (chapter 34) suggests that teachers have a responsibility to refuse replicating structures of institutionalized power. This is a notion with which I completely agree, and I believe that teachers who facilitate a shift from teacher-centered to learner-centered pedagogy are increasingly necessary. Both the motivation to engage with music and the obligation to do so must come from students, with the teacher serving as a guide.

As I have described elsewhere (Dorfman, 2013), I believe that the teacher's role in a technology-based music environment is to help students past obstacles they encounter on their paths to solving musical problems. Technological tools enable us to guide students toward new solutions to old problems. With the right tools—those that allow students to act in musically sophisticated ways—teachers can serve as model creative thinkers to help students overcome their creative conundrums. This represents a shift from the authoritarian model of the teacher as the sole possessor of knowledge that she instills in students. Teachers of technology-based music share in the pursuit of creative outcomes and help students engage in the hard work of novel creation and revision. Rather than the top-down approach of old, this places teachers and students on a more even plane and allows stakeholders to share in the power associated with making decisions and guiding educative experiences.

Further, one potential of technology of which we have only begun to scratch the surface lies in its ability to make the outcomes and benefits of music education more easily visible and instantly recognizable. In ensemble-based music education, the typical outcomes are formal events—concerts at which loved ones gather to hear the product of a long period of preparation. While concerts will (hopefully) never go away, music education with technology, particularly that which works toward students composing their own music, allows individual or small groups of students to demonstrate their learning in less ceremonial ways. These products can be shared online or via less time-and-space-dependent media. Services such as SoundCloud, Nimbit, and even the CD (though perhaps not for much longer!) are means to distribute work that students do, and to make the products of educational efforts apparent for various stakeholders. As Elliott and Silverman (2015) have recently suggested, teachers should be comfortable using these services and other applications as distribution platforms and as learning tools—their students and programs will likely benefit from teachers' comfort in doing so.

Conclusion

As a university professor, I frequently conduct workshops and special sessions for teachers; these sessions are usually focused on a particular type of technology or software. So that I can understand my students a bit better, at the beginning of these sessions I always ask the attendees to introduce themselves and tell me and their fellow attendees about the potential uses they have for the technologies we are studying. At a recent weeklong workshop on a particular DAW package, the teaching levels and concentrations ranged from elementary general and instrumental music to high school vocal music, and even included an information technology specialist who wanted to learn the software for his personal and freelance uses. While their needs were varied, it was enlightening and challenging for me to think about a single piece of software—albeit an extremely deep and varied one—in terms of its uses for diverse entry points. This was a valuable exercise and one that I need to engage in constantly.

It is important for teachers, and those invested in music technology, to take stock of their own entry points into music, as Lum has in chapter 34, and as I have done in the introduction to this chapter. Most of us will quickly realize that the ways we have approached music throughout our lives are multifaceted. While developing expertise in a particular kind of music is still desirable, so is the ability to function in varied types of musical contexts, of which technology is merely one. By attending to the ways we interact with music, and the foundational experiences that have shaped those ways, perhaps we can begin to understand our own creative tendencies and how technology might support them.

Certainly people can be creative without computers and digital technology, but technology may help us to understand the kinds of support mechanisms that help people to be creative. The potential of technology is to provide immediate feedback, to allow choice of creative style (linear/nonlinear, broad/detailed), to encourage collaboration and communication, to support multiple versions and revisions of work, and to provide practically infinite storage of ideas. Each of these alone would serve as powerful support for people engaged in creative activities; taken as a whole, they represent major shifts in our approaches to creative work. And while we can see evidence of each of the potentials of technology in relatively common educational practice, the extent to which they might be integrated in the future is virtually limitless.

References

Boyle, J. D., & Radocy, R. E. (1987). *Measurement and evaluation of musical experiences.* New York: Schirmer Books.

Dorfman, J. (2013). *Theory and practice of technology-based music instruction.* New York: Oxford University Press.

Elliott, D. J., & Silverman, M. (2015). *Music matters: A philosophy of music education* (2nd ed.). New York: Oxford University Press.

Gordon, E. E. (2004). *The aural/visual experience of music literacy: Reading and writing music notation.* Chicago: GIA.

Gudmundsdottir, H. R. (2010). Advances in music-reading research. *Music Education Research, 12*(4), 331–338. doi: 10.1080/14613808.2010.504809

Mayer, R. E. (2009). *Multimedia learning* (2nd ed.). New York: Cambridge University Press.

Mayer, R. E. (Ed.). (2005). *The Cambridge handbook of multimedia learning.* New York: Cambridge University Press.

National Association of Music Merchants. (2012). *A new duet: Emerging tech syncs with tradition to convert aspiring music makers into musicians.* Retrieved January 5, 2017, from http://www.harrisinteractive.com/vault/2012_NAMM MusicandTech.pdf.

Runfala, M., Etopia, E., Hamlen, K., & Rozendal, M. (2012). Effect of music instruction on preschoolers' music achievement and emergent literacy achievement. *Bulletin of the Council for Research in Music Education, 192,* 7–27.

Schleuter, S. L. (1974). Use of standardized tests of musical aptitude with university freshman music majors. *Journal of Research in Music Education, 22*(4), 258–269.

CHAPTER 38

TECHNOLOGY AND MUSIC COLLABORATION FOR PEOPLE WITH SIGNIFICANT DISABILITIES

DONALD DEVITO

Up until the twenty-first century, technology in music education tended to be in the form of equipment utilized in the classroom to play and watch examples of music performance. Examples are the CD player for listening and the VCR or DVD player for watching recorded performances of local ensembles, classical orchestras, and movies about famous musicians. While these were utilized to enhance the curriculum by viewing and listening to a piece of music from a given style or genre, simple technology exists today to greatly transform the ways interaction with music and musicians takes place in the classroom. The opportunity through Skype, Google Hangout, and simple online communication technology to create real and enduring international connections in the classroom curriculum are widely available. The iPad is increasingly coming to be considered a music resource and, especially with children with disabilities, an adapted music instrument. The same device can be plugged into a projector and, with Skype or other readily available online communication programs, literally open the world of music to a classroom by engaging with musicians from all areas and professions. The ramification is that technology is no longer just a CD player, and music literature in the classroom is no longer Americanized versions of traditional world music offered in state-adopted textbooks. The manners in which these engagements take place are wide and varied.

The Sidney Lanier Center School, where I teach, serves students with moderate to severe disabilities between the ages of 3 and 22. The description of our program that follows includes each course by grade, including supplemental music ensembles. The music program highlights a global music education approach that utilizes Skype to link our music classes with universities and community-based programs around the world. The efforts of this music program have been published in peer reviewed journals, book chapters, and international music education conferences since 2003. Institutions linked with the Sidney Lanier music program include, among others, the University of Roehampton and the University of London in the United Kingdom, the China Conservatory, the Universidad de Londrina in Brazil, Limerick University in Ireland, Queensland University in Australia, the Special Education Centre music program led by Arthur Gill in Gujarat, Pakistan, Group Laiengee in Guinea, West Africa, the Notre Maison orphanage led by Gertrude Bien Aime (founder of one of the few programs that serves children with disabilities in Haiti), Queensland University (Australia), University of Windsor (Canada), the University of Minneapolis-Duluth, Pacific University, James Madison University, Jackson State University, the University of South Florida, and Syracuse University. This multicultural real-time approach to music and special education has opened doors and adapted learning experiences for our students for over a decade.

The online collaboration began by combining the Sidney Lanier students with Dr. Emma Rodriguez Suárez's pre-service music educator course at Syracuse University. was referred to as the Virtual Classroom Disabilities Project and combined the Orff approach with accommodation for special learners (Suárez & DeVito, 2012). The online sessions incorporated the four processes of the Orff Schulwerk: (1) imitation, (2) exploration, (3), improvisation, and (4) music literacy. Online sessions occurred first during a drum circle at the school in Florida and then the university students selected the activities within the four processes they wished to teach to the individual student they were assigned. Sessions were done individually so the university student could get to know the student they were paired with in a greeting session and then adaptation of the lesson to be taught was done by the music major to suit the needs of their student. This sequence allowed for individualized accommodation which is something that will be vital when they are teaching in the field. Although the entire interactions were online, the college students reported detailed affective responses for themselves and their students in the educational online experiences and a motivation to work with special learners when they are full time teachers (Suárez & DeVito, 2012).

A current project in development was brought about by an invitation I received to present on this program at the "Listen" Conference, held by the organization Sabreen in the West Bank, for Palestinian educators looking to develop music education as a course in public schools. A variety of presenters shared the latest approaches in music education development. As a result of teaching in local schools and this session, Skype collaboration was discussed between two special schools of varying populations that cater to children with disabilities, one in Bethlehem, the other in Jerusalem, and the Sidney Lanier Center in Florida. The first school in Bethlehem was the Ephpheta Paul VI Pontifical Institute, for the audio-phonetic rehabilitation, and the educator who accompanied me in the workshop was Luab Nayef Hammoud, a creative and dedicated music teacher at the school. One grant project idea for the Jerusalem school was to share music education online and then have the activity culminate with an in-person visit by children at the school in Jerusalem, who would visit their colleagues in Florida. The inclusion of this developing project in this book is an early and very meaningful distinction for of all the special schools involved.

What Can Technology Do for Music Education?

Technology can break down barriers between cultures, create previously unthought-of educational and interaction opportunities in the music classroom, and enhance the quality of life for underserved populations (Suarez, E., DeVito, D., Kleber, M., Akombo D., Krikun, A., & Bingham, S. (2010). Consider this example: The Sidney Lanier Center School links with Dr. Adam Ockelford of the University of Roehampton and Dr. Graham Welch of the University of London through Skype to create facilitation of the newest approach developed in analyzing behavioral response frameworks to music called Sounds of Intent (Vogiatzoglou, A., Ockelford, A., Welch, G., & Himonides, E. (2011) training in which is shared with the Notre Maison orphanage in Port-au-Prince, Haiti. This project included the Society for Education, Music and Psychology Research, through University CollegeLondon professor Dr. Graham Welch, which provided a projector for group instruction, allowing the orphanage to include in the Skype experience their entire orphanage at one time, and GoTalk communication technology, for facilitating a nonverbal communication approach for children with severe speech language challenges at the orphanage.

One utilization of the Go Talk was programming the device with recorded Haitian music for the children to indicate their musical preferences by selecting the switch with the corresponding music examples. This approach has been explored in a number of its facets over the years, with the Sidney Lanier Center utilizing Skype in the music education classroom. A variety of in-person collaborations, trainings, and shared music education activities have developed as a result of this approach.

After several Skype activities with Notre Maison, I went in person to develop a music and special education program there and followed it up with enhanced Skype interactions with the students at Sidney Lanier upon my return. Documentation of this led to being involved with Projects for Haiti, which invited Gertrude Bien Aime—the director of Notre Maison, with whom we collaborate online—and me to travel together to Cap-Haïtien in the north of the country to train 400 Haitian teachers in the concept of music and special education inclusion developed at Notre Maison. The culminating activity resulted in motivating hundreds of Haitian teachers to volunteer in adapted arts activities in their communities with children with disabilities. A second result was the publication of their feedback to the Projects for Haiti training (which involved 12 other educators from the United States) in The First Educational Publication of the Haitian Teachers Association on their approaches to developing inclusion in Haiti, edited by myself and Phil Mullen for Project for Haiti creators Bertrhude Albert and Priscilla Zelaya (DeVito, D. & Mullen, P. (2015).

All 400 Haitian educators are listed as contributing authors to this book, which is translated into Creole and distributed throughout Haiti. The book also has an ISBN which allows it to be internationally accessed and sold as an electronic document benefitting the teachers association through online sales.

Considering the fact there was no electricity at the training site, simple technology in the form of: cell phone video to record discussions, songs and lessons by the Haitian teachers; photography of homework assignments related to each day of training; and a laptop with Microsoft Word were utilized to achieve this original document. One cumulative effect is the enhancement of their Haitian teachers academic status in their communities and the further utilization of arts education to enhance the academic knowledge and quality of life for underserved populations.

All this from the development of music education and Skype technology, triangulated between the efforts of major universities in the United Kingdom, public schools in the United States, and community-based programs in low-resource areas such as Haiti. Further communication on this development and shared experiences is being continued, with the Haitian teachers using simple technology, such as cell phones, that allow them to access Facebook to post and share their needs and experiences with this collaborative project.

Technology can bring disparate groups together and reverse typical roles of instruction. In this example, the integrated elementary special education majors of Dr. Jian-Jun Chen-Edmund's Music for the Elementary Classroom course at the University of Minnesota Duluth are Skyping to a combined grade 3 to 5 class of "hard to reach" students with significant behavioral needs at the Sidney Lanier Center School. These particular students were sent from their home zoned schools to the Sidney Lanier program due to the level of additional assistance needed to function in an organized school setting. With the university's mission of fostering diversity and inclusion, this collaboration provided an additional opportunity for future educators to interact with minority students and students with disabilities.

In one particular lesson, the role of teacher and student was reversed as the students demonstrated hip hop dancing combined with Motown era music of the Jackson 5. This fusion of styles was choreographed by the students prior to the lesson in their music class and perhaps more importantly was requested as a lesson component by the students themselves. The complexity of the dancing was evident and the long distance and live teaching interaction would not have been possible without online technology and the Smart Board utilized for the large classes in both locations.

Music education organizations such as the International Society for Music Education (ISME) can help to foster contacts between schools and institutions that result in creative curriculum development (Higgins, L., Akuno, E., & DeVito, D 2016, Vol 5). Catherine Pestano, a U.K. community music and ISME member saw a Skype presentation at the ISME conference and was motivated to initiate the first Music "TeachMeet" with international music practitioners such as myself providing feedback to local educators. Pestano contacted music educator Mike Rathbone and thanks to a technology grant from We Got Tickets in partnership with the organization Youth Music, initiated the "Border Crossing" project between a Pupil Referral Unit for students with emotional and behavioral difficulties (EBD) at a local school in Croydon, Surrey, U.K., and the Sidney Lanier students in the United States (Rathbone, M., & Pestano, C., 2013).

Performances of Jambo Bwana given by the secondary students with intellectual and physical disabilities in the U.S. for Rathbone's primary aged classes with behavioral needs were recorded live, then synced with original raps and music by his U.K. students to create a unified music composition. Logic software was used to edit and loop the recordings from the Skype sessions and then combined with the musical accompaniment and vocals. Underlying goals that were accomplished in the U.K. included discussions on disability. developing an understanding of all types of people and experience with new music software. Some of the music was added using eight colored Koby Nanopads, each assigned to a separate note on the pentatonic scale, meaning that up to eight students could play and record simultaneously. It was also possible to assign each pad to a different drum sound to make up a full drum kit. The U.S. students received enjoyable peer music experiences, a global perspective, and an application of their performance recordings using new software.

The project documented that providing students with meaningful international social and arts inclusion enhances their understanding of the world. To me, the fact this new music could be composed collaboratively with children of varying exceptionalities on different continents provides an example for utilizing technology that can be replicated in almost any region or population. The Border Crossing project was later published in the U.K. Music Educators journal Music Mark (Rathbone & Pestano, 2013).

In What Ways Has Technology Forced Us to Reevaluate Definitions of Musicality? Of Musicianship?

Musicality has changed with technology as the venue of music making has changed. This has been accounted for in several developments with the advent of technology, namely, in relation to (1) venue, (2) the definition of audience, and (3) the manner in which music performance is shared and transmitted. With online technology, music is no longer primarily taking place in a single room or recital hall. It is being shared in between vast interconnected distances in real-time settings, with the transmission of music and its affective experiences no longer taking place solely in a theatre or established hall. With this change in venue comes a change in acoustics and in the styles of music being transmitted, allowing for real-time blending of cultural styles and genres.

The definition of audience is also changing, with a much more varied integration of participants, from school students to professional musicians of greatly varied cultures and backgrounds. The manner in which music performance is shared and transmitted is also changing, with the advent of iPads as a recognized musical instrument. In fact, informal music integration can take place in handheld devices, with the differences in time zones becoming less of a factor, with friends and colleagues simply adjusting their schedules or sharing spontaneous music activities during free time, as opposed to set class scheduling or evening performances.

Of Who Is and Is Not a Musician

There seems to be a common perception that people with disabilities receive music therapy, while those without receive music education. With advancements in the iPad and other related technology that only require the movement of a finger to generate musical sounds, the ability to participate as an active musician in an ensemble is being redefined in some truly amazing and positive ways. In many cases, students with severe disabilities who are in a traditional public school have often been kept out of higher-level performing ensembles because of their special needs (often through auditions, inability to read written music, absence of a mandate requiring their participation, and families not understanding that their children have the skills necessary to participate).

Thanks to examples of ways that students can gain these necessary skills, such as the global online program at the Sidney Lanier Center, students are having inclusive online experiences with ensembles from around the world and at all levels of accomplishment. Some of this collaboration in the aforementioned Sidney Lanier Center online program include Gary Day, an expert in iPad music adaptation who is currently affiliated with the University of Roehampton in London. He has already shared traditional Irish music with the students and instructs on the utilization of iPads in shared music experiences. Another is community musician and More Music artistic director Pete Moser who Skyped from Liverpool to share a collaborative session of Beatles music with the Sidney Lanier Center students

In What Ways Has Technology Transformed Our Understandings of Creativity?

The manner in which technology is utilized can transform our current understandings of creativity by opening the door to a variety of media that previously was not available. Currently, hardware can be developed that is specifically designed to minimize the hand-eye coordination needed to create musical sounds, while pairing with software that opens up a variety of methods for creating music. Specifically, music can be enjoyed by the performer alone, or adapted to include others in a concert environment. One such example was the development of a personal sound-generating device by Mallory Moyer, a University of South Florida student, who created a device that looks much like an oil painting, with deep grooves and designs, but becomes a personalized music instrument at most, or a sound-generating device at least, for people with severe disabilities when they touch sensors with varying sounds on each groove.

The project required a study of music therapy, design, and technology, which helped her to create a musical instrument with universal playability. The musical design of this instrument includes fun graphics and a touch sense technology, allowing anyone to play and have fun. The goal was for people with disabilities to achieve the therapeutic benefits of playing a musical instrument. Throughout the project she introduced her instrument to a wide variety of people: musicians, doctors, and people with disabilities.

My music program had a rewarding and meaningful role in this project. I provided suggestions for the project, and my students approached the instrument for the first time at the school and had a wide range of reactions that revealed the potentials of this technological approach. The project also revealed the potential advantages of having a touch-sensitive instrument on hand in the classroom. It also opened up discussions on the possibilities of having instruments that are customizable, tailored to the individual. Choice of sounds, colors, graphics, size, and placement of the touch-sensitive graphics could be considered. The most profound realization that arose from this initial project was the recognition of its potential for growth. Mallory intends to continue to develop the project and work with our school in the future. She will collaborate with creative minds to make new prototypes and instruments based on universal playability and music therapy and hopes to continue to utilize these insightful resources as she and her project grow.

Another example of the new possibilities of inclusion was the use of iPads by Sounds Global founder Phil Mullen and myself at the first inclusive music performance held at the China Conservatory in Beijing, coordinated by Professor Xie Jiaxing in 2012, and including adults with learning disabilities from the Anhuali Community Center. This concert utilized iPad apps that created a variety of acoustic and modern sounds that fit within a hands-on pitched and unpitched percussion ensemble made up of conservatory students and local adults with learning disabilities from the Anhuali Community Center (Harrison & Mullen, 2013). A few months after the concert, conservatory translator Creek Martin continued this connection by Skyping lessons on basic Chinese terms and music to my students and utilizing an Ipad to show visual aides to represent the concepts she was teaching.

What Are Some of the Untapped Potentials in This Area?

One of the key untapped potentials in this area is people's increasing ability to create and design their own apps for computer and personal device technology. I would guess that in the near future people's ability to do this will become commonplace. What sounds represent music and what constitutes an instrument are being redefined as technology develops in the fields of music education and performance over time. Originally, the format of music reception was through recordings, which could be listened to individually or shared at dances and restaurant jukeboxes. Technology is not only changing the context in which music is shared but the utilization of music performance and interaction by people who may no longer be referred to as "nonmusicians." How we define who is a musician and who is considered to have the prerequisite music skills and expertise to be referred to as a performer will increasingly change, or in fact become irrelevant, with the advent of personal music technology that brings music opportunities to all populations, regardless of special needs, around the world.

References

DeVito, D. & Mullen, P. (2015). *The First Educational Publication of the Haitian Teachers Association*. Gainesville: Projects for Haiti.

Harrison, E & Mullen, P (2013). *Reaching Out: Music Education for Hard to Reach Children and Young People*. Wilts: Music Mark.

Higgins, L., Akuno, E., & DeVito, D (2016). International Cooperation in Music Teaching and Learning: Perspectives from the International Society for Music Education, *European Perspectives in Music Education, 5*, 117–128.

Rathbone, M., & Pestano, C., (2013). Border Crossing: A Skype creative music project with the USA at an EBD primary PRU. *Music Mark Magazine, 1*(1), 26–29.

Suárez, E., DeVito, D., Kleber, M., Akombo D., Krikun, A., & Bingham, S. (2010). *Discovering abilities through harmony: An interdisciplinary music approach for students with disabilities with the Sidney Lanier Center*. ISME CMA, and NACCM. ISME CMA XII: Harmonizing the diversity that is community music activity, 87–92. Retrieved January 5, 2017, from https://issuu.com/official_isme/docs/2010_cma_proceedings.

Suárez, E. R., & DeVito, D., (2012). Music education preparation for teaching students with disabilities. *The Orff Echo, 44*(3), 26–29.

Vogiatzoglou, A., Ockelford, A., Welch, G., & Himonides, E. (2011). Sounds of Intent: software to assess the musical development of children and young people with complex needs'. *Music and Medicine, 3*(3), 189–195.

CHAPTER 39

PROSUMER LEARNERS AND DIGITAL ARTS PEDAGOGY

SAMUEL LEONG

Lum and Freedman (chapters 34 and 35 in this book) clearly show how key aspects of traditional approaches to music teaching and learning have been transformed by technological developments. I agree with Lum that "technology has created an unprecedented fluidity in the teaching and learning of music" (chapter 34, p. 365).... "Agency provided to the learner through technology has considerably weakened the power of music teachers in controlling mastery over time and space" (chapter 34, p. 360). I also agree with Freedman that music education's primary focus should go beyond having students become readers and writers of standard music notation to focus on the "." Both writers have also pointed out how technology has challenged traditional views of music, musicality, and creative work in music. Their chapters have stimulated me to offer three "Further Perspectives," having to do with prosumer learners, digital arts pedagogy, and technology-enhanced music curriculum.

Prosumer Learners

Both Lum and Freedman have reminded me that learners are the subject of the educational endeavor. In a technology-driven world, it is easy to be more concerned about things such as Web 2.0 tools and technological applications in music teaching and lose the central focus that the learners are at the core of teaching and learning. Learners include the "teachers" and everyone from the very young to seniors in increasingly diverse music education settings. Technology has expanded the range of possible learning contexts, enabled the provision of lifelong learning opportunities, and assisted in catering for learners with different academic abilities and behavioral needs, while resetting the traditional classroom-based schooling and herd-driven consumer culture to a life-wide participatory culture. With instant and 24/7 connections to the world via technologies, the educational ecosystem has greatly changed. Today's learners are prosumers who live, work, and play in the "age of the prosumer" (see Leong, 2003; Toffler, 1980). Prosumers advocate the do-it-yourself approach to musical creation by individuals and groups, made possible by increasingly powerful and more affordable technologies, with support rendered by the new forces of social networks, crowdsourcing, and crowdfunding. Prosumer learners are "proactive individuals [who] not only seek out as much information as possible about products and services but are eager to pass it on to others" (Euro RSCG Worldwide, 2011, p. 4). They share traits of the producer-cum-consumer in seven ways: they need inspiration, want personal attention, do their research before commitment, demand faster and better products or services, are not afraid to question the status quo, "buy" on their terms, and value-focused rather than brand loyalty (Maseko, 2012).

With access to smart digital devices and digital media networks, prosumer learners act as both consumers and producers who post, share, comment, critique, publish, and produce individually, as a group, or as creators of their own groups. This marks a shift from the traditional top-down static flow of information to a more dynamic process of circulation via multiple network systems. The shift is also from the traditional push-music culture, in which a few produce for many, to a networked pull-music/on-demand culture.

Whereas listeners in the past had only simple control options, such as play, pause, skip, and volume, today's prosumers can respond via comment sections beneath news articles and YouTube videos, express their thoughts on all kinds of topics via online media, and even participate in a forum to suggest how to improve a particular site. Network media such as YouTube, Spotify, and SoundCloud can empower them to develop closer relationships with educators, musicians, other learners, and the wider global community. Education systems today should tap into the potential of prosuming activities that can contribute to cultural, social, educational, and economic values via the new media. The voice and creativity of the consumer-learner need no longer be marginalized, and the collective creativity and collaboration of all learners can enrich the quality of music education for everyone.

Jam2Jam (Brown & Dillon, 2007) is an example of prosumer learning platform that presents opportunities for creative improvisatory collaboration and positive interaction with music and that integrates listening, composing, and performing. This platform blurs the domains of production and consumption, enables remote participation, and facilitates learners' gaining of meaning through musical reflexivity and decision-making in response to musical stimuli from others within an evolving musical-cultural setting and structure. This approach exemplifies a new kind of inclusive and experiential technomusicking (see Small, 1998), where music is regarded as process rather than object, and as a social and cultural activity through which learners explore their identity and their relationship with others. Technomusicking can also take place by using applications such as Smule's Ocarina and Guitar!—in which prosumer learners can play a musical instrument without having to acquire the necessary performance techniques—or access T-Pain Video to create their own music videos with control of vocal effects and one-click uploading to YouTube. These new technologies abolish the technical barriers that prevent many avid music lovers from being engaged in the rich array of creative activities that music offers. Unfortunately, traditional computer usage in music education has primarily supported traditional pedagogy: creating lesson materials, including worksheets, drills, recording, storage, and playback.

Given the prevalence of digital technologies in a world inhabited by prosumers learners, today's music education should feature digital arts pedagogy that engages learners with the computer and computer devices as tools, media, and musical instruments, while at the same time "amplifying" musicianship effectively through technology-enhanced teaching (see Brown, 2007, 2012; Brown & Nelson, 2013). In addition, today's music education should develop learners' competencies in the manipulation and creation of images, text, audio, video, animation, sound, and visual effects, as well as in the processes of imagining, sourcing, sampling, designing, editing, composing, presenting, and disseminating.

Digital Arts Pedagogy

Today's computers are bundled with software that provides the means to create music clips, films, and even video games. And a wide array of free software and apps are available. Today's technologies are helping prosumers create their own homemade videos, video game "mods" (i.e. modifications), independent music projects, and crowdsourced music video projects. Indeed, we are seeing a convergence of the old and new, of producer and consumer, of professional/celebrity and layperson, and of formal and informal. The learning landscape has transformed in such a way as to feature open learning environments, open educational resources, web communities, smart streams, augmented reality, synchronous technologies, and mobile technologies. Despite the potential and promises of technologies and digital learning platforms, contemporary education should note that most social networks and media sharing tools such as Facebook and YouTube operate without clear pedagogical goals (Snickars & Vonderau, 2009). More research is needed in this area, and new pedagogies that cater for prosumer learners using multimedia technologies need to be developed. Nonetheless, three emerging pedagogical trends have been observed: (1) technology has opened up learning, making it more accessible and flexible—the classroom is no longer the unique center of learning, based on information delivery through a talk session; (2) technology enables power sharing between teacher and learner—toward more support and negotiation over content and methods, and a focus on developing and supporting learner autonomy, such as new social media, peer assessment, discussion groups, even online study groups but with guidance, support, and feedback from content experts; and (3) increased technology use, not only for teaching but also to support and assist students and to provide new forms of student assessment. Indeed digital arts pedagogy cannot be overly dependent on technology to do the job but needs to properly utilize digital tools that can support user autonomy, access to peer-to-peer networks, connectivity to open communities, and social interactivity.

My students have learned to produce their own indie magazine that covers topics of interests from their program of study. Indeed, digital technology can enable the collection, storage, and assessment of learners' online activities and contributions, including a permanent record of their online discussion, academic and creative e-portfolio outputs, and peer-mentor feedback and assessment. This can enhance the process of reviewing, monitoring, and assessing the quality of student learning. An e-progress mapping project at my institution encourages learners to take ownership of their learning, set goals toward desired graduate outcomes, map their learning journeys, construct their individual knowledge bases, and collate evidence to substantiate their learning outcomes. This is proving very useful for course and program review, internship placements, and a better understanding of the teaching and learning issues of prosumer learners. The project has also helped the teaching team realize the importance of harnessing the power of technologies to customize learning activities and develop the digital arts literacy of our students: the ability to locate, organize, understand, evaluate, analyze, and create information using digital technology.

One of the most important challenges for digital arts pedagogy is embracing the broad range of networked digital devices that includes wearables, smartphones, tablets, phablets, laptops, and desktops and developing learners' knowledge, skills, attitudes, and behaviors in using them. Mobile learning is now encouraged in addition to blended learning. Another challenge is to come to terms with the digital medium itself—which requires learners to imagine, think, design, construct, develop, and share "products" that are digital in nature. While digital production and creativity involves "products" with origins in different media, such as music, art, video, or writing, nondigital media require praxis (see Regelski, 1998; Elliott, 2005) that involve creative skill sets that are linked to cultural conventions and traditions. Hence, digital arts pedagogy and assessment need to consider the distinctions and interrelationships between the creative skills required when producing a nondigital product versus those required to realize a digital product linked to nondigital practices, as well as the transferability between the two. As the digital medium provides interactions with other digital citizens, digital arts pedagogy and assessment should also consider how communication and sharing during the creative process could have impacted the product outcome. One of the issues that has arisen from my experiences with assessing multimedia presentations of discipline-based music projects is the lack of musical understanding and concomitant musical creativity in otherwise creatively designed presentations involving images, text, animation, video, and music. Three other issues relate to this one: (1) assessing students' creative skills as author/composer versus editor/producer of a creative product; (2) their demonstrated knowledge of historical, local, and popular cultures versus a fusion of cultural elements; and (3) novelty/originality versus the reuse of found material (with appropriate acknowledgment).

Technology-Enhanced Curriculum

Technologies that support music teaching and learning have come a long way from educational television, audio, and VHR cassette tapes, play-along CDs, VHS and VCD instructional videos, portable personal minidisc recorders, film strips, and Hypercard and floppy disks. Today's new learning ecosystem, afforded by new technologies, empowers learners and educators to foster the skills and dispositions needed by the twenty-first-century workforce, instill innovative and entrepreneurial thinking, and enhance global, glocal, and local awareness. Technologies should liberate teachers and learners from the constraints of rigid timetables, fixed rows of desks and chairs, hierarchical formal curricula, and the shackles of assessment. Multiple activities should be designed as mediators of learning—not only for gaining literacy and numeracy but also for searching, designing, performing, creating, improvising, and playing. The learner should become more adaptable and flexible, able to negotiate, navigate, and manipulate the oceans of knowledge, where knowledge itself is fluid, multifaceted, and open to interpretation.

The technology-enhanced curriculum should enable learners to chart their own unique learning pathways, adjust their paces of learning, and find ways forward that caters for their individual learning styles, strengths, interests, and cultural differences—supported by technologies such as webcasts, webinars, video conferences, digital learning platforms, and educational games. Some important pedagogical possibilities offered by technologies for music education can be found in Gouzouasis and Bakan (2011).

Technology-enhanced music curricula should move beyond drill and practice, worksheets, and written assignments to feature a more socially engaged and creative approach that truly engages all prosumer learners. Prosumer learners love learning through games, which can help them learn a number of transferable skills, such as processing information from multiple sources and making decisions quickly, deducing the rules of a game from actual experience instead of being told, creating strategies to overcome obstacles, making sense of complex systems via experimentation, and learning to collaborate with others (see Prensky, 2003). Learners can undergo game-based musicianship training using programs such as SingStar, in which learners sing along with music in order to score points for their pitch accuracy. Younger learners can benefit from game-based learning, such as that found in the New York Philharmonic Orchestra's KidZone's Game Room, the BBC's Maestro game site, and the Mrs. Dennis 110 Free Music Education Apps site. A collaborative project between upper primary teachers used Guitar Hero as a context for learning between Easter and June, in which everything the students did was based around the game (Bray, n.d.). These included dance lessons performed to rock music, creative writing activities inspired by Guitar Hero characters, planning a world tour in geography, creating band logos and merchandising in art and design, financial education on marketing, profit, and loss, and other activities.

The technology-enhanced curriculum could have learners design and create new musical sounds, new instruments, new combinations of performance ensembles, even conveying new ways of expression and interpretation through the production of music recordings. But, as music educators, we should be aware of the implications that music might have become "computer data" in a postperformance era, where performance is regarded a "secondary substitute" while "the authentic version is the recorded version" (Thibeault, 2012a, p. 6; also see Thibeault, 2012b).

Three initiatives at my institution are supporting the realization of a technology-enhanced undergraduate music curriculum. Two "mobile labs" with 33 sets of laptop computers on customized trolley each can be rolled into a class to facilitate interactive learning activities involving the entire class, small groups, and/or an individual. Students' musicianship training is enhanced through rehearsing and performing in the i-Orchestra, an ensemble that uses synthesizers and computers to simulate the sounds of an acoustic orchestra. New learning approaches using mobile devices, such as the iPhone and iPad, are being trialed in aural training and other aspects of the music curriculum. The inclusion of 3D printing in the music curriculum is currently being considered.

As technology and learning sciences continue to advance, they offer exciting new possibilities for a new era of digital-enhanced music teaching and learning approaches that more effectively engage today's prosumer learners in a globalized interconnected world. This calls for boldness and creative risk taking to cocreate a viable and sustainable technohuman educational future beyond entertainment, gadgetry, and gimmickry.

References

Bray, O. (n.d.). Imaginative ICT: Guitar Hero in the classroom. *Teach Primary*. Retrieved January 5, 2017, from www.teachprimary.com/learning_resources/view/imaginative-ict-guitar-hero-in-the-classroom.

Brown, A. R. (2007). *Computers in music education: Amplifying musicality*. New York: Routledge.

Brown, A. R. (Ed.) (2012). *Sound musicianship: Understanding the crafts of music*. Newcastle upon Tyne, UK: Cambridge Scholars Press.

Brown, A. R., & Nelson, J. (2013). Digital music and media creativities. In P. Burnard (Ed.), *Developing creatives in higher music education: International perspectives and practices* (pp. 61–74). Abingdon, UK: Routledge.

Brown, A. R., & Dillon, S. C. (2007). Networked improvisational musical environments: Learning through online collaborative music making. In J. Finney & P. Burnard (Eds.), *Music education with digital technology* (pp. 96–106). London: Continuum.

Elliott, D. J. (Ed.). (2005). *Praxial music education: Reflections and dialogues*. New York: Oxford University Press.

Euro RSCG Worldwide. (2011). *Prosumer report: The second decade of prosumerism*. Paris: Author.

Gouzouasis, P., & Bakan, D. (2011). The future of music making and music education in a transformative digital world. *UNESCO Observatory E-Journal*, 2(2). Retrieved January 5,

2017, from http://education.unimelb.edu.au/about_us/specialist_areas/arts_education/melbourne_unesco_observatory_of_arts_education/the_e-journal/volume_2_issue_2.

Leong, S. (Ed.). (2003). *Musicianship in the 21st century: Issues, trends and possibilities*. Sydney: Australian Music Centre.

Maseko, D. (2012, October 31). The rise of the pro-active consumer "prosumer" [Blog entry]. *Enlyne Branding*. Retrieved January 5, 2017, from http://www.enlynebranding.com/news/theriseofthepro-activeconsumerprosumer.

Prensky M. (2003). Digital game-based learning. *ACM Computers in Entertainment, 1*(1), 21. doi: 10.1145/950566.950596.

Regelski, T. A. (1998). The Aristotelian bases of praxis for music and music education as praxis. *Philosophy of Music Education Review, 6*(1), 22–59.

Small, C. (1998). *Musicking: The meanings of performing and listening*. Hanover, NH: Wesleyan University Press.

Snickars, P., & Vonderau, P. (Eds.) (2009). *The YouTube reader*. Stockholm: National Library of Sweden.

Thibeault, M. D. (2012a). *A critical pragmatic approach to technology in music education: Response to John Kratus' "Transitioning to Music Education 3.0."* Paper presented at 2011 Music Education Conference, Michigan State University. Retrieved January 5, 2017, from https://www.ideals.illinois.edu/handle/2142/42591.

Thibeault, M. D. (2012b). Music education in the postperformance world. In G. McPherson & G. Welch (Eds.), *The Oxford handbook of music education* (Vol. 2, pp. 517–529). New York: Oxford University Press.

Toffler, A. (1980). *The third wave*. New York: Bantam Books.

CHAPTER 40

A PLURALIST APPROACH TO MUSIC EDUCATION

JAMES HUMBERSTONE

Music that is made with technology has the potential to revolutionize music education and redefine musicianship and creativity. By adopting a genuinely pluralist approach to music education, creative processes (Burnard, 2012a) with technology at the core can engage and motivate students, advocate for music's continued relevance in the curriculum, and help music educators retain and refine traditional pedagogies and systems. Technology can save music education from its falling status in worldwide curricula (Cox & Stevens, 2010).

Adopting such a pluralist approach is profoundly difficult for music educators, because we come from a self-perpetuating system. We are part of the significant minority of students who choose to study music to the end of high school. For almost all of us, that means learning to sing or to play a (Western) instrument to a relatively high standard (Green, 2008). We joined an even smaller minority to study music at a tertiary level, and then a subgroup of that minority, becoming qualified music educators. We become teachers, and we teach music based on our own success—the same Western instrumental or vocal traditions, maybe with a bit of popular music thrown in just to show we can. Fewer than 10% of our students will choose to study this kind of music right through high school (Humberstone, 2014), and so the next minority generation is born.

Hein's experience, outlined in chapter 22 of this book, is profoundly different, and because of that, his model provides the exact answers (to the Provocations) that evade most of us who are stuck in this cycle. He proposes a genuine paradigm shift in terms of the understanding of musicality, musicianship, and the identity of the musician. In this Further Perspective, I will make a case for the coadoption of Hein's paradigm, drawing on the inspirational examples of professional transformation outlined by practitioners such as Freedman (chapter 35) and Dorfman (chapter 37), who already live and breathe such a plurality in their working lives. I will argue that Western music education systems have already lost relevance in nearly all of our students' lives, but that we do not need to take a "baby with the bathwater" approach to making musical creativity as important and enjoyable for them as it should be.

The Twenty-First-Century Learner and Music Education

Our children live in a "flattened" world (Friedman, 2007) where they are, even at young ages (Steiner-Adair & Barker, 2013), able to access information, entertainment, and relationships and to project a "version" of their self-identity (Gardner & Davis, 2013). Craft (2012, p. 176) identifies arising contrasting and extreme views of this recent transformation as seeing "childhood at risk" (e.g., Turkle, 2011) versus "childhood empowered." This contrasting discourse of a single situation is reflected in the wide range of viewpoints found in this handbook, and in research on the impact of new technologies on children's lives. Contrast, for example, the popular research of Sugata Mitra, which conclusively shows that Internet-enabled audodidactism works (Mitra, 2005; Mitra & Dangwal, 2010; Mitra et al., 2005; Mitra & Rana, 2001), with Dimitri Christakis's work proving the negative impact of screen time on young children's brain development (Christakis, Zimmerman, DiGiuseppe, & McCarty, 2004; Zimmerman & Christakis, 2005). I encourage my students to be "enthusiastic cynics" in terms of technology in education; to approach new technologies and cultural change optimistically but to think critically about their educational value. As I will show, such a metamodern approach (Vermeulen & van den Akker, 2010) is appropriate for established music teachers who are engaging in new music for the first time, too.

Whichever worldview one adopts, it cannot be denied that we live in a time of tremendous change when children have at their fingertips almost limitless choices for learning. David Price, one of the original designers of the United Kingdom's "Musical Futures," suggests that learning has already "gone open" (2013, p. 5). Research has not yet measured whether more people are learning to play an instrument from freely available Internet resources such as YouTube than from qualified music teachers, but anecdotes tell me that such a hypothesis may well be true. What we *do* have is a critical appreciation of how YouTube is used in both formal and informal music education (Waldron, 2011, 2013a, 2013b, 2016) and an understanding that informal music learning contributes to a greater uptake of music in schools and can lead to greater student involvement in both informal ensemble performance and formal music training in and out of school (Hallam, Creech, & McQueen, 2011; Hallam, Creech, Sandford, Rinta, & Shave, 2009).

The depth of resources open to the bedroom musician (Green, 2002), which may be used in a traditional didactic manner or in a more collaborative manner (Thorgersen & Zandén, 2014), means that students can and do become composer-performers of the music that most interests them, an experience which may not be mirrored in classroom composition experiences (Ruthmann, 2008). Bedroom musicians do not choose to become violinists, to learn an instrument so that they can join a higher-level concert band, or to seek out training in vocal blending so as to get into chorus. And they do not just learn, as may be supposed, sporadically or shallowly: consider the number of bedroom-trained musicians, like French DJ Madeon (Hugo Leclercq), who are now performing professionally and, in his case, producing for Lady Gaga after some of that bedroom-produced work went "viral" on social media (http://youtu.be/lTx3G6h2xyA): "I had a very chaotic and complicated education history. I've always been more attracted to self-learning. I didn't learn English or Music at school—I preferred to do it myself" (Popjustice, 2012, para. 26).

And there's the rub. Our students actually *love* studying music, but generally not the music we teach in schools (exceptions being music programs designed to engage students, usually through popular music cultures, rather than to impart information *about* music (e.g., Green, 2008; Rusinek, 2008). The music *we* teach persuades only 5–10% of our students to study music right through high school, despite the fact that they "report that listening to music is their favourite activity" (Rose & Countryman, 2013, p. 49). Hein's account of his own experience mirrors exactly this: "I abandoned formal music as soon as I was able." *Formal* music.

Technology, then, provides a platform for informal learning, and for learning about the musical genre and associated skills in which the student is most interested. The literature on the impact of new technologies shows that the integration of mlearning *does* improve motivation and engagement in students (Bebell & O'Dwyer, 2010; Kinash, Brand, & Mathew, 2012). Therefore, technology not only holds the *content* to reel our students in, but in itself is a motivator to get them started and earn their trust once more.

The Barrier Does Not Really Exist: Teacher's Perceptions of Difficulty

Freedman (chapter 35) proposes solutions to the problems that Lum (chapter 34) says teachers face. The barrier of professional learning, for example, is dismissed by Freedman's trust that her students will work out what they need. The "I'll teach music, and *we'll* work out the technology" approach is not a new one, and has been applied since long before the technology in question was computers or tablets (Quinn, 2007).

The use of technology is also often linked to composition (Burnard, 2007) rather than performance, and as Strand (2006) has shown, trained music educators who identify as performers or directors are less likely to include composition tasks in their curriculums. Strand also showed that teachers perceive lack of access to technology and their own comfort teaching composition as barriers to including this kind of creativity in either generalist classrooms or ensembles. Reese (2003) found that most teachers are not trained in how to teach composition in a classroom context, and more recently

Bernhard (2013) observed similar reservations in preservice music teachers about teaching improvisation.

This perceived double barrier, of technology that teachers do not have funds to buy, training to use, or time to learn (Gall, 2013), combined with curriculum demands in aspects of musical practice that they feel least able to teach—specifically composition—then clashes with teachers' own likely musical-cultural background, as presented in the introduction, and understanding of what musical literacy is.

Successfully promoted research that has entered the lexicon of popular thought about twenty-first-century education, such as Prensky's divisive generational description of digital natives and immigrants (2001a, 2001b), has perpetuated the idea of the technology barrier that teachers lacking confidence claim holds them back (Martinez & Prensky, 2011), when "even" mature teachers are able to be digital culture leaders (Helsper & Eynon, 2010). Such generalization about the generations also lacks cultural subtlety (C. Brown & Czerniewicz, 2010). Conversely, Freedman had the confidence to know that her musical training was valuable in music class, whatever the context, and the courage to follow her own advice: "Today, if you want to learn something about a piece of software, take a course, read a book, look it up on the Internet, or ask a kid!" (2013, p. xv). Prensky is right to suggest that teachers need to learn to speak the new tongue, but the debate about student-centered, experiential education has been going for more than a century, and this new "language" is one that several generations are making up together. There *is* no right or wrong accent, and Prensky's more recent suggestion that the future debate should be about digital wisdom (2012) is apt. Similarly, the introduction of many successful bring-your-own-device (BYOD) programs prove that technology will soon not be something schools have to *afford* but that students will *bring*.

More helpful ideas gaining some traction are those based around giving teachers frameworks for the introduction of technology, such as technological pedagogical and content knowledge (TPACK) (Voogt, Fisser, Pareja Roblin, Tondeur, & van Braak, 2013). Bauer (2013) has explored ways in which TPACK can be implemented in music education and further discussed this in his recent book reviewing the intersection of technology, culture, and music pedagogy (2014).

Is It All About Pop Music?

Embracing new technologies and their related music cultures does not mean giving in to teaching semesters full of Taylor Swift and Justin Bieber. Nor does it mean giving up the tried and tested that pedagogies teachers know from experience *do* engage and educate their students. A pluralist music education means just that: many musics, many pedagogies, and not necessarily "integrated" just for the sake of it (Green, 2008, 2014).

The barrier for music teachers is not the technology, nor "having" to teach music they don't necessarily personally like. The barrier is in coming to understand that the categorizations and descriptive language we have developed to explain music are irrelevant to many new musics. Just in the same way that it would be pointless teaching serial composition through a prism of formal harmony, or Balinese gamelan in relation to the rules of fugal writing, much modern music does require a redefinition of musical terminology, of musicianship, and of what we mean by *musician* or someone who is *musical*.

Hein (chapter 36) explores questions of identity and terminology. In reference to the latter, he writes: "the inclusion of popular dance music would significantly enrich the curriculum, particularly in areas traditionally neglected: groove, timbre, and space." "Groove" and "space" are not terms found in music syllabi around the world, nor are they terms often taught to preservice music teachers. Just as one shouldn't assume that a young person is a digital expert (if not native), one can't assume that a young (classically trained) music teacher is an expert in contemporary popular music. Rose and Countryman (2013) also found that musical terms meaningful to young people included "groove" and "delicious" timbres (p. 55) and that they had the ability to discuss in a sophisticated manner and socially share "delight in patterns of sound" (p. 56).

Our students are deeply interested in the physical, psychological, and cultural effects of music, but if we do not learn, first, to accept as valid their own understanding of music, and second, to attempt to understand the language they use to describe that understanding, we cannot fully engage them, and we cannot understand the music *they are engaged in*. Trying to make new music fit prescribed, art music categorizations imposes value systems and "causes teachers to extract simple analyses that over-generalize" (Rose & Countryman, 2013, p. 49). And even *if* new terminology can be adopted, recategorized, and understood, it is important to remember that by itself, "formal musical knowledge is inert and unmusical" (Elliott, 1995, p. 61).

Syllabus-mandated categorization of the "elements" of music is also a problem at a curriculum level, where teachers are encouraged to tick boxes of components of musical learning. In countries of many diverse cultures and musical backgrounds, musical education is broken down into the separated experiences of (1) performance, (2) listening, and (3) composition, in that order of importance. Performance is often the only experience of music education in which students partake (Cox & Stevens, 2010). Listening means listening and responding to music in most generalist classrooms—*merely* listening (Reimer, 2004)—but sometimes also includes the development of the skills of traditional musicianship: theory, aural, and analysis. Composition is the third and final "experience," because as discussed, most teachers do not feel trained to teach it.

Even where this holy trinity of musicking (Small, 1998) may be beautifully balanced, as in Moore's "perfect music experience" (2003, p. 195), the electronic music of which Hein speaks blows apart these categorizations. Without elaborating on whether the act of improvisation fits neatly into the experience of performance or composition, or somewhere between them, the arts of DJing and producing draw on those discrete experiences *simultaneously*.

Consider a live DJ set. Before it begins, the DJ has collected or created sounds, and has rehearsed combining these in ways that she and her listeners will sound interesting. On top of these samples the DJ may create or improvise with synthesized sounds or samples using MIDI controllers or dedicated hardware. Effects are not added as part of the mixing or production process but as options for live performance nuances, such as a low pass filter sweeping up toward a musical climax (the "drop"). Some DJs perform live on acoustic instruments. Different genres of electronic or sample-based music pose different expectations of tempi, timbres to be employed, keys and harmonic devices to be used, and structures to be followed (Snowman, 2014), and every DJ carefully plans how to manipulate the listener by following—or not—these expectations (Burnard, 2012a).

Some of these preexplored combinations will be reproduced exactly in the live set, and may even be precombined as new tracks, ready to go; others are options for the DJ to explore during the set: they could be samples that may or may not be "dropped in," performed live from a MIDI controller, or scratched-in from vinyl. Live sounds may be captured in real time and retriggered from a loop pedal or laptop. Climaxes may be lengthened through the exploration of audio effects when the DJ feels the audience is going with her, or a sudden decision to go to "the drop" when more excitement needs to be created. All of this is performance. All of this is composition. All of it is improvisation. And all of it is listening, analyzing, and exploring the structures of the genre within a live context.

Just as much contemporary music may not fit into the categories we use to teach about music, the creative process does not divide neatly across strict boundaries of DJ, composer, or songwriter (Hickey, 2003; Odena, 2012). Some composers audiate (imagine the sounds in their head) first, then write out compositions; some explore at an instrument, and some do indeed use technology as part of their creative process. Research has shown that mature composers who compose *in* notation software are less creative and provide less detail (Peterson & Schubert, 2007), yet many of Australia's most esteemed and senior composers compose directly in computer programs, using the playback feature for immediate feedback (M. Hindson, personal correspondence, December 30, 2013; Hindson, 2015; Humberstone, 2015).

Contrast this mix of audiation and "tinkering" with the processes of "serious" art music composers who work in soundscape design, and one realizes that the DJ processes discussed above are no longer unique to any one genre. Lisa Whistlecroft describes her digital exploration of sound as the *beachcomber-bowerbird approach to composition* (2013), working exclusively with sounds recorded in the field in a DAW (digital audio workstation). Compare this to Madeon's method (he composes first at the piano, then develops each sound from scratch through synthesis; Leclercq, 2013) and the compositional processes are reversed, and through the use of technology. Neither is "better." Neither is "the right way."

Happily, this range of practices more closely mirrors the postmodern world teachers and students inhabit, where the "high" and "low" arts not only coexist but pastiche and borrow from one another and even from themselves (Hutcheon, 1988). Indeed the remix form established by electronic music is now impacting the "classical" tradition (for example Max Richter's *Vivaldi Recomposed* or Steve Reich's *Radiohead Remix*), and the side-by-side (r)evolutions in both art and popular electronic music technologies over the past 40 years may offer a musicological equivalent to what Vermeulen & van den Akker (2010) have described as "metamodernism." Outside the classroom in the "real" musical world, pluralism means that culture can be separate and together, disconnected and connected, all at once. Similarities in all music, even the most aleatory (Gill, 2003), such as repetition of sounds that neurologically trigger amygdala activation (Levitin, 2006), are the starting points. It's not all about popular music. It's about *all* music.

Concluding

In this chapter I have attempted to show that while the challenges of educating in the twenty-first century are very real, the claimed barriers to engaging students through technology-based music are not a valid excuse. Conversely, redefining musicianship and musical experiences and understanding new musical genres and cultures *is* a challenge for "classically" trained music educators. I have suggested that a pluralist approach to music education, one that embraces *all* musics and *all* creative practices, is one way to not "throw the baby out with the bathwater" in terms of tried and tested pedagogies, but also to ensure that our students benefit from the revolution that music technology and its related genres brings our classrooms. I have shown that boundaries between "high" and "low" art are increasingly meaningless in modern society, and that all music is created with skill and ingenuity: in fact, that even the most clichéd popular "trash" is created very self-consciously in our metamodern world.

Hein suggests that one key to engaging our students is helping them identify as producers. Hein is right. But to do this, we teachers have to begin by legitimizing those creative practices. We have to learn *why* those musics are more popular, more engaging, and more successful than the repertoire we may currently be teaching. We have to take them seriously—and when we do find ourselves teaching Taylor Swift and Justin Bieber, or electronic dance music made with software we don't understand, we have to remind ourselves that "inspired by a modern naïveté yet informed by postmodern skepticism, the metamodern discourse consciously commits itself to an impossible possibility" (Vermeulen & van den Akker, 2010, p. 5). A wry smile may suffice.

References

Bauer, W. I. (2013). The acquisition of musical technological pedagogical and content knowledge. *Journal of Music Teacher Education*, 22(2), 51–64. doi: 10.1177/1057083712457881.

Bauer, W. I. (2014). *Music learning today: Digital pedagogy for creating, performing, and responding to music*. New York: Oxford University Press.

Bebell, D., & O'Dwyer, L. (2010). Educational outcomes and research from 1:1 computing settings. *Journal of Technology, Learning and Assessment*, 9(1), 16.

Bernhard, H. C. (2013). Music education majors' confidence in teaching improvisation. *Journal of Music Teacher Education*, 22(2), 65–72. doi: 10.1177/1057083712458593.

Brown, C., & Czerniewicz, L. (2010). Debunking the "digital native": Beyond digital apartheid, towards digital democracy. *Journal of Computer Assisted Learning, 26*(5), 357–369.

Burnard, P. (2007). Reframing creativity and technology: Promoting pedagogic change in music education. *Journal of Music, Technology and Education, 1*(1), 19. doi: 10.1386/jmte.1.1.37/1.

Burnard, P. (2012a). *Musical creativities in practice*. Oxford: Oxford University Press.

Burnard, P. (2012b). The practice of diverse compositional creativities. In D. Collins (Ed.), *The act of musical composition: Studies in the creative process* (pp. 111–138). Farnham, UK: Ashgate.

Christakis, D. A., Zimmerman, F. J., DiGiuseppe, D. L., & McCarty, C. A. (2004). Early television exposure and subsequent attentional problems in children. *Pediatrics, 113*(4), 708–713.

Cox, G., & Stevens, R. (2010). Introduction. In G. Cox & R. Stevens (Eds.), *The origins and foundations of music education: Cross-cultural historical studies of music in compulsory schooling* (pp. 1–12). London: Continuum International.

Craft, A. (2012). Childhood in a digital age: Creative challenges for educational futures. *London Review of Education, 10*(2), 18. doi: 10.1080/14748460.2012.691282.

Elliott, D. J. (1995). *Music matters*. New York: Oxford University Press.

Freedman, B. (2013). *Teaching music through composition: A curriculum using technology*. New York: Oxford University Press.

Gall, M. (2013). Trainee teachers' perceptions: Factors that constrain the use of music technology in teaching placements. *Journal of Music, Technology and Education, 6*(1), 5–27. doi: 10.1386/jmte.6.1.5_1.

Gardner, H., & Davis, K. (2013). *The app generation: How today's youth navigate identity, intimacy, and imagination in a digital world*. New Haven, CT: Yale University Press.

Gill, R. (2003). Kindergarten debrief. In MLC School (Ed.), *The creative classroom* [DVD] (Disc 1). Sydney: MLC School.

Green, L. (2002). *How popular musicians learn: A way ahead for music education*. Aldershot, UK: Ashgate.

Green, L. (2008). *Music, informal learning and the school: A new classroom pedagogy*. Aldershot, UK: Ashgate.

Green, L. (2014). *Hear, listen, play! How to free your students' aural, improvisation, and performance skills*. New York: Oxford University Press.

Friedman, T. (2007). *The World is Flat*. New York: Picador.

Hallam, S., Creech, A., & McQueen, H. (2011). *Musical futures: A case study investigation*. London: Institute of Education, University of London.

Hallam, S., Creech, A., Sandford, C., Rinta, T., & Shave, K. (2009). *Survey of musical futures*. Retrieved January 5, 2017, from www.musicalfutures.org.

Helsper, E. J., & Eynon, R. (2010). Digital natives: Where is the evidence? *British Educational Research Journal, 36*(3), 503–520. doi: 10.1080/01411920902989227.

Hickey, M. (Ed.). (2003). *Why and how to teach music composition: A new horizon for music education*. Reston, VA: MENC.

Hindson, M. (2015, March 30). Variety is the spice of life. *About music*. [Public lecture series.] University of Sydney.

Humberstone, J. (2014). *Why music education should lead all education in the 21st century*. Retrieved December 21, 2014, from http://musicaustralia.org.au/2014/11/why-music-education-should-lead-all-education-in-the-21st-century/.

Humberstone, J. (2015). *Defining creativity for a more pluralist approach to music education.* Paper presented at the ASME XXth National Conference 2015. Music: Educating for life, Parkville, Victoria: Australian Society for Music Education.

Hutcheon, L. (1988). *A poetics of postmodernism: History, theory, fiction.* London: Routledge.

Kinash, S., Brand, J., & Mathew, T. (2012). Challenging mobile learning discourse through research: Student perceptions of Blackboard Mobile Learn and iPads. *Australasian Journal of Educational Technology, 28*(4), 16.

Leclercq, H. P. (2013). *I'm Madeon, let's talk production!* Retrieved April 4, 2014, from http://www.reddit.com/r/edmproduction/comments/186kli/im_madeon_lets_talk_production/.

Levitin, D. (2006). *This is your brain on music.* London: Atlantic Books.

Martinez, S., & Prensky, M. (2011). Point/counterpoint—Is the digital native a myth? *Learning and Leading with Technology, 39*(3), 6.

Mitra, S. (2005). Self organising systems for mass computer literacy: Findings from the "hole in the wall" experiments. *International Journal of Development Issues, 4*(1), 71–81.

Mitra, S., & Dangwal, R. (2010). Limits to self-organising systems of learning—the Kalikuppam experiment. *British Journal of Educational Technology, 41*(5), 672–688. doi: 10.1111/j.1467-8535.2010.01077.x.

Mitra, S., Dangwal, R., Chatterjee, S., Jha, S., Bisht, R. S., & Kapur, P. (2005). Acquisition of computing literacy on shared public computers: Children and the "hole in the wall." *Australasian Journal of Educational Technology, 21*(3), 407.

Mitra, S., & Rana, V. (2001). Children and the Internet: Experiments with minimally invasive education in India. *British Journal of Educational Technology, 32*(2), 221–232. doi: 10.1111/1467-8535.00192.

Moore, B. (2003). The birth of song: The nature and nurture of composition. In M. Hickey (Ed.), *Why and how to teach music composition: A new horizon for music education* (pp. 193–207). Reston, VA: MENC.

Odena, O. (Ed.). (2012). *Musical creativity: Insights from music education research.* Farnham, UK: Ashgate.

Peterson, J., & Schubert, E. (2007, December 5–7). *Music notation software: Some observations on its effects on composer creativity.* Paper presented at the inaugural International Conference on Music Communication Science, Sydney: Australia.

Popjustice. (2012). *"I say one funny weird thing and that becomes the headline." Madeon is basically a genius. Here's a long chat with him.* Retrieved December 21, 2014, from http://www.popjustice.com/interviewsandfeatures/madeon-interview-2012/102676/.

Prensky, M. (2001a). Digital natives, digital immigrants, part 1. *On the Horizon, 9*(5), 6.

Prensky, M. (2001b). Digital natives, digital immigrants, part 2: Do they really think differently? *On the Horizon, 9*(6), 6. doi: 10.1108/10748120110424843.

Prensky, M. (2012). *Brain gain: Technology and the quest for digital wisdom.* New York: St. Martin's Press.

Price, D. (2013). *Open.* London: Crux.

Quinn, H. (2007). Perspectives from a new generation secondary school music teacher. In P. Burnard & J. Finney (Eds.), *Music Education with Digital Technology* (pp. 21–29). London: Continuum.

Reese, S. (2003). Responding to student compositions. In M. Hickey (Ed.), *Why and how to teach music composition: A new horizon for music education* (pp. 211–232). Reston, VA: MENC.

Reimer, B. (2004). Merely listening. In L. R. Bartel (Ed.), *Questioning the music education paradigm*. Waterloo: Canadian Music Educators' Association National Office.

Rose, L. S., & Countryman, J. (2013). Repositioning "The Elements": How students talk about music. *Action, Criticism, and Theory for Music Education, 12*(3), 20.

Rusinek, G. (2008). Disaffected learners and school musical culture: An opportunity for inclusion. *Research Studies in Music Education, 30*(9), 15. doi: 10.1177/1321103X08089887.

Ruthmann, S. A. (2008). Whose agency matters? Negotiating pedagogical and creative intent during composing experiences. *Research Studies in Music Education, 30*(1), 43–58. doi: 10.1177/1321103x08089889.

Small, C. (1998). *Musicking: The meanings of performing and listening*. Hanover, NH: Wesleyan University Press.

Snowman, R. (2014). *Dance music manual: Tools, toys and techniques*. Burlington, VT: Focal Press.

Steiner-Adair, C., & Barker, T. H. (2013). *The big disconnect: Protecting childhood and family relationships in the digital age*. New York: HarperCollins.

Strand, K. (2006). Survey of Indiana music teachers on using composition in the classroom. *Journal of Research in Music Education, 54*(2), 154–167. doi: 10.2307/4101437.

Thorgersen, K., & Zandén, O. (2014). The Internet as teacher. *Journal of Music, Technology and Education, 7*(2), 233–244. doi: 10.1386/jmte.7.2.233_1.

Turkle, S. (2011). *Alone together: Why we expect more from technology and less from each other*. New York: Basic Books.

Vermeulen, T., & van den Akker, R. (2010). Notes on metamodernism. *Journal of Aesthetics and Culture, 2*. doi: 10.3402/jac.v2i0.5677.

Voogt, J., Fisser, P., Pareja Roblin, N., Tondeur, J., & van Braak, J. (2013). Technological pedagogical content knowledge—a review of the literature. *Journal of Computer Assisted Learning, 29*(2), 109–121. doi: 10.1111/j.1365-2729.2012.00487.x.

Waldron, J. (2011). Conceptual frameworks, theoretical models and the role of YouTube: Investigating informal music learning and teaching in online music community. *Journal of Music, Technology and Education, 4*(2-3), 189–200. doi: 10.1386/jmte.4.2-3.189_1.

Waldron, J. (2013a). User-generated content, YouTube and participatory culture on the Web: Music learning and teaching in two contrasting online communities. *Music Education Research, 15*(3), 257–274. doi: 10.1080/14613808.2013.772131.

Waldron, J. (2013b). YouTube, fanvids, forums, vlogs and blogs: Informal music learning in a convergent on- and offline music community. *International Journal of Music Education, 31*(1), 91–105. doi: 10.1177/0255761411434861.

Waldron, J. (2016). An Alternative Model of Music Learning and 'Last Night's Fun': Participatory Music Making in/as Participatory Culture in Irish Traditional Music. *Published in Action, Criticism, and Theory for Music Education, 15*(3), 86–112.

Zimmerman, F., & Christakis, D. (2005). Children's television viewing and cognitive outcomes: A longitudinal analysis of national data. *Archives of Pediatrics and Adolescent Medicine, 159*(7), 7. doi: 10.1001/archpedi.159.7.619.

CHAPTER 41

AUGMENTING MUSIC TEACHING AND LEARNING WITH TECHNOLOGY AND DIGITAL MEDIA

EVAN S. TOBIAS

The Provocation Questions in this part of this book imply that technology may help music educators move beyond static notions of musicianship, musicality, creativity, teaching, and learning. Both of the Core Perspectives demonstrate how technology might support educators in being more inclusive of varied types of musicians, musicianship, or creativity. Hein's discussion (chapter 36) of digital production and associated ways of being musical highlights how technology affords educators ways of broadening the musical genres and practices addressed in their programs. Lum provides a compelling argument for educators to consider the potential of technology to redistribute power in ways that open spaces for students to be expressive (chapter 34). Freedman describes how music educators can use technology to address Western music concepts typically addressed in "music theory" contexts in ways that students might find engaging (chapter 35).

All three of these Core Perspectives emphasize the importance of music learners having opportunities to create music and describe how technology might play a role in supporting such engagement. In this Further Perspective, I address the way technology enables people to augment aspects of, blur boundaries between, and connect across musical experiences in ways that might expand conceptions of musicality, musicianship, and who might be considered a musician. I propose three such approaches to leveraging technology: augmenting music and musical engagement, augmenting musical sharing and concert experiences, and augmenting performing and ensembles.

Augmenting Music and Musical Engagement

Given the shrinking size and increasing sophistication of technology with Internet access (consider that a small mobile device has the computing power that was once only available in a large desktop computer), music educators might imagine future possibilities for digital media to augment music with layers of information, meaning, or engagement. The following vignette explores what might be possible now or in the future:

> Diana shifts her focus between the deceptive cadence she just played and other cadences contained throughout the music, all displayed on a small lens over her eye overlaying her view of a beautiful lake. She is standing on a wooden dock, clarinet in hand, engaging with music she enjoys while reveling in the beauty surrounding her parents' house 15 feet from Macinaw Lake. As she considers relationships of the phrases and cadences in the music, she adjusts her personal computing system (PCS) to display the phrase she just performed with other cadences, allowing her to play and listen to what the music would sound like had the composer chosen a different route. She performs several options, switches back to the original version, and records herself playing through the first section. she then says out loud "upload recording and title 'section one second try by lake' to my streaming account number three and share with my networks and the public," with which her PCS complies. She also instructs the PCS to tag the recording with the composer's name and to tag all of the deceptive cadences accordingly.
>
> Several minutes later, Diana watches a duck floating on the lake in front of her while listening to audio of people's reactions to her recordings. The responses were posted by people who subscribe to her streaming account, though others may find her recording via related searches in the future. She listens to a friend's recording of the same piece, noticing a different approach to the phrasing, and instructs her PCS to post "I like how you leaned into the B natural and crescendoed." She plays through the phrase again before having the PCS display options of available harmonic analyses onto the music. She views several of the analyses, noticing that some are inaccurate. "Uploads of student work from a theory class," she wonders to herself. She instructs the computer to copy and embed aspects of different musical analyses as optional layers to her music and hides or displays them in varied combinations, creating a metaanalysis that she finds meaningful.
>
> Diana plans to organize an interdisciplinary recital with her friends to stream in a virtual world. As part of her process she wants to identify varied connections to the music she will perform and work with designers to create multimedia experiences for people to explore in relation to the music. At some point she will research and curate information on the composer and music. When engaging with music in such ways, she uses a "project" tag in conjunction with related

> terms. She plays through her music several more times before switching off her digital music visualization, and begins to improvise with the birds and wind while looking across the lake.

In this vignette Diana plays music outdoors with an acoustic instrument while leveraging wearable or perhaps embedded technology to display and augment music. While such technologies may not be widely available yet, the concept of displaying music digitally and augmenting it with additional information and possibilities, ranging from harmonic analyses to alternative content, is already available, albeit with more basic technology, such as mobile devices. In discussing augmenting in this context, I am referring to the act of adding layers of experiences, content, or information to music. This process is not specific to technology; however, digital media and the Internet expand the ways people might augment music. For instance, while one could augment music by annotating sheet music with a pencil, digital media might enable one to annotate music digitally, connect among annotations by others across the world, organize annotations by related "tags," and link to related information, ranging from analyses to video tutorials. Such processes call for music educators and learners to consider music from holistic and connected perspectives.

Teachers and students might ask questions such as (1) What information might be related and added to this music? (2) How does this music relate to other music? (3) How does this music relate to social, cultural, historical, and musical contexts? (4) What meaning do I (and others) make or what perspectives do I (and others) have in relation to this music? (5) What connections does this music have to my current or future life and interests? (6) How might people engage with this music? Educators' and students' answers to these questions might inform their use of technology and digital media to augment music and related engagement. In this way, students might combine thinking in sound with the rhizomatic nature (Deleuze & Guattari, 1987) of hypertext to extend the ways they engage with music (Miller, 2011). Lum similarly suggests that technology affords rhizomatic engagement with music, particularly in terms of the fluidity and multiplicities of musical experience. These perspectives suggest that envisioning and identifying connections and relationships is an aspect of contemporary musicianship and a way of being musically creative.

Augmenting Musical Sharing and Concert Experiences

Technology might also be used to augment how we share music with others, expanding beyond traditional Western concert experiences. From this perspective, one might be creative in imagining and realizing interesting and meaningful ways for people to engage with music. This might be considered a critical skill for entrepreneurial musicians interested in addressing diverse audiences in varied ways. Some musicians and ensembles have experimented with integrating into the concert environment social media aspects such as live chatting (Tobias, 2008), Tweeting (Lewin & Wise, 2012) and having people vote on aspects of the music, such as which repertoire might be included in a program (Clifford, 2009). To what extent do music educators help students envision and experiment with diverse approaches to sharing music with others? This vignette portrays how one might leverage technology and digital media to augment musical sharing and concert experiences:

> Linda looks out at the people sitting and standing in front of her, noticing some scanning QR codes she provided on the paper concert notes and interacting with related content online ranging from information about Eric Satie to discussions of his music. After quickly glancing at a video screen with images of people attending her recital via the Internet, Linda focuses and begins playing her arrangements of Satie's Gymnopedies. A peer records her performance and streams the live video and audio feed online.
>
> Linda has decided against charging people for access to her recording, instead posting it on YouTube after the recital. To increase her chances of potentially earning advertising revenue, she shared links to the video on Twitter, Facebook, her personal blog, and other social media, using related hashtags such as #Satie and #saxophone. She figured that some people might share the link with their own networks and considered that more people might listen to her performance on YouTube than attended her recital.
>
> On her blog, Linda encouraged people to share versions of Satie's *Gymnopedies* that they enjoyed and included a playlist of her favorite performances, along with links to music that she thought set a similar mood. Since the Gymnopedies are in the public domain, Linda posted her arrangements and audio recording on platforms such as musopen.com and the International Music Score Library Project. Her music educator friend planned to have his students create music incorporating her audio recordings. Linda is interested to see how people might remix and transform her performance of the Gymnopedies. She is also in the process of collaborating with a local dance studio to have young people create their own related choreography, which she plans to share on her blog and project digitally at her next public performance. She is also considering hosting a collaborative performance event in which she would perform and stream Gymnopedies live online, inviting others to improvise with her.

Almost any content, once in digital format, can be shared, networked, and circulated (Jenkins, Purushotma, Weigel, Clinton, & Robison, 2009; Manovich, 2002) in the ways outlined in this vignette. This type of engagement is also captured in Lum's description of a student who might want to become a YouTube sensation. Music educators and learners might leverage technology and digital media to share music and connect with others in ways that expand beyond or augment traditional presentational paradigms in which music is performed live for an audience present in a physical setting. This can involve people engaging with musicians and music before, during, and after a performance. In this way, technology might facilitate the design and realization of music sharing events extended across time and location while offering the potential for people to interact and engage with music in a participatory manner.

While some might see this as a way of connecting with others through music and musical experience, others may conceptualize such engagement from an entrepreneurial stance, making it part of their careers. While this in no way means eliminating traditional concert events or opportunities to engage with music without technology, it may open spaces to imagine playful, engaging, and meaningful ways to expand on concert experiences. Music education programs might address such uses of technology and related issues, ranging from what it means to perform live to the expansion from a paradigm of presenting music to a paradigm of sharing music with others.

Augmenting Performing and Ensembles

> On Wednesday evening at 7:45 p.m., two music programs are well under way in their respective performance locations. Gene Samson and Tessa Gonzales both include traditional acoustic ensembles and the use of technology in their music programs. Their approaches differ, however, in the ways they structure their curriculums and the types of practice they include. After Gene's concert band completes its portion of the concert, the stage curtain closes, and a student appears holding an iPad. She begins sliding her finger on the iPad's surface, and large speakers emit sound on either side of the stage. Another student steps onto the stage, joining in the music. Soon, a small group of students are performing an arrangement of a popular song on iPads, each student appearing on the stage as their parts combine.
>
> Thousands of miles away, Tessa Gonzales looks across her school's stage at students with woodwind, brass, and string instruments, MIDI control surfaces, laptops, and tablets, a rhythm section, a digital DJ setup, and a mixer. Several of the microphones in front of the students playing acoustic instruments lead to a mixer and to a laptop running software that enables live sampling and manipulation. This is the school's first hybrid ensemble, which has emerged out of Tessa's efforts to merge aspects of formerly separate classes and ensembles.
>
> Tessa looks at her setlist and lingers for a moment on each selection. The performance will start with students' original music, drawing on multiple influences, with an emphasis on electronic dance music. The students will then perform a group improvisation, which segues into the finale, which combines aspects of Afrobeat and hip-hop. Tessa scans across the people sitting in seats, wondering

> how many will dance during the finale. She cues a student who then presses a small rubber pad, sending a loud and aggressive kick drum pattern through the sound system, shattering the silence and making several parents flinch. The ensemble joins in, melding acoustic and digital sonorities, while a student samples her peers' performance, adjusting knobs and sliders, and sometimes slamming rubber pads with hyperbolic gestures matching the sound pounding from the speakers. After finishing their original music, the group pauses and begins improvising in ways that explore sensitivity, texture, and contrasts in timbre.

As this vignette highlights, music educators can maintain acoustic paradigms of performance in contexts that are technology-specific or explore expanded approaches to performing in hybrid contexts that support a convergence of acoustic, electric, and digital instruments. The use of MIDI or Open Sound Control (OSC) controllers and music applications or software to create and perform music, a practice sometimes referred to as "controllerism," offers music educators and students opportunities to include sample-based and live production approaches to creating and performing. Hein discusses how such aspects of music and musicianship connected to digital production and popular music contexts afford expanded ways of engaging with and thinking about music. With applications such as Cycling 74's Max, music educators and students can generate instruments and interactive systems to use in their performing. Augmenting performing and ensembles with such technology and digital media calls for a broader and more diverse set of musical practices that take into account idiosyncratic aesthetic and logistical issues.

Hugill's (2012) inclusion of sound diffusion, networked performance, notions of liveness, interactivity, and live coding as aspects of musicianship offers music educators opportunities to consider expanding beyond acoustic or Western classical performance paradigms. Music educators can augment music programs by including technology and digital media, along with related practices and issues, in ways that extend or even transform how students can be musical. This has intriguing implications for the ways that musical learners and doers can be creative by thinking through and engaging with sound. While music education scholarship is beginning to address some of these issues in terms of musicians' practices (Burnard, 2012; Savage, 2005) and in classroom contexts (Finney & Burnard, 2007), potential exists for additional work and research in this area.

Augmenting Music Teaching and Learning

The three approaches to augmenting music teaching and learning through technology and digital media discussed in this chapter have potential to expand how music educators and their students conceptualize and enact musicianship. Leveraging technology and digital media to augment music and musical engagement situates the act of connecting and identifying musical relationships as a form of musicianship and a way of being a musician. This invites related pedagogical and curricular adjustments.

By augmenting musical sharing and concert experiences, music educators and learners might add to or expand on presentational paradigms of music. This means including newer ways of engaging with people through music and understanding the potential trajectories of music as it circulates across space and time. Augmenting ensembles and performing with technology and digital media might provide students with more diverse ways of performing and creating music. Music educators might consider how one can be creative and musical in identifying and leveraging musical connections or relationships, sharing music with others, and creating and performing music with or through technology and digital media.

While augmenting music programs with technology and digital media can maintain much of what is traditional in K–12 settings, music educators might also consider the ways such change has transformative potential. The use of technology and digital media to augment music programs in ways that lead to new musical possibilities for students and teachers may reveal some of technology and digital media's most untapped potential. Augmenting music teaching and learning in such ways calls for music educators to consider if and when it is appropriate to treat acoustic and digitally mediated musical contexts as separate and distinct or as hybrid and convergent (Tobias, 2012, 2013). This may require ongoing exploration and negotiation of what it means to be musical, creative, and musically educated. Developing related theory and praxis necessitates that music educators and learners consider and address technology and digital media thoughtfully, critically, and imaginatively. This may help us open our eyes and ears to technology and digital media's possibilities and to ramifications of what it means be musical.

REFERENCES

Burnard, P. (2012). *Musical creativities in practice*. Oxford: Oxford University Press.

Clifford, S. (2009). Texting at a symphony? Yes, but only to select an encore. Retrieved January 5, 2017, from http://www.nytimes.com/2009/05/16/arts/music/16text.html.

Deleuze, G., & Guattari, F. (1987). *A thousand plateaus: Capitalism and schizophrenia*. Minneapolis: University of Minnesota Press.

Finney, J., & Burnard, P. (Eds.). (2007). *Music education with digital technology*. New York: Continuum.

Hugill, A. (2012). *The digital musician* (2nd ed.) New York: Routledge.

Jenkins, H., Purushotma, R., Weigel, M., Clinton, K., & Robison, A. J. (2009). *Confronting the challenges of participatory culture: Media education for the 21st century*. Cambridge, MA: MIT Press.

Lewin, N., & Wise, B. (2012, March 28). Attraction or annoyance? Orchestras invite audiences to use their smartphones. WQXR. Retrieved January 5, 2017, from http://www.wqxr.org/-!/articles/conducting-business/2012/mar/28/attraction-annoyance-orchestras-invite-audiences-use-smartphones/.

Manovich, L. (2002). *The language of new media*. Cambridge, MA: MIT Press.

Miller, V. (2011). *Understanding digital culture*. Thousand Oaks, CA: Sage.

Savage, J. (2005). Information communication technologies as a tool for re-imagining music education in the 21st century. *International Journal of Education and the Arts, 6*(2).

Tobias, E. S. (2008). *Liveblogging a concert experience*. Retrieved January 5, 2017, from http://evantobias.net/2008/09/19/liveblogging-a-concert-experience-3/.

Tobias, E. S. (2012). Hybrid spaces and hyphenated musicians: Secondary students' musical engagement in a songwriting and technology course. *Music Education Research, 14*(3), 329–346.

Tobias, E. S. (2013). Toward convergence: Adapting music education to contemporary society and participatory culture. *Music Educators Journal, 99*(4), 29–36.

CHAPTER 42

POSSIBILITIES FOR INCLUSION WITH MUSIC TECHNOLOGIES

DEBORAH VANDERLINDE

INCLUSION for learners of all abilities remains a challenge for music educators. Elementary and secondary general music teachers engage with large class loads with full inclusion expectations for musical engagement and social interaction. The challenges and strengths that promote or impede learning are unique to each learner; this requires teachers to reconsider barriers to learning or what might be added to the classroom environment and curriculum for every child to have opportunities to create musical meaning.

Teachers can enable success for all learners when applying principles of universal design for learning.[1] These principles are key to fostering success and for supporting teachers as they consider ways to accommodate and modify their music curriculum. Underlying the goals for universal design is the goal of seamless access—to the classroom environment, to the music and music curriculum, and to peer interaction. These avenues of access are challenging for all learners, but especially problematic for learners with exceptionalities.

Universal design for learning is a commitment to differentiation on every level, presenting and engaging with content in varied and flexible ways. That is, teachers might provide multiple ways for students to engage with content and to express what they know with multiple avenues for applying and assessing understanding (aural, oral, singing, verbally, visually, enactively, and kinesthetically, with iconic representation, with manipulatives and assistive technologies). These principles are best practices for all teaching, but essential for learners with exceptionalities. Without them, some learners will remain on the periphery of musical and social engagement, fostering a sense of exclusion and lack of opportunity for musical expression. Issues of joint attention pervade the challenges among learners with special needs, particularly children on the autism spectrum. This refers to the limited ability to jointly attend to a shared experience or object; eye contact and empathy for the experience of the other is difficult, compounded by a lack of interest in the other's actions.

With these concerns in mind, mobile technologies have much to offer as a form of assistive technology for learners with exceptionalities. iPads and other touch screen devices are appealing to learners—not necessarily because of novelty or "cool" factor; children with special needs often seem indifferent about this. However, the ease of access to the device, the intuitive nature of interaction, and the colorful and visual applications are very attractive. The touch screen allows actions that are smooth, and, for children with physical impairment, the lack of a mouse, keypad, or multiple functions to "get going" is critical. While some apps like GarageBand take some practice and require multiple steps, there are many apps with fewer functions that look and sound sophisticated musically and offer an avenue for immediate satisfaction as well as increased proficiency. Here lies another key to the attractive nature of the iPad—the notion of immediate success. Learners with exceptionalities rarely experience success quickly; for some activities, success is never attained. It can be debilitating when learners sense their own "difference" and they may easily lose heart when this becomes their daily school experience. To be able to produce and express musical ideas in a way that is socially normative and personally meaningful is empowering. App selection requires the same differentiation needed for other musical activities; with an understanding of the child's strengths and challenges, including the child's IEP and curricular goals, teachers can—usually with trial and error—find apps that meet musical and social goals.

Social inclusion will not happen immediately but teachers may find that using iPads with appropriate apps will eventually draw children into the whole classroom experience. For lessons where students are learning to sing and play songs together, a wide range of classroom instrument apps offer learners a similar experience—playing a song on an instrument alone (if needed) or jointly with a peer (if possible). Anecdotally, I have observed children with exceptionalities work alone, only to move closer to or alongside a peer, and finally interacting with a peer working on the same app or with a new, intriguing app. Gradual and self-initiated joint attention fosters greater agency and longer lasting success for learners who struggle with social interaction. I have also observed learners who require initial assistance gently take over the iPad but willingly allow the teacher or helper to sit alongside to observe or engage when invited. These seemingly minor occurrences are often big steps for learners who may otherwise refuse to engage with others or who may act out simply because of being uncomfortable with the close proximity of another. The attractiveness and shared joy in iPad engagement seems to offset these challenges.

Some Suggestions

Here, I offer suggestions for apps that have been successful in my own teaching and research (Blair, 2012, 2013). This is a very small sample intended only to stimulate your

FIGURE 42.1 Singing Fingers.

own thinking regarding the learners in your classrooms and the ways you might creatively design lessons to provide more inclusive experiences.[2] Students may begin with these apps individually (possibly with headphones), but all of them are conducive to more collaborative music making with a partner or small group: playing an improvisation or composition together, engaging in call and response, creating musical layers with grid-style apps, creating musical layers with touch and draw iconic representation apps, creating soundscapes, and creating layers in an "add on" fashion using apps like Soundrop or Bebot. Many of these ideas provide multiple entry points for a wide range of musical ideas and musical proficiencies while students are collaboratively and successfully engaging in an authentic musical process and product.

Singing Fingers (fig. 42.1): learners with autism may have limited verbal capacity. This app has been useful for some students as it requires creating vocal (or other) sounds that are recorded when a finger touches and moves across the screen.[3] Once recorded, learners can replay the sounds by touching the screen. The silly factor of this app may be the initial attraction, but students will quickly discover other ways to create more complex musical ideas.

Seuss Band offers songs that can be played with increased difficulty (fig. 42.2).[4] Using icons that fall across the screen, children tap the instrument keys to play along with a background track. Initial goals for our children with autism were to focus on the task and to develop tracking skills. We soon found that as they sustained attention (big accomplishment), they would anticipate musical ideas and later improvise musical ideas within the song. Joint attention was fostered as students played this app like a game with a teacher or paraprofessional (only when it was their choice to do so). "Watch me, look what I can do!" was an accomplishment for many, and we celebrated this small step toward mutual engagement.

Soundbrush and Articulator offer touch and draw soundscapes (fig. 42.3).[5]

Wavebot and Melodica are learner-friendly apps that use a grid format for developing musical layers and loops (fig. 42.4).[6]

Soundrop and Bebot use novel sound sources and are intriguing apps for engaging in individual or collaborative improvisational music making with a peer (fig. 42.5).[7]

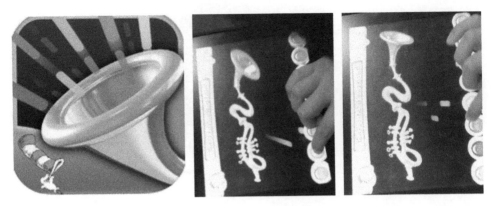

FIGURE 42.2 Seuss Band.

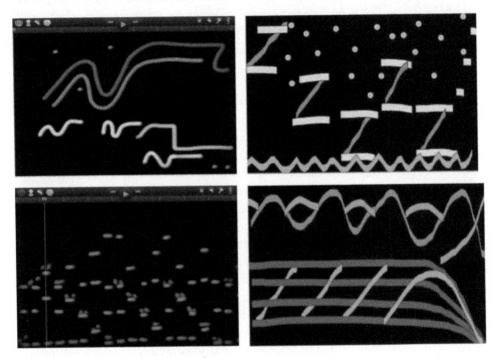

FIGURE 42.3 Soundbrush and Articulator.

MadPad is well suited to foster joint attention (fig. 42.6).[8] The inherent function of the app requires recording video sound bites and creating a "pad" of sounds on which one can compose or improvise music. Initially, we found that students would engage others to assist in the recording of music or playing music while they served as videographer. Students can play the music alone but often have musical ideas that require more "hands," thus furthering engagement with others. Gains toward successful peer interaction can be fostered with skillful scaffolding and gentle support. The app has layers

 **Wavebot**  **Melodica**

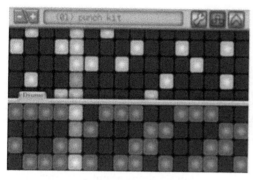

FIGURE 42.4 Wavebot and Melodica.

Soundrop   **Bebot**

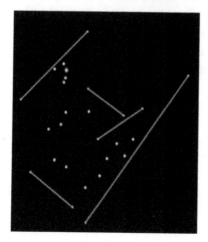 

FIGURE 42.5 Soundrop and Bebot.

FIGURE 42.6 Madpad and Songs for You and Me.

of sophistication that are easily accessible; thus, sustained engagement with an activity or musical composition is fostered for students who self-select to pursue their musical ideas. Sustained engagement and self-regulation are key Individualized Education Plan (IEP) goals for many learners with exceptionalities; this app is particularly suited for these goals when students find these musical processes interesting.

Songs for You and Me is an iBook series that includes illustrated songs with iconic representation (fig. 42.6).[9] The songs are presented in multiple ways—phrase by phrase, in video format, and with videos of the songs played on iPad apps or classroom instruments. These multiple modes of representation provide layers of instructional material. An underlying goal, however, is to offer learners time with the songs outside class (in their own classrooms or on home devices) in order to gain competence and confidence with the music, thus fostering better social inclusion and musical engagement once in the classroom. These public domain songs are aimed at younger children (grades K–5) in situations where music class time is limited and the use of the iBooks outside class to gain proficiency would be especially helpful.

The iPad offers unique possibilities for learners with exceptionalities, not only as an assistive technology but also as an authentic musical instrument. The intuitive nature of the touchscreen and ease of flow of functions creates a setting where musical processes become highly accessible for all learners but is of particular value for learners who struggle with decoding notation, who have cognitive or processing delays, who struggle physically with playing acoustic instruments, or for whom collaborative music making is a challenge. The authentic ways that musicians formally and informally use mobile technologies generates genuine musical contexts for children as they, too, engage in digital music making.

NOTES

1. See www.udlcenter.org and www.cast.org.
2. A more comprehensive list of apps can be found on the "Children with Exceptionalities" website: "Awesome Apps" link under "Resources for All" tab; https://sites.google.com/site/exceptionalitiessrig/home/resources/related-website.
3. See http://singingfingers.com/.
4. See https://itunes.apple.com/us/app/dr.-seuss-band/id474940131?mt=8.
5. See http://www.getsoundbrush.com/ and http://www.artikulatorapp.com/.
6. See https://www.youtube.com/watch?v=8wUzqAkde5E(Wavebot) and https://itunes.apple.com/us/app/melodica/id313573137?mt=8(Melodica).
7. See http://www.normalware.com/(Bebot) and https://itunes.apple.com/us/app/soundrop/id364871590?mt=8 (Soundrop).
8. See http://www.smule.com/apps. This YouTube video was helpful for our students and could be downloaded for their own use: https://www.youtube.com/watch?v=ZlubpvztxWY.
9. See https://itunes.apple.com/us/book/songs-for-you-and-me/id821772963?mt=11. Teacher resources: http://musicalmaps.weebly.com/songs-for-you-and-me.html.

REFERENCES

Blair, Deborah VanderLinde. (2013, October 30). *MadPad: Creative, collaborative and critical thinking in a digital musical context.* Digital showcase session presented at the annual meeting of the Association for Technology in Music Education, Cambridge, Massachusetts.

Blair, Deborah VanderLinde. (2012, June 9). *iPads in music education.* Presentation at eCornucopia 2012: Creativity through Technology Conference, Oakland University.

3 B

Core Perspectives

CHAPTER 43

GETTING IN THE WAY?

Limitations of Technology in Community Music

GILLIAN HOWELL

ARE there some conditions in which music technology is inappropriate or ineffective? This is the question I address in this chapter, drawing upon my experiences as a community musician who leads groups of people—all ages and abilities, experienced players and first-timers—to create and perform their own music. Based in Australia, I work in diverse environments around the world—communities, schools, hospitals, prisons, and conflict-affected settings—leading practical and collaborative projects that by necessity must respond in some way to the realities of the local environment.

These "local environment realities" lead me to question the suitability or value of music making based around contemporary music technologies for some community music projects. While the extraordinary capabilities of computers, music creation and recording software, synthesizers, and computer-generated sound-makers, tablets, smartphones and their apps, and platforms that facilitate remote linking and file-sharing across great distance each can offer incredibly powerful solutions and learning opportunities to some community music projects, there are a number of limitations that are worth examining in more detail.

In my experience, technology can indeed offer solutions and create new possibilities, but it can just as easily create problems, and thus be as much a burden as a resource. Many technologies, particularly computer-based technologies, have elements of unpredictability that can disrupt or derail workshop flow and focus. Programs freeze, updates suddenly need to be installed (which may not be a quick process), operating systems crash, or shut down, or need to be rebooted. Work may be lost, and group music making will definitely be paused while solutions are found. Individuals may require more personalized attention to assist them with their project. It may be that the group leader or person in command of the technology does not have sufficient knowledge to troubleshoot or solve the problem on the spot, requiring access to specialized technicians, or a plan B for that particular class or workshop that can be quickly executed. These interruptions halt the momentum and flow of the music-making experience and endanger the sustained engagement of the group.

Technology also begets technology. Amplified instruments need leads, extension cords, power boards, and amps. Singers need microphones to boost their sound in order to be heard above louder instruments. Individuals surreptitiously increase their output so that the overall volume produced by the group gets progressively louder. Sooner or later a mixing desk (and someone to operate it) becomes necessary. Soundproofed rooms become essential in order to lessen disturbance to others in neighboring spaces. And so on—all of which makes the group music-making endeavor less portable, less spontaneous, and more prone to interruptions to the flow of simply playing music together.

In this chapter, I argue that choices about appropriate music learning technologies are context-dependent. This includes the intended goals or rationales for the music activity—community music projects often have stated aims focused on well-being and positive social experiences, especially in those projects that target vulnerable or marginalized groups. This chapter considers questions of environment, resources, sustainability, and participant-centered practice. It examines the value of technology in relation to what the physical and human environment can sustain, the immense value of direct and seamless pathways into active music making, and the importance of the feeling of playing music. Finally, I highlight the importance of the facilitator's interests, experiences, and skills in deciding how or what technologies to utilize in music education and music participation projects.

Sustainable Choices That the Environment Can Support

One of the primary limitations of the use of music technologies in community music activities is that which is imposed by the environment and workshop context. A leader may be keen to utilize particular technologies, but the environment's capacity to support this choice is a critical factor. Let us consider, for example, how integral electricity is to many contemporary technologies and music-making platforms. In some parts of the world, a safe and reliable source of electricity cannot be guaranteed or even assumed. Alternatively, electricity and the technologies themselves may be present—access to new technologies is indeed spreading through many parts of the developing world—but the skilled technicians needed to support any kind of ongoing or sustainable engagement may not yet exist, to say nothing of affordable Internet access.

I illustrate this point with a vignette from a community music residency I undertook in a remote community in Timor-Leste (East Timor). One of the world's newest and poorest countries, Timor-Leste spent 500 years under the neglectful colonial yoke of Portugal. Neighboring Indonesia invaded in 1975, and for 24 years subjected the East

Timorese people to brutal military occupation. After a UN-sponsored independence referendum in 1999, the Indonesians withdrew, leaving a vengeful trail of terrifying violence and mass destruction in their wake. Timor-Leste has been the recipient of more than a decade of international nation-building and development interventions. It has seen gradual but steady progress in many areas of human development since 1999, but access to reliable electricity, clean water, healthcare, nutrition, and sustainable livelihoods remains a distant goal for many rural communities (United Nations Development Programme, 2013).

My community music residency took place in a remote rural town where there were several unavoidable technological limitations on resources. Electricity supply was limited. It was only available during the evenings, with frequent blackouts and some days when it never came on at all. Charging up the various devices I used each day was a nightly ritual. Internet access was extremely expensive—well beyond the means of the average local person, with the cost of an hour's access the equivalent of a day's income for many—and available via a mobile USB modem, or at a single computer in one of the local shops. Few people owned their own computers—at least, one that was fully functioning, or one that they would happily let multiple people use.

My project goals were to engage young people and local musicians in some kind of collaborative music project. Rather than being focused on imparting specialized knowledge and skills, this was a participant-centered and environment-sensitive community music project, responsive to the needs, interests, and capacities of the community and building on the musical skills and knowledge already present in the group. It was an example of what Higgins (2012) calls an "active intervention" approach to community music (p. 4), where a music leader engages a group of people in hands-on music making through facilitation and collaborative methods. In this community music context, the extreme environment dictated all of the technological choices I could make.

The first time children came to my house in Lospalos to play music I had very few instruments to share with them. I had three sets of alto chime bars (pitches C to C, no chromatic notes), a pair of bongo drums, an empty "water gallon"—one of those large filtered water cooler containers ubiquitous in Western offices—and some bamboo sticks. I imagined that a few children would play and the rest of us would sing. About eight children came along to the first jam, so there was never long to wait before getting to play an instrument. But once other neighborhood children heard the sounds coming from our veranda their curiosity and excitement was piqued, and we soon had 30–40 children coming every day. They would rush to grab the instruments that they wanted to try (many of these children had not been to school and so were not good at the social rules of turn-taking, or taking care of equipment) and sit so close to the people holding the instruments that they were barely able to play them. We needed more instruments.

We started with bamboo. A neighbor agreed to sell me some from her property. I borrowed a machete; my partner hacked the tall stalks down, and we carried it in armfuls back to my house with the children's help. It was green and slender, and splintered easily. We cut it into short lengths and experimented with blowing the different lengths to find different pitches (like blowing across the top of a bottle) and tapping pairs together to make claves. At the local market I spotted a plastic home wares stall, where I bought buckets of all shapes and sizes. We turned these upside down to make drums, and they proved a good accompaniment to the sounds of the empty water gallon.

We made agogo bells from empty plastic water bottles, pumped full of air by inserting a tire valve in the lid. In a country like Timor-Leste, where you cannot drink the tap water and there is no rubbish disposal system or recycling service, empty 1.25 liter bottles pile up quickly. When pumped full of air with a bicycle pump, the side of the bottle gave a bell-like "ping" sound when tapped. We then fixed two well-matched contrasting pitches together with gaffer tape, and *voila*: an agogo bell!

At a workshop at the local convent and orphanage, I noticed a pile of thick bamboo, much older and drier, in a pile in the garden. "Do you need all of that old bamboo?" I asked. They were happy for me to take some, and these much thicker, wider-bored pieces of bamboo had far greater potential for instrument making. We discovered that one of our neighbors had traditional instrument-making skills. He showed us how to make a bamboo log drum that he called a *kakalo*, using a machete as the primary tool. We held a Saturday working bee, and the older children and teenagers helped make a set of 12 kakalos.

All technologies give us the potential to make some kind of change in the material environment. If we understand "technology" as being something that extends the capacity of the human body or mind to make, create, and manipulate something, then musical instruments are also technologies, for they enable exploration, expression, and amplification of musical ideas, even in the hands of first-time players.

In this community, bamboo, bottles, buckets, and a machete were the music technologies that were a sustainable, suitable match for a humid and remote environment with little access to many of the resources taken for granted in more developed parts of the world. These "old" technologies were available locally. If things broke, they could be replaced easily. They made satisfying sounds. The children's attraction to the tuned instruments I had brought with me from Australia (a donation that was later given to the local school) was considerable, but they loved the kakalos. They were excited to share something unique from their culture with us and were delighted in turn by our interest in making them and in taking them with us in workshops throughout the community (Howell & Dunphy, 2012). Our enthusiasm for their traditions generated renewed pride and confidence in their cultural knowledge.

Could new technologies have been bought and donated to the community for them to use in this project, and then to keep forever as community assets? Sadly, the issue is not purely one of limited financial means. Valuable items require ongoing care and housing, and this must be sustained locally without the need for external support or provision. I learned that storage space is in short supply in poor communities. Often, suitable storage space that is dry and secure does not exist; even when it does, problems remain. Who will assume (or accept) responsibility for the valuable items? Such a role can be unattractive in a country grappling with the legacies of long-term authoritarian rule, where being the person in charge once meant being the scapegoat if things went wrong. Another possible scenario is one of nepotism and potential abuse of power, where only those people closest to the "keyholder" are given opportunities to use the donated equipment.

There are also ethical points to consider. What does it mean when you bring new items into a community where they cannot be sustained? Could this create a greater sense of loss or deficit than was present before said items were introduced? Choosing technologies that the environment can support is crucial if the benefits derived through the course of the project are to be sustained in the longer term, beyond the span of the artist's visit.

In addition, extreme environments can make extreme demands on equipment. Humidity, bad roads, and a collective social culture in which all assets should be shared can all conspire to make less robust technologies like computers have very short lives. When things break down, technicians are unlikely to be available to perform repairs.

Finally, I have described this approach to community music making as one that is participant-centered, as well as environment-sensitive. I therefore needed to consider the priorities and goals of the participants, who were mostly young people aged 5–20. The children in this town had very few opportunities for engagement in hands-on music making and adult-led creative music making (Howell & Dunphy, 2012). Creating sustainable experiences also meant establishing foundational skills and fostering creative and curious minds that would serve the children long after I had gone. In an environment where people survive on the most basic of resources and enjoy little access to the outside world, perhaps the most important thing to develop and nurture was this curiosity, through positive experiences of oneself as a learner and discoverer.

Removing Obstacles and Prioritizing Immediacy

The following vignette illustrates a workshop process with a group of children who were newly arrived in Australia and spoke a language other than English. The needs of this group were considerable. Many of the children in this group were refugees and came to school in Australia with very few prior schooling experiences to guide them. They experienced many conflicting feelings of insecurity and fear about learning, extreme optimism and hope for their new lives, and uncertainty about the behavior expected of them. (The cultural rules of Australian schooling are a world away from the hypervigilant survival skills they learned in refugee camps.) This was a confusing mix, made potentially volatile when combined with the traumatic experiences many of them had survived and the transitional nature of the school community, with new students constantly arriving and others moving on to mainstream schools (Howell, 2011)

The primary goal of the music program was to make and play music, but there were additional goals related to collaborative learning, social cohesion and trust, increased individual confidence to explore and learn new things, and the continued acquisition of expressive and receptive language. This workshop, a weekly occurrence as part of a long-term music residency in the school, aimed to support these children in making the necessary adjustments to the culture and expectations of school life in Australia.

> The children enter the room. Some bound in, full of excitement and eagerness. They already know the workshop leader, having been students in the school for a term or more. Others are shyer, more cautious—they are the newest arrivals, and everything in this school is unfamiliar to them. We form a circle, children and adults together. We play some introductory games that encourage each person to say his or her name, or make a sound or gesture, in turn. There is space for every voice.
>
> We sing a song together. The children who are old hands already know it, but the leader reminds them of the words, saying each line slowly and in rhythm and asking them to say it back in rhythm. The newer children watch the mouths of the leader and children moving. Some will make some of the sounds, others will just be absorbing all the visual and aural information, making sense of it for themselves but remaining quiet at this time.
>
> We add body percussion. The leader starts a syncopated rhythm, and the children join in when they are ready. The newer children take part in the body percussion even if they aren't yet singing. One of the body percussion moves makes the children laugh. Each time it comes around in the song they laugh harder. By the end of the song some of them are falling about on the floor, they are laughing so hard. We all collapse down in a tumble, laughing and grinning. It's playful, a bit silly, a bit of fun. It's also a musical warmup that establishes ensemble skills and risk-taking, and lets the children know that it is okay to take things lightly. This is not a place of severity or punishment but of trying, and exploring.
>
> We move to instruments. The children choose what they want to play. They love playing instruments more than anything else in music classes. We reestablish the rhythm from the earlier song, but now on instruments. The leader suggests notes to play on the tuned instruments, using the alphabet letter names that the children learn in class, and removing the bars from the instrument that aren't needed.
>
> There is very little verbal instruction given in this class. We don't need it. The newer children can see what is going on and join in by copying the other children. Grooves and rhythms are learned aurally. Instrument techniques are learned when required in the context of the music. Later, one of the newest children will tap absent-mindedly on a drum, revealing a rhythm that she learned months or years before in another country, reflecting the music of her cultural group. The leader will imitate it and encourage it forth, later incorporating it into one of our class compositions. Later, too, we will pass a drum around the circle, inviting each child to play a solo. This reveals all sorts of extraordinary talents and knowledge. As confidence in English grows, we will write songs

together that describe, with matter-of-factness, sometimes sadness, and often with humor, their shared experiences of transition, and the hopes they hold for their new lives in this country.

This vignette describes a musical environment in which any potential obstacles to group music making have been avoided or removed. The music making begins the moment the children enter the room. There are no special instructions to listen to, nor equipment to choose (or fight over) or move to. Everyone is working together, and the workshop flows seamlessly from the playful warmup games to establishment of foundational skills in rhythm, introduction of specific musical materials, and to instrumental music and jamming. The atmosphere is positive and encouraging, and there is no "dead" time. Everyone is focused and actively involved.

A workshop like this is as much about promoting individual and collective well-being as it is about making great music and developing new musical skills. Like many "intervention-based" community music projects, the project's goals include what Cunha and Lorenzino (2012) describe as the "secondary" outcomes or byproducts of collective music making (p. 73). These relate to the social, cultural, affective, and cognitive aspects of human experience that occur in parallel with, or as a result of, group music making.

Well-being is multifaceted. It comprises many dimensions of how we feel within ourselves, about ourselves, and about who we are in relation to a wider social group. The reasons people choose to take part in community music activities relate to these feelings of well-being, and include the desire to develop new musical knowledge and experiences, to build social networks and feel bonded with others, and to be part of something that contributes positively to the wider community (Gridley, Astbury, Sharples, & Aguirre, 2011).

Consideration of what in the music-making experience generates these feelings of well-being is therefore important. Ruud (2012, citing his earlier work from 1997) has identified four dimensions of well-being that collective music making can positively impact in relation to overall quality of life: *vitality* (feelings of happiness, relaxation, being "present in the moment" and free of other worries); *social bonding and belonging* (feeling connected to others); *agency* (having voice and visibility, confidence in one's powers, and being a contributor); and *meaning* (where the music making has value and importance beyond itself). These dimensions are in evidence in the above vignette. We see a group of children arrive and immediately fall into music-making experiences. Their energy is focused through the seamless flow of activities, and the sense of exploration and inclusion supports them to relax, have fun together, take calculated risks, and experiment with new ideas and skills. There is laughter in the process, and acceptance of the different contributions and levels of participation each child offers. These are characteristics that point to the presence of sustained vitality in the group.

The workshop focus and flow also promotes feelings of belonging and social connection. The groups' members remain in synchrony with one other throughout the workshop, enjoying a mutual focus of attention, and entraining their emotional energy just as the rhythms of the music entrain and fall into place (Stige, 2012). The participants are "in tune" and "in sync" with each other, an experience that is deeply satisfying, and can serve to increase the group's levels of emotional energy and make them feel supported, more courageous, and a part of a community. Stige (2012) suggests that these feelings of connection, emotional support, and community are akin to those experienced in other "interaction rituals," such as religious ceremonies.

Increased agency develops as the group gains confidence in its abilities to play and create music that sounds satisfying and connects them to the musical worlds they inhabit outside the school environment. Over time, they gain confidence in their capacity to learn, to contribute, and to present their work in visible and audible ways. This is a common experience for many music learners but particularly significant when applied to groups that might otherwise be marginalized or socially invisible.

From the detail included in the vignette we can speculate about the meanings that these musical experiences may hold for the children. The inclusion of musical material from their cultures and traditions in the group composing activity may serve to build confidence in their identity at a time when they feel under pressure to assimilate with a foreign culture. The songwriting process may support them to find meaningful ways to describe and share their experiences with others. For these children, coming from backgrounds of war and division, the experience of working collaboratively with people from so many different cultural backgrounds and religions may have additional significance, demonstrating that such crosscultural collaboration is possible, even desirable, and enjoyable. In this particular vignette, the workshop leader chooses to go down the non-technological pathway in order to avoid any interruptions to the workshop's flow and focus. This takes the group directly into the music-making experience, and the attendant benefits to well-being that this immediacy and clarity can engender.

This is not to claim that dimensions of well-being are not present in community music projects that utilize music technology. The health-promoting experience of increased agency is evident in many music activities built around sampling, sequencing, and synthesizing softwares. These technologies enable individuals to be active producers of their own art, using accessible tools and platforms that enable users to participate in real-world, authentic cultural creation and exchange.

However, these technologies also draw the user's focus toward the hardware, the software, and in particular the screen, rather than to others in the room. The opportunity to be in synchrony with others in nonverbal ways through the real-time music-making experience is greatly reduced, if present at all. In many music softwares, the user interface is designed primarily for single, independent users, and while collaborative platforms do exist, they are not the primary purpose of many of the most widely used sequencing, sampling, recording, and composing tools. There are of course exceptions to this generalization. Hullick's (2013) sonic art ensemble Amplified Elephants have cultivated a strong group identity and sense of community through their work, honed through their emphasis on group-devised work and public performance. Nor do I wish to suggest that technology itself is the problem. There are many technologies in use in these sessions, but they are in the background, and secondary to the activity itself. The navigation of new technologies is not the focus, and it therefore cannot disrupt or derail the music-making experience.

Finding the most direct, seamless pathway into practical music making experiences is essential in community music groups. The momentum it creates contains and holds the group in a friendly and safe space, and helps the group members to relax and place their trust in the leader's skills. In other words, by removing obstacles and potential or unpredictable interruptions, the facilitator can put in motion the activities that contribute important experiences of well-being for participants. Interruptions that halt proceedings and pull the facilitator's attention away from the group will disrupt or destabilize the feelings of safety and community that are being cultivated by the leader.

The Feeling of Playing

For most community music groups, making music is the primary goal. The participants may be motivated by the opportunity to learn more about or participate in a style of music that holds particular meaning or interest for them, but even more, they may be attracted by the opportunity to play. Music making is not just about the sounds that ensue, although these are a core part of the great satisfaction many people derive from music participation. It is also about the feeling of playing music.

I question the extent to which music making that is strongly mediated through music technology can satisfy the desire for the feeling of playing music. Recently I worked as a visiting artist in a primary school with a group of 15 children aged 10. Despite the project goal being built around the use of music technology, a participant-centered approach revealed what they really wanted to do. This vignette describes the evolution of a project that began with a music technology focus but evolved into something quite different:

> It's my fourth week in this school. I've been invited to be artist-in-residence at a well-to-do inner-city primary school in Melbourne, Australia. It is a collaboration with the Mandarin language teacher, with the aim of using iPad technologies to make quirky, child-designed language-learning aids. I'll guide the children in recording sounds and language samples within the school environment, and we'll use various sampling programs to create the language-focused musical soundscapes, embedding Mandarin vocabulary and phrases into original melodies, harmonies, and beats to create "ear worms" that will help users to memorize the words. We have all the technology that we need, and I am excited to get started.
>
> By week 4 we have lost some momentum. My Mandarin-language collaborator has fallen sick, the iPads aren't yet ready (something to do with them being new, needing to be "registered" on the school system, and loaded with the appropriate software, which I, as a visitor, cannot instigate), and the alternative machines (netbooks) are

slow and clunky, with oversensitive trackpads and tiny screens. No one is having fun with the netbooks.

Meanwhile, we have access to a great music room, with a drum kit, a keyboard, several guitars, and loads of percussion. So while we wait for the Mandarin teacher to get better and the iPads to be available, the children and I have begun to play. We've made a rock band, learning to count and play together, and inventing riffs and grooves. The children correct and support each other, showing initiative, engagement, and ownership. They are filled with enthusiasm. Every week when I arrive in the school the children race up to me, demanding, "Are we doing the rock band today?"

It's taken me by surprise that these children are far more engaged by the immediacy and physical challenges of instrumental music making than by the more abstract experience of working with sound recording and sequencers. Sure, that's interesting, they seem to be saying, but give us a chance to play these real instruments over holding that box, pressing buttons, and listening through headphones any day!

This vignette raises two points in particular for discussion. The first is that, given the choice, the child participants in this project were far more interested in playing drums, guitars, keyboards, and glockenspiels than in using technology to make music and soundscapes. There could be several reasons for this preference. One is that the instruments were familiar, housed as they were in the music room, albeit stored under dust sheets and rarely used in the classroom music program. Clearly, the children had an interest in these instruments that was not yet close to being sated. They hadn't had enough turns playing instruments!

Music making with instruments and voices is very much a multisensory, embodied experience, and I believe this was a strong part of the attraction of using the instruments for these children. There is a compelling and absorbing quality of immediacy and responsiveness; as the player makes the gesture, the sound *sounds*. The air molecules inside the instrument and around the room move in turn, and the body responds in sympathy. The energy generated is palpable even to those in the room who are not involved in the music making. This contributes to what participants may identify as the exhilarating and heady "buzz," the "vibe," or the "energy" of the music-making experience. This multisensory, multidirectional, intangible feedback loop is one of the hard-to-pin-down but thrilling ingredients in collective music making with instruments and voices.

The more a group music-making experience is mediated by music technology, the more the multisensory feedback is lessened. Music making with new technologies is more abstracted from the physical world and embodied experiences, with much of the sound happening through headphones, or amplifiers and speakers, and generated through the computer rather than through physical human sound-making gestures. While the concentration or on-task focus of participants may be similarly intense, the "buzz," the "vibe," or the "energy" that emanates from embodied human musicking is simply not as discernable when the sounds are being generated electronically.

Furthermore, the pleasurable experience of being in synchrony with others, discussed in the previous section, is lessened. Such synchrony is audible, but it is reinforced and intensified through the eye contact, physical proximity, and visual and aural aspects of sound generation, and the entrainment of pulse, breath, and energy that is part of a live, in-person, vocal or instrumental collective music-making experience.

Music technologies still continued to play a minor role in the project above. What is a rock band, after all, without an electric guitar or an amplified keyboard? We used online music notation software (www.noteflight.com) to score and rehearse the compositions they created, and we documented the process with video diaries. However, playing music that they had created together on instruments that were both new (as far as skills and experience were concerned) and familiar (in that they had real-world credibility) was the clear preference of this group.

Technology, the Music Facilitator, and Keeping It Real

A further point for discussion from the second vignette recalls my observation at the start of this chapter that the use of technologies can increase the scope for things to go wrong. However, rather than highlighting again the unpredictable characteristics of certain technologies, I now focus on the skills, experiences, and interests of the music facilitator, and the support offered by the environment, as the final limitations to consider with regard to the place of music technology in community music projects.

The vignette above describes a well-resourced, wealthy, inner-city school that prided itself on the opportunities provided to its students. All the equipment that was required for this project was (in theory, at least) there, listed in inventories and willingly made available to the project. However, things still went wrong that halted the momentum of the workshops and the overall project. Use of the technology required further supports (technical support expertise, access to networks, the sustained interest from key staff members) that could not be relied upon.

Furthermore, the project moved me into new creative territories. I was familiar with the technologies involved, having used them to make many works in the past, but despite having the requisite knowledge to successfully lead the project described in this vignette, I was nevertheless halted when a lack of support in the school environment arose. Interestingly, the change in direction was a return to music-making ground that, for me as facilitator, had far fewer unpredictable elements, and required far less dependence on supports outside the project. The project regained its momentum, which indicates the difference that familiar ground, with tools and external supports that are predictable for the facilitator, can make to a community music outcome.

When workshops flow, when participants feel safe, supported, encouraged, and inspired, and when exciting and engaging musical outcomes emerge from the process, there is a skilled facilitator at work. Skilled facilitators and the knowledge they possess are critical factors in community music outcomes, musical or otherwise. If the facilitator is working outside his or her sphere of experience, knowledge, and local influence or authority, this will have an impact on the workshop. If external supports are not available, the potential for unpredictable elements to derail the workshop process increase. Hullick (2013), describing his role as mentor to Amplified Elephants, emphasizes the importance of artists mentoring others in "the art that they know" (p. 223) rather than following particular trends, formulas, or manuals. Of course, new directions can and should be trialed. But if this is to happen, extended time frames and external technical supports that allow the new approach to evolve become critical factors, for both the artist-facilitator and the community music group.

What draws us as community music facilitators and leaders toward particular tools and approaches? For some, there may be pressure (direct or perceived or self-imposed) to keep up to date with the latest new resources and to use music technology with our project participants. However, community musicians and music educators should not be pressured to engage with music technology just because it is there, especially if they are sufficiently skilled in practical music making to draw participants into satisfying and inspiring music experiences using resources like instruments, voices, and bodies. Musicians in all contexts should not underestimate the power of what we have at the grassroots of musical practice and engagement.

The pressures to get on board are not imagined: in community music, the music leader is often required to be everything to everyone, from engaging challenging groups with different needs to practical facilitation of a wide range of musical genres. However, as leaders we have to ensure we know our materials or we risk losing the trust and engagement of the group, which, once lost, can be very difficult to win back. Community music participants sign up to make music. There may be many other benefits related to health and well-being that ensue, but it is the music that attracts them in the first place. Therefore, it has to happen in ways that make them feel satisfied, or intrigued, or excited, or inspired, or safe, for them keep coming back.

Finally, there is a risk that, in our eagerness to embrace the solutions that music technology offers, we lose sight of what the goal is, which is frequently centered on live, socially rich, and responsive musical interactions. For all that watching a concert live online is an amazing and enjoyable thing to do, it remains second best to actually being there. Remote linking community music projects do a wonderful job of enabling musical interactions to take place between people who are geographically distant, and offer a platform for meaningful contact with skilled musicians and other enthusiasts. But as one participant in an online old time and bluegrass community explained, online engagement did not replace the "relationships and bonds that form as a result of 'live' jam sessions and lessons" (Waldron, 2012, p. 99). The remote linkup is beneficial, and better than no contact at all, but let us not forget that what would be really ideal would be for the same experiences to be available in person! The technology offers valuable solutions, but it is not a complete substitute for the real face-to-face, in-person physical experience.

Conclusion

In this chapter I have raised questions about where and when there may be limitations to the use of technology in music learning and music participation contexts. I propose that environments can pose limitations, as can facilitators in the skills and experience they possess. I also highlight the multiple benefits to well-being that collective music making can have, and the importance of these in participant-centered community music endeavors. Not least of these is the significance of the multisensory, emotional, and social feelings of playing music with others.

The limitations I observe are not insurmountable. They are not inherent within the technology-in-workshop context itself. Variables such as the amount of time the group has to work and explore together, the makeup of the group in terms of interests, abilities, and commitment, and the relationships between goals, intentions, time frames, and resources (available skills and knowledge as well as material resources such as gear and space) will each play a part in determining the content of a workshop and the place of technology within it.

Technology in community music can offer students and music learners powerful tools for the creation of their own work, and platforms for connecting with others who are geographically distant. However, this potential does not mean technological solutions are always suitable substitutes. The benefits of face-to-face musical engagement with instruments requiring an element of physicality for the sounds to sound are substantial, and in a music context where a central reason for the project's existence is related to dimensions of well-being, these benefits need to be given due weight. Before launching into the latest technology on offer, perhaps the facilitator should think deeply about how suitable and sustainable it is for the music environment in which she is working, whether it is a good match for the needs of the group and her own skills, and whether she is sure she can't get a more potent shared outcome through engaging in "old technologies" like voices, bodies, and instruments. Sometimes the tools you need might already be there, forgotten, in the back of the classroom cupboard.

References

Cunha, Rosemyriam, & Lorenzino, Lisa. (2012). The secondary aspects of collective music-making. *Research Studies in Music Education*, 34(1), 73–88.

Gridley, Heather, Astbury, Jill, Sharples, Jenny, & Aguirre, Carolina. (2011). *Benefits of group singing for community mental health and well-being*. Carlton, Australia: Victorian Health Promotion Foundation (VicHealth).

Higgins, Lee. (2012). *Community music: In theory and in practice*. Oxford: Oxford University Press.

Howell, Gillian. (2011). Do they know they're composing?: Music making and understanding among newly arrived immigrant and refugee children. *International Journal of Community Music*, 4(1), 47–58.

Howell, Gillian, & Dunphy, Kim. (2012). Toka Boot/The Big Jam: Making music in rural East Timor. *International Journal of Community Music, 5*(1), 27–44.

Hullick, James. (2013). The rise of the Amplified Elephants. *International Journal of Community Music, 6*(2), 219–233.

Ruud, E. (2012). The new health musicians. In R. MacDonald, G. Kreutz, & L. Mitchell (Eds.), *Music, health, and wellbeing* (pp. 87–96). Oxford: Oxford University Press.

Stige, B. (2012). Health musicking: A perspective on music and health as action and performance. In R. MacDonald, G. Kreutz, & L. Mitchell (Eds.), *Music, health, and wellbeing* (pp. 183–195). Oxford: Oxford University Press.

United Nations Development Programme. (2013). *Rise of the South: Human progress in a diverse world*. New York: UNDP. Retrieved 25 January, 2014, from http://hdr.undp.org/sites/default/files/reports/14/hdr2013_en_complete.pdf.

Waldron, Janice. (2012). YouTube, fanvids, forums, vlogs and blogs: Informal music learning in a convergent on- and offline music community. *International Journal of Music Education, 31*(1), 91–105.

CHAPTER 44

MEANINGFUL AND RELEVANT TECHNOLOGY INTEGRATION

MICHAEL MEDVINSKY

TECHNOLOGY is pervasive in today's world. We integrate it into every aspect of our lives: when we interact with our families, perform tasks at work, and pursue our hobbies. Nowhere is this integration more intriguing than in our classrooms. In today's education climate it is hard to imagine learning without technology. With appropriate integration, technology transforms learning, thinking, and sharing. Some music educators are just beginning to explore the possibilities and potential for supporting this shift in musicians' learning. Just as learners are developing new ways of experiencing the world, teachers must also develop new understandings of how to integrate technology into learning experiences. As both learners and teachers navigate their way through these new experiences, we must create an environment where learner-musicians have a "voice" in designing their own learning processes and choices in the ways they show their musical understandings, whether digital or analog. In order to inspire the most profound technology-based learning outcomes, it is important for music educators to provide learner-musicians with opportunities to listen to, perform, and create music in a culture where technology is transparently integrated into their everyday practices.

As a school music educator myself, I have always made technology a core element of my classroom. Currently, I teach middle school general music and guitar in a suburban school district in Michigan where the learners' daily musical experiences are supported with different technologies. Previously, I taught kindergarten through high school general music in a rural school district where I developed and established a secondary general music program, called Musician's Workshop, in which all the participating musicians experienced music within a blended learning environment. Regardless of where I've been employed, in all my classrooms we have used technology to support our musical practice, including solo performing, ensemble performing, arranging, composing, chording, improvising, listening critically and analytically, and reflecting.

I strive to enable learner-musicians to be thinkers and to use sound as a vehicle for personal expression. Unfortunately, I have personally experienced and observed lesson designs that create obstacles to the music learning process, such as:

- Teachers expecting inexperienced learner-musicians to work from staff notation before they are ready
- Situations where the learner-musicians conceive music that is more complicated than they have the experience to perform
- Learner-musicians being asked to create music without essential prior experiences that would enable their success
- Learner-musicians never being invited to reflect on listening examples or performances
- Teachers never gaining an understanding of how contemporary artists perform
- Music teachers lacking an understanding of how instruments and performance practices have evolved

The seamless integration and transparency of contemporary technology can enable learner-musicians to engage with music in ways that eliminate or resolve these kinds of issues. My desire to provide my students with the most enabling music learning environment is why I have committed to integrating music technology into the learning and teaching experiences in my classroom, and it is why I continue to search for and investigate the strengths of emerging technologies.

Educational Philosophy

My approach to teaching music is rooted in a constructivist view of learning (Brooks & Brooks, 2001; Fosnot, 2005). From this perspective, I see myself as a facilitator and problem setter and the learners as peer teachers and problem solvers. By including my students in my thought process and engaging them in a hands-on approach to musical understanding, I help them focus on the process of learning, and together we construct an understanding of our thought process and the ways we think in music rather than about music (Wiggins, 2015).

As a learning experience designer (Wiggins & McTighe, 2005), I scaffold learner-musicians as they navigate their ways through new experiences inside their zones of proximal development (Vygotsky, 1978). All students' zones of proximal development are different, due to prior experiences, and may change within different situations where they may be more or less experienced. I endeavor to design meaningful experiences in which learners will have multiple entry points, allowing each individual learner-musician to continue growth toward independent musicianship (Boardman, 2002; Wiggins, 2015). When approaching experience design framed by the use of technology, I carefully consider the ways technology supports the learner. These considerations ultimately guide the ways in which certain technology is integrated into each learning experience. Some guiding questions I use to frame my thinking are:

- What (if any) technology is suited to support each learner?
- How can the technology connect the musicians outside the walls of the classroom?
- Does the technology support the musician in creating music previously unachievable?

As I began experimenting with digital music on my own, I noticed my role in the classroom transforming. My experimentation with what was possible began to affect the ways I was designing experiences for my students. I began by using a MIDI keyboard, a digital audio workstation (DAW), an audio interface, and recording microphones—all of which changed the ways my students were creating and capturing music in class. I introduced another technological element when I noticed that my students beginning to own mobile devices. These devices allowed me to augment my students' acoustic musical creating and performing with creating and performing on digital instruments.

Learner-musicians continue to perform and create on acoustic instruments while also having the option of using digital as well. This choice has fostered a blended learning and performing environment. The option of creating and performing on acoustic or digital instruments is an important force in our classroom. Providing musicians with the opportunity to make their own musical decisions and amplify their artistic intents in the best way they see fit is empowering. Each musician has her or his own way of using technology, digital or mobile, to support her or his musicianship. We also began using mobile apps to support our musical thinking and to reflect upon our experiences in class. Apps such as Socrative, Three Ring, Poll Everywhere, and Popplet allow me to assess my learner-musicians' thinking in real time and to adjust my instruction accordingly. These emerging technologies push my thinking as a music educator.

Some Potentials of Technology

Learner-musicians who progress through traditional Western music education perform and re-create music that others have composed; these processes are the core of the classroom experience. This tradition is perceived by many practitioners as "tried and true": *Why should we change the way learner-musicians are experiencing music in our classrooms? Can we be musicians through processes other than singing or playing acoustic instruments?* One way for technology to enhance today's music classroom is to provide avenues for learner-musicians to explore music in nontraditional ways. New genres and digital instruments have played an integral role in extending the definition of "musician." This extension still includes performing and creating, but it also provides opportunities for listening and engineering.

Becoming digitally literate in the twenty-first century provides new opportunities for musicians to be able to use technology to organize sound, create and innovate, share ideas in a global community, and think creatively about performers' musical interfaces. People are creating new instruments, MIDI devices, Internet devices, and peripherals. We are experiencing a renaissance in musicians who can create new or modify existing instruments. Musicians are inventing custom instruments that are personalized to the ways they perform. They think beyond the musical performance and audience engagement. Musicians are no longer limited by the traditions of historical instrument construction. They think about the ways in which they interact with their musical interfaces. This ideation used to be limited to elite groups of wood lathers, electricians, engineers, and instrument makers. Today, the mindset of inventing and redesigning instruments is being reimagined as a folk tradition, something that is changing the ways musicians think about creating and expressing with sound.

Lasse Gjertsen, an animator, musician, and videographer, is exploring the potential of music technology. He composes and performs original pieces on instruments he claims he cannot play. He produces his pieces by organizing video clips of himself playing one sound at a time in a stop-motion video, which culminates in intricately performed pieces. Gjertsen expresses his musical creativity in ways that he may not have had the ability to do without the support of music technology (e.g., Gjertsen, 2006). Curtis Thorpe, a musician and inventor, is another music technology innovator. He created a MIDI digital harp guitar (e.g., Thorpe, 2013) by modifying an acoustic guitar with wood, quarters (coins), and a MaKey MaKey board. The quarters were designed to act as triggers for the bass notes of the harp guitar. Thorpe created an instrument that enabled him to perform with a unique sound on a unique instrument.

Today's musicians are performers who also need to think as producers, sound engineers, and inventors. Mike Tompkins is a producer, engineer, and YouTube musician who creates unique solo a cappella music. He re-creates popular music with just his voice, Pro Tools, vocal processors, and loop machines. His videos have become viral sensations: some have been viewed more than 85 million times (e.g., Tompkins, 2006). Clearly, the ways in which musicians are creating and performing music has evolved as we have entered the twenty-first century.

While music teachers and learners should, I believe, continue to engage in traditional musicianship and musical activity, authentic musical processes are changing. The ways in which teachers design experiences in our music classrooms should also reflect the innovative technological possibilities available to help augment twenty-first-century musicianship. Yesterday's cutting edge technological advances, such as CDs, thumb drives, and interactive white boards, are almost obsolete today. Technology is accelerating, and innovation is happening so quickly that, without the proper tools, educators may be left behind. Music educators sometimes become too comfortable with the experiences they provide for the learners who are progressing through their music programs. They tend to teach the way they were taught, and they use the same music literature they were exposed to as learner-musicians. This presents problems because contemporary learner-musicians come to our classrooms with life experiences and musical identities that may not connect with traditional music or the way it is taught. They may not find traditional musical experiences relevant to their lives. It is our role as music educators to design experiences that bring learner-musicians into the unknown, but we must begin where they are when they arrive. Many are already familiar with digital musicians, loop-based software, MIDI instruments, and digital effects processors. They may not be familiar with these terms, but they are likely familiar with the sounds and style of the music that uses them.

How do music educators keep up with the progression of popular music, the ways musicians are creating music, and the evolution of instruments? One way is to connect with musicians and music educators through social media. Networks such as Twitter, Facebook, Instagram, Pinterest, and blogs can serve as tools for keeping current with trends in music and music education, thus allowing music educators to share resources and expertise in a connected digital community. Communicating about classroom planning and practice through social media can be a form of reflective practice that enables teachers to improve their own teaching while becoming facilitators of innovative learning experiences for their students. Connected educators support each other through searching for new articles, blog posts, conversations, and connections. This trust and support may empower teachers to try new approaches in designing learning experiences. My personal learning network has introduced me to innovators in the field of music education and connected me with contemporary musicians who are thinking differently and reinventing the ways people are performing and interfacing with music. These connections empower me to rethink the experiences I am designing for my students. In turn, I model ways to be a connected educator and network ideas to the learners in my classroom. Musicians are changing the ways they are collaborating and composing is through sites such as BandHub, Soundtrap, Indaba Music, and Soundation. It is through experiences such as these that students and teachers are able to begin a shift in the paradigm.

Technology Integration

Technology integration is most effective when it provides a transparent scaffold within a musical learning experience. The technology provides a differentiated experience from learner to learner, such that music learners can have their voices and choices in deciding which technology best suits their needs and how to use it to support their musicianship. However, it is important not to concentrate too much on the technology itself. The music must remain the focus of the learners' experience, with the technology becoming a vehicle for their musicianship. Technology simply provides musicians with multiple pathways to express, problem-solve, and show their understandings of learning goals, thus fostering divergent thinking.

I have experienced a shift in the culture of my classroom. I find that learner-musicians engage deeply in experiences that connect to the music that is the soundtrack of their lives. It is relevant, current, and forward thinking; they see themselves as innovators and creators of new music. It is a new mindset, and technology is at the forefront of this seismic shift. "Digital musicians must find a distinctive musical "voice." They may build on what has gone before, so they may sidestep existing work. Either way, they become a new kind of musician: one who originates and performs, who creates and produces, and who harnesses the potential of technology in new and exciting ways" (Hugill, 2012).

When music learners are given these choices, they begin to take more ownership of their learning and to construct their own understandings within an inquiry-based learning environment. Creating opportunities for musicians to show their understandings of musical concepts in their own ways nurtures an environment where they feel valued and honored. This model also fosters divergence in the ways different musicians interpret and create music. While some may choose technology, some may prefer creating music with acoustic instruments, and still others may opt for a cappella. The important idea is that all are acceptable. We, as educators, just need to ensure that all are available. Making music on acoustic instruments has been accepted as a way of being a musician for a long time; it is making music through technology that we must now also consider valid.

While it is tempting to just dabble with technology in the musical classroom, it cannot be simply an extra "add-on." Learning to use technology as a musician should be one of the core processes in the classroom. Technology should be ubiquitous, transparent, and ever-changing; it must constantly evolve along with learning. Educators should not simply plan "Technology Tuesdays" or tell students "when you are finished with your work, you may play a game on the iPad." Because learners are not inherently born with an understanding of how to use technology, they need to engage in experiences that foster their understandings of its appropriate uses. If the only way learners use technology in a classroom is to play games after they are finished with a lesson, they will come away believing that is all music technology can be. Instead, technology must support music learners' engagement in new musical experiences in new ways.

App as Scaffold

One of the most intriguing ways teachers may use technology in the music classroom is to identify related apps and pair them with learner-musicians as a means of scaffolding new experiences. For example, as an extension of ideas and inspiration derived from listening to "Hyperactive" a piece by Lasse Gjertsen (2006), my class decided to explore organizing recorded sounds. When we began considering the difference between sound and music, we used a few apps in succession to provide digital scaffolding while solving this musical problem. We began with Vidrhythm, by Harmonix, which records vocal samples and organizes them into layered rhythmic pieces. The learner-musicians had their choice as to which styles within the app they felt best represented them. Then they recorded their sounds, and Vidrhythm made all the musical decisions while composing the piece. Eventually, the learner-musicians shared their pieces with one another as a class performance. They then used their music as anchor pieces and analytical listening examples to gather ideas about how to create a piece like the one they created through the app. Throughout the listening process, they gathered characteristics of the rhythmic motives and syncopation, phrasing, form, and any other elements of the piece they found interesting. They then transferred their understandings to compositions, using Smule's app MadPad. MadPad samples environmental sounds and arranges them on a device screen that is similar to a drum pad interface. MadPad does not arrange the sounds, as Vidrhythm does; instead, the musicians perform and create loops from the samples they have recorded. They can also use this app to create texture and form changes during performance and recording.

When learner-musicians engage in this type of composing experience using these two apps, it provides a scaffold for processes of digital composition. The progressive release of digital support is what makes Vidrhythm a scaffold for MadPad. This is by no means a replacement for teacher or peer scaffolding; it is but another layer of technological support. It is the design of musical experiences supported by technology that creates this digital scaffolding.

Creating and Performing on Glass

The interface between performer and audio has changed since the introduction of touch-and-react glass technology. Today, a single musician can perform and create multilayered music with one device with much more ease than with the cumbersome keyboard workstations of the past. Many apps enable creating and performing music through the glass interface, and I have found these apps most useful in my classroom. In particular, the way we view GarageBand for the iPad has made it one of the most transformative pieces of software created for Internet devices. The depth of the software and its multiple entry points for musicians of different experience levels are what make this software so impressive.

One idea we explored in class with GarageBand's instruments and smart instruments is the idea of scales. Learner-musicians explored why some notes fit into a chord progression better than others. How are these notes organized? I created a simple chord progression (C–Am–F–G) using GarageBand's smart guitar and the autoplay knob, which allows you to select a variety of built-in strumming patterns. The learner-musicians had to use their understandings of whole and half steps while staying within the chord structure. They used one of GarageBand's melodic instruments set to notes to improvise during the chord progression. Some chose instruments they were not familiar with, like the guitar fretboard or violin neck. The struggles began. The fretboard is laid out chromatically and in fourths. I asked them to figure out the melody to "Heart and Soul" (which is supported with the chords I originally recorded). For most of those who chose guitar, it was difficult to gain an understanding of the fretboard and the intervallic relationships between strings. A few were able to problem-solve and deduce parts of the melody, but others continued to struggle with understanding the fretboard. We then set the scale to major within the software. This limits the notes to the diatonic pitches of the key. With the obstacles of string intervals and fret relationships removed, the learner-musicians were able to engage with the sound, and all were able to easily perform the melody on glass over the accompaniment. This experience created a doorway in that allowed students to first engage with the sound and then pitch relationships on string instruments (Wiggins, 2015). The relationship between pitches is what makes musical scales meaningful. Many learner-musicians are confused by the constraints of the whole and half steps of different key signatures and neglect to consider the sound that connects to the pitch choices. The GarageBand app helps students to better understand these pitch relationships. The smart instruments and drum machine features of GarageBand also provide support for musicians with varied prior experiences to perform and create together. This enables learner-musicians to create with the understanding that the organization of sound remains the central focus. Self-expression continues to be core of the experience supported by technology.

Device as Controller

I have recently upgraded my DAW from GarageBand to Logic Pro X. The depth of this DAW has revealed new timbres, arpeggiators, and vocal effects previously unknown to the learner-musicians and me. Prior to this upgrade, I found myself at the controls while recording original class songs; I was duplicating tracks, fading in and out, cutting and pasting loops, and adding effects to tracks during class. The learner-musicians were peripherally engaging in the process through the visuals projected onto our interactive white board, but they were not actively engaged in the process. I felt I was doing all the work, which in turn meant that I was doing all the learning. The musicians were engaged in the role of recording artists while I took on the role of producer-engineer.

I began to consider ways of engaging the learner-musicians in the composing and recording process as well as the sound engineering and postproduction of their music. Unfortunately, we were all working from one laptop computer. When I began researching possible solutions, I found that Logic X had developed a Logic Remote app for the iPad. The app is designed to utilize the iPad as a peripheral controller and provide new ways to record, mix, and even perform instruments in Logic Pro X from anywhere in a room—turning your iPad into a keyboard, drum pad, guitar fretboard, mixing board, and transport control. Logic Remote is designed to navigate inside Logic projects. Learner-musicians can gain control of the recording remotely, where they can trigger Logic Pro X key commands. The app also allows them to customize buttons. During our recording sessions, the musicians were thinking about the commands we used most, and they mapped those commands to customizable buttons on the glass. A favorite feature among the learner-musicians in my classroom has been the addition of the arpeggiator plugin to any instrument. As a music educator, these apps have helped deepen my understanding that there are many ways of being musical, which in turn has broadened my design of musical experiences.

Experiencing audio is a unique way of being musical, and the use of iPads deepens this experience by encouraging the students to be more accountable for the music they produce. This shift in ownership has had quite an impact on our class culture. The learner-musicians have always taken ownership of their songs, but there has been a distinct shift in the ownership of production. This shift became apparent when I gave the iPad to a student as we were composing together as a class. The learner-musicians had stopped to discuss the harmonic progression of a section in their piece. They decided that the chords needed to change at a different point in the melody. I asked the "engineer" to erase the track and take it back to the beginning. His reply was, "Mr. M, I'm one step ahead of you." Device as peripheral controller has enabled learner-musicians to take control of more than the process of composition. The learner-musicians in my classroom have developed a broader scope through which to approach a new musical experience. The technology provided this shift; it is now up to us to integrate these opportunities into the experiences we create. As these kinds of experiences become routine in your classroom, you may find that the learner-musicians get there before you do.

Mirroring Devices

AirServer—an AirPlay receiver for Mac and PC—and Apple TV are two pieces of technology that have amplified the way musicians share ideas and present projects in the classroom. Each device serves different purposes. AirServer mirrors a class set of Internet devices to a screen, and Apple TV can mirror one device, but my classroom has benefited most from cloud streaming music and videos.

Many of the musicians' experiences in my music classroom are rooted in listening, performing, and creating, all of which are enhanced by the use of AirServer and Apple TV. The learner-musicians are often creating in small collaborative groups, and technology is always one option they can choose to use as an instrument, recording device, or for accompaniment. When there are many devices in use during their creative process, we use AirServer to mirror all the devices in use to the screen. This enables learner-musicians to scaffold one another with visual aids without interrupting their own creative processes. If there is something musicians see that they want to learn more deeply, they know whose group to visit. The technology enables them to take control of their learning. Often I find myself visiting groups to learn more about an app or a feature I have not known about because I have seen them manipulating the app on our screen.

When musicians present a final iteration of a composition or arrangement, they mirror their device to our classroom Apple TV for the entire class to see. Having a visual representation of the sounds may support the learners' listening experiences. This digital listening map provides visual representations of changes in texture, instrumentation, form, phrase, and loop length. As other musicians listen to and see their peers' music, they often edit and revise their pieces. The class becomes a collaborative place for modeling critical thinking, creativity, and the process of creating and expressing through music. These are times where technology enables educators and their students to create a community that builds caring relationships and high levels of collaboration among its members.

While these technologies are a boon to my classroom, they don't always function as well as I would wish. The difficulty that I have found when wirelessly mirroring devices to a large display screen is latency. There is a distinct lag from the time you touch the surface of a device to the time the instrument realizes the sound through the speakers. This creates a problem when performing or recording live. As an experienced musician, I am able to anticipate the beat and touch the glass early to make it sound in time, but less experienced musicians find this troublesome. Another solution is to quantize after recording. However, this does not always correct the rhythm to what was intended. When performing or modeling using a device, I use a dongle connected to a VGA cable and an audio cable wired directly to a sound source to eliminate latency for a simultaneous performance.

Equity and Resources

Many districts are facing economic hurdles, insufficient funding of arts programs, and downsizing of support departments such as instructional technology, and I have not been spared these difficulties in my own career. Most of the technological additions I have made to my music rooms have been by my own design and funding. However, beginning in upper elementary or middle school, there are solutions to these economic hurdles. A "bring your own device" (BYOD) model can supplement the lack of technology in classrooms. In a BYOD environment, in addition to the technology provided by the school, learners have the option of bringing their own devices to class. This puts more devices in your classroom but can also lead to platform inequity. Some devices run Android operating systems, while others run iOS. This is not an issue if you can use web-based technology, but it is important to keep in mind that some mobile apps are not cross-platform.

Another dilemma is that public schools are not always equipped with the hardware to handle many devices connected to the Internet at one time. There may be a cap on the number of devices one can connect at a time, which leads to some learners engaging in a blended learning experience and some not. This does not affect my classroom to a great degree, since many of the apps my students use do not require the Internet. Most of our daily experiences of listening, performing, and creating are done offline. The times that are difficult to navigate are when musicians share reflections online or upload finished compositions or performances to a web-based service. I have solved this problem by creating a separate wireless network within my own classroom. The learner-musician's devices connect to my room's wireless signal, which allows all our devices to connect at once.

Closing Observations

I think that it is important for learner-musicians to have a choice about whether or not to use technology to express their musical ideas. Some may start with acoustic instruments and then decide to use a digital means of producing their sound, and others may start with mobile devices and decide to move to acoustic instruments. It is imperative that the culture of the class gives them voice and choice. Musicians have unique ways of expressing their inner thoughts and feelings through their organization of sound and silence. Musical experiences need to be designed so that there is enough room for their musical voices to be expressed in their own way. Whether acoustic, or through the use of digital instruments, mobile devices and apps, or music created in the browser, the choice must remain that of the musicians in your classroom. It is their voices and choices that represent their musical identities.

Musicians continue to create new ways of expressing and creating music. For music education to keep up with the innovation and evolution of music, educators must keep current with emerging technologies. When designing experiences for listening, performing, and creating, we must provide opportunities for musicians to create with these technologies. This creates a classroom culture where musicians become autonomous in their approaches to showing their musical understandings. The technology becomes ubiquitous, becomes transparent, and must constantly evolve along with learning.

References

Boardman, E. (Ed.). (2002). *Dimensions of musical teaching and learning*. Reston, VA: MENC.

Brooks, J. G., & Brooks, M. G. (2001). *In search of understanding: The case for constructivist classrooms* (3rd ed.). Alexandria, VA: ASCD.

Fosnot, C. T. (Ed.). (2005). *Constructivism: Theory, perspectives, and practice* (2nd ed.) New York: Teachers College Press.

Gjertsen, L. (2006). *Amateur* [Video]. Retrieved November 7, 2006, from https://youtu.be/JzqumbhfxRo.

Gjertsen, L. (2006). *Hyperactive* [Video]. Retrieved May 8, 2006, from https://youtu.be/o9698TqtY4A.

Hugill, A. (2012). *The digital musician* (2nd ed.) New York: Routledge.

Thorpe, M. (2013). *DIY MIDI harp guitar—because it's there—Michael Hedges cover* [Video]. Retrieved March 24, 2013, from https://youtu.be/Mw9sRPqWoZA.

Tompkins, M. (2006). *Mike Tompkins* [YouTube channel]. Retrieved July 21, 2006, from https://www.youtube.com/user/pbpproductions.

Vygotsky, L. S. (1978). *Mind in society: The development of higher psychological processes.* Cambridge. MA: Harvard University Press.

Wiggins, G. P., & McTighe, J. (2005). *Understanding by design* (2nd ed.). Alexandria, VA: ASCD.

Wiggins, J. (2009). *Teaching for musical understanding* (3rd ed.). New York: Oxford University Press.

Wiggins, J. (2015). Musical agency. In McPherson, G. E. (Ed). *The child as musician: A handbook of musical development* (2nd ed.), pp. 102–121. New York: Oxford University Press.

CHAPTER 45

THE CONVERGENCE OF NETWORKED TECHNOLOGIES IN MUSIC TEACHING AND LEARNING

JANICE WALDRON

I am bimusical—a classically trained saxophonist and wind band conductor and an informally "self-taught" Irish traditional musician on tin whistle, Irish flute, and Uilleann pipes. Entering graduate school in 2002, I didn't think about incorporating the half of my musical self that identifies as "Irish trad musician"—and the music learning and teaching practices that accompany it—into my future career as a university music educator-researcher. I had been a successful band director for 20 years, and I was going to back to school to become a music educator of future instrumental educators (re: band directors). Playing Irish trad "on the side" was an enjoyable and, on occasion, profitable hobby, but I never envisioned it occupying a place in my professorial research career.

In retrospect, 2002 was an important year for another reason, although I didn't realize it at the time. Putzing around on the Internet curing the summer before graduate school began, I discovered the Irish trad bulletin board and forum Chiff and Fipple (C and F) and began regularly checking and posting messages to a group of like-minded enthusiasts—my "peeps," some of whom I already knew offline. Together with "chiffers" met online, we chatted, swapped stories, playing tips, and tunes, and organized informal sessions where we met and played together in the "real" world. Pre-Facebook, before social media was a "thing," this was pretty revolutionary and exciting stuff—I could now interact daily with a community that before, due to a lack of geographical proximity, I couldn't always easily connect with in "real" life. Being a "baby boomer" born in the late 1950s, I'm definitely not a "digital native," but checking the "chiff" board along with my e-mail quickly became part of my morning routine.

Besides the chat boards, C and F had another intriguing feature: the "clips and snips" forum, where one could record oneself playing and then upload the result for commentary from other "chiffers." Recording and uploading a tune was a pretty tedious procedure at the time, complicated by numerous technical issues, but it was pretty exciting nonetheless; board members from around the world would then take the time to listen and make suggestions on the "clips and snips" commentary board—there was even a dedicated "learning with 'clips and snips'" how-to thread. I didn't think of it at the time as "informal music learning." I didn't know there was an actual label for what I and others were doing—it was just an enjoyable experience I did in my spare time, a nice addition to live session playing when I couldn't get my IrTrad playing "fix."

Fast forward to the fall of 2002, with me sitting in my first doctoral philosophy class. The professor of the class was raving about Lucy Green's just-published book *How Popular Musicians Learn*, passing it around the seminar for us to peruse and discuss. I remember looking at the first few pages and thinking "What's the big deal? Everyone I know in the IrTrad world (including myself) learns this way." That was the "aha" moment that placed me on this research path. Serendipity really can't be overrated.

The one thing Green (2002) did not discuss in her book was using Internet resources and online collaboration for informal music learning; nor did it occur to me to explore or include a discussion of them in my subsequent and various graduate school studies of informal music learning, even while I continued to be an active member of several online IrTrad communities. That epiphanic moment didn't come until the introduction of YouTube in 2006, when I noticed people posting all kinds of YouTube hyperlinks of well-known IrTrad musicians playing and teaching onto the various forums; I myself used several videos to learn new tunes and improve my playing. This was even more revolutionary than uploading and listening to audio recordings, because as a player, I could see and "read" other musicians' fingers as they performed; observational learning combined with aural/oral music learning became possible any time I felt like playing without leaving home. Furthermore, YouTube allowed me unlimited access to performances by legendary IrTrad artists, many of them my "musical heroes," from all over the world—a radical development regardless of genre, but particularly so for those of us who play nonmainstream musics.

I knew I was onto something when I presented my first conference paper on online communities, YouTube, and music learning. After my presentation, a skeptical music education researcher raised her hand and emphatically pronounced: "I just don't believe people learn music that way" (i.e., informally, using Internet resources and YouTube). I think I've had the last laugh on the topic!

Context

"Networked technology," defined as the convergence of the Internet and mobile phones with social networks, has revolutionized the ways people communicate, collaborate, and learn, with ramifications that continue to unfold. The purpose of this chapter is to consider the implications of networked technology use—for us, as educators, and for our students, "digital natives"—for music teaching and learning. To contextualize this discussion, I first draw from new media scholars theorizing networked technologies' history, evolution, and functions, because those researchers' views come from a perspective primarily concerned with what the theoretical a prioris of networked technologies hold for the general populace, and not just educators (let alone music educators). For example, many new media scholars maintain that more important than what any single technology does is how we use technology overall, the effect(s) that use has on us, and the relationships we form through it and with it (Turkle, 2011). With that in mind, in the second part of this chapter I will explore the evolution and the implications of networked technology use for current and future music education research and practice.

As a profession, music educators tend to focus on technology as a knowable "thing," that is, hardware and software with their "practical classroom applications" and what they can do for our students and us in the moment—and not the greater epistemological issues underlying their use. How will we, the born "analog," engage musically in a meaningfully relevant way with a generation of students—"digital natives"—who have grown up technologically "tethered"? How will these different "ways of knowing," facilitated by the omnipresence of networked technologies, challenge and perhaps change our beliefs and practices about music learning and teaching, now and in the not-so-distant future? To position these questions, I next present two contrasting new media researchers' views on the development of networked technologies and how this evolution has subtly changed our expectations and acceptance of their usage as a "new" normal in daily living.

Three Decades of Technology—Use, Change, Convergence, and the Rise of the "Digital Native"

Thirty years ago, social scientist and technology researcher Sherry Turkle (1984, 1995, 2011) began exploring how people perceived their relationships to various technologies, among them the newly available personal computer. In the early 1980s, "technology" was commonly understood to be software and hardware not the Internet; as it was generally unavailable, the average person remained unaware it existed. That quickly changed when Web connectivity became widespread in the 1990s; personal computer usage rapidly evolved, along with the Internet's swift growth, and transformed what, a few years previously, had been an intimate experience—"the person alone with his/her machine"—to a communal one, with personal computers acting as "portals" for people to "lead alternate lives in virtual worlds" (2011, p. xi). Gradually, "technology" became an all-inclusive word for software, hardware, the Internet, and (the now quaint-sounding term) cyberspace—referring to places "out there" on the web where one could "go" and interact with other users—instead of a "thing" like a piece of hardware or a software program. This represented a major paradigm shift in the way researchers, scholars, and users viewed technology, because the term now included possible location(s) as well as a product.

As the ways in which people used their computers changed, so did the focus of Turkle's research; shifting away from examining the one-to-one relationships individuals had with their computers as "second selves," that is, the personal computer as a reflection of its user's identity, to exploring the bonds "that people formed with each other using the computer as an intermediary" (2011, p. xi). Though she had been an early champion of the positive implications of "virtuality," Turkle became increasingly disillusioned with the roles that computers, along with their promise of constant connectivity, assumed in the 1990s. She asked: "why is it so hard for me to turn away from the screen?" adding the observation that "it is striking that the word 'user' is associated mainly with computers or drugs" (1995, p. 30). However, she reiterated that, despite these troubling implications, "the computer offers us new opportunities as a medium that embodies our ideas and expresses our diversity" (p. 31). But she also recognized that people's interactions with computers could result in unintended and ambiguous effects, because intentions of use do not reside within the computer but are instead determined by the ways people interact with, perceive their relationships to, and develop expectations of what their machines can or should be able to do over time: "we construct our technologies, and our technologies construct us and our times. We become the objects we look upon but they become what we make of them" (1995, p. 46). She wrote: "people think they are getting an instrumentally useful product, and there is little question that they are. But now it is in their home and they interact with it every day. And it turns out they are also getting an object that teaches them a new way of thinking and encourages them to develop new expectations about the kinds of relationships they and their children will have with machines" (p. 49). There are several key points to take away from Turkle's arguments. The first and most important one is that technological determinism is a false modernist construct of where technology leads, because what any given technology is capable of becomes clear only during the process of its use, including how it is used and what it is used for. The act of use itself changes the technology, the person interacting with it, and the expectations that such exchanges have for a culture and society. Simply put, a machine's makers and programmers cannot possibly predict what can be produced with a connected computer and a creative individual or individuals manipulating it.

Second, Turkle's argument that (1) children develop new expectations of what their machines can and should do, and (2) computer use changes the way people think is particularly prescient today; writing in 1995, Turkle was discussing people's relationships to their newly acquired personal home computers. Her argument is even more fitting now, given the omnipresence of smartphones and the expectation that "there's an app for everything," resulting in digital natives for whom life without networked technologies is both unthinkable and untenable.

Along with the notion that there is a technological solution for every niche and problem (or if there is not one now, there will be), Turkle questions the significance of what being technologically "tethered"—that is, always available via one's devices—means and what has become lost in the process:

> [we] have changed as technology offers us substitutes for connecting with each other face-to-face ... we are offered a whole world of machine-mediated relationships on networked devices. As we instant message, e-mail, text, and Twitter, technology redraws the boundaries between intimacy and solitude. Teenagers avoid making telephone calls, fearful that they "reveal too much." They would rather text than talk. Adults, too, choose keyboards over the human voice. It is more efficient, they say. Things that happen in "real time" take too much time. ... Yet suddenly, in the half-light of virtual community, we may feel utterly alone. (2011, p. 12)

And further: "over time, a new style of being with each other becomes socially sanctioned. In every era, certain ways of relating come to feel natural. In our time, if we can be continually in touch, needing to be continually in touch does not seem a problem or a pathology, but an accommodation to what technology affords. It becomes the norm" (p. 177). And herein lies the problem: there is a steadily growing disjuncture between how we, as music educators, teach, not having grown up "tethered," compared to our students, for whom this is the norm. These digital natives, as Turkle states, are "the first generation that does not necessarily take simulation to be second best. As for online life, they see its power—they are, after all risking their lives to check their messages—but they also view it as one might view the weather; to be taken for granted, employed, and sometimes endured" (p. 172). Like Turkle (2011), new media researchers Rainie and Wellman (2012) trace the intersection of networked information technologies over the past three decades, coining the term "the Triple Revolution" to describe the relatively recent convergence of the Internet and mobile smartphones with social networks. They offer a far less dystopic view of what networked technologies hold for people and society than Turkle's (2011), recognizing that, as in all paradigm shifts, there will be some negatives, but that these are inevitable when change occurs at a fundamental level. They are also emphatic that the gains far outweigh the losses; for example, they argue that the convergence of networked technologies has created unprecedented educational opportunities, which I will discuss from a music learning and teaching perspective later.

In their research, Rainie and Wellman focus on how the Triple Revolution works in "relationships, families, work, creativity, and information," explaining that the way social technological networks function has profoundly transformed the way we connect in person and electronically (2012, p. xi). In this, their arguments are similar to Turkle's (2011); further they stress that "humans determine how technologies are used [and regardless of technological advancements], that one thing holds true: The Internet and mobile phones have facilitated the re-shaping of people's social networks, enabling them to be larger and more diverse. And they have re-configured the way people use their networks to learn, solve problems, make decisions, and provide support to one another (pp. xi–x). Texting via mobile phones is perhaps the most striking example of how the intersection of technology with social networks has changed the ways the general populace communicate with one another on a regular basis, particularly among teenagers and young adults. According to a 2011 Pew Internet survey, teens and twenty-somethings are the heaviest text users of all, on average sending and receiving 1,500 texts per month (Rainie & Wellman, 2012). As a demographic, they prefer texting to talking—due to both its simplicity and the privacy it provides while communicating—so much that the principal author of the Pew Internet report, Amanda Lenhar, concluded: "if teens are a leader for America, then we are moving to a text-based communication system" (as cited in Rainie & Wellman, 2012, p. 89). Intertwine the prominence of texting with readily available smartphone apps, and what and how the average young adult now communicates and expects of technology becomes the new normal, consistent with Turkle's (1995) supposition discussed above.

Digitally native teens, however, are not the only ones maintaining multiple social contacts and information searches online simultaneously while performing the tasks of day-to-day living. For better or worse, people of all ages have quickly adapted to technological multitasking, traveling back and forth between what Michael Castells calls "the space of flows," similar to Turkle's notion of "tethering" but with a temporal twist, in that networked technologies now make possible "near simultaneous communications [that] can be consummated at any moment, including times when people are walking down the street, standing in line, or driving" (as cited in Rainie & Wellman, 2012, p. 102). Within this "space of flows," Rainie and Wellman identify three distinct ways to conceptualize "place" in blurred temporal and physical space. These are:

1. Connected presence, in which "people can update aspects of their lives without having to wait for the next time they see each other in person. There is less backlog of information";
2. Absent presence, in which "extensive ICT use means that people can be physically in one place while their social attention and communication focus is elsewhere"; and, its converse;
3. Present absence, whereby "those who are not physically there can be incorporated into the group conversation," for example, as in teleconferencing. (2012, p. 102)

While there are pros and cons to each of these three connective states—a loss of awareness of being "in the moment," for one—for many people, in particular digital natives, being technologically tethered is their reality; it represents a different "way of being" from the "way of being" that people experienced before the advent of networked technologies (Rainie & Wellman, 2012).

For music educators, it is important to recognize that we—and more important, our students—now live in this newly socially sanctioned technological reality, along with its "space of flows," if music teaching is to retain its relevancy in formal institutions now and in the near future. As with all paradigm shifts, some practices will become obsolete; others will expand, while new ones will be created. I will discuss some examples and possibilities based on current music education literature in the next section.

Music Education Literature and Networked Technologies

Although research on technology and music education has been around since the 1970s (Webster, 2002), until recently much of the work done in the field has concentrated on the "nuts and bolts" of how to use technology in the music classroom, and not on the larger epistemological implications that networked technologies have for music teaching and learning. Further, music education scholars have not, in general, considered that how we conceptualize and use networked technologies challenges and could change what we do and how we do it at the most basic level—in this case, music teaching and learning.

In the last five years, exceptions have quickly begun to emerge from the literature, in particular work by music education scholars and practitioners who are researching Internet use and music learning and teaching. Research in this area tends to fall into one of two categories, the first being work by music education scholars investigating effective uses of networked technologies as a delivery system for music learning when geographical physical access to music teaching is impractical or impossible. Most often the goal in this type of research is to discover how to most closely replicate the traditional music studio or classroom experience using networked technologies (Brändström, Wiklund, & Lundström, 2012; Kruse & Veblen, 2012; Pike & Shoemaker, 2013; Riley, 2009). Typically in these studies, a traditional one-on-one master-apprentice approach to music learning is examined, with comparisons to how technology could be used to deliver instruction similar to that which takes place in conservatory and classroom settings. In other words, the research tends to focus on how to duplicate, through networked technology use, a linear approach to music learning and teaching already in place in formal institutions and conservatories. Most often, what is not explored is how learning through networked technologies challenges and could change the ways in which people conceptualize music learning and teaching in toto. Technology's value lies in how closely it replicates the delivery of the pedagogical status quo, not in challenging the pedagogical status quo.

The second growing body of research in the field explores, along with investigating how people learn music through online collaborative interconnectivity, how networked technologies could change peoples' fundamental beliefs about both what constitutes and what it means to "teach and learn music." Based on evidence that the Net generation already socializes, works, learns, and collaborates through networked technologies, both outside school and in formal institutions, Leong (2011) advocates for a "new music education future that is strongly linked with the global knowledge economy in the digital and conceptual age" (p. 233), concluding that "music education practices need to devote more attention to developing learners' digital literacies, analytical and critical thinking, and the other twenty-first-century skills with reference to the realities of the cultural and creative industries. Curriculum priorities should focus on nurturing the creative, reflective and intuitive abilities of learners, utilizing web-based and game-based pedagogies, developing communities of learners as well as teaching them how to make use of the vast information and knowledge bases that are available and growing exponentially" (p. 236). Examples of music education research in this second emerging category are wildly diverse in their study foci, but reflect of many of the ideas and areas mentioned by Leong here. What they have in common is that, through the use of networked technologies, they implicitly challenge or expand upon already established perceptions about music learning and teaching in different mediums and areas. This second body of work includes research on (1) open source music learning software development (Dillon, 2012); (2) online music communities and online and convergent music communities of practice (Mercer, 2012; Partti & Karlsen, 2010; Ruthmann & Hebert, 2012; Saluvuo, 2008, 2006; Tobias, 2012; Waldron, 2009, 2011a, 2011b, 2012, 2013; Waldron & Bayley, 2012; Waldron & Veblen, 2008); (3) autoethnographies of online music learning (Kruse, 2012; Savage, 2011); (4) collaborative music learning through vlogging (Cayari & Fox, 2013); (5) YouTube videos for music learning and teaching (Cayari, 2011; Kruse & Veblen, 2012; Waldron 2011a, 2011b, 2012, 2013; Waldron & Bayley, 2012; Webb, 2007); (6) music learning through collaborative video gaming (Tobias, 2012); (7) collaborative digital media performance and online collaborative technologies for music learning (Brown & Dillon, 2012); and (8) participatory music making facilitated through user-generated content and networked technologies (Draper, 2008; Ruthmann & Hebert, 2012; Tobias, 2013; Waldron, 2013). Probably the most significant common denominator among these studies is that informal music learning practices figure prominently but with digitally networked twists that are considerably different from those practices first identified by Green (2002) in her landmark research on the same. Because their use is implicit in many of these studies, "informal music learning practices" are often not named or identified as such.

In the age before the digitally networked one, Green's (2002) original definition of "informal music learning" was limited to a group of practices whereby the learner "picked up" popular music aurally/orally from recordings, friends, peers, and family through hard artifacts—CDs, tape cassettes, and vinyl—and personal contacts in "catch as catch can" localized scenarios. People wanting to learn a specific type of vernacular music were confined to and bound by resources physically around them, having to find artifacts and resources on their own, a sort of musical learning "treasure hunt." Musical collaboration could only occur with those one knew personally in a physical setting; further, "informal music learning practices" occupied physical spaces that lay firmly outside music classrooms and conservatories. They are now, however, recognized as an established legitimate mode of learning, and music educators are exploring ways to incorporate them into school music environments (Green, 2008; Younker, Wright, Linton, & Beynon, 2012).

I label these original "analog" practices "informal music learning 1.0." Due to the omnipresence of networked digital technologies, people now have instant access to what could be thought of metaphorically as "informal music learning 2.0." Because learners can quickly locate online resources along with a networked community of learners for support in almost any musical genre, this represents a significant leap forward (or sideways and upside down) from what Green first postulated in 2002. Specifically, I define "informal music learning 2.0" as group of practices whereby the learner:

1. Can "pick up," via networked technologies, resources, tools, user-generated content (such as YouTube videos and MP3s) 24/7 in any number of diverse musical genres; hard artifacts are no longer necessary, and cloud storage makes for easy safekeeping and collaboration
2. Can locate and join a networked online community of learners for support, information, discourse, and collaboration; because online music communities often overlap with corresponding offline ones, this can lead to "real life" opportunities for connecting, playing and performing, and music learning and teaching

Like earlier informal music learning practices, digitally networked informal music learning practices still lie largely outside the classroom, but with a significant difference: a good percentage of our students, as tethered digital natives, are already familiar and comfortable with them.

Conclusion

As the basis of constructivist learning, integrating what students already know into formal contexts is not a new idea; what differs now is that "digitally native" students are likely to have more expertise in and knowledge of the possibilities that networked technologies hold for music making then the average music educator. And this should be potentially good news for teachers, because it means they need not necessarily be experts in every new technological development (which would be nearly impossible and an unrealistic endeavor anyway). Instead, music educators can use their students' networked technological expertise to advantage by being open to incorporating student-brought knowledge of "informal music practices 2.0" into the classroom. Teachers then can offer students what teachers do when at their best: provide expertise in moderating, facilitating, and guidance.

This offers a "win-win" scenario for both teachers and learners; students become musically engaged in meaningful relevant ways, and educators, in turn, learn from and with their students while simultaneously creating collaborative places and "spaces of flow" for the processes of music making. In this way, networked technologies do not replace music teaching but instead offer opportunities for changing and enhancing practice in unanticipated ways. We need only be open to what those opportunities might or could be.

References

Brändström, S., Wiklund, C., & Lundström, E. (2012). Developing distance music education in Arctic Scandinavia: Electric guitar teaching and master classes. *Music Education Research*, 14(4), 448–456.

Brown, A., & Dillon, S. (2012). Collaborative digital media performance with generative music systems. In G. MacPherson & G. Welch (Eds.), *The Oxford handbook of music education* (pp. 549–566). New York: Oxford University Press.

Cayari, C. (2011). The YouTube effect: How YouTube has provided new ways to consume, create, and share music. *International Journal of Education and the Arts*, 12(6), 1–28.

Cayari, C. & Fox, S. (2013). *The pedagogical application of collaborative video logs*. Presentation to CIC conference, University of Illinois, Champagne-Urbana., April 2011.

Dillon, S., Myllykoski, M., Rantalainen, A., Ruthmann, S. A., Thorgersen, K., & Väkevä, L. (2012). Open source. *Journal of Music, Technology and Education*, 5(2), 129–132.

Draper, P. (2008). Music two-point-zero: Music, technology and digital independence. *Journal of Music, Technology and Education*, 1(2-3), 137–152.

Green, L. (2002). *How popular musicians learn: A way ahead for music education*. Aldershot, U.K.: Ashgate.

Green, L. (2008). *Music, informal learning and the school: A new classroom pedagogy*. London: Ashgate.

Kruse, N. (2012). Locating the "Road to Lisdoonvarna" via autoethnography: Pathways, barriers, and detours in self-directed online music learning. *Journal of Music, Technology and Education* 5(3), 293–308.

Kruse, N., Harlos, S., Callahan, R., & Herring, M. (2013). Skype music lessons in the academy: Intersections of music education, applied music and technology. *Journal of Music, Technology and Education*, 6(1), 43–60.

Kruse, N., & Veblen, K. (2012). Music teaching and learning online: Considering YouTube instructional videos. *Journal of Music, Technology and Education*, 5(1), 77–87.

Leong, S. (2011). Navigating the emerging futures in music education. *Journal of Music, Technology and Education*, 4(2-3), 233–243.

Mercer, A. (2012, Spring). Rural high school students studying music via the Internet are shaping local music communities. *Canadian Music Educator*, 38.

Partti, H., & Karlsen, S. (2010). Reconceptualising musical learning: New media, identity and community in music education. *Music Education Research*, 12(4), 369–382.

Pike, P., & Shoemaker, K. (2013). The effect of distance learning on acquisition of piano sight-reading skills. *Journal of Music, Technology and Education*, 6(2), 147–162.

Rainie, L., & Wellman, B. (2012). *Networked: The new social operating system*. Cambridge MA: MIT Press.

Riley, P. (2009). Video-conferenced music teaching: Challenges and progress. *Music Education Research*, 11(3), 365–375.

Ruthmann, A., & Hebert, D. (2012). Music learning and new media in virtual and online environments. In G. MacPherson & G. Welch (Eds.), *The Oxford handbook of music education* (pp. 567–584). New York: Oxford University Press.

Technological determinism. (n.d.). Retrieved January 17, 2014, from http://en.wikipedia.org/wiki/Technological_determinism.

Tobias, E. (2012). Let's play! Learning music through video games and virtual worlds. In G. MacPherson & G. Welch (Eds.), *The Oxford handbook of music education* (pp. 532–548). New York: Oxford University Press.

Tobias, E. (2013). Toward convergence: Adapting music education to contemporary society and participatory culture. *Music Educators Journal, 99*(4), 29–36.
Turkle, S. (1984). *The second self: Computers and the human spirit.* New York: Simon and Schuster.
Turkle, S. (1995). *Life on the screen.* New York: Simon and Schuster.
Turkle, S. (2011). *Alone together: Why we expect more from technology and less from each other.* New York: Basic Books.
Waldron, J. (2009). Examining the old time virtual music "community of practice": Informal music learning on the Internet. *Journal of Music, Education, and Technology, 2*(2–3), 97–112.
Waldron, J. (2011a). Conceptual frameworks, theoretical models, and the role of YouTube: Investigating informal music teaching and learning in online music community. *Journal of Music, Technology and Education, 4*(2-3), 189–200.
Waldron, J. (2011b). Locating narratives in postmodern spaces: A cyber ethnographic field study of informal music learning in online community. *Action, criticism, and theory in music education, 10*(2), 32–60.
Waldron, J. (2012). YouTube, fanvids, forums, vlogs and blogs: Informal music learning in a convergent on and offline music community. *International Journal of Music Education, 31*(1), 91–105.
Waldron, J. (2013). User-generated content, YouTube, and participatory culture on the Web: Music learning and teaching in two contrasting online communities. *Music Education Research, 15*(3), 257–274.
Waldron, J., & Bayley, J. (2012). Music teaching and learning in the Online Academy of Irish Music: An ethnographic and cyber ethnographic field study of music, meaning, identity, and practice in community. In D. D. Coffman (Ed.), *Proceedings from the ISME 2012 Seminar of the Commission for Community Music Activity* (pp. 62–66). Retrieved, December 15, 2013, from http://issuu.com/official_isme/docs/2012_cma_proceedings?viewMode=magazine&mode=embed.
Waldron, J., & Veblen, K. K. (2008). The medium is the message: Cyberspace, community, and music learning. *Journal of Music, Technology and Education, 1*(2), 99–112.
Webb, M. (2007). Music analysis down the (You) tube? Exploring the potential of cross-media listening for the music classroom. *British Journal of Music Education, 24*(2), 147–164.
Webster, P. (2002). Computer-based technology and music teaching and learning. In R. Colwell & C. Richardson (Eds.), *The new handbook of research on music teaching and learning* (pp. 416–439). New York: Oxford University Press.
Younker, B. A., Wright, R., Linton, L., & Beynon, B. (2012, Spring). Tuning into the future: Sharing initial insights about the Music Futures pilot project in Ontario. *Canadian Music Educator 53*(4), 14–18.

3 B

Further Perspectives

CHAPTER 46

NARCISSISM, ROMANTICISM, AND TECHNOLOGY

EVANGELOS HIMONIDES

It is exciting to having been given the opportunity to participate in such a well-structured anthology of expertise, critical thinking, and creative intuition as this book presents. It is equally exciting to have been given the freedom to employ a somewhat nonmainstream vernacular, relative to that of usual academic writing, in order to foster a novel critical discourse about key issues that the writing team is passionate about. I find myself usually rehearsing technology's potential roles within particular educational and commercial contexts, but I am very excited that in this instance I have been invited to offer a short commentary on issues arising from other contributions to this book that I have found to be interesting, controversial, or challenging. I will present a number of these, and will conclude by offering my personal opinion about the implications of technology's role (alas, "the role" once again) in music and music education.

It is, to begin with, intriguing to see Waldron's presentation of her "bimusical" self (chapter 45), comprising a "formal" side and a "self-taught" one. Waldron, a seasoned figure in academia with a substantial portfolio of written and spoken public output, as well as an often exploited skill at performing sharp metacognitive assessments, provides a great example of the institutionalized divides between different musical identities, expertise, pasts, or "trajectories," if you will. Funnily, it was not my own "schizomusical" background—as a classically trained choral conductor and music theorist, in tandem with a self-taught bluesy jazz guitarist (or jazzy blues, depending on the "judge's" own gestalt and musical niche) and vocalist—that stimulated my thinking about this frequently reported duality; it was Waldron's account. Obviously, Waldron, myself, and numerous academics and practitioners the world over do have to market themselves at given junctures, and do have to specialize in particular fields in order to make their livings, sustain their families, and pursue their careers. On the other hand we are past the time where we should need to be terribly protective, secretive, or apologetic about any facet of our music experience that dwells outside the expected existential horizon. Popular musics are, gladly, well-established disciplines, the study of which is continually supported and enriched by multimodal research enquiry. Similarly, informal, nonformal, and other-than-formal learning, whatever these terms might mean to the people who use them, are also established notions that appear to be under continual scrutiny, internationally.[1] Why do I, therefore, still have the impression that our other-than-formal side is sometimes presented (often by ourselves) either as a "guilty pleasure," a "cute quirk," a "niche lite," or perhaps an "interlude" from the "proper stuff"? Is it simply "presented" this way, or is learning actually "perceived" in this way by some of us who do it? My somewhat dubious differentiation of perception and presentation is by no means implying ill intent or deception; it probably echoes countless examples that I can recall where I have personally presented my other-than-formal side as a guilty pleasure or quirk. In my own case, this usually has been either in order to feel more relaxed, to fraternize, or simply to be able to communicate with people whom I thought had limited to no understanding (interest, or respect!) of something I have been extremely passionate about.

But the taxonomy of things that are either in conflict or celebrate our raisons d'être would not be complete if I didn't rehearse another interesting and very familiar to me example in Waldron's text: her "What's the big deal?" reaction when introduced to an academic book that attempted to uncover how popular musicians learn. I recall reacting to my introduction to that particular text with very similar surprise and consternation. Within the few passages that I could actually identify with as somewhat approaching my own past learning experience as a popular musician, I remember being in a continual state of astonishment when reading about the fact that one could be so excited to discover such novelty in other people's learning journeys. To me, and the hundreds of people who were part of my extended network of popular musicians, this was simply how things either were, or at least should be. I don't know whether part of my discomfort might have been aroused because of the particular musical genre that I was passionate about, or the people who gave birth to it. But I have a vivid recollection of experiencing a synaesthetic drawing of mental parallels to old documentaries and encyclopedias that present the exciting findings of the curious elite of the British Empire during their wondrous journeys and their studies of the noble savages. Although sometimes skillfully presented, the material feels somewhat "not right." It is important for me to highlight that I certainly don't mean to refer here to the multiple reasons one can usually identify in these old reports for feeling extremely uncomfortable about them (such as racism, exploitation, and cruelty). But besides these profound issues, there are also some more "refined" issues that potentially manifest in works like the aforementioned book about how popular musicians learn, like condescension, superficiality, and ignorance. These more refined issues appear somehow to form what perhaps the late vocal pedagogy pioneer Richard Miller (2000) presented as some of the "pedagogic pollutants" in the ambient instructional air that today's singers breathe. I will try to assess whether such pedagogic pollutants might inform some of the dialogues on the use of technologies in education and music education and whether similar pollutants might exist within the overarching fields of music and technology.

I intend to do so because I find myself experiencing comparable inner "field disturbance" when I come across similar accounts wearing the "technologist's" hat.[2]

In this setting, technology is often being presented or accounted for as a knowable "thing," as Waldron intuitively argues, or as a set of heuristic remedies, similarly presented by myself and Purves (Himonides & Purves, 2010; Himonides 2012). Furthermore, I have argued (2012) that, in most cases, our conventional "understanding" or "agreement" about the threshold past which the non-technological (or "analogue") ends and the technological begins is arbitrary, and perhaps elusive. Furthermore, we can also observe a number of different "junctures," or time-bins, even when referring to the technological side of things. For example, Howell (chapter 44) talks about "new" music technologies; Waldron rehearses the prenetworked and networked technological contexts (and informal music learning 1.0 and 2.0); and Medvinsky (chapter 44) alludes to "contemporary" technology. How well defined are these, and does it matter anyway?

I believe that it does, and greatly so. The reason that I believe this is that the definition of these junctures, within ourselves, our microcosm(s), our macrocosm(s), and the greater society(ies) and world, is consequently tightly interwoven with our gauging of ourselves, and others (e.g., our students), somewhere on the digital immigrants versus natives (Prensky, 2001) continuum. This gauging, as a result, will inform our general "attitudes" (see Himonides, chapter 59 in this book) toward technology and, even further, our ethos, modus operandi, and potential influence on other people's musical and other-than-musical development. I believe that this type of classification is flawed, philosophically, conceptually, and praxially. Several writers have rightfully challenged Prensky's use of this type of classification as somewhat exaggerated, scaremongering, alarmist, or even naïve. I would prefer to characterize such a classification as simply unnecessary.

What shines under this light is, once again, that the biggest part of the "civilized" world is very much attracted to, reliant on, or simply "stuck" onto the worship of the tyranny of labels. We might come across people who identify themselves as popular musicians, classical musicians, digital natives, digital immigrants, qualitative researchers, quantitative researchers, constructivists, behaviorists, phenomenologists, pragmatists, objectivists, and so on. For those of us who are educators, we might also come across students from a plethora of musical backgrounds, and perhaps from families who celebrate some, or other, or absolutely none of the above philosophical stances. How could we possibly compartmentalize, classify (or pigeonhole, if you wish) the students in our classes according to all their existential "facets," in order to provide appropriately tailored developmental pathways for them? And how could we use technology appropriately in order to tailor-fit their experiences?

In line with what I argue in chapter 59, I believe that it is "attitudes" that need to be adjusted; in this case, not our attitudes toward technology but our attitudes toward diversity, originality, and the celebration of creative expression and learning needs that do not necessarily align with our own. Here is where the notion of narcissism could be used as a paradigm. We tend to appreciate and value our own trajectories. We were raised in a particular way, we formed an understanding of the various bits around us in a particular way, we have developed ways of understanding new concepts in a particular way, we have become musicians in a particular way, we sustain our musicianship in a particular way, we use technology in a particular way, we teach and we create in particular ways. At the same time, we have particular ways in which our self-esteem is formed, our confidence is built, and our aspirations are cultivated. Particularly for those parts of our past trajectories in which we assess that we have been quite successful, it is not uncommon for us to possess a strong sense of causality. A causal ascription for success (Meyer, 1980) can potentially harbor doubts about the effectiveness of other possible ways (or trajectories) that other people might have followed. Interestingly, a lack of empathy toward others is a very common trait with narcissistic personality disorder (U.S. National Library of Medicine, n.d). Perhaps equally interesting is the fact that according to published evidence (e.g., Kohut, 2013), people with narcissistic personality disorder often display elitist, snobbish, contemptuous, or patronizing attitudes. One cannot but notice that such attitudes are not infrequent among musicians, be it performers, educators, theorists, philosophers, or combinations of these.

Complementary to the paradigm of narcissism that I offer here is the paradigm of our romantic attachment to whatever tools or instruments we utilized during our perceived-to-have-been-successful developmental paths. This attachment can involve anything from instruments, tools, spaces, methods, and affordances to simple commodities such as food, drink, clothing, and accessorizing. Musicians can have romantic recollections about their favorite performance shoes, or pencils, or metronomes, or music stands, or manuscript paper. Similarly, technologists can have affectionate memories about particular programming languages, hardware interfaces, operating systems, storage devices, sequencing packages, or anything else that they have formed a positive association with.

These conditions frequently seem to infuse our *ethe* and *praxes* as educators and researchers. Examples of educational software interfaces that bear a freakish resemblance to the artwork in the textbooks that the developers (or their educational consultants) were taught with when they were children are not just numerous, they are the majority. There is usually a great disparity between the graphical user interfaces that children are invited to use within their formal learning contexts and those that they choose to use at home for gaming and general "edutainment." What they are expected to use at school is dumbed down, both visually and functionally, solely because this is what the people who designed the educational software think they would have expected to experience when they were children learners. It would be helpful for such disparities to be rectified. I believe that this will gradually happen because of the additional strength of what Waldron metaphorically calls "informal music learning 2.0," that is, networked digital technologies, and what I simply call *social technologies*.

Finally, I would like to highlight what I believe to be an often overlooked issue. Technology is often viewed as a tool, or as sets of tools, that can be used in order to "enable," "facilitate," or "enhance" educational (in the present case, music education) curricular aims and objectives, some of which haven't really changed in the past century or so. I see this view, too, as particularly counterproductive. It is through technology that a new "music education" can be envisaged, celebrated, and experienced. Technology can facilitate, enable, and foster a new praxis in music education. We simply need to embrace our technological selves, free from stereotypes, labels, and romantic attachments, and free from prejudice about the unfamiliar. We need to stop worrying about *what* technology we want to use *where* and focus on *how* we are going to use it and *why*. This will result in a critically formed framework of effective practice that will help us celebrate our development with and through music in new and very exciting ways. So let's not reminisce about how great things used to be; the best is yet to come!

Notes

1. I personally belong in the "camp" that sees learning belonging on a continuum, instead of discrete philosophical (or praxial) pillars.
2. Presented more extensively elsewhere in this book.

References

Himonides, E. (2012). The misunderstanding of music-technology-education: A meta perspective. In G. McPherson & G. Welch (Eds.), *The Oxford handbook of music education* (Vol. 2, pp. 433–456). New York: Oxford University Press.

Himonides, E., & Purves, R. (2010). The role of technology. In S. Hallam & A. Creech (Eds.), *Music education in the 21st Century in the United Kingdom: Achievements, analysis and aspirations* (pp. 123–140). London: IOE.

Kohut, H. (2013). *The analysis of the self: A systematic approach to the psychoanalytic treatment of narcissistic personality disorders*. Chicago: University of Chicago Press.

Meyer, J. P. (1980). Causal attribution for success and failure: A multivariate investigation of dimensionality, formation, and consequences. *Journal of Personality and Social Psychology*, 38(5), 704–718. doi: 10.1037/0022-3514.38.5.704.

Miller, R. (2000). *Training soprano voices*. New York: Oxford University Press.

Prensky, M. (2001). Digital natives, digital immigrants. *On the Horizon*, 9, 5, 1–6.

U.S. National Library of Medicine. (n.d.). *Narcissistic personality disorder—National Library of Medicine—PubMed Health*. Retrieved April 7, 2015, from http://www.ncbi.nlm.nih.gov/pubmedhealth/PMHT0024871/.

CHAPTER 47

PEDAGOGICAL DECISION-MAKING

RYAN BLEDSOE

TECHNOLOGY comes in many forms—computer software, hardware, do-it-yourself materials and projects—which makes the implementation of technology challenging in a creativity-driven music classroom. To express some of these challenges, I present vignettes from my experiences with students and other teachers that have lead me to focus my attention more on the pedagogies I implement than on considerations of the actual technologies. To close, I offer principles that I use to guide my own implementation of technology in the music classroom.

TEACHER CONFIDENCE WITH TECHNOLOGY

When I arrived at school today I noticed a small, rectangular package in my mailbox. Inside the package were four white plastic cubes: the AudioCubes. My excitement quickly turned to anxiety as I wondered what to do with them. I have the technology, but what comes next?

Teacher confidence with technology presents a challenge to the implementation of technology. With computer- and non-computer-related technologies evolving almost daily, I empathize with the feeling of finding it impossible to keep up with new tools. In 1968, Babbitt advised: "there's no reason in the world why the music educator should have to be a technological expert. Very few composers are. I don't have to know how to build an oboe to write for it, and I don't have to know how to build a synthesizer to write for it" (p. 132). Music educators might not need to be experts in the technology they use, but it is essential that they develop expertise in helping the students they work with to explore technologies that might help them access or express their own creative ideas. More important, today's students should learn to teach themselves to use new technologies, as the technologies we teach today will not be the same in the future. The musical tools used outside the school setting have changed drastically in the past century, so it seems strange to teach students the same music-making tools over the course of one's entire career.

When approaching projects that include technology, I remind myself that my confidence, or lack thereof, is not necessarily reflective of student confidence; students will sometimes have more confidence than me and sometimes less. My confidence level regarding technology can, however, prevent students' access to technology while in the music classroom. In our current educational practices, the teacher acts as the gatekeeper who chooses which technological tools are available and acceptable in the classroom. Whether or not this control is ethical, it exists. Perhaps teachers can position themselves as colearners who learn to use new tools alongside students rather than excluding new technologies.

Student Exploration

The sixth graders spent the past month making music with Victorian synthesizers (a speaker head attached to a 9-volt battery). Many of them had not used this technology before, so we began with a period of exploration. Our first class began with a short introduction to "twitching" the speaker to make a sound. The students spent the rest of class freely exploring the Victorian synthesizers while documenting the sonic discoveries they made using the device and other objects in our classroom. They filled the speakers with pennies, scratched the battery terminals with screws, and incorporated classroom instruments into their circuits. We continued our exploration of the Victorian synthesizers for the next four class periods, when it became clear that they were ready to share their discoveries with each other and move on to making pieces of music to share with the class (which some students had already begun).

Teachers' confidence with creative processes, specifically that of exploration, challenges the implementation of technology. Creativity researchers find that playful activity, where children have opportunities to play freely with their materials, and exploration time prior to creative task engagement can positively affect creativity (Amabile, 1996; Berretta & Privette, 1990; Pellegrini, 1984). Similarly, play and time for exploration are important in learning to use new technologies (Jenkins, Purushotma, Weigel, Clinton, & Robinson, 2009; Stager & Libow, 2013; Papert, 1993). In the vignette above, I felt that students would be more creative musically after exploring the sonic possibilities of the unfamiliar Victorian synthesizers, so I challenged them to find as many uses for the instruments as possible. During this time, students playfully discovered many different sounds they could create with the Victorian synthesizers and by using them with other instruments and materials available in the classroom. I also sought to make this time more exploratory by engaging students in an open question dialog (Pellegrini, 1984) about how they used the Victorian synthesizers and their potential uses. In a music technology educational setting, Greher and Heines (2014) emphasize the importance of providing space for students to "freely explore the material . . . [especially when] introducing new concepts either musically or computationally" (p. 30). Exploration of both music compositional techniques and technological tools are equally important in a creativity-driven music classroom. Similarly, without the time to explore *musical* ideas, students may not develop the techniques necessary to express their ideas in sound. An appropriate balance of time, tools, and techniques (Stauffer, 2001) requires that teachers accept and routinely act on the importance of play and exploration. Without the time to explore their tools, students may not develop the knowledge of tools or the techniques they need to express their ideas in sound.

Creativity and music education researchers have demonstrated that teachers who include less structured time in their classrooms foster creativity (Amabile, 1996; Burnard, 2000; Greher & Heines, 2014; Stauffer, 2011). In the vignette above, students had opportunities for student-led exploratory learning time with a wide range of musical tools, both the new Victorian synthesizer and our everyday classroom instruments, from which they chose instruments with which to create their pieces of music. A more structured scenario might include the teacher guiding students through tasks and modeling ways to use new technology. While providing more structure aids in teaching specific techniques, the musical product may be more similar to an etude or study of a certain technique rather than a personal expression of an idea through sound (Burnard, 1995; Stauffer, 2013). Outside music education, Martinez and Stager (2013) describe the value of students acting as makers, tinkerers, and engineers in the classroom in a less structured and more exploratory manner than is usual. In what ways could students be makers, tinkerers, or engineers in the music classroom? Future research in music education toward this discourse might be fruitful.

When Is Music?

Today the sixth graders shared the pieces that they worked on over the past few weeks. Before their performance, each group of students practiced one last time. I'm quite fond of one piece, and it stays in my head for hours after class each time I hear it. Going into the performances, I had no doubt that this group was well prepared for the performance. I didn't feel this way about all of the groups. One group in particular seemed like a mess. Their piece sounded unorganized, and I wondered whether these students had any intention in their music and the sounds they used. Their sound sources were microphone feedback and a contact microphone scraping against sandpaper, which they ran through a looper. These sounds were not pleasing to my ear. I wondered what the principal would think if she walked in and heard this piece.

After school, I read through the students' reflections and listened to the recorded performances from earlier in the day. The corresponding reflections from the boys in the group I worried about revealed that they were relating the experience of having a heart attack through their music—one of the boys in the group had recently lost his father to a heart attack. They wrote that the unorganized nature of their piece reflected the uncertainty his family felt and the irregularity of his father's heartbeat. What a powerful piece of music. I have to remember that this is their music making and not mine.

Teachers' conceptions of when music happens can hinder the implementation of technology. If the conception is that music happens when students perform pieces written by experienced composers for specific instruments, then occasions for music making, especially with technology as in the vignette, are limited. Not valuing all forms of music that people create may affect not only which technological tools the educator makes available for music making in the music classroom but also what the music educator accepts as "music" when it happens. Do we value certain instruments and styles of music over others? When does "creating music" happen—only after years of training and learning to read standard notation, or when children make something meaningful to them? How do these teacher conceptions affect students' music classroom experiences? As is told in the vignette, I discredited the musical value of the students' work because I did not like the sounds they used or the organization of their work. Their music did not make sense to *me*. My worst fear was that the principal would walk in and perhaps judge *my* teaching based on what *I* thought might have been perceived as something unmusical occurring in the music classroom. In the end, however, I was reminded of the greater importance of the *students'* self-expression and *their* intention, which moved me, over my own tastes.

In the vignette, I allowed my preconceptions of what music should sound like get in the way of the students' sonic expression of ideas and emotion. Although I disliked the sound objects they used—specifically the use of feedback and a contact microphone on sandpaper—their ability to express their intentions behind the sounds helped me to acknowledge my own musical narrow-mindedness. Electronic music makers emphasize the sound object (Roads, 1996) as the most fundamental unit of music rather than the note. A sound object can be a recorded found sound, feedback, a certain wave type, a sound made using granular synthesis, a processed acoustic instrument, and so on. The emphasis of the sound object over the note can be problematic for music educators who place value on a different way of understanding music: through notation centered around a specific set of 12 pitches and with timbres from an established bank of acoustic instruments.

Risk and Failure

I am so frustrated with the AudioCubes! Today the sixth grade students were all gathered around the cubes, ready to experiment with them and make sounds, when the software repeatedly crashed. After restarting the computer three times, I gave up. I sat with the students and we vented together about how unpredictable technology could be. What a disaster!

Two weeks later I asked the students to respond to questions about what they wished the AudioCubes could do, in addition to responding to what they actually did (when they worked properly). The students had so many great ideas, such as making an interactive wall out of them, using them to record responses for people who cannot speak, or connecting the cubes to your brain so that your moods could control the music they played. The students appeared to be inspired to think creatively about the AudioCubes because we struggled with them. It's as if the students want the AudioCubes to do more because they seemed to do so little. This turned out to be a great introduction to a new musical controller. I wonder how they would design their own controllers?

Trying or learning something new involves taking risks. Taking risks involves opening oneself to the possibility of failure. Teachers may be failure and risk averse for many reasons: wanting lessons to be perfect for their students, wanting flawless performances, motivating students, wanting student experiences to be successful, avoiding wasted time. As demonstrated above, experiences with technology may include the risk of failure, but even then learning can still occur. In fact, researchers in technology (Gee, 2007; Jenkins et al., 2009; Papert, 1993) and music composition (Cohen, 2002; Priest, 2006; Stauffer, 2013; Tarnowski, 1999; Wiggins & Medvinsky, 2013) recognize risk taking and failure as part of the learning and creative processes that provide students from early childhood through early adulthood with the experiences they need to grow. Although the AudioCubes caused frustration for me and the students, they were able to look at the tool's shortcomings in such a way as to discover what they desired in a musical tool to help meet their needs as people who create music. What disposition would I have modeled if I had abandoned the experience with the AudioCubes after they did not work as I had planned? Educators need to be conscious of how their reactions to failure affect their students' reactions to failure and willingness to take creative and technological risks (Greher & Heines, 2014). Becoming less failure and risk averse may encourage students to be more innovative and may create a disposition for exploring the unknown more often.

Curriculum

The students are learning so much from their experiences with inventing their own handmade instruments and sounds. Some students are designing and making electronic music controllers using microcontrollers and various electronic switches. Others are recording the sounds of classroom instruments and found sounds, manipulating them in GarageBand and other software, tailoring the sounds to their needs. I think they need another two weeks to follow through with these projects, but the sixth-grade curriculum document says I need to cover Gregorian chant and teach them to read the bass clef before the semester ends. What am I going to do?

The final challenge to the implementation of technology is the curricular imperative. Music educators may wrestle with stated or implied directives to "cover" specific skills and knowledge as mandated by the school, district, or state, depending on student age or the class or ensemble. People think differently about music depending on the tools they use to create music, especially when using electronic tools (Higgins & Jennings, 2006; Martin, 2012; Savage, 2005; Savage & Challis, 2001). These various ways of thinking about music are not necessarily reflected in music curricula. Real-world experiences with music will help students to develop a deeper understanding of music through questions and ideas important to the musical practices of today (Wiggins & McTighe, 2005). Students should encounter these ideas and questions as musicians rather than being taught about them (Boardman, 1989, 2002; Wiggins, 2009). Experiences as *real-world* musicians help to build students' confidence and empower them as actual musicians when they engage with music in ways that are familiar to them because they have encountered them in what they see and hear outside school.

Principles

I identify three principles that form the foundation for my own implementation of technology in music classrooms with music students of all ages where student-created music is the ultimate goal. I prefer to call them "principles" rather than "strategies" because, from my perspective, they underlie all of students' music-making experiences, not just those with new technology. These principles are (1) treat all musical tools as equal in the classroom, (2) provide access to both acoustic and electronic instruments in all classroom settings, and (3) provide opportunities for exploration and play to occur with all instruments.

Many music makers outside the school setting, in the "real world," do not discriminate between acoustic and electronic instruments. Instead, music makers choose an instrument because they have acquired technique on it or because it offers something to their music that helps them to express their ideas (Bledsoe, 2014). If educators conceptualize electronic instruments as creative tools that are just as useful as acoustic instruments, then students have opportunities to learn to use all of these tools, and which ones they prefer, so as to best achieve their musical goals. As an important component of creativity, time for exploration should be allocated with all musical tools, both acoustic and electronic, for students to discover first-hand the constraints and possibilities a tool provides to them. Following these principles, implementing them along with creativity-driven pedagogies, empowers students to be real-world music makers in the music classroom.

As I reflect on the experiences of implementing technology in the classroom, I initially worried about what to do with the technology, and things did not always go as I planned. In the end, the students engaged with music making that was meaningful to them, which is such a valuable part of being music makers. Pedagogical decisions, such as offering less structured classroom time to explore the possibilities of new technology and making risk taking and failure part of the creative and learning processes, provided students with a space for creating with a variety of music tools. More personal decisions, such as expanding my perception of when music happens and working through my technological anxieties, opened up opportunities for students to create with acoustic and electronic tools in all of our classroom activities.

REFERENCES

Amabile, T. M. (1996). *Creativity in context: Update to the social psychology of creativity.* Boulder, CO: Westview Press.

Berretta, S., & Privette, G. (1990). Influence of play on creative thinking. *Perceptual and Motor Skills, 71*(6), 659–666.

Bledsoe, R. N. (2014). *Electronic music makers: Perspectives and pathways.* Unpublished manuscript.

Boardman, E. (Ed.). (1989). *Dimensions of musical thinking.* Reston, VA: Music Educators National Conference.

Boardman, E. (2002). *Dimensions of musical learning and teaching: A different kind of classroom.* New York: Rowman & Littlefield Education.

Burnard, P. (1995). Task design and experience in composition. *Research Studies in Music Education, 5*(1), 32–46.

Burnard, P. (2000). How children ascribe meaning to improvisation and composition: Rethinking pedagogy in music education. *Music Education Research, 2*(1), 7–23.

Cohen, V. (2002). Musical creativity: A teacher-training perspective. In T. Sullivan & L. Willingham (Eds.), *Creativity and music education* (pp. 218–237). Toronto: Canadian Music Educators Association.

Gee, J. P. (2007). *What video games have to teach us about learning and literacy.* New York: Palgrave MacMillan.

Greher, G. R., & Heines, J. M. (2014). *Computational thinking in sound: Teaching the art and science of music and technology.* New York: Oxford University Press.

Higgins, A. M., & Jennings, K. (2006). From peering in the window to opening the door: A constructivist approach to making electroacoustic music accessible to young listeners. *Organised Sound, 11*(2): 179–187. doi: 10.1017/S1355771806001464.

Jenkins, H., Purushotma, R., Weigel, M., Clinton, K., & Robinson, A. (2009). *Confronting the challenges of participatory culture: Media education for the 21st century.* Cambridge, MA: MIT Press.

Stager, G., & Libow, S. (2013). *Invent to learning: Making, thinkering, and engineering in the classroom.* Torrance, CA: Constructing Modern Knowledge Press.

Martin, J. (2012). Toward authentic electronic music in the curriculum: Connecting teaching to current compositional practices. *International Journal of Music Education, 30,* 120–132. doi: 10.1177/0255761412439924.

Martinez, S. L., & Stager, G. S. (2013). *Invent to learn: Making, tinkering, and engineering in the classroom.* Torrance, CA: Constructing Modern Knowledge Press.

Papert, S. (1993). *Mindstorms: Children, computers, and powerful ideas.* New York: Basic Books.

Pellegrini, A. D. (1984). The effects of exploration training on young children's associative play. *Imagination, Cognition, and Personality, 4*(1), 29–40.

Priest, T. (2006). Self-evaluation, creativity, and musical achievement. *Psychology of Music, 34,* 47–61.

Roads, C. (1996). *The computer music tutorial.* Cambridge, MA: MIT Press.

Savage, J. (2005). Working towards a theory for music technologies in the classroom: How pupils engage with and organize sounds with new technologies. *British Journal of Music Education, 22*(2): 167–180. doi: 10.1017/S0265051705006133.

Savage, J., & Challis, M. (2001). Dunwich revisited: Collaborative composition and performance with new technologies. *British Journal of Music Education, 18,* 139–149. doi: 10.1017/S0265051701000237.

Stauffer, S. L. (2001). Composing with computers: Meg makes music. *Bulletin of the Council for Research in Music Education, 150,* 1–20.

Stauffer, S. L. (2011). *Theories of creativity.* Unpublished manuscript.

Stauffer, S. L. (2013). Preparing to engage children in musical creating. In M. Kaschub & J. Smith (Eds.), *Composing our future: Preparing music educators to teach composition* (pp. 75–109). New York: Oxford University Press.

Tarnowski, S. M. (1999). Musical play and young children. *Music Educators Journal, 86*(1), 26–29.

Wiggins, G. P., & McTighe, J. (2005). *Understanding by design.* Upper Saddle River, NJ: Pearson Merrill Prentice Hall.

Wiggins, J. (2009). *Teaching for musical understanding* (2nd ed.). Rochester, MI: CARMU Press.

Wiggins, J., & Medvinsky, M. (2013). Scaffolding student composers. In M. Kaschub & J. Smith (Eds.), *Composing our future: Preparing music educators to teach composition* (pp. 75–109). New York: Oxford University Press.

CHAPTER 48

EQUITY AND ACCESS IN OUT-OF-SCHOOL MUSIC MAKING

KYLIE PEPPLER

DISCUSSIONS around music technology and learning often center on the capabilities of a particular tool to impact individual learning. What musical concepts can it convey? How does the learner interact with it? How can we assess improvement in conceptual understanding and performance through experiences with the tool? What we miss from this focus, as several others have also argued in this book, is that musical learning is rarely, if ever, about a learner operating a tool in isolation. Rather, the communities—the physical places and spaces—that come together around these tools are central to how youth shape their ideas, interests, and identities in music.

Attention to the communal function of music is particularly important when we consider access and equity in music learning in out-of-school spaces. Music is inherently social: it is meant to be performed, listened to, and even danced to. Therefore, it is critical that we don't lose track of the notion that social influences and spaces heavily mediate the development of musical identities. In other words, in the rush to embrace new technologies, often overlooked in the learning process is the value of shared physical and communal spaces. However, youth need not only physical access to high-quality, professional grade software and hardware to record, mix, and share their original work (e.g., high-quality microphones, digital audio workstations (DAWs), soundproofed rooms, etc.) but also access to communities where they can jointly engage in shared music making. Consequently, informal or nonformal after-school spaces—such as libraries and other community centers—are now serving this nearly insatiable need, particularly in traditionally underresourced areas. These kinds of communal spaces extend access to high-quality equipment and software, as well as providing social settings in which youth can make and share music, and are playing an important role in the musical development of youth.

Out-of-School Music Learning Communities

One of the major findings to emerge from a series of ethnographic observations is that today's youth are using technologies like the DAWs described above to assume increasingly public roles as musicians, performers, and producers and are sharing their work through social media platforms (Ito, Baumer, & Bittanti, 2010). Prior ethnographic research suggests that whereas little creative production and reflection occurs when young people work alone at home on their computers (Giacquinta, Bauer, & Levin, 1993; Sefton-Green & Buckingham, 1998), well-designed social learning environments encourage youths to explore new kinds of musical learning more than they would be inclined to on their own (Peppler, 2014).

Many out-of-school spaces support social music learning by featuring a private recording studio that youth can reserve to record and mix their own albums. These studios are typically a room off to the side of the main communal area, with a microphone, a keyboard and/or piano, a computer with music software, and a glass window that allows others to see into the space while a dedicated group rehearses, records, and produces their work. In most out-of-school centers, the music studio is one of the most frequently used areas, and several spaces boast stories of youth who have "made it" in some fashion through the albums created there. The studios generally do not have a dedicated music staff, so youth often learn how to use the recording equipment or play on the instruments by observing or working with more experienced peers. This type of peer-to-peer learning is a trademark of informal learning communities, which support learning that happens in a casual or haphazard way through the use of dedicated "zones" for musical production or participation. That is, they have available resources, including tools, materials, and adult mentors, but they lack a formal organization and, importantly, rarely feature technology dedicated to more traditional musical skill acquisition, such as interval identification or music engraving. By contrast, content learning is an unplanned or unintentional byproduct when youth congregate based on a shared interest, and they unexpectedly gain insight through activities that were otherwise designed to be "just for fun." An organizing principle of many informal communities is that if youths engage in music making, learning will happen without additional and intentional curricular scaffolding.

One long-standing example of a large network of informal learning communities that support music making is the Computer Clubhouse Network, which aims to give youths, especially those in economically disadvantaged communities, opportunities to become fluent with new technologies (Kafai, Peppler, & Chapman, 2009). With the realization that youths who were disengaged from formal schooling were still interested in creative spaces in which to explore, design, and share work, the Computer Clubhouse Network adopted a constructionist approach to ensure that participants had opportunities for discovery and personal expression (Papert, 1980). As Mitchel Resnick, one of the founders of the Computer Clubhouse Network, says, "if [youth] are interested in video games, they don't come to the Clubhouse to play games; they come to create their own games.... In the process, youth learn the heuristics of being a good designer; how to conceptualize a project, how to make use of the materials available, how to persist and find alternatives when things go wrong, how to collaborate with others, and how to view a project through the eyes of others" (Resnick, 2002, p. 34). Likewise, clubhouses in the Computer Clubhouse Network offer similar kinds of experiences of making and sharing music, generally striking a balance between structure and freedom in youths' music learning. Sample songs and albums are shared widely and are distributed on the Computer Clubhouse Network. These projects give youths multiple entry points into the production process, as well as a sense of what is musically possible.

Another example of an informal community that supports music learning is YOUmedia and the associated Learning Labs, which consist of a rapidly growing network of more than 30 public libraries, museums, and community-based organizations across the United States. The spaces are dedicated to youth for their explorations of digital media. Within these settings, designers allow for "hanging out, messing around, and geeking out" (Ito et al., 2010) and make a special effort to develop youths' critical thinking, creativity, and digital media skills through hands-on activities. YOUmedia Chicago, for example, provides an open, 5,500-square-foot meeting space on the ground floor of the Chicago Public Library's downtown center, which serves close to 200 youths per day. At this YOUmedia site, any youth with a valid Chicago Public Library card has free access to equipment, including still and video cameras, rhythmic video games, instruments, laptops, and professional grade software. With the support of mentors from Chicago's Digital Youth Network as well as librarians, young people create rap music, spoken-word pieces, documentaries, and other digital media and art forms. The design of the YOUmedia learning space encourages individual and collaborative work and provides a safe and open area where youth can hang out and observe work created by their peers. YOUmedia youth also have access to a version of Remix World, a social learning network where young participants can share and reflect on their work, including their original tracks produced in the YOUmedia music studio.

Some out-of-school centers augment their musical material resources with more direct forms of musical instruction, an approach to out-of-school programming referred to as "nonformal learning communities" (Hirsch, 2005; Cole & Distributed Literacy Consortium, 2006). Nonformal communities differ from informal learning spaces because they emphasize goal-oriented learning and organized programming determined and implemented by adults. Proponents of these types of space believe that these learning communities can support large numbers of young people while also allowing individual students to pursue their interests at their own pace (Moll, Amanti, Neff, & Gonzalez, 1992). It is important to note that many learning communities—informal or nonformal—tend to be hybrid in nature. For example, the Computer Clubhouse Network includes physical spaces but also has a virtual Intranet that connects young people across the globe and allows those within the network to share and comment on one another's work.

Deepening Interest via "Progression Pathways"

There are several substantial hurdles between initial interest or experience in music and the type of long-term engagement that is sought after in more formal settings. Toward this end, a number of approaches have been developed to help youth cross the divide between initial interest and long-term engagement, approaches referred to by some as "progression pathways" (Price, 2006). YOUmedia offers a progression pathways approach in some of their programming, where they initially provide a place for youths to explore new programs and tools informally and then let them sign up for workshops in digital music production, digital video production, radio and podcasting, and the spoken word. This allows youth to first "mess around" and then "geek out" with the new technologies. Beyond the accommodation of different types of engagement with music and other forms of new media, the success of YOUmedia is largely based on having professional artists work directly with youths. Nichole Pinkard, founder of YOUmedia, and colleagues have observed that having adult mentors who are active musicians and professional artists is an important starting point for youths' creative engagement, but that these artist-mentors must also undergo professional development to ensure that they have the right combination of technical skills (a presence in Web 2.0 and social networks, knowledge of new media tools, and an understanding of a mentor's role), cultural capital (the ability to relate to youth as well as the credibility in their art form), and pedagogical knowledge (how to teach project-based learning strategies, critique youths' work, and incorporate new media literacies). The mentors then serve as brokers, inviting youth to pursue their passions and supporting their development in the arts.

Another exemplary model of progression pathways is the Musical Futures program in the United Kingdom (Price, 2006). The Musical Futures curriculum, which emerged from an exploration into why teens were not pursuing music education despite their obvious passion for music in their everyday lives, offers a series of models and approaches that teachers can adapt to their own settings and instructional styles. The emphasis is on guiding and modeling, rather than direct instruction; experienced students act as peer leaders. For example, youth incorporate their favorite pieces into the curriculum, teaching each other how to play simple parts on their instruments, either through notation or by ear, depending on varying levels of prior musical ability. Musical Futures is connected to a virtual space called NUMU, where participants can publish, share, and critique one another's work. Research from the Musical Futures project has conceptualized developmental pathways, based on speaking to young people about their aspirations and observing their responses to projects. As a result, this research identifies four typical participant archetypes in music (Price, 2006, p. 4):

1. Refusers—those with little or no inclination to engage with music other than as consumers. Refusers are perhaps the most complex group to understand because

they carry cultural and social baggage that often keeps them from participating. Skilled administrators of the curriculum, with strong interpersonal skills and persistence, can usually persuade these young people to shed their reluctance and defensiveness.

2. Waverers—those who have an interest in music but are not sure what they want to do or how to participate. Conventional music skills may often be rudimentary, and confidence can be fragile. Participants are more likely to feel comfortable finding "their own way in" to projects. Having a negative experience with performance can turn Waverers into Refusers.
3. Explorers—those who have acquired some skills and confidence but have not yet found a good match for their interests. Youth making the transition from primary to secondary education often fall into this category.
4. Directors—those who have already accessed a range of opportunities and are developing performance and rehearsal skills. They are confident among peers and motivated, with a clear sense of musical direction. They often form their own groups and are eager to extend their depth and range of skills.

By identifying young people's interests and skills, adult mentors, professionals, and the youths themselves are able to coconstruct progression pathways. The general progression pathways model both (1) outlines distinct stages toward deeper engagement, and (2) suggests the need to coordinate across multiple settings so as to deepen participation over time (including schools, informal learning, nonformal learning, and online social networks). As a result, the Musical Futures pathways suggests different activities

Table 48.1 Musical futures—categories of engagement and suggested activities

A: Refusers	B: Waverers	C: Explorers	D: Directors
Music Taster Workshop: Arrange a number of taster sessions in different musical styles and genres for young people, as a means of establishing what extra-curricular work might be successful and might appeal to students.	**Songwriting and technology club:** An after-school club that involves re-mixing and using technology to create music and videos.	**Songwriting project:** Introduction to chords, riffs, scales, lyric writing, melody, group skills and performance.	**Professional recording sessions** to produce an album.

targeted to each group of participants, ranging from music "taster" workshops to professional recording sessions. The intent of these activities is to meet youth wherever their current interests in music lie and get them excited about the possibilities of making and performing music. Note that the activities recommended by Musical Futures all position learners to actively reformulate their conceptions of music through the production of meaningful artifacts. Thinking about different strategies to engage learners at various points along the continuum is developmentally appropriate. Strategies that might engage Directors who want more opportunities to create professional quality work (such as providing supplies or resources for their latest projects) probably won't work for Refusers who aren't yet interested in the arts. Instead, Refusers would first need an engaging introduction, such as rhythmic video games like Rock Band, that are novice-friendly and allow them to play popular music.

Mixing formal, nonformal, and informal learning spaces, for example in the way described in the progression pathways model, is a necessary and important part of the individual's whole learning system, forming what scholars have referred to as an ecological approach (Sefton-Green, 2006) to interest-driven learning. New efforts in interest-driven music learning should build on this work, seeking to determine its applicability to other spaces, including virtual communities. For example, is it possible to use a similar progression pathway solely in online communities? Is the distinction between virtual- and physical-space-based learning important to youths today? If so, what are the major differences? This stance encourages us to design out-of-school learning communities more intentionally, as well as to change the way we study and see music learning at work in these environments.

REFERENCES

Cole, M., & Distributed Literacy Consortium. (2006). *The fifth dimension*. New York: Russell Sage Foundation Press.

Giacquinta, J. B., Bauer, J., & Levin, J. E. (1993). *Beyond technology's promise: An examination of children's educational computing at home*. New York: Cambridge University Press.

Hirsch, B. (2005). *A place to call home: Afterschool programs for urban youth*. New York: Teachers College Press.

Ito, M., Baumer, S., Bittanti, M., boyd, d., Cody, R., Herr, B., Horst, H. A., Lange, P. G., Mahendran, D., Martinez, K., Pascoe, C. J., Perkel, D., Robinson, L., Sims, C., & Tripp, L. (2010). *Hanging out, messing around, and geeking out: Kids living and learning with new media*. Cambridge, MA: MIT Press.

Kafai, Y. B., Peppler, K., & Chapman, R. (Eds.). (2009). *The computer clubhouse: Creativity and constructionism in youth communities*. New York: Teachers College Press.

Moll, L. C., Amanti, C., Neff, D., & Gonzalez, N. (1992). Funds of knowledge for teaching: Using a qualitative approach to connect homes and classrooms. *Theory into Practice, 31*, 132–141.

Papert, S. (1980*). Mindstorms: Children, computers, and powerful ideas*. New York: Basic Books.

Peppler, K. (2014). *New creativity paradigms: Arts learning in the digital age*. New York: Peter Lang.

Price, D. (2006). *Supporting young musicians and coordinating musical pathways.* London: Paul Hamlyn Foundation.

Resnick, M. (2002). Rethinking learning in the digital age. In G. S. Kirkman, P. K. Cornelius, J. D. Sachs, & K. Schwab (Eds.), *The Global information technology report 2001–2002* (pp. 32–37). New York: Oxford University Press.

Sefton-Green, J. (2006). *New spaces for learning: Developing the ecology of out-of-school education* (Hawke Research Institute Working Paper Series, no. 35). Magill, South Australia.

Sefton-Green, J., & Buckingham, D. (1998). Digital visions: Children's "creative uses" of multimedia technologies. In J. Sefton-Green (Ed.), *Digital diversions: Youth culture in the age of multimedia* (pp. 47–79). London: UCL Press.

CHAPTER 49

TECHNOLOGY, SOUND, AND THE TUNING OF PLACE

SANDRA STAUFFER

TAKE a moment to imagine your home, the place where you live. If you could not see, how would you know you were there? If you could *only* hear, how would you know that you were either "home" or "not home"? Could you recognize home by sound alone? How? If you moved to a new location, what sounds would you want to hear, or not hear, to feel "at home"? In other words, how would you go about making a space *sound* like the place of "home"?

These questions have to do with the relationship between sound and place. Sound is one of the ways in which humans experience the world, recognize place, and distinguish one place from another. Even when we are not conscious of the sounds around us, we are conscious of the sonic qualities of place and know where and even who we are because of those sonic qualities. Sound shapes our experience of place and our sense of self. Conversely, we are adept at using sound to shape place and to identify ourselves. But why, and how? Why is place important? How is sound part of place? Why and how do humans use sound to make places and identify themselves and for what reasons? How are technologies of all kinds, including sound-making and sound-inhibiting technologies, part of the making of place? And what might the relationship of place, sound, and technologies mean for music teaching and learning?

ABOUT PLACE

We tend to think of place in terms of physical location or space. Place philosophers, however, think of place as something more fundamental to human experience and to our sense of being in the world. To illustrate, consider again the place you call "home" relative to space, location, and time. You can measure the space of home in feet or meters and describe the dimensions or adequacy of the storage space. You can identify the location of home using an address, using latitude and longitude coordinates, or by pointing to a map. You can also measure the age of your home in days or years, identify when it was built, identify when you moved in, or describe the sequence of additions and renovations you or others may have made. None of those descriptions, however, or even all of them together adequately account for what makes the place where you reside "home." Place, or sense of place, has to do not only with space, location, and time but also with who and how we are in the world, with experiences and practices, with actions and interactions, with sociality and culture, and with memories and anticipations of the future. When Dorothy intones "There's no place like home" in *The Wizard of Oz*, she's referring to something much more than Kansas.

SOUND AND PLACE: AN EXAMPLE

Ethnomusicologist Steven Feld suggests that place always has an acoustic dimension, and he argues for "acoustemology"—the study of "sonic sensibilities, specifically of the ways in which sound is central to making sense, to knowing," and to place (1996, p. 97). Although Feld grounds his argument for "acoustic knowing" in part in extensive study among the Kaluli people in the rainforests of Papua New Guinea, one need not be in a rainforest to think about sound and place in human experience. Consider, for the span of a few paragraphs and the time it takes to read them, several examples of the complex interplay of sound and place in human experience.

I grew up in a place where cicadas (known as locusts in some places) began their characteristic buzzing in late August, just before school was about to start. I enjoyed school and still do. In my youth, the sound of cicadas signaled return to a place I enjoyed (school) in the place (rural southeastern Pennsylvania) where I lived. But I live now in a different location (urban central Arizona), and when I hear cicadas begin their buzzing in July, I think "too early" and "not yet" and "climate change." The sound of cicadas in July seems out of place relative to my earlier experience, or at least out of time, even though the sound still signals a return to the actions and interactions of the place "school," as well as a possible warning about the health of the larger place "planet," since I am now more conscious of the environment than I was in my youth.

If I do heed the sound of the neighborhood cicadas and go to "school" (now a school of music at a university—a different place altogether) in late July, the building will likely seem strangely quiet or even still, a sonic quality I recognize as the absence of the people and interactions that typically occur when classes are in session. The quiet may be unsettling, too unsettling, and I may leave. If I choose to stay and read or write in the place of my office with the door open, which I prefer, I can be fairly certain that, if I have turned them off, the telephone will not ring and the email will not ping, and that incidental sounds beyond my immediate visual frame are unlikely to disrupt my concentration unless they are out of place—a sudden loud crash, for example, or someone yelling.

By late August, when faculty and students have returned to campus, the place of the school of music will sound and feel different—vibrant, exciting, alive—at least for me, but perhaps not for everyone. For the custodian down the hall, more sound may signal more or different kinds of work. The musical sounds in the spaces of this school of music may not be equally inviting to all who pass through its entrance doors. I know, for example, that I am more likely to hear a violin than a vihuela, more likely to hear Western art music than country music, more likely to hear a rehearsal than a jam session, unless I know where to go in the building and when, or how to find the spaces where those sounds occur. I am also likely to hear and participate in the sounds of conversation, including discussions about which musical sounds are present or absent and why, and what the absence of certain musical sounds means in a place called "school of music." When all of the sounds of the new academic year become too loud for reading or writing at my desk, I may retreat to the library, which is similarly busy but with different kinds of sounds. Or I may choose to return to my home, where I can read and write to the familiar chirping of birds and the occasional barking of the neighbor's dog, until I become aware that an air conditioner doesn't sound quite right, prompting a different set of actions and feelings altogether.

The extended example above illustrates some of the ideas underpinning place. First, we are all multiply situated in a complex web of nested and overlapping places that is specific to each of us, yet shared with others in complex ways. Second, each of us understands place or has a sense of place that is particular to that person's own perspective and experience (the place of one's self), yet places are also social constructions, made and remade through actions and interactions with others. We understand place *with* others. Third, places are not static; rather, places and sense of place evolve over time and through human actions and interactions, including actions and interactions that may become habits or practices, as well as actions and interactions that may be unique or just-once events. Fourth, sense of place is embodied. We acquire our evolving sense of place through and in our bodies, in the sensations, perceptions, and actions that are part of place and that are shaped by place, including the place of ourselves. Fifth, places simultaneously include and exclude through the very actions and interactions that make places what they are. Places can both embolden and silence. Sense of place may free people to invent or may force compliance and make docile bodies of even those who are most attracted to a particular places. Places are rarely neutral. Finally, sound—musical or otherwise—is inseparable from the sense of place and being in place.

All of the ideas above are thoroughly explored in the literature of place (e.g., Casey, 2009; Cresswell, 2004; Malpas, 1999, 2007) except the last one—the idea that sound is inseparable from sense of place. The place literature focuses on sight, movement of bodies through space, and direction and dimension as part of the understanding of place. Yet sound precedes sight and, in large part, the ability to move in space. Before infants are born they hear the sounds of bodies and voices as well as the sounds of the environments they will soon experience directly. Sound is fundamental to the sense of place and the sense of who, where, and how one is in the world. Humans tune in to sound, tune out some sounds in favor of others, and over time, consciously or unconsciously, use sound to tune place (Coyne, 2010).

Sound Technologies and the Tuning of Place

The phrase "tuning of place" comes from Richard Coyne, who uses this metaphor to describe how people use digital media and technologies to adjust continually to the environment and to each other. In the twenty-first century, the word "technology" tends to conjure images of some kind of current or battery-driven and chip-enabled device, and, for the purposes of this short essay, one that makes a sound. Yet wherever and whenever humans have existed, they have used whatever means they have had at their disposal to make or suppress sounds for their own reasons and purposes (Peterson & Bledsoe, 2014). Whatever they used were the technologies of their time. In other words, for the purposes of this essay, I define "sound technology" as any tool used to create, separate, reproduce, intensify, or cancel sound—no current, chip, or battery needed. Both an acoustic guitar and an iPad are sound-making technologies. Both an office door and noise-canceling headphones are sound-reducing technologies. In both examples, one technology may be more or less satisfying than the other, depending on the individual and the circumstance. Still, the action of using either device—the acoustic guitar or the iPad, the office door or the headphones—effects an adjustment or sonic tuning of place. But why tune place at all?

Where are you reading this essay? What sounds are you experiencing as you read? Asking the latter question draws (perhaps unwelcome) attention to sonic qualities of the environment of which you were subconsciously aware before the question was asked. You know which places are best suited to your ability to read with attention and concentration, if that is what you have chosen to do, and your knowing of "reading places" includes both explicit and tacit understanding of sound. For example, if you are reading at home, you know without thinking which sounds to ignore and which ones require attention. You also know, through experience and practice, whether and how and how much you can adjust the sounds of different kinds of spaces using technologies of all kinds to suit your "reading place" preferences, and what degree of control you have of those technologies. For example, if you prefer to read with music playing in the background, you likely have more than one technological device available to produce that sound, as calling in live musicians to play while you read is both inconvenient and expensive. You may have more control of which music, how loud, which source (device), and location (another room?) in your own home than in the local coffee shop. If you go to the local coffee shop to read, you likely have less control of whatever technologies are producing sounds, if any, and if a few musicians arrive, set up, and begin to play, then you may continue reading, stop reading and listen, or leave. In some ways, then, technologies of all kinds allow humans to use sound to curate spaces and thereby make places.

In addition to curating particular physical spaces using sound, the evolution of electronic and digital technologies has enabled the curating of sounds themselves in ways unimaginable two centuries or even two decades ago. The act of curating sound, once the province of specialists (librarians, DJs), is now available to almost anyone, and this ability to curate has blurred the lines of public and private spaces. For example, you likely own a small digital device on which you have collected music you prefer or find useful in some way, and you have likely organized that music in some way that satisfies or pleases you. That digital device affords the opportunity for you to create a private music listening place almost anywhere, including public spaces where quiet is required or where music curated by others may be playing. Technologies can constrain as well. If you use that same small device as a telephone, then you may be constrained by the sounds it emits to get your attention by the ringtones and other signal sounds selected for you by the programmers of the device, or, if you have the know-how and the programming of the device permits it, you may have curated the sounds it produces by rearranging or even deleting the preprogrammed ones, by adding new sounds, or by tagging certain sounds to specific callers or functions that satisfy your ideas of sonic space and place. You have also likely learned to operate the device to avoid disrupting public spaces with sound, and you have probably frowned at those who are less adept or forgetful about silencing private devices in public spaces. You may have imagined or even participated in an event in which the sounds of private devices are part of a public and communal activity.

Sound and Place, Teaching and Learning

What, then, are the implications of sound and place for music teaching and learning? In 1992, in the introduction to a book for teachers, R. Murray Schafer commented on the changing "soundscapes" of the world, noting: "sounds are multiplying faster than people as we surround ourselves with more and more mechanical gadgetry" (p. 8). Schafer suggested that we become conscious of sounds, asking:

> Why do we focus on certain sounds and merely overhear others? Are some sounds discriminated against culturally so that they are not heard at all? . . . Are some sounds filtered out or rendered inconspicuous by others? And how does the changing acoustic environment affect the kinds of sounds we choose to listen or ignore? . . . Aside from the physiological dangers of a noisier [environment], how is our hearing psychologically affected by these changes? Is there a way of filtering out unwanted sound and still allowing the desired message through? Or does sensory overload finally beat us into a state of dopey submission or frazzled desperation? (pp. 7–9)

Schafer argued for a renewed sense of sound and listening, and for developing abilities to make "conscious design decisions" about sound.

Being conscious of sound is crucial, but Schafer's argument, the arguments of Coyne in *The Tuning of Place*, and the ideas of place philosophers suggest that something more is required—*being critical*. Here I use "being critical" not to suggest the role of the professional music critic, but rather a community of learners engaged in critical questioning and conversation about sounds (musical and otherwise) and places, and the relationship of sound and place. For example, what does the inclusion or exclusion of certain kinds of sounds (or music) in certain kinds of spaces communicate about place and belonging? (Consider what sounds, made by what people, are "allowed" in the place called "school." Why?) How are actions and sound related, and how do they function in certain spaces? (Consider how the act of reading a book signals a sound-place relationship. Does reading on an airplane create a micro place that silences conversation? Where and when and how did you learn that "reading" signals quiet?) How are places sonically designed, whether consciously or unconsciously, to attract some people and keep others away, and why are they designed that way? (Consider the differences in the sonic spaces of apparel stores and restaurants.) What does sonic design have to do with personal identity? How does the ability to acquire sounds from anywhere in the world, and to send sounds to anywhere in the world, impact local places? How do sounds, music, sound tags, and other sonic phenomena shape our experiences? Why bother to think about the sound-place relationship?

These and similar questions can be framed so as to focus on technologies of all sorts. What technologies are used to produce, inhibit, or curate sound in a space? Why these technologies and not others? Who has access to the technologies, the control of the technologies, and why? And if access includes know-how, where and when does the learning of sound technologies occur, who is involved, and why, and which technologies are used (or not)? As places and technologies evolve, these kinds of sound consciousness and critical questioning about sounds and spaces should be part of any environment that claims to be a place for music education.

References

Casey, E. (1997). *The fate of place: A philosophical history.* Berkeley: University of California Press.

Casey, E. (2009). *Getting back into place: Toward a renewed understanding of the place-world* (2nd ed.). Bloomington: Indiana University Press.

Coyne, R. (2010). *The tuning of place: Sociable spaces and pervasive digital media.* Cambridge, MA: MIT Press.

Cresswell, T. (2004). *Place: A short introduction.* Malden, MA: Blackwell.

Feld, S. (1996). Waterfalls of song: An acoustemology of place resounding in Bosavi, Papua New Guinea. In S. Feld & K. H. Basso (Eds.), *Senses of place* (pp. 91–135). Santa Fe, NM: School of American Research Press.

Malpas, J. (1999). *Place and experience: A philosophical topography.* Cambridge, UK: Cambridge University Press.

Malpas, J. (2007). *Heidegger's topology: Being, place, world.* Cambridge, MA: MIT Press.

Peterson, J., & Bledsoe, R. N. (2014, January 30–February 1). *Musical hacking: Creating inexpensive digital instruments.* Presentation to the Arizona Music Educators Association Conference, Mesa, AZ.

Schafer, R. M. (1992). *A sound education.* Indian River, Ontario: Arcana Editions.

though
PART 4
COMPETENCE, CREDENTIALING, AND PROFESSIONAL DEVELOPMENT

4 A

Core Perspectives

CHAPTER 50

TRADITIONS AND WAYS FORWARD IN THE UNITED STATES

JAY DORFMAN

Music teacher preparation must account for the combination of executive and expressive skills, the artful weaving of prescribed methodology with student-centered pedagogy, and many other aspects of music education to achieve the goal of preparing well-rounded, thoughtful teachers for the profession. Scholarly self-examination of our work as we integrate technology into teacher education, and thus into music classrooms, is essential for improving the ways in which we approach that integration. In many ways, scholarship in music education echoes the concerns of other subjects. Technology integration, in particular, is a component of music teacher preparation for which researchers have examined concerns similar to those of our general education counterparts. Teacher educators are obligated to continually reexamine the foundations of practice so that we can maintain the integrity and aesthetic basis of our subject while integrating new, research-based approaches that may benefit our students. We must regularly confront "the tacit nature of our practice" (Keast & Cooper, 2012, p. 66) and preserve self-awareness regarding the curricula we implement.

I began teaching technology-based music classes as a high school teacher in the late 1990s, and the possibilities for basing music learning on technology have fascinated me since. My perspective is that of an American schoolteacher who transitioned into the role of scholarly researcher, though I maintain an active role as a technology-based music teacher at the university level, and occasionally in K–12 classrooms. I have studied, using various lenses and methods, interactions between students and technology, and teachers' attitudes toward and practices with technology integration. Based on those experiences, in this chapter I offer my views regarding the interplay between music teacher education, technology skill development, pedagogical knowledge and implementation, faculty modeling, and long-lasting influences of each. While this chapter is based on much previous research in music education and technology, the conclusions I will draw here are my opinions based on my years of experience as a teacher and researcher, not on any new data collection or analysis. Central to the argument of this chapter is my concern that students in music teacher preparation programs must acquire both the skills to use technology and the pedagogical techniques to base their future teaching on it. Through self-study of programs and courses, we can determine the effectiveness with which those two related sets of skills are reaching our future educators.

There are practical concerns to be considered as teacher education curricula are examined, but it is similarly important that we address the theoretical foundations of integrating technology into music teacher preparation, and later into music instruction. Berry (2007) articulated several justifications for teacher educators engaging in self-study: (1) to check for consistency between practical approaches and philosophical beliefs; (2) to scrutinize individual aspects of preservice preparation; (3) to develop models of critical reflection; and (4) to evaluate our practice in ways that are more valuable than typical, institutional evaluations might be. By inspecting the traditions of thought and practice in preservice music teacher technology-based learning, perhaps we can begin to promote self-study, and to "encourage the kinds of flexible understanding our youth and our teachers need to engage creatively with technology" (Caillier & Riordan, 2009, p. 490). Rather than a comprehensive review of the issues that plague preservice programs as they integrate technology, I hope that this chapter will represent an initial step toward outlining a few elements of technology integration into teacher education that are ripest for scrutiny. I close this chapter with suggestions for research that may help us to examine rooted approaches to technology integration, and some concerns that may help propel this subgroup of our profession.

Technology and Traditions of Music Teacher Education

For several years, teacher education scholars have recognized that technology must be infused into the content and pedagogy classes typically found in teacher preparation curricula (Darling-Hammond, Banks, Zumwalt, Gomez, Sherin, Griesdorn, & Finn, 2005). Within the milieu of music teacher preparation, traditions divide the kinds of courses students take into standard types. As I will describe here, courses often serve specific, isolated purposes. As a result, technology skill development and its related pedagogical techniques can be left in a problematic, undefined position. In this section I explicate the lack of definition that results from program structures, and suggest resolutions to the problem.

Techniques versus Methods

Having studied and worked in several music teacher preparation programs, I am concerned about the place of technologically enhanced studies that occur in them. My specific concern is regarding the models that exist for integrating technology, and the fact that technology often does not fit easily into a mold that has existed for a long time. Examining most traditional music teacher training curricula in the United States will lead quickly to a generalization that there are essentially two kinds of classes that students are expected to take, as follows. (1) Techniques classes are those in which students learn performance skills on unfamiliar instruments. Students typically take a sequence of classes in woodwinds, brass, strings, percussion, and voice, and sometimes in classroom and folk instruments. Depending on the rigor of these courses and the inclinations of their instructors, they may not focus on practices related to the teaching of these instruments; instead they focus on providing students with performing experiences to develop beginning proficiency. (2) Methods classes, which students typically take in the later years of their program, focus on pedagogy, usually of large-group instruction, and frequently in the settings of band, orchestra, choir, and elementary general music. Since techniques classes are taken earlier, methods classes assume functional skills with instruments. Methods classes often include peer-teaching activities, which are designed to simulate the classroom environment and to let students practice teaching, relying on each other to mimic the performance of younger students. Methods courses also may carry an expectation that students spend time in the field, observing teachers in their classrooms, and perhaps participating in limited teaching activities.

While other models of music teacher education curricula certainly exist, and new structures are beginning to emerge, the traditional dichotomy between techniques and methods classes leaves technology studies in a state of limbo. For those programs where music technology topics reside in a stand-alone course, the approach to that course may be unclear. The course might be taught in a manner similar to the techniques approach, in which students are expected to acquire skills but are not introduced to pedagogical techniques nor given opportunities to practice them. Conversely, the course might be taught as a methods course in which the focus is on pedagogical technique but perhaps at the expense of technological skill development. In either case, the concern is that students will receive only partial preparation for technology-based music teaching.

Reenvisioning Technology Experiences

Solutions to the problematic in-between state where technology courses often sit do exist; they rely on music teacher educators to reframe how students might engage in technology experiences, as I have suggested previously (Dorfman, 2012). One solution is to distribute technology skill development and pedagogy throughout the larger curriculum where appropriate. Both methods (instrumental, elementary general, etc.) and techniques (woodwinds, brass, etc.) classes could contain streams that address technologies that are connected to the content of the course.

Some problems arise from this model. First, by including technology as a component of other courses, technologies that are not directly related to traditional performance areas may be ignored. Second, this model requires a great deal of collaboration between faculty members to determine which technologies will be addressed in each part of the curriculum. Third, faculty members have to possess the necessary skills to introduce technologies to their own students, which may be difficult to acquire.

A second solution is to maintain the model of a stand-alone technology course, but to reimagine its structure so that students gain both technology skills and pedagogical experiences. This is difficult because of the limited time allotted for the course (often far less than for any other single area of the curriculum) and often results in dilution of both skills and pedagogy; however, if this is the only possible solution because of practical concerns, such as credit distribution and teacher licensure requirements, then this approach is more desirable than simply ignoring either skills or pedagogy.

A final solution is to allot an equitable amount of time to developing technology skills and pedagogy; that is, as instrumental and choral segments of the curriculum do, perhaps curricula should offer both a technology techniques class and a technology methods class. Doing so would allow students to develop skills for using technology for musical purposes and the pedagogical orientations, approaches, and practices for teaching music with and through technology. Further, investing in study of technology from both practical and pedagogical perspectives allows the technology to become transparent, and for aspiring teachers to critically evaluate technology and the experiences that may result from its integration (Kirby, McCombs, Barney, & Naftel, 2006; Weiland, 2008).

Technology Skill Development

The notion of skill development in music technology might raise some eyebrows because of the inherent link between skill development and behaviorist approaches to education. This is a particularly problematic notion because it relates to some technology-based teaching and learning scenarios that feature drill-and-practice models found in many of the most popular computer-assisted instruction products. Zeichner (1993) reminded us more than 20 years ago that skill development approaches are intrinsically linked to "innumerable attempts during the 1920s and after to break down and analyze the teaching task into its component parts and to build a teacher education curriculum around these bits and pieces" (p. 4). Modern music pedagogy with technology can be designed to lead students through engaging, thought-provoking activities for which they might call on imaginative and creative processes.

Despite the criticism of skill development, we should remind ourselves that technology is a material, similar to a text book or sheet music, that we use to convey musical ideas or reach musical goals; it is not the musical content itself. Similarly, the development of technology skills is not the same as the development of teaching skills. Many technologies are designed with discrete components, and learning to use them requires experience (to one degree or another). Developing technology skill is a different thing entirely from developing teaching skills. Developing technology skills, and contextualizing technology within musical and social growth, is an important component of developing a complete professional teacher (Sykes, Bird, & Kennedy, 2010).

As in other segments of music teacher education, there are technology-related skills that preservice teachers should acquire. Miller and Lambert (2012) showed that the influence of undergraduate and graduate institutions on technology skill development in arts students was significantly greater than that of high schools; this type of evidence should inspire music teacher educators to feel empowered to include technology skill development as an important component of their curricula. Some recent scholarship has shown that there is a consistently agreed-upon set of technology skills that university music students should be able to understand or demonstrate (Webster & Williams, 2011, 2012, 2013); perceptions of the importance of those skills remain relatively stable for a sample of music teachers at various stages of their careers, including preservice teachers (Dorfman, 2015). Regardless of the skills that teacher educators or preservice teachers perceive as important, development of those skills should be based on sound educational models that draw upon best practices of sequencing and scaffolding content.

It should be noted that the type of learning in which preservice teachers (or anyone, for that matter) learn to use the functions of technology (software or hardware) is often referred to as training. I distinguish between educating and training, and hope that the experiences provided in preservice music teacher programs leans further toward educating than toward training. With abundantly available online resources, along with countless brick-and-mortar facilities, just about anyone can obtain training that contains procedural knowledge. They can learn step-by-step processes for accomplishing particular tasks with technology. Educating, however, encourages students to be thoughtful and creative with—and perhaps critical of—the technologies they use. Technology education, as I view it, subsumes training, and compels students to think about inventive ways that technology might be used for teaching. Education in technology encourages future teachers to consider the content first, and to explore ways in which the technology might enhance learning experiences that surround the content (Roblyer & Doering, 2012).

Still, preservice teachers need to learn the procedural elements of technology. When music teacher educators provide training so that preservice teachers can acquire skills with music technology, they should strongly consider the sequence of presentation of those skills and, as in the learning of any sophisticated task, consider how moving from simple to complex can help students succeed. The open-ended nature of many music applications can be overwhelming, so presenting tasks in a logical sequence can help to reduce students' anxiety and bring some order to potential chaos. For example, when learning to create scores in notation software, a logical sequence would be to begin with relatively simple scores, such as single line melodies, piano-vocal arrangements, and those without many advanced symbols. Once students master placement of the common symbols that occur in modest scores, they might advance to more sophisticated ones.

Similar considerations for simplicity and complexity might be made when selecting software itself. The market is replete with software programs that accomplish the same tasks, though their levels of sophistication vary. When learning techniques of MIDI sequencing, teachers should question whether a complex piece of software such as Logic or Pro Tools is necessary or GarageBand or Mixcraft would suffice. A key consideration is the usefulness of software that students learn in their preservice years once they enter the P–12 classroom. Decisions as to the software we employ, and the sequences we use to learn software, should be made carefully and intentionally, with an eye toward usefulness in the field and many other factors.

In my technology training classes, a portion of skill development is tied to faculty and program expectations of students in other parts of their preservice curriculum. While technology can be an effective means of teaching music, in reality, most of the students I teach are more likely to work in traditional settings, such as band, choir, orchestra and elementary general music, than to teach purely technology-based music classes. Part of my responsibility, then, is to help them develop skills that can be used in those settings. For example, when students take instrumental methods courses, they are often asked to prepare materials for beginning instrumental lessons. Technological skills such as sequencing MIDI tracks can support these activities. Recognizing the links between technology classes and the other core components of music teacher training can help us create activities that are relevant; they allow students and faculty to draw close links between classes that might otherwise seem disconnected.

An Emerging Model of Pedagogical Knowledge

Until quite recently, few theoretical models have existed that helped both music teacher educators and future music teachers to conceptualize the difficult interconnections between music, pedagogy, and technology, and the specific kinds of knowledge associated with each of those domains. Berry (2007) described lack of theoretical work in teacher education in general, but the description seems applicable to technology-based music instruction: "for many teacher educators, the difficulties associated with researching personal practice lie not so much in recognizing the complexities inherent in their own work (these they readily see) but in finding ways of representing that complexity to others. Because so little of the 'swamp' has been mapped, it is hard to know how to proceed" (p. 31). The "swamp" of technology-based instruction had been sorely in need of a map for some time. The theoretical framework known as technological pedagogical and content knowledge (TPACK) is influential and its significance is likely to continue growing. It is an extension of Shulman's (1986) model of "pedagogical content knowledge," which suggests that teachers possess special kinds of knowledge of their subject areas and the ways in which students might best learn content. The theory of TPACK (Koehler & Mishra, 2008; Mishra & Koehler, 2006) suggests that technology adds a layer of complexity to both content and pedagogy, and causes interesting intersections in the process. (Fig. 50.1 shows how the authors depict the model.)

Complete explanations of the TPACK model are available elsewhere (e.g., Bauer, 2013, 2014; Dorfman, 2013); for the purposes of this chapter, I will focus on the most complex of the intersections, the center section that depicts the TPACK construct itself. Essentially, TPACK for music captures the ways in which technology influences the musical content that teachers know, and the ways in which they teach that musical content to students. This model has emerged both as an approach to preparing teachers for entering the technology-based music class and as a means for measuring their preparedness or level of accomplishment.

The outer dashed circle of the model suggests that the model considers various contexts. Contexts might include constructs such as socioeconomics, gender balances, grade levels, students' and teachers' expertise with the content, and many other factors. Here, I will assume the context of TPACK to be music teaching in general, while acknowledging that music teaching scenarios vary greatly.

The TPACK framework provides us with a way of thinking about how to navigate the swamp of technology-based music teaching. As we prepare teachers, the model helps us to remember that content, pedagogy, and technology are distinct; when these three elements overlap, more sophisticated teaching may result (Dorfman, 2013). An assumption that can be extracted from this model is that teacher knowledge, in all the permutations the model suggests, eventually is applied to the pedagogical choices that teachers make in their classrooms. If this is the case, then the model further supports my previous

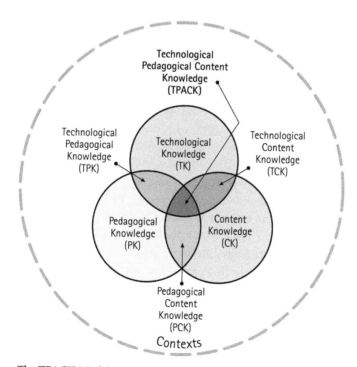

FIGURE 50.1 The TPACK Model. Reproduced by permission of the publisher, © 2012 by http://tpack.org.

assertion that preservice preparation for technology-based music instruction must include emphases on both skill development and pedagogy, so that teachers can shape their technology-based music instruction practice. Only by doing so can we properly position future teachers to build sophistication in their teaching.

The Roles of Faculty

Mere exposure to technology does not guarantee that preservice teachers will later be successful implementers of technology into their teaching. We must consider the limitations of technology, and we should not assume sustained access, support, or changes in teaching practice as a result of technology experiences in the preservice program (Wang, Spalding, Odell, Klecka, & Lin, 2010). The onus falls on music teacher educators to design technological learning opportunities and to model behaviors. These faculty members play a pivotal role in the development of technology skills and pedagogical understanding for their preservice students. In my estimation, the primary difficulty related to furthering the use of technology in the music classroom is music teacher education faculty's hesitation to embrace technology and emphasize it as an effective means for teaching music. By modeling reasonable uses of technology in their own practices, music teacher educators may have a profound influence over students' thinking about using technology in their later teaching.

Perhaps the most prevalent sophisticated technology available on most college campuses is the online learning management system; popular examples include Blackboard, Moodle, Edmodo, and Canvas. While these systems vary in design and feature sets, most include the ability to post class documents, host discussion boards, manage student grades, host media, and manage communications between members of classes. Learning management systems are becoming more popular in the P–12 environment as well, and teachers frequently rely on the assessment and communication portions of these applications for their daily work. By modeling good uses of learning management systems, music teacher education faculty can set an example of technology uses for their preservice students. While the software is sophisticated, most learning management system packages are relatively simple to use, and help faculty to accomplish tasks they would do anyway. Using academic technology is merely one example of modeling technology use that will set an example for future teachers.

Faculty might also consider the ways that their own collaborative behaviors can influence the integration of technology into their programs. Faculty can make decisions regarding how best to include technology in the curriculum. Structuring technology experiences as a stand-alone class, as opposed to distributing them throughout various classes, should be a faculty decision, and faculty who determine the approach that their program takes will adjust according to the outcomes or consequences of that determination. Faculty can collaborate on this decision and implement what they think is the best strategy for their program. If they choose to distribute technology experiences throughout multiple courses, then they should collaborate further to determine which skills are most appropriately addressed in each component of the curriculum. Collaborative acts further model the structures often found in innovative technology uses at the P–12 level. Personally, I remain torn on this issue, because neither approach (nor any permutation thereof) has been shown in the research or pedagogical literature to be superior or inferior to any other. My experiences have been only in the isolated class model, though I remain open to experimentation with other models. I will return to this issue in my suggestions for research at the end of this chapter.

An important influence that music teacher education faculty can have toward the goal of technology integration in their students' later work is to encourage reflection. In my teaching, I regularly engage students with technological tools and methods for reflecting on their own work. For example, when students in instrumental methods classes participate in microteaching activities, they are video recorded. Those recordings are uploaded to a streaming video service such as YouTube, and the students are required to access the video and create reflections, often using a blog post or a journal entry in the university's learning management system (Blackboard). I use the same techniques for microteaching experiences in my technology classes. I also have my students watch video recordings of teachers in technology-based music classes and write guided reflections about their observations. In these ways and others, I encourage my students to think deeply about the ways in which technology, content, and pedagogy interact; the elements of teaching and learning with technology that are particularly positive; and the elements of teaching and learning with technology that need improvement. Darling-Hammond and colleagues (2005) suggest that "technology provides tools that aid in reflection and improvement" (p. 188); learning technology skills, and using technology to improve teaching, therefore, creates a cyclical experience of learning and reflecting that can be viewed repeatedly and at slower speeds than those found in "the rapid pace of classroom life" (Rosaen & Florio-Ruane, 2008, p. 723).

Reflection may also help students to draw connections between the contents of technology studies and ideas they learn and practice in other portions of the preservice curriculum. Particularly important is reflection on preservice activities in the arts, which enables students to consider the focus of their practice on incorporating values of aesthetic education into their teaching, rather than designing courses that are merely technical (Holzer, 2013).

As teachers who set examples for preservice students, faculty members must practice "reflection-in-action" (Schön, 1987) and must be transparent about their own pedagogical experiences and decisions. Zeichner (1993) suggests that "reflection is treated as a social practice rather than merely as a private activity" (p. 11). Clarity around the process of reflection would certainly benefit preservice teachers. Reflective learning and teaching will help preservice teachers to develop their own pedagogical theories and decisions; faculty members should encourage critical examination of teaching and learning experiences so that preservice teachers can develop their distinct pedagogical styles.

Finally, faculty members can seek opportunities to simulate real technology-based music classes or to actually immerse their students in technology-based music environments as much as possible. As mentioned earlier, a methods approach to technology preparation would suggest that preservice teachers engage in activities such as peer teaching, field observation, and, if possible, teaching in the field. Only through these types of experience can preservice teachers gain valuable familiarity with situations they may encounter in their careers.

A mythology has developed that students enter undergraduate programs with technology skills that far surpass those of faculty members, or so-called digital immigrants (Prensky, 2001). More accurately, while the undergraduate students I encounter are savvy with particular consumer technologies, such as particular functions of their cell phones, they generally lack depth of experience with applications for music teaching and learning. Put another way, they know what they need to know to make particular technologies function for their personal habits, but they do not always consider the educational applications of those technologies. Certainly these are generalities, but I have seen them consistently throughout my college teaching career, and some research literature supports the notion (e.g., Margaryan, Littlejohn, & Vojt, 2011).

Because of this phenomenon, in my teaching I draw a distinction between preservice teachers and in-service teachers in terms of their needs. Preservice music teacher education, I believe, needs to include introductions to technology skills that are specific to music teaching, and varied opportunities to practice those skills. At the preservice stage, technology skills should be taught and practiced with an eye toward their application in the music classroom, or for uses that are connected to classroom needs. Even the most pedestrian applications—for example, office-type software or music player software—can be approached from an educational perspective and framed in ways that will be beneficial for the music teacher.

At the graduate level, I propose that the emphases should be different from those at the preservice level. Graduate music education curricula typically include coursework in which students engage in reading theoretical perspectives of critical issues in music education and in which they learn to consume and critique research in the field. Technology-focused coursework at the graduate level should adopt this same stance. Students at the graduate level should be exposed to theoretical constructs that may influence how they think about technology-based music teaching. This approach can certainly include the TPACK framework, and might also include theories of creativity, constructivist learning, social learning, and others. As students are exposed to theories, they should be encouraged to recognize items of technology that appeal to theoretical constructs inherently, and to seek out opportunities to approach technology-based learning from a theoretically informed perspective.

While exposure to theoretical perspectives, and practice informed by them, is important, maintenance of skills remains necessary, and is complicated by the ever-changing nature of technology. Koehler and Mishra (2008) wrote of this volatility: "the knowledge required to learn to use digital technologies is never fixed. Technology changes quickly, causing hardware and software applications to become outdated every few years. One has to continually keep up with the changing demands of new technologies. . . . Moreover, the rapid changes often happen in piecemeal fashion, which leads to users having to work with a variety of versions of software and hardware, some of which may be incompatible with each other. . . . The instability of digital technologies requires that teachers become life-long learners who are willing to contend with ambiguity, frustration, and change" (p. 8). Technology does not stand still, so teachers must continually renew their technological knowledge through formal professional development and informal exposures to software and hardware. Recent research has shown that music teachers are more likely to pursue formal technology professional development when the motivation to do so is their own (Dorfman & Dammers, 2015). Given these findings, professional development providers should continue to offer learning opportunities that are accessible to teachers, and should include a variety of topics in those opportunities. Perhaps the most significant contributor to this effort in the United States has been the Technology Institute for Music Educators which, since 1995, has been offering professional development workshops through various universities. The weeklong workshop model that is typical of Technology Institute for Music Educators courses has been shown to significantly increase several indicators of technology comfort and use for teachers (Bauer, Reese, & McAllister, 2003). Regardless of the source of professional development, I encourage school administrators to support teachers in their quest for skill development in technology.

Research Needs for Moving Forward

In accord with Zeichner's (1999) call for research about teacher education in general, I propose that music teacher education has several research needs that might propel the incorporation of the technology component of our field. In accord with Berry's (2007) suggestions regarding the purposes of self-study of teacher preparation referenced earlier, it is critical that we examine and modify the ways technology is approached in preservice music programs. Doing so would help us align practices with beliefs, and develop reflective critiques that are valuable. I suggest the following possible ways forward for researchers.

Technological competency must be a cornerstone of program development as programs are evaluated. Researchers should consider evaluating programs from the perspective of their inclusion of technology. The long-held (and previously mentioned) question of the effectiveness of the stand-alone technology class, as it compares to other models, may be addressed through program evaluation. Further distinguishing between the techniques and methods course models might be accomplished through program evaluation, though it is possible that researchers could propose effective alternatives. This type of research may also shed light on the types of technology integration that are most appropriate in particular branches of the music teacher education curriculum, and offer models for replication.

Policy studies might provide insight into effective ways to include technology in preservice music curricula. With the recent focus of the teaching profession on measurable performance indicators (Shuler, 2013), teacher education faculty must be certain that their curricula prepare teachers for increasingly data-driven schools where they will be held accountable to measured benchmarks for their students' progress. With more specific standards that reference detailed outcomes of learning, the types of skill that preservice music teachers are required to develop may be further nuanced. While the relevance of skill development maybe be promoted, this might impact the creativity and freedom with which faculty can design courses. The research community should examine the impact of state and federal policies on music teacher education curricula, and the mandates that will influence their students' future integrations of technology into P–12 classrooms.

I have previously drawn distinctions regarding technology content that might be addressed at the preservice versus in-service levels. No researchers have yet examined this distinction, or the usefulness of content that might reside in courses at one level or the other. Music teacher education faculty design courses based on what they suspect might be useful, or—perhaps worse—based on the newest gadget or software to hit the market. Systematic examination is needed to determine what types of in-service technology learning experiences are most relevant or desired, and in what ways those experiences relate to classroom implementation. This type of information may help to clarify the preservice/in-service distinction, if indeed a clear line exists. Also at the in-service level, a decade has passed since the most substantial investigation (Bauer, Reese, & McAllister, 2003) into the effectiveness of professional development for music teachers in technology. With new, informal, and personalized professional development models now appearing—including online synchronous and asynchronous learning—it is important to examine the effectiveness and desirability of these new models. Doing so could shape the ways in which skills are learned for the next decade.

Finally, with the rise of the TPACK model, we are in a position to observe, and perhaps accurately measure, students' preparation for technology-based music instruction. While some have questioned the validity and usefulness of the model (Brantley-Dias & Ertmer, 2013), it remains the most sophisticated and all-encompassing theoretical perspective from which to paint an accurate picture of preservice teachers' preparedness for teaching in a technology-based environment. Similar to researchers in other fields, music education researchers should conduct studies in which we examine the usefulness of TPACK for conveying the complexity of technology-based music teaching to preservice students, and in which they use the notion of TPACK to assess development in the practices of technology-based teaching. Since TPACK is a model that is intended to cut across disciplines, researchers could examine the validity of its constructs specific to music education (Cavanagh & Koehler, 2013; Shinas, Yilmaz-Ozden, Mouza, Karchmer-Klein, & Glutting, 2013). Studies in music education that draw on the TPACK framework might also examine the extent to which technology can be an effective means for teaching music, and the types of preservice technology experiences that translate particularly well into in-service classroom applications.

References

Bauer, W. I. (2013). The acquisition of musical technological pedagogical and content knowledge. *Journal of Music Teacher Education*, 22(2), 51–64. doi: 10.1177/1057083712458593.

Bauer, W. I. (2014). *Music learning today: Digital pedagogy for creating, performing, and responding to music*. New York: Oxford University Press.

Bauer, W. I., Reese, S., & McAllister, P. A. (2003). Transforming music teaching via technology: The role of professional development. *Journal of Research in Music Education*, 51(4), 289–301.

Berry, A. (2007). *Tensions in teaching about teaching*. Dordrecht: Springer.

Brantley-Dias, L., & Ertmer, P. A. (2013). Goldilocks and TPACK: Is the construct "just right?" *Journal of Research on Technology in Education*, 46(2), 103–128.

Caillier, S. L., & Riordan, R. C. (2009). Teacher education for the schools we need. *Journal of Teacher Education*, 60(5), 489–496. doi: 10.1177/0022487109348596.

Cavanagh, R. F., & Koehler, M. (2013). A turn toward specifying validity criteria in the measurement of technological pedagogical content knowledge (TPACK). *Journal of Research on Technology in Education*, 46(2), 129–148.

Darling-Hammond, L., Banks, J., Zumwalt, K., Gomez, L., Sherin, M. G., Griesdorn, J., & Finn, L.-E. (2005). Educational goals and purposes: Developing a curricular vision for teaching. In L. Darling-Hammond & J. Bransford (Eds.), *Preparing teachers for a changing world: What teachers should learn and be able to do* (pp. 169–200). San Francisco: Jossey-Bass.

Dorfman, J. (2012, November 15–18). *Reinventing music education technology courses*. Paper presented at the annual conference of the Association for Technology in Music Instruction, San Diego, CA.

Dorfman, J. (2013). *Theory and practice of technology-based music instruction*. New York: Oxford University Press.

Dorfman, J. (2015). Perceived importance of technology skills and conceptual understandings for pre-service, early-career, and late-career music teachers. *Symposium of the College Music Society*, 55. Retrieved January 5, 2017, from http://bit.ly/1FswGr9.

Dorfman, J., & Dammers, R. J. (2015). Predictors of successful integration of technology into music teaching. *Journal of Technology and Music Learning*, 5(2), 46–59.

Holzer, M. F. (2013). Many layered multiple perspectives: Aesthetic education in teaching for freedom, democracy, and social justice. In D. K. Lee & N. M. Michelli (Eds.), *Teacher education for democracy and social justice* (pp. 131–148). New York: Routledge.

Keast, S., & Cooper, R. (2012). Articulating our values to develop our pedagogy of science teacher education. In S. M. Bullock & T. Russell (Eds.), *Self-studies of science teacher education practices* (Vol. 12, pp. 65–84). Dordrecht: Springer.

Kirby, S. N., McCombs, J. S., Barney, H., & Naftel, S. (2006). *Reforming teacher education: Something old, something new*. Santa Monica, CA: RAND Corporation.

Koehler, M., & Mishra, P. (2008). Introducing TPCK. In AACTE Committee on Innovation and Technology (Ed.), *Handbook of technological pedagogical content knowledge (TPCK) for educators* (pp. 3–29). New York: Routledge/American Association of Colleges for Teacher Education.

Margaryan, A., Littlejohn, A., & Vojt, G. (2011). Are digital natives a myth or reality? University students' use of digital technologies. *Computers & Education, 56*(2), 429–440. doi: 10.1016/j.compedu.2010.09.004.

Miller, A. L., & Lambert, A. D. (2012). Comparing skills and competencies for high school, undergraduate, and graduate arts alumni. *International Journal of Education & the Arts, 13*(5). Retrieved January 5, 2017, from http://www.ijea.org/v13n5/.

Mishra, P., & Koehler, M. (2006). Technological pedagogical content knowledge: A framework for teacher knowledge. *Teachers College Record, 108*(6), 1017–1054.

Prensky, M. (2001). Digital natives, digital immigrants. *On the horizon, 9*(5), 1–6.

Roblyer, M. D., & Doering, A. H. (2012). *Integrating educational technology into teaching*. Boston: Pearson.

Rosaen, C., & Florio-Ruane, S. (2008). The metaphors by which we teach: Experience, metaphor, and culture. In M. Cochran-Smith, S. Feinman-Nemser, & D. J. McIntyre (Eds.), *Handbook of research on teacher education: Enduring questions in changing contexts* (3rd ed., pp. 706–731). New York: Routledge/Taylor & Francis and the Association of Teacher Educators.

Schön, D. A. (1987). *Educating the reflective practitioner: Toward a new design for teaching and learning in the professions*. San Francisco: Jossey-Bass.

Shinas, V. H., Yilmaz-Ozden, S., Mouza, C., Karchmer-Klein, R., & Glutting, J. J. (2013). Examining domains of technological pedagogical content knowledge using factor analysis. *Journal of Research on Technology in Education, 45*(4), 339–360.

Shuler, S. C. (2013). From retrospective to proactive: Creating the future that students need. *Arts Education Policy Review, 115*(1), 7–11. doi: 10.1080/10632913.2014.847353.

Shulman, L. S. (1986). Those who understand: A conception of teacher knowledge. *American Educator, 10*(1), 9–15, 43–44.

Sykes, G., Bird, T., & Kennedy, M. (2010). Teacher education: Its problems and some prospects. *Journal of Teacher Education, 51*(5), 464–476. doi: 10.1177/0022487110375804.

Wang, J., Spalding, E., Odell, S. J., Klecka, C. L., & Lin, E. (2010). Bold ideas for improving teacher education and teaching: What we see, hear, and think. *Journal of Teacher Education, 61*(3), 3–15. doi: 10.1177/0022487109349236.

Webster, P. R., & Williams, D. B. (2011, October 20–22). *Music technology skills and conceptual understanding for undergraduate music students: A national survey*. Paper presented at the annual conference of the Association for Technology in Music Instruction, Richmond, VA.

Webster, P. R., & Williams, D. B. (2012, November 15–18). *Refining a national survey on music technology competencies: Active ways to engage students*. Paper presented at the annual conference of the Association for Technology in Music Instruction, San Diego, CA.

Webster, P. R., & Williams, D. B. (2013, October 31–November 3). *Defining undergraduate music technology competencies and strategies for learning: A third-year progress report*. Paper presented at the annual conference of the Association for Technology in Music Instruction, Cambridge, MA.

Weiland, S. (2008). Teacher education toward liberal education. In M. Cochran-Smith, S. Feinman-Nemser, & D. J. McIntyre (Eds.), *Handbook of research on teacher education: Enduring questions in changing contexts* (3rd ed., pp. 1204–1227). New York: Routledge/Taylor & Francis and the Association of Teacher Educators.

Zeichner, K. (1999). The new scholarship in teacher education. *Educational Researcher, 28*(9), 4–15. doi: doi:10.3102/0013189X028009004

Zeichner, K. M. (1993). Traditions of practice in U. S. preservice teacher education programs. *Teaching and Teacher Education, 9*(1), 1–13. doi: http://dx.doi.org/10.1016/0742-051X(93)90011-5.

CHAPTER 51

TECHNOLOGY AND INVISIBILITY IN MUSIC TEACHER EDUCATION

GENA R. GREHER

In my entire life as a student, I remember only twice being given the opportunity to come up with my own ideas, a fact I consider typical and terrible.

—Eleanor Duckworth, *The Having of Wonderful Ideas*

I am a music teacher educator in the United States who came to this field after 20 years in advertising as a music producer/music director. I have seen firsthand how music can enhance a message, alter the mood and feel of a film, sell a product, foreshadow events, and transport the listener's imagination. I have experienced the role of technology from all facets of production, from its most mundane uses to its ability to help shape and create sonic experiences that were previously unimaginable to prior generations of musicians and their audiences. Technology in the right hands can expand our creative options as musicians. What sets one production apart from the next comes down to the ideas generated by the people at the helm of the technology. Ideas matter. Creating the environment to encourage the free exchange of ideas as well as encouraging students to think about music and sound and its infinite possibilities is what drives me as an educator.

As a music teacher educator in an institution of higher education, one of my objectives is to help music education students broaden their conception of teaching music to be inclusive of all areas where people interact with and create music beyond the traditional paradigm of band, orchestra and chorus. Because the students at my institution earn their teaching certification at the graduate level, as opposed to most programs where certification is earned at the undergraduate level, I have built a number of school-university partnerships into the curriculum for our undergraduates. The objective for these partnerships is for my students to have as much time in the classroom applying what they are learning with children, in a variety of schools with a diversity of students and school cultures, well before their student teaching practicum. What I have learned over the years through these various partnerships informs my thinking on music teacher education, including the role that technology can and should play in music education, as well as the importance of understanding and connecting with one's students.

In the United States there are many requirements imposed on higher education faculty and music education students from the university and state levels regarding content, learning standards, and professional dispositions that shape most of our teaching and assessments. It is, therefore, often easy to lose sight of the fact that our ultimate goal as music teacher educators is to provide our students with the types of skills, learning habits of mind, and musical abilities that will have relevance to the students these music teachers will eventually teach. It is this future generation of students, who after all will be the "end-users" of all of our efforts. As technology is increasingly playing a larger role in our society, my goal for this chapter is to define a role and a place for technology in the music teacher education curriculum with benefits for students and teachers, as well as teaching and learning. The ubiquity of technology has done little to change what we teach and how we teach it. What should technology's purpose be in music teacher education and how do we get there? While my views are shaped by my experiences in the United States, I believe the ideas put forth in this chapter will find traction in other parts of the world as well.

The current discourse on the skills needed for a "twenty-first-century" education outside music education puts a great deal of emphasis on technology and media literacy as the means by which teaching and learning will be transformed (Cuban, 2006; Dede, 2010; U.S. Department of Education, 2010). Much of the discourse in the United States regarding education reform centers around preparing students for the technology sector of the workforce, with a strong focus on the twenty-first-century skills that are needed for STEM subjects. In thinking about where and how music technology fits into the education, training, and certification of music teachers, a number of thorny issues immediately come to mind. Not the least of them is whether or not music technology should have a place of its own in the music teacher education curriculum and licensing track, right up there with the more traditional emphasis on instrumental, choral, and general music. In theory this would safeguard that music technology is more than an afterthought. Though at the present time the dominant belief system still privileges the performance track in the music education hierarchy in the United States, with general music occupying a lower status in both the school music curriculum and within music teacher education programs. This is in spite of the fact that many states in the United States require a comprehensive K–12 instrumental/vocal/general license, where music educators need to be well versed in all three areas of concentration. The study of music technology within a school setting, if it exists at all, will generally fall into a subset of the general music area. At the secondary level, performance ensembles are the dominant modes of emphasis in music education, with a general music class offered as the catchall for the nonperformers looking to satisfy a music and arts elective. At this level, general music may include digital audio (if there is a computer lab); keyboarding; guitar; or theory classes. One would be hard pressed to find any infusion of technology in an ensemble setting at this level, due to the emphasis on traditional music ensembles in most schools.

On the other hand technology, if used imaginatively with a focus on creative musicking, can enhance all areas of music teaching and learning and should be embedded within all manner of musicianship, pedagogy, methods, and performances classes. So why compartmentalize music technology in its own discreet silo, as is often the case in our public schools? Might we need a combination of both approaches when thinking about how to provide preservice and in-service music teachers with not just technological skills, but the ability to synthesize and apply what they are learning in order to create rich music learning environments with these skills? What might that look like in terms of how we educate and credential teachers? In my experience, I have found it instructive to first examine what the role of technology currently is, and what its role ought to be in education in general, and music teacher education in particular. How might we infuse technology into the curriculum in a manner that is relevant to both our collegiate students and their future students?

Technology as Content Provider

In thinking about education in general, Seymour Papert (1987), seeing technology's potential for transforming teaching and learning, identified two opposing poles in education. At one pole, emphasis is placed on the development of the child and the child's construction of knowledge. This is considered a child-centered approach. Papert coined the term "constructionism," whose goal was to "to teach the most learning for the least amount of teaching" (1993, p 139). He believed that by being posed real-world problems to solve, students gain the knowledge they need through the making of tangible objects, making the learning process more concrete for students. At the other pole, what Papert terms "instructionism," importance is placed on the delivery of content—a more curriculum- or teacher-centered approach (Papert, 1993).

With computers and technology becoming less expensive and more readily accessible, along with the capability for increased storage and retrieval of vast amounts of data, it is easy to see how one might adopt a view of technology education as merely a conduit to developing the skills necessary to access and acquire information. It is now possible to easily access endless amounts of information, directions, news, music, school-related research, through one's digital devices, anywhere and anytime. In Papert's view this is a dangerous stance, one that will not engage or transform students' learning experiences. Papert (1987) stated that technology should be integrated into the lives of children with a more constructionist focus, such that the technology itself would become "invisible." By invisible, Papert means that the goal of the use of technology in the classroom is for the children to focus on what they are creating, rather than thinking about the tool they are creating with. This more constructionist approach is a pathway for the students to become engaged learners, critical thinkers, and problem solvers. Sadly, much of technology education is centered on learning how the tools work, with little emphasis on what one might do with them, or, in the case of music teacher education, what one might do with them and how that could potentially transform the teaching and learning experience.

The two opposing poles come into sharp focus with regard to music teacher education. As suggested by Allsup and Westerlund (2012), the history of child-centered education is fraught with tension, resistance, and contradictions. The authors point out that music education has a long tradition of skill building and developing musical expertise that would seem to fly in the face of allowing students to construct their musical knowledge. Not too many years ago, when I was observing an instrumental instruction class, I noticed the students noodling around on their instruments trying to figure out a tune that was popular on the radio. When the teacher walked into the class, one group of students wanted to demonstrate the tune they had figured out how to play on their instruments. Rather than support and encourage the students to explore their instruments, the teacher immediately shot them down and directed them to their exercise book—the implied message being that the notes on the page were all that mattered. This stance by the teacher instantly negated not only the students' interests but also the tremendous amount of aural skills capability the students would have demonstrated.

At the time of writing, the United States is moving toward a more nationalized curriculum known as the Common Core. While our National Association for Music Education (NAfME) has in the past established national standards for music education, each state still has its own set of curriculum guidelines, because educational authority in the United States resides at the state and not the federal level. In many cases, state standards are based on the current National Standards, with many state curricula subdividing the nine National Standards into a string of subsets and strands. Each state's having its own sets of guidelines makes comparing data between one state and another nearly impossible. A goal among many policy-makers is now to create a "common core" for all areas of education, including the arts.

Perusing the new proposed National Core Music Standards for Music Education (National Coalition for Core Arts Standards, 2012), one is hard pressed to find any mention of technology until the secondary level, and then it only appears as an elective, along with theory and composition and harmonizing instruments, such as keyboard or guitar. While these are performance ensemble guidelines, the learning "strands" focus on creating, performing, and responding. Creating, performing, and responding can undoubtedly manifest themselves through all manner of music making, and provide a broad enough umbrella under which we can and should develop our musical goals. It is an umbrella under which technology can most certainly play an important role; technology can also be used to reach a wider group of students than is currently the case

(Williams, 2007). However, it seems to be assumed in the proposed National Core Arts Standards that the route to developing musical expertise is through learning to sing or play an instrument. Among many music educators, music technology is apparently not perceived as providing such a route to the development of students' musicianship.

Back in 1994 (MENC, 1994) when the current National Standards for Music Education were developed, technology was not on the agenda. In 1999, the Opportunity-to-Learn Standards for Music Technology were established (MENC, 1999). These standards deal mostly with hardware, software, and lab setup issues, with a nod to the importance of including music technology in music education. These standards, while welcomed on many levels, laid the groundwork, whether intentionally or not, for our current tools-focused approach to music technology education. Surprisingly, given all the recent changes in digital technology in both hardware and software, there have been no updates to these standards. These standards clearly skew toward the information provider pole and how to acquire and employ the right tools for the access needed. In fact, a recent visit to the NAfME website (www.nafme.org) yielded a dropdown menu for "standards" that outlines the National Core Arts Standards, in which music technology appears noticeably absent. However, typing "technology standards" into the search box brought one to a page with multiple options (http://www.nafme.org/?s=technology+Standards) and with several links to pages created in 2009. There is one link to "Core Music Standards"; once you click on it and scroll down you can find a "technology" link that will bring you to a rubric for creating and composing using analog and digital means. One really has to go searching to find it.

The near absence of music technology from the National Core Arts Standards and the NAfME website is not in the invisible way that Papert intended, but it is absent in the sense of being rendered insignificant. Only through an exhaustive purposeful search of the NAfME website will you find digital media at all; in this case they use "digital citizenship," which is from an old post about guitar education. This leaves the implicit message that any study of music outside the sphere of traditional performance ensembles is considered of marginal value. It is little wonder that if there is a course in music technology for the music educator or a professional development session for credentialed teachers, a great deal of the focus will be on learning about the software and hardware that music educators might find useful. There will be little emphasis, however, on the pedagogy or practical applications of this technology and how its use might impact teaching and learning.

On another front, the guidelines of the National Association of Schools of Music (as presented in the *NASM Handbook*) place the emphasis for music education on educating preservice teachers in the areas of general music, vocal/choral, and instrumental music, with language that encourages knowledge of the appropriate digital and emerging technologies but doesn't actually require it. The NASM is the accrediting body for schools of music and postsecondary music programs throughout the United States. The most recent *NASM Handbook* for the professional baccalaureate degree states: "graduates are expected to demonstrate competencies in the common body of knowledge and skills expected of all who hold a professional undergraduate degree in music, including, but not limited to, performance, aural skills and analysis, composition and improvisation, repertory and history, and technology" (National Association of Schools of Music, 2014–15, p. 189). This statement would appear to lean toward the more comprehensive approach toward incorporating technology, but will this approach invite some unintended consequences? There is a perception that since technology is pervading all aspects of our lives, a separate course on how to use technology might seem redundant. But just as all music students play instruments or sing and perform in ensembles, we still require quite a number of pedagogy classes in those areas; a pedagogy course or two on teaching music with and through technology would not seem out of line. This thinking begs the question: are we finally at that post-music-technology instruction moment when, thanks to our students growing up in the computer culture (Turkle 1995), we have achieved Papert's (1987) goal of "making the technology invisible"? Or is it that, in our rush to render music technology an implicit component of a well-rounded music education, we are truly making music technology invisible and, by extension, inviting its more negative connotation: scrubbing music technology from the forefront of our thinking? Those students and instructors who are interested in working and teaching with technology will pursue those interests regardless of the standards, while those with no interest or comfort level can now safely continue to sidestep the issue. In the degree program at my institution, technology-based projects are infused throughout the curriculum, starting in an immersive introduction to technology in music education course, which is then followed up by infusing technology projects throughout the sequence of methods and pedagogy classes. The point of this structure is to model and to help our music education students connect the dots between our in-class activities and how they can create similar projects with their students. While there is often initial resistance from many of our in-service and preservice teachers, I have discovered that when they get involved in a project that is personally meaningful to them, students will drop their defensive guards surrounding technology. There are still those who continue to maintain a very parochial attitude toward ensembles and draw a line in the sand where technology is concerned. While they might allow that technology is probably beneficial in the general music realm, in their minds it clearly has no place in the ensemble world.

Perhaps those associations that are charged with developing educational guidelines, such as the NASM and NAfME on the national level, as well as the state level associations, are conflating the pervasive existence of these technological tools in our lives both in and out of school—whether the tools are workstations, laptops, tablets, or any other digital device—with the existence of a sound pedagogy for thinking and learning with these tools (Mishra & Koehler, 2006). The logic stream for this thinking is somewhere along these lines: technology is now prevalent in most schools; the teachers are being trained in how to use the technology; the students are gaining access to the technology, and they use it outside school anyway. Most problematic is the belief that students will learn how to use technology in their other subjects, so let us not take away from precious rehearsal time where the real music learning happens.

The More Things Change

Cuban (2006) cites a host of technological transformation points leading up to the wholesale adoption of laptops for education, yet the innovation needle has not moved much, nor has much changed pedagogically. In Cuban's study, the teachers were using the technology more for administrative purposes than for enhancing their teaching or their students' learning. We are now firmly ensconced in thinking about iPads and other tablet devices as an educational game changer. While the technology is still in its early stages with regard to the research available on the use of these tablet devices in the classroom, there are now a number of articles regarding school districts adopting iPads with the expectation that such adoption will lead to transforming their schools and transforming students' learning (Pasnik, 2007; Simpson, 2011; Pegrum, Oakley, & Faulkner, 2013). Much of the discussion in these articles centers around the iPad's ease of use, and its ability to deliver content via eBooks and publisher-designed curriculum. Through many of these devices, teachers have the ability to track students' progress as they work through the various activities. Given the current emphasis from curriculum developers promoting the use of tablet devices as a means to transform education, one has to wonder what aspects of education will be transformed. Just putting a new device in a teacher's hand loaded with all the cool new curricular materials in and of itself is not going to be a pedagogical game changer as much as it will be a content delivery game changer. It will also propel the evaluation and assessment trend taking hold throughout the "edusphere" to a whole new level.

Unlike most of the rhetoric regarding educational reform, where students are generally left out of the discourse, the tablet sales pitch places the benefits to the students clearly front and center. Tablet devices are touted for the ease with which all students will be able to do research and create collaborative presentations, as well as the potential of its lower price point, as compared to desktop or laptop computers, to decrease the digital divide for those students who do not have computers in their homes. As with Cuban's (2006) research into how teachers used, or, more to the point, didn't use laptops in their teaching, there is now a great deal of buzz around tablet computing's potential to supply vast amounts of information; there is less buzz concerning how teachers are reconfiguring their pedagogy and actually transforming learning. If we only think about technology as a content delivery system, we may be missing the greater potential of incorporating technology in music education and, by extension, we may not be giving experienced and novice teachers the appropriate professional development experiences in using this technology in a way that might actually transform their teaching.

I am an early adopter and champion of iPads in the classroom. My students and I have experienced increased engagement levels in the middle school learners we work with. Just as Savage (2007) observed in his research, we are observing positive benefits and outcomes when we and the teachers we work with incorporate technology in our lessons. I was an early adopter of iPads not because they were the next new cool device but because of the creative potential they can unleash in our students and in us as teachers, and the intuitive nature of most apps. A tablet device has tremendous potential to promote collaborative exploration, problem-solving development, and creative music making—not to mention that its ease of use is an important factor, in that many of the apps are designed to be intuitive with few directions needed. I'm sure as I write this, some next new thing is already on the horizon.

Ease of Use and Other Mismatched Assumptions

Apple has been at the forefront of making technology intuitive by designing a range of creativity tools that are affordable and accessible to the average person. In a 2007 white paper regarding the benefits of the iPod for teaching and learning (Pasnik, 2007), the major focus was on students' ability to take their learning, or perhaps more to the point, the teacher's teaching, outside the classroom, where they can access, share, and collaborate with others regarding the content on the device. It was being promoted as a device with the potential to allow students to listen to lectures anytime, anywhere. At that particular time it was portrayed to the larger educational community less as a device that would promote the construction of knowledge than as a device to acquire and track knowledge.

In their study regarding iPads, Pegrum, Oakley, and Faulkner (2013) noted that interviewees commented that the play-based nature of iPads would be appropriate for early childhood education, but they questioned the appropriateness at the upper levels, where experimentation is less valued. The notion of experimentation being perceived as less valued at the upper levels of school is problematic to be sure, and unpacking that thinking could be a chapter of its own. The authors also noted how these devices could be viewed as an alternative to traditional books, due to the large storage capacity of the hardware for eBooks. As in other studies regarding technology's use in schools, many teachers reported feeling overwhelmed and underprepared (2013). The authors suggest the focus should be on pedagogy before the technology in professional development, as well as noting that technology-based professional development should be targeted and context specific.

This disconnect with music technology in music teacher education is not limited to the United States. In a study conducted in the United Kingdom, access to technology was often perceived as a barrier for preservice music teachers (Gall, 2013). The expense of music software made it difficult for students who did not have access to the software at home to have enough time exploring and familiarizing themselves with the technology. The technology in schools was perceived as being often out of date and unable to support the software packages that were being used in collegiate training. Gall's (2013) research also found that computers in schools were frequently located in a variety of places and not always in a dedicated classroom. Mentors often were not well versed in the latest technology and reluctant to incorporate it in their lessons. This view is echoed in studies by Doering, Hughes, and Huffman (2003) and McGrail (2006) and is a sentiment I often hear from the students and teachers I work with. While many of the teachers I work with are in technology-rich schools with laptop carts and iPad carts, it is rare for the music teachers to gain access to them. They are often forced to make do with a couple of old computers scattered around their classroom or to use their one laptop connected to a SMART board. As long as a mismatch continues between what and how learning takes place and what is assessed, the handheld mobile technologies will be seen as a distraction (Pegrum et al., 2013).

The Role of Technology in the Classroom

Papert (1993) suggests that having children appropriate knowledge that is personally meaningful, as opposed to learning facts because they need to pass a test, should be the goal of education. To that end, when we move our thinking from a mindset of technology as a content delivery system to one in which a technology-rich environment provides the means for students to imagine and create using sounds and their intuition as a springboard, we can begin to bridge the disconnect between music education practice and the multiple ways that our students understand music on their own outside school. Yet, given the level of enculturation into the field of school-based music education that begins as early as elementary school, for a new teacher to embrace multiple ways of being musical beyond the ensemble paradigm means confronting a sense of disequilibrium and tension for many music education students as well as seasoned music teachers.

In a recent graduate seminar, after being immersed in designing and creating instruments and musical compositions with and through technology, my students discussed their own engagement levels as well as how they imagined students would react to this activity. They were able to make the interdisciplinary connections into other core areas as well as reflect on what their students could learn musically from such an activity. They spoke about how this approach to technology could teach science concepts, support the teaching of mathematics, and support literacy development. In addition, they were eager to discuss the variety of thinking processes needed to engage with these materials, making connections to procedural and creative thinking. They appreciated the hands-on learning aspect of what we were doing, the fact that no two groups of students came up with the same idea, and having to logically think through ideas to see what would and wouldn't work. What was interesting is that most of the students found a bit daunting the exploratory aspect of first trying to figure out what they could do with a particular piece of technology we were working with. Just about all the students believed that their students would need step-by-step directions. Most of my students also felt that if they were to use this technology it would only be with high school students. In actuality, when we introduced the same technology to a group of middle school students, we used the same exploratory approach, and the students were completely engaged and coming up with some great musical ideas. They were problem solving and making musical decisions.

While the discussion that followed the activity was quite positive, the subsequent web-based discussion about their beliefs on what the role of technology should be in music education revealed an ingrained mindset that making music through technological means is somehow less valid than making music with what they referred to as "real" instruments. Comments such as that technology "could be the icing on the cake," "can be employed as a supplementary activity," or "shouldn't be our main tool in the real-life classroom" were common, whereas their more positive thoughts on the role of technology in the classroom leaned toward the wealth of content that can be accessed through technology in order to enhance their lessons.

My students' reactions were similar to findings by Gall (2013) and Savage (2007) regarding teachers' perceptions of music technology and some of the benefits and challenges of using it. On the plus side, since my students were working in groups, they saw a tremendous benefit in being able to bounce ideas off one another and learn from each other. They appreciated the hands-on exploratory approach that was presented to them and the ability to create something they were personally invested in. They saw the benefit in creating open-ended activities so that students could attach their personal "stamps" to an assignment. While one group actually explored the potential for incorporating this technology into an ensemble setting, most groups were clearly of the belief that music technology was more suited to a "music tech" class. They could not envision a role for this type of technology use in an ensemble setting. A few made the argument that music has existed for centuries with just acoustic instruments. According to one student's journal, though the activity was clearly fun, the student believed that in order to be comfortable using this type of technology with a class, more training and time allotted to exploration would be needed. Paradoxically, this student made a good case for why we probably need more contact and experiences with technology so that we can internalize the technology enough to focus on our pedagogy.

It is interesting that these very same students who relished the creative challenges in our classes through many technology-based projects still viewed these types of activities as the exception rather than the rule. They defaulted to shoehorning technology into the preexisting music education paradigm rather than reimagining what music classes might look like in the twenty-first century. Their grounding in the performance ensemble traditions, reinforced by much of the music teacher education curriculum, gave them a narrow lens through which they viewed those moments when we were creating with musical materials as mere enhancements to a "real" music-making activity, rather than envisioning musical exploration as the foundation upon which they could build their music curriculum. For the most part these assumptions about what constitutes a quality music program then get reinforced when they commence their student teaching. They may see teachers using SMART boards in classrooms that may even have a few electric keyboards and computer stations scattered around the room, but otherwise the music classroom, and the practices within it, pretty much resembles what they experienced in their own schooling.

It Takes a Community of Learners

Citing Dewey's frustration with the binary mindset with regard to the either/or of a traditional teacher-centered approach to education versus a child-centered *progressive* approach, Allsup and Westerlund (2012) continue to make a case for the interplay of both. They suggest that achieving this will take a willingness on teachers' part to reconstruct their role: rethinking the student–teacher dynamic. The technology question seems to raise similar issues with regard to the tension it creates for many music teachers who think that by adopting technology one is giving up the rich traditions of performance-based music education—that somehow one mode of music making will supplant another. Considering the small percentage of students at the secondary level of schooling who participate in school music, there would seem to be room for both approaches. There needs to be room in the curriculum to encompass the variety of ways in which students can be musical. The long-ingrained music education traditions enjoyed by the performance and general music tracks will continue to thrive, even if we allow technology more space and more exposure in the curriculum.

As Lipscomb (2012) and Bauer (2007) suggest for music education, as echoed by Dede (2010) for general education, there needs to be a rethinking of the interplay among the higher education community, the larger school community (including administrators, teachers, and students), and the technology suppliers and developers. Carving out a space for sustained professional development should be a priority for music teacher educators. If we wait to find suitable state-of-the-art sites where our students can do their fieldwork, with that rare technologically savvy classroom music teacher, we become complicit in the fact that nothing will change. Our graduates will continue to feel overwhelmed and underprepared to incorporate technology into their own curriculum and will continue to fail to make connections to the musical world of their students. We, as music teacher educators, need to create those opportunities by inviting the school community into our classrooms, creating a larger technology learning community. We need to be the architects of what Wenger (1998) espouses as a "community of practice." If we don't do it, who will?

Figure 51.1 depicts the reciprocal learning currently taking place in one of our technology-based school-university partnerships (Greher, 2011). While this may seem more a theoretical than an actual learning model, I can attest that this is a model I have been cultivating for several years now, and our music education students and our local music teachers and public school students are indeed a learning community. We are all learning from each other, fitting Wenger's (1998) notion of a community of practice. In addition, in order to offer technology-based professional development workshops, local

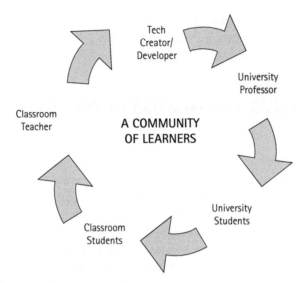

FIGURE 51.1 The Technology Learning Community

teachers are invited into my classroom to explore new technologies with my students on a regular basis.

As part of our weekly scheduled field-based experiences connected to either the intro to music technology class or the general music methods classes, our undergraduates, under the supervision of a graduate student, myself, the classroom teacher, and occasionally a music technology creator, develop lessons incorporating technology in some form into their lessons. Before taking ideas into the classroom, the college students gain experiences in fully immersing themselves in a specific technology application and explore the possibilities, challenges, and affordances. Then, working closely with the goals of the classroom teacher, the college students strategize ideas for how best to fulfill specific music learning objectives incorporating technology. As they work with the classroom students, they are modeling ideas for the classroom teacher. It is a constant process of refining and tweaking of concepts. For instance, our students recently developed a unit incorporating listening skill development based around the songs our classroom teacher partner would like her students to know about, including knowing the songs' chord progressions, bass lines, lyrics, and melodies. They did this in consultation with Eric Rosenbaum, the creator of a new music app in development. In addition to us learning about how middle school students are interacting with the app—what features are intuitive, what still needs more work, and what educational potential the app may offer—these middle school students are demonstrating to all of us their ability to re-create tunes they are learning, create variations, improvise, and create new tunes. The college students are eager each week to devise new activities; the classroom students, including those who are considered "at-risk," are eager to explore and make music. It is a process not just of only using technology in their lessons but of constantly seeking opportunities for incorporating technology where its use is appropriate and where it may spark an interest in the classroom students. These opportunities can help the public school students expand their creative and musical thinking and tap into their intuitive musical selves and possibly spark an interest in learning more about music they may not have had before.

The college students and the classroom teachers often notice a heightened interest and engagement within the classroom. The classroom students are more willing to communicate their ideas and collaborate. Through these experiences the college students are learning to work with and problem-solve technology platforms they may or may not be familiar with and are coming to understand the need to know who their students are so as to design projects better. Most important, they are learning firsthand the strengths and weaknesses of each program, in order to better design appropriate lessons around a particular software application

Will this approach change the mindset of all music education students and music teachers regarding technology's role in music education? Perhaps not. However, greater percentages of preservice students and teachers in the community are feeling more confident in their abilities to employ technology in their lessons with the explicit goal of engaging more students in music-making activities. In fact, many of the teachers are now writing grants and working with their districts to seek out funding sources to upgrade technology in their schools.

Teaching involves more than just providing content. Teaching involves designing a learning environment, inviting learners' engagement, and enabling interaction and scaffolding within the environment. As with more traditional approaches to music teaching, creating these music technology experiences for students is still teaching music education students about lesson planning, organization, and the importance of understanding the relational aspects of teaching.

While this model of creating a learning community takes a bit more planning, time, and revisions than a typical lecture-based class, the benefits are many. The music education students, along with the classroom teachers are using their imaginations more, thinking about the process as well as the product, and empowering students in their own learning. The classroom teacher gets fresh ideas, mentoring, and the extra time needed to internalize the technology in a less formal, less threatening manner. Most important, they get to observe how these activities are impacting their students. The college students get mentoring in the classroom, and the classroom students get the opportunity to create and work on personally meaningful activities with the college students whom many regard as their peer. The technology creators and developers get feedback on what is and isn't connecting with the end user, and will most likely get new ideas on how their device or software application can further impact the music-learning process. The college professor, through engaging in a very loose, informal peer-to-peer learning loop with all the parties involved, is learning what is and is not working out in the field in order to better refine the educative process. Most important, the construction of knowledge is at the forefront of this process. While it is not explicitly taught, it is teased out through each stakeholder's own experiences and observations.

References

Allsup, R. E., & Westerlund, H. (Eds.). (2012). Through simple discovery. In A. de Quadros & A. J. Palmer (Eds.), *Tanglewood II: Summoning the future of music education* (pp. 191–208). Chicago: GIA.

Bauer, W. I. (2007, 25–29 June 2007). *Transforming music teaching via technology*. Paper presented at the Tanglewood II Satellite Symposium; The effect of technology on music learning, University of Minnesota.

Cuban, L. (2006). 1:1 Laptops transforming classrooms: Yeah, sure. *Teachers College Record*. Retreived January 5, 2017, from http://www.tcrecord.org/Content.asp?ContentID=12818.

Dede, C. (2010). Transforming schooling via the 2010 national educational technology plan. *Teachers College Record*. Retreived January 5, 2017, from http://www.tcrecord.org/Content.asp?ContentID=15998.

Doering, A., Hughes, J. E., & Huffman, D. (2003). Preservice teachers: Are we thinking with technology? *Journal of Research on Technology in Education, 35*(3): 342–361.

Gall, M. (2013). Trainee teachers' perceptions: Factors that constrain the use of music technology in teaching placements. *Journal of Music, Technology & Education, 6*(1): 5–27.

Greher, G. R. (2011). "Music technology partnerships: A context for teaching and learning. *Arts Education Policy Review, 112*(3): 130–136.

Lipscomb, S. (2012). Enhancing opportunities through technology. In A. de Quadros & A. J. Palmer (Eds.), *Towards Tanglewood II: Summoning the future of music education* (pp. 221–245). Chicago, IL: GIA.

McGrail, E. (2006). "It's a double-edged sword, this technology business": Secondary English teachers' perspectives on a schoolwide laptop technology initiative. *Teachers College Record, 108*(6): 1055–1079.

MENC. (1994). *National standards for arts education: What every young American should know and be able to do in the arts*. Reston, VA: Author.

MENC. (1999). *Opportunity to learn standards for music technology*. Reston, VA.

Mishra, P., & Koehler, M. J. (2006). Technological pedagogical content knowledge: A framework for teacher knowledge. *Teachers College Record, 108*(6): 1017–1054.

National Association of Schools of Music. (2014–15). *National Association of Schools of Music handbook 2013–14*. Reston, VA: Author.

National Coalition for Core Arts Standards. (2012). *National Core Arts Standards: A conceptual framework for arts learning*. http://nccas.wikispaces.com.

Papert, S. (1987). *A critique of technocentrism in thinking about the school of the future*. Paper presented at the Children in an Information Age: Opportunities for Creativity, Innovation, and New Activities Sofia, Bulgaria.

Papert, S. (1993). *The children's machine: Rethinking school in the age of the computer*. New York: Basic Books.

Pasnik, S. (2007). iPod in education: The potential for teaching and learning. One in a series of iPod in Education white papers. Sponsored by Apple Inc.

Pegrum, M., Oakley, G., & Faulkner, R. (2013). Schools going mobile: A study of the adoption of mobile handheld technologies in Western Australian independent schools. *Australasian Journal of Educational Technology, 29*(1): 66–81.

Savage, J. (2007). Reconstructing music education through ICT. *Research in Education, 78*, 65–77.

Simpson, K. (2011, May 31, 2011). More Colorado schools turning to iPad to improve education. *The Denver Post*.

Turkle, S. (1995). *Life on the screen: Identity in the age of the Internet.* New York: Simon & Schuster.
U.S. Department of Education (2010). *Transforming American education: Learning powered by technology.* Office of Educational Technology.
Wenger, Ettienne. (1998). *Communities of practice.* Cambridge, UK: Cambridge University Press.
Williams, D. B. (2007, 25–29 June 2007). *Technology in music education.* Paper presented at the Tanglewood II Satellite Symposium; The effect of technology on music learning, University of Minnesota.

CHAPTER 52

AUTHENTIC APPROACHES TO MUSIC EDUCATION WITH TECHNOLOGY

JONATHAN SAVAGE

BIOGRAPHY

I have always had an interest in music technology. Right from the early days of an Atari ST in the mid-1980s, I have been fascinated by what music technology can do for the processes of musical performance and composition. Although my own musical training was fairly conventional, when I started teaching in a school I quickly realized that technology could provide an alternative "way in" to music making, and I sought to integrate various technologies within my teaching. I also began studying for a doctorate that reflected on the uses of technology that experienced composers were making in an electroacoustic musical environment. I drew on my observations of, and conversations with, these composers and designed a range of curriculum projects that built upon them. Since then, I have researched how music education can be developed through the use of technology and sought to inspire others to do the same. The challenges and joys of working with young teachers each year are inspiring. However, one thing is crucial to me. Music technologies provide a range of tools; they are means to an end; they are not *the* end in and of themselves.

Robert Benchley, the famous American columnist, once quipped: "there are two kinds of people in the world: those who divide the world into two kinds of people, and those who don't." I was reminded of this quotation when approached to write this Core Perspective that addresses, among other questions, whether "music technology" should be taught as an independent subject or should be integrated across the music curriculum. It does seem to be the case that human beings are prone to like to categorize things; music education is no exception. But, as I will show, this can lead to many unhelpful, although perhaps unintended, consequences. This is an important question, but in many senses—in the United Kingdom's schools and colleges at least—it has been answered in recent years through educational policy and resulting curriculum frameworks. However, this does not mean that the questions are not worthy of further consideration.

Given this mix of experiences, this chapter begins with an examination of the current curriculum framework in the United Kingdom and analyzes some of the consequences of this separation between music and music technology. Following this, I will present two alternative visions of musicianship, drawn from the work of two writers, and argue that a more coherent approach to the development of music technology skills within a music teacher's pedagogy is urgently needed.

Music Education with Technology in the United Kingdom

The National Curriculum for England still includes music as a subject. This means that it should be taught in all state-maintained primary and secondary schools. The National Curriculum is still structured around "Key Stages," and a new revised National Curriculum was implemented in schools in September 2014 (Department for Education, 2013).

For the last 16 years, students have been able to study for a separate Advanced Level (A Level) qualification in music technology; more recently, during the early 2000s, vocational qualifications in music, with a significant, if not entire, course content devoted to music technology, recording, production, and other technologically mediated musical activities, have been available to students from the age of 14 onward. Alongside these specialist courses, references to music technology and its use have continued to appear in National Curriculum Programmes of Study in various forms. Students have also been able to study music in more traditional qualifications such as General Certificate of Secondary Education and A Level music courses.

The fate of music technology within the revision process surrounding the National Curriculum has been mixed. From September 2014, having been present in the previous Programmes of Study, it is now entirely absent from the Key Stage 2 program of study (for children aged 7–11). At Key Stage 3 (for children aged 11–14), the National Curriculum makes mention of music technology in one sentence: "[students] should use technologies appropriately and appreciate and understand a wide range of musical contexts and styles" (Department for Education, 2013).

The second key element of music education is the General Certificate of Secondary Education (GCSE), the principle mechanism for the assessment of students at age 16. Despite a reprioritization of subjects, this has meant that music has been placed in the lowest tier (and will be ignored by many students, parents, and schools), though students may still have access to a GCSE in music course in many schools. But, as I will show, numbers of students taking this examination have fallen by 11% over 3 years, and the trend is only going to continue on a downward spiral.

At Key Stage 5 (A Level for students aged 16–18) qualifications have remained in music and, as a separate qualification, in music technology. These qualifications are accompanied by a range of vocational qualifications in music, such as the Business and Technology Educational Council (BTEC) First (for students in the Key Stage 4 age range) and BTEC National (for students at Key Stage 5). Both these BTEC qualifications contain a significant amount of music technology–related content.

What is quickly apparent from a cursory overview of the curriculum and qualification frameworks within the United Kingdom is that students' experience of music technology begins as part of a broader music curriculum experience within the school but quickly becomes specialized around the age of 14 into discrete qualifications that focus, to a greater or lesser extent, on the specialist skills associated with music technology.

An analysis of the number of students taking these examinations is informative. Figures drawn from government sources (Cambridge Associates, 2013a, 2012a, 2011a) show clearly that the number of students studying for an A Level in music and an A Level in music technology has fallen by around 35% over the last 3 years. While the gender balance in relation to A Level music slightly favors boys (16% more boys than girls studied for this examination in 2012), the differences in terms of gender within the A Level music technology intake are stark. Between 2010 and 2012, for example, 85% of the entrants for this examination were male.

In respect to the "vocational" curriculum offered through qualifications such as the BTEC First and BTEC National examinations, a similarly marked difference in gender has been noted (Edexcel, 2012, 2013). Across the entire portfolio of examinations (i.e., in every subject) offered by Edexcel (the awarding body), male students outnumber female students 52% to 48%. However, within the music courses that, as stated, have a core element of music technology, the gender imbalance is significant. Over the last 3 years, within the Level 2 BTEC First, male students account for around 63% of the total student entries; at Level 3 (BTEC National) they account for 80% of the total student entries.

Against this backdrop, the GCSE music examination has been taken by around 11% fewer students in 2012 than in 2010 (Cambridge Associates, 2011b, 2012b, 2013b). The gender imbalance has narrowed from around a 10% gap (with boys outnumbering girls) in 2010 to a 3% difference in favor of the boys in 2013.

I have dwelt on these figures for a number of reasons. First, and obviously, the separation of the study of music technology from the study of music in the United Kingdom's curriculum and examination framework has created a significant gender imbalance. While the boys studying a formal qualification in music or music technology always outnumber the girls studying for the same qualification, the more music technology content that is included within the examination itself, the wider the gap between male and female students (at its largest to a massive 80% difference with the Level 3 BTEC qualifications).

Second, the decline in students taking GCSE and A Level examinations has been marked, while during the same period the total number of students taking vocational qualifications, such as the BTEC courses referred to here, has increased by around 22% from 2012 to 2013. What are perceived as academic qualifications on one hand have diminished, while those relating more closely to technological skills, popular musical styles, and perceived employability have expanded. I make no value judgments here, but the change in fortunes is stark. However, this changed significantly beginning in September 2014, when certain qualifications of this type were downgraded by the government and no longer count in the assessment of a school's performance tables. This will result in large numbers of schools opting out of these more vocational courses and pushing students into the standard GCSE programs that they offer.

On balance, these changes reinforce the folly of categorizing musicians and musical activity by type or technology. Within the United Kingdom, the move in the 1990s to distinguish music without technology (in a general sense) from music with technology has resulted in such a highly unhelpful, gendered, and rigid delineation of musical content, activity, and technologically mediated practice to such an extent that it is difficult to see any way this situation could be reversed. It has had a major impact on the way individual students have seen their musical skills. Many of them are the ones who are training to be teachers today.

To explore the issues here in a little more detail, I would like to shift my focus to the work and practice of an individual musician. In what follows, I will present two alternative "visions" of a musician who uses technology. The first of these is Hugill's vision of the "digital musician" (Hugill, 2008).

Challenging the Notion of the Digital Musician

In his book *The Digital Musician*, Hugill (2008) suggests that digital musicians are a discrete class of musician who exhibit certain characteristics that separate them from other musicians who (he acknowledges) may make use of digital technologies on odd occasions within their work). Hugill acknowledges that the classification of musicians into "types" is bound to be problematic. Yet this does not stop him from embarking on a stringent approach to the classification of musicians by various labels, including musical "dimensions" or "resources." For instance, he starts with those musicians who work in established traditions, such as classical or folk music, whose principal preoccupation, he argues, is "pitch." These musicians generally play an acoustic instrument as their main practice, and they travel a prescribed career path that involves the gradual evolution of technique and musicianship. The criteria for recognizing virtuosity are clearly established, and there is broad agreement among listeners about the level that an individual musician has attained (p. 3).

Second, Hugill cites "rhythm" or "beat" as being a primary starting point for popular musicians. These musicians tend to show a relative lack of interest in pitch when compared to the first type, although of course they do not ignore it completely. Bands without some sort of percussion section are rare, and when they do appear, the instruments often emphasize rhythmic content in order to compensate (2008, p. 3).

A third type of musician starts with timbre. These musicians, Hugill suggests, are harder to pin down, for the simple reason that many timbre-focused artists do not consider themselves to be musicians at all, or do not use the word "music" to describe what they produce. Into this category can be placed the majority of those working in electronic and electroacoustic music, but also sonic art, sound art, sound design, and various forms of radiophonic and speech-based works. By dealing with sounds rather than notes, these musicians have changed the nature of music itself, raising questions about what is "musical" (2008, p. 4).

Digital musicians, for Hugill, are not defined by their use of technology per se. A classical pianist giving a recital on a digital piano is not a digital musician, nor is a composer using music notation software to compose a string quartet. Hugill's definition of a digital musician is as follows:

> A digital musician is one who has embraced the possibilities opened up by new technologies, in particular the potential of the computer for exploring, organising and processing sound, and the development of numerous other digital tools and devices which enable musical invention and discovery. This is a starting point for creativity of a kind that differs from previously established musical practice in certain respects, and requires a different attitude of mind. These musicians will be concerned not only with how to do what they do, or what to do, but also with why to make music. A digital musician, therefore, has a certain curiosity, a questioning and critical engagement that goes with the territory. (2008, p. 4)

Furthermore, Hugill goes on to specifically distinguish what makes a "digital musician" different from other musicians. The skills that a digital musician requires, he argues, are as follows (p. 5):

> Aural awareness (an ability to hear and listen both widely and accurately, linked to an understanding of how sound behaves in space and time);
> Cultural knowledge (an understanding of one's place within a local and global culture coupled with an ability to make critical judgements and a knowledge of recent cultural developments);
> Musical abilities (the ability to make music in various ways—performance, improvisation, composition, etc. using the new technologies);
> Technical skills (skill in recording, producing, processing, manipulating and disseminating music and sound using digital technologies).

The notion of a digital musician is a flawed one. It makes little if any sense in my view. It is about as useless as defining a musician as being principally concerned with pitch, rhythm, or timbre. Classifications of this type are worse than simplistic and reductionist;

they are dangerous and divisive and, as I have described above, have resulted in an unhelpful demarcation in terms of the curriculum within the United Kingdom. They have also led to a crisis in confidence in music teachers' perceived skills (on which more below).

In contrast to Hugill's characterization of the digital musician, I would like to explore an alternative model drawn from the work of one of the most interesting electroacoustic composers whom I have had the privilege of working with: John Bowers.

The Improvising Musicians

Bowers's *Improvising Machines* (2003) explores the improvisation of electroacoustic music from various standpoints, including its musicological, aesthetic, practical and technical-design dimensions. Within it, detailed ethnographic descriptions of the author's own musical performances over a period of a number of years at different concerts across Europe are described, and the various pieces of hardware and software through which these were facilitated are analyzed. For anyone with an interest in electroacoustic music, musical improvisation, the human-design interface, and the wider adoption of digital technologies, it is a fascinating account.

Bowers presents an argument that electroacoustic music is an indigenous "machine music." He explores his own experience as an improviser in this idiom, giving special attention to observable variations in the forms of technical interactivity, social interaction, and musical material that existed across the various musical performances that he gave with fellow performers. Toward the end of this chapter (2003, pp. 42–51) he identifies four issues that inform his writing in later chapters (notably the exploration and development of a technical design aesthetic in chapter 3) but are of particular interest me here in relation to my discussion of how music education with technology could be more broadly defined, integrated, and promoted within the curriculum and examination frameworks.

Contingent Practical Configurations

The music has arisen in relation to these contingencies in such a way that, from an ethnographic point of view, it should not be analytically separated from them (Bowers, 2003, p. 43). Bowers defines "contingent practical configurations" as the technologies used, the musical materials and forms explored, the performance practices employed, and the specific settings and occasions of, as well as the understanding generated from, musical improvisation with technology. Contingencies of this type are "topicalized" within the performance itself. They are integral to it and shape the resulting musical statements, interactions, and expressions. Improvised musical conduct of the sort

Bowers describes is a space in which contingencies are worked through in real time and in public. The specific contingency of a technology-rich musical improvisational conduct is embodied in the relationship between human beings and their machines. You cannot have one without the other. Specifically, "it is in our abilities to work with and display a manifold of human-machine relationships that our accountability of performance should reside" (p. 44).

Variable Sociality

For Bowers, *variable sociality* is the different social interactions and relationships that are worked out through musical performance: "the sociality of musical production is an important feature of improvised electro-acoustic music. Publicly displaying the different ways in which performers can position themselves with respect to each other and the different ways in which technologies can be deployed to enforce, negate, mesh with, disrupt, or otherwise relate to the local socialities of performance could [again] become the whole point of doing it" (2003, p. 45). As with any musical practice, within "machine music" the social, interactional relationships that Bowers and his fellow musicians enjoyed varied over time. There was a deliberate playfulness. Different alternatives were experimented with, variably and often interchangeably, within the course of a specific musical performance. Social norms could be disrupted at a particular point, perhaps due to technical issues (perhaps the cables were not long enough or the monitoring was lopsided) or other factors (the audience began to leave or the music was too loud and complaints were received from others in the locality). The social dimensions of musical production are highly important. They need to be understood and explored as an integral part of the aesthetic, not as a separate issue.

Variable Engagement and Interactivity

Just as performers can variably relate to each other, they can variably engage with the technologies and instruments before them (Bowers, 2003, p. 45). Linked to the above, Bower's concept of *variable engagement and interactivity* facilitates a consideration of the different and varying relations that performers have with their instruments and technologies. In particular, he identifies a number of different patterns of engagement for the musical performer and for the listener.

Bowers's 12-hour musical performance in Ipswich (at which audience members were given a free can of baked beans in return for their attendance!) utilized a range of mechanized musical production technologies that, at particular points, automatically set new parameters for musical statements or even drew on new source materials from the performers' laptop computers. The pattern of engagement from the performers' point of view was one of initiation, delegation, supervision, and intervention. This process meant that it was not always necessary for one, or both, of the musical performers to be physically present within the space for the whole of the 12 hours.

Other more conventional forms of musical production within Bowers's performance events utilized conventional instruments that required some kind of human incitement to action (striking something, manually triggering a sample, etc.). The pattern of engagement here would be one of physical excitement/incitement and manipulation.

Different forms of engagement have different phenomenologies associated with them. How one listens, hears, or responds intellectually or is physically moved all effect and affect one's engagement and interaction with sound and its means of production.

Musical Materials

"To construct workable and intelligible performance environments, I have made various distinctions between these musical materials in terms of their real-world sources, the media by which they are conveyed, the manipulability of those media, the kinds of gestures and devices, which are used to realize those manipulations. From time to time, all of those features are seen to be bound up with identifiable forms of social organization between co-performers, and those forms of interaction have musical-formal aspects to boot. I have tried to reveal these interconnections through ethnographic description of the performance situation" (Bowers, 2003, p. 48). Bowers's sophisticated organization of musical materials draws on a range of existing methodological structures for electroacoustic composition. While he is at pains to emphasize the differences here, his account is illuminating when placed alongside his analysis of Pierre Schaeffer's acousmatic composition (and allied practices), Simon Emmerson's distinction between aural and mimetic discourse, and Denis Smalley's spectro-morphological categorizations. These all provide a frame for dialogue and discussion about the sounds that Bowers and his coperformers produced during their improvisations and, importantly for us, about how they reflected on and justified the musical "product" that resulted at the various concerts.

Central to this discussion (Bowers, 2003, pp. 48–50) is the question of how an overall musical structure of "syntax" can emerge from an improvised performance practice. Drawing directly on Emmerson's (1986) work on musical syntax, Bowers writes:

> Improvised forms are naturally immanent, ad hoc-ed moment-by-moment on the basis of what has gone before and projecting opportunities for what might come following. In the language I hinted at above, multiple threads of significance may link up several of the elements in play. There may still be singularities and other "unattached" offerings. The threads may be thin or may be densely interwoven (steady with the metaphor now!). We may have a sense of "a piece" or a collection of "moments" or

some point in between. These are some of the immanent forms, of abstracted syntax, one can hear generated by electro-acoustic improvisors. (2003, p. 50)

Authentic Music Technology within Music Education

Improvising Machines presents an illuminating narrative about the processes and products of Bowers's improvisational conduct within the context of electroacoustic music. It contains a blend of musicological features, technical considerations, and reflective comments, underpinned throughout by a rigorous approach to ethnographic and critical analysis. What can it teach us about music education and how it can be enriched through the use of music technology in an authentic way?

All Music Education Takes Place within a Rich Context of Contingent Practical Elements

The contingent practical context of music education is fundamental and integral to Bowers's process of creating music with technology. It is only through a strong commitment to exploring the intricate relationships that develop that a true (or at least a defensible) understanding of what music education really is can be created.

This raises a number of pertinent questions. To what extent are we able to map out the contingent practical elements that are at work within a particular context or process of musical instruction? The type and location of these elements might be diverse, extending from the classroom, where learning might be initiated, to the students' home environments, where it continues and develops, and from their conversations with their friends at school to conversations they have online with others about their work; they may include formal elements, such as the units of work within which the musical learning is contextualized, to informal elements, such as the album a student listened to yesterday. They will undoubtedly include the quality of relationship that students have with their teachers, their peers, and their instrumental teacher or other admired role models. They will also include a whole range of digital technologies.

Understanding these elements is important if we are to truly understand and know how students' musical learning has developed. Only by developing a rich understanding of the broad context within which students' work has been produced can you begin to understand why they have made their particular musical choices. This understanding is not helped by simplistic and reductionist categorizations of the type presented within the concept of the "digital musician." These only serve to atomize our understanding and prevent us from considering in appropriate detail the real essence of what it means to be a musician and a music educator.

Music Education Always Takes Place within a Rich Technological Context

Technologies are integral to all music making, digital or otherwise. I would argue, philosophically, that there is not much difference between the development of the sustain pedal on the pianoforte in the nineteenth century and the latest Boss guitar effect pedal in 2014. Technologies of any type can help enforce the social order, or they can negate it; they can facilitate a meshing of ideas and responses, or they can helpfully or unhelpfully disrupt them.

The rich technological context of music education extends beyond our choices of instruments and their uses in educational settings. The broad arrays of technology that mediate our students' lives implicate, fundamentally, their engagement with us, as teachers, and with music more broadly. One cannot escape this reality, and it is ridiculous to imagine that one can.

Working Productively with Technologies: Initiation, Delegation, Supervision, and Intervention

As I have discussed, making music with any technology, digital or otherwise, is a complex business. It builds on numerous contingent practical elements and configurations that are mediated through a process of variable sociality. It demands that a student be able to diagnose and work within a range of approaches for musical engagement and interactivity in an integrated and holistic way. Trying to disassemble skills, concepts, or processes as technological or nontechnological is nonsensical. As teachers, it is vital that we understand these processes; policy makers develop examination frameworks that facilitate them in an integrated way.

Different technologies demand different approaches. We need to encourage our students to be flexible, to embrace and respond, intuitively and fluently, to the emerging streams of sound that these technologies produce. As teachers, we can initiate something. We can start our students off in a direction. But following an initiation, there is a delegation. For effective musical learning to take place, we need to allow our students to take ownership of their creative ideas. They need space and autonomy, time to explore, to experiment, to work with their machines and obtain outcomes that are of value to them. Delegation might involve handing over significant control to a technology, for a time, to see what emerges. The key here is to consider the human endeavor in equal measure with the technological input. It is the student who will add and express value in a technological utterance.

If students are not to continue their work indefinitely, there will come a time when the teacher has to exercise a legitimate supervisory role. Perhaps the time is up for that piece of work, a new direction needs to be taken, or the deadline for submission is near. With supervision comes intervention. Intervention might mean a day of reckoning, a formal assessment or examination. However, it could just mean a moment of reckoning or accountability, a pointing in a new direction—a tack as it were—before the students are off again.

Initiation, delegation, supervision, and intervention; here is just one potential approach to music education that is in tune with the ways of working with digital technologies inspired by Bowers's *Improvising Machines*.

Conclusion: It's All about the Music, Not the Technology

Music is not a universal language, but within the specific musical utterances of a particular genre or style one can begin to recognize gestures, shapes, sentences that carry meaning. The use of all technologies, digital or otherwise, needs to be firmly contextualized within music itself. Technologies are authenticated within the context within which it they are used. For teachers, the key is to find a way to integrate music technology into musical activities, games, curricula, and conversations with their students in a way that facilitates their students' creativity and engagement with music itself.

Within the United Kingdom, teachers and their students will work, for the foreseeable future, within examination and curriculum frameworks that are seeking to divorce musical skills and processes from those categorized as being concerned with "music technology." This is a system that prioritizes certain forms of knowledge in a simplistic and unhelpful way, for example, the rewards of studying for an A Level in "music," as opposed to "music technology" are more favorable (e.g., in accessing a course of higher education). As I have shown, it also creates artificial and unhelpful barriers in terms of the gendered discourse surrounding music itself. These barriers are very difficult for individual teachers and students to overcome.

However, the study of music and the provision of music education within the context of individual teachers' work with their students is a location where it may be possible to begin to chip away at some of these barriers. Teachers have a responsibility not to buy into the narrative that music technology is only for some students, perhaps those who cannot access music in the "proper" way or who are male! They need to realize that the skill sets that they need if they are to implement a broad and appropriate range of music technology within their work are their own responsibility, and not something that should be hived off to a technician or support staff. Most important, their conceptual models for music education and how it is organized must be built upon an understanding of an authentic musicianship that embraces technology, of any shape and form, and sees it as integral to musical expression. Music technology is too important to be categorized as being solely within the domain of the "digital musician" and left at the doorstep in the experiences of so many others. Artificial categorizations only divide; what music education needs to develop first and foremost are students who have a rich and authentic form of music expression, regardless of the tools they choose to use.

References

Bowers, J. (2003), *Improvising machines*. Retrieved January 24, 2014, from http://www.ariada.uea.ac.uk/ariadatexts/ariada4/.

Cambridge Associates. (2011a). *Uptake of A-level subjects 2010: Statistics report series no. 28—revised*. Retrieved January 27, 2014, from http://www.cambridgeassessment.org.uk/Images/109918-uptake-of-gce-a-level-subjects-2010.pdf.

Cambridge Associates. (2011b). *Uptake of GCSE subjects 2010: Statistics report series no. 35—revised*. Retrieved January 27, 2014, from http://www.cambridgeassessment.org.uk/Images/109925-uptake-of-gcse-subjects-2010.pdf.

Cambridge Associates. (2012a). *Uptake of A-level subjects 2011: Statistics report series no. 42*. Retrieved January 27, 2014, from http://www.cambridgeassessment.org.uk/Images/109931-uptake-of-gce-a-level-subjects-2011.pdf.

Cambridge Associates. (2012b). *Uptake of GCSE subjects 2011: Statistics report series no. 44*. Retrieved January 27, 2014, from http://www.cambridgeassessment.org.uk/Images/109933-uptake-of-gcse-subjects-2011.pdf.

Cambridge Associates. (2013a). *Uptake of A-level subjects 2012: Statistics report series no. 55*. Retrieved January 27, 2014, from http://www.cambridgeassessment.org.uk/Images/150182-uptake-of-gce-a-level-subjects-2012.pdf.

Cambridge Associates. (2013b). *Uptake of GCSE subjects 2012: Statistics report series no. 57*. Retrieved January 27, 2014, from http://www.cambridgeassessment.org.uk/Images/150205-uptake-of-gcse-subjects-2012.pdf.

Department for Education. (2013). *National curriculum in England: Music programmes of study*. Retrieved January 27, 2014, from https://www.gov.uk/government/publications/national-curriculum-in-england-music-programmes-of-study/national-curriculum-in-england-music-programmes-of-study.

Edexcel. (2013). *BTEC results day 2013*. Retrieved January 27, 2014, from www.edexcel.com/btec/news-and-policy/.../BTEC_results_day_2013.ppt.

Edexcel. (2012). *BTEC results day 2012*. Retrieved January 27, 2014, from www.edexcel.com/btec/Documents/BTEC-Results-Day-2012-External.ppt.

Emmerson, S. (Ed.). (1986). *The language of electroacoustic music*. London: Macmillan.

Hugill, A. (2008). *The digital musician*. London: Routledge.

4 A

Further Perspectives

CHAPTER 53

TECHNOLOGY IN MUSIC INITIAL TEACHER EDUCATION

MARINA GALL

It has been fascinating reading the three Core Perspectives in this part of this book. Dorfman and Greher write from the perspectives as teacher educators in the United States; Savage considers music teacher education in England. My experience is also within the English context: I am responding as an ex–secondary school music teacher and mentor to novice teachers who has spent the last 15 years working as a music teacher educator. In this short chapter I will reflect upon some of the issues raised by the American authors, and will build upon the ideas presented by Savage concerning the English system of initial teacher education for secondary music teachers.

In 2006–2009, while working with European colleagues on a project mapping music teacher education across Europe, I became aware that the English system of training secondary teachers in association with higher education institutions differs from almost all others in Europe (Institute für Music Pädagogik Wien, n.d.). One might term this system a "3 + 1" approach: students complete their 3-year undergraduate studies in a specialist subject and then select their teacher education pathway, which is usually 10–11 months long. The "one-year" Postgraduate Certificate in Education course (PGCE) is only offered at certain higher education institutions in England; thus, many students move to different universities for their education studies, which are required to include at least 120 days of teaching practice in schools (Department for Education, 2016). Many music undergraduate courses have no education component whatsoever (Crow, 2008), and thus the majority of those who select this route into teaching engage with a system that, until the final year of four, separates the acquisition of subject knowledge from that of pedagogical and pedagogical content knowledge (Shulman, 1987). A benefit is that students have a considerable length of time to decide on their vocations; the huge disadvantage for those opting to become teachers is that they have less than one year to develop pedagogical awareness and skills in relation to all aspects of music learning and teaching.

There is an additional complexity when one considers the place of music technology within the PGCE. In his chapter on teacher education in Boston, Massachusetts (chapter 50), Dorfman notes that undergraduates entering his course have limited awareness of the technologies that support musical learning. In England, undergraduate courses differ widely, and not all require engagement with music technology; as a result, students can begin their teacher education year with no skills at all in this area (Gall, 2013). Helping student teachers develop personal music technology skills and the confidence to use technology in classroom settings within one year—at the same time that they are acquiring other subject knowledge and competences, honing learning and teaching skills, and completing masters level written assignments—is a difficult undertaking.

Furthermore, while a National Curriculum for information and communications technology (ICT) in initial teacher education, including details of requirements for each secondary subject, existed at the start of this century (Teacher Training Agency, 2002), the latest English teaching standards make no mention at all of technology (GOV.UK, 2011). Although I feel that a list of specific technological skills to be developed and software and equipment to be mastered would be too prescriptive for a creative subject such as music, I agree with Dorfman that preservice teachers should be compelled to engage with technology, in a range of ways, prior to qualifying. In my article on research into English trainee teachers' notions of the factors that inhibit their use of music technology in school I suggest that, in the United Kingdom, music technology skills development should be mandatory within all undergraduate courses to enable student teachers to begin their one-year teacher education course with at least some skills (Gall, 2013). I have designed the 10-month PGCE music course on which I teach in such a way as to ensure that every trainee teacher gains experience in creating and teaching projects that include music technologies. The student teachers are also required to reflect upon these experiences both individually, as part of mandatory assignments and with the PGCE music cohort as a whole. Similar to what Dorfman describes, they too use "technological pedagogical content knowledge" (Mishra & Koehler, 2006) as a framework for thinking (Gall, 2017).

From my own perspective as a teacher educator, it is interesting to note that, despite the differences in the school curricula that form the backdrop to teacher education work in the United States and in England, comparable issues arise when one considers technology. Dorfman and Greher both discuss the difficulties of preparing preservice teachers for a school music curriculum that places a strong focus upon vocal and instrumental ensemble work and relatively little emphasis on technology. In England, although technology has been included in the Music National Curriculum since the mid-1990s (Department for Education, 1995) and the government has continually expressed concern about its ineffective use and underuse (Office for Standards in Education, 2004; Office for Standards in Education, 2012), it is relatively uncommon within school music for younger secondary school children (Gall, 2013). Given the importance of practical school experience of technology-based learning environments for trainee teachers, highlighted by both Greher and Dorfman, the paucity of good modeling by experienced English music teachers in lessons for 11- to 14-year-olds is problematic (ibid.). Moreover, even novice teachers who are competent and confident in their own ICT skills, and who have creative ideas for supporting musical learning with technology, can be hindered in their school practice. This is a result not only of the dominant culture of the many secondary school music departments in which technology is sidelined or used inappropriately with younger secondary school students but also—noted too by Greher in relation to U.S. schools—of music teachers' limited access to appropriate hardware and software (Gall, 2013), and of mentors' perspectives and school policies that fail to embrace skills and experiences that pupils gain outside school.

Reading the chapters by Greher and Dorfman, I wonder if perhaps the job of English music teacher educators in fostering trainees' enthusiasm about ICT is somewhat easier than that of our U.S. colleagues because of the increasing importance of music technology within school examination syllabi. Savage notes that, of late in England, the numbers of those who are studying both the music and the music technology A Level courses for 16- to 18-year-olds have diminished. At the same time, however, a recent government report indicates that there has been an increase in students opting for music technology rather than the more long-standing "more traditional" music course (Office for Standards in Education, 2012). In my experience, the awareness that teaching using technology at an A Level may well be required of them impacts positively upon trainee teachers' attitudes toward the acquisition of personal and pedagogical music technology skills. Regarding my own PGCE course, some trainees begin feeling underconfident or uninterested in technology yet, on entering schools, very quickly recognize that ICT competence and pedagogical understanding and skills are essential to their teacher toolkit. This is in strong contrast with Greher's experience in Massachusetts. Dorfman, also working in Massachusetts, questions the need for trainee teachers to learn how to use complex sequencing software. In England, preservice teachers view this as vital, since the ability to teach the A Level music technology examination course—which largely entails work with professional-standard software such as Logic or Cubase—is increasingly included as a requirement in job specifications.

Readers from outside England might view the existence of both a music and a music technology end-of-school examination course as positive, but Savage raises key concerns associated with presenting music technology skills as separate from other musical competencies. (For more discussion on the subject see also Savage, 2012). Of these, one that has been the focus of my own thinking for some time, concerning both teacher education and classroom work, is that of the gender disparity in relation to technology. This is discussed in depth in Victoria Armstrong's *Technology and the Gendering of Music Education* (2011), which suggests ways in which males are often privileged within technology-based school composition work. Data related to my own teacher education course from 2007 to 2014 (see Gall, 2015) indicates that many females enter their PGCE year with no personal experience of using music technologies, and that even within a course that focuses strongly on ICT within music education, such novice teachers often remain underconfident in this area when they enter their first teaching jobs. Since school students' perceptions and attitudes are influenced strongly by gender (Green, 1997), my concern is that unless gender issues surrounding technology are addressed, the status quo will be maintained. Furthermore, in England, the burgeoning of interest in music technology as an end-of-school examination subject could lead to a considerable number of women being disadvantaged when applying for music posts in schools.

To this point, I have focused upon the "traditional" teacher education pathway for aspiring secondary school teachers, which is one of the most long-standing of those currently on offer in England. However, across the globe there has been growing emphasis on developing routes into teaching that offer more extended practice periods in schools, including training that occurs almost exclusively within schools (Ellis & Orchard, 2014). Here I use the term "training" advisedly, and for reasons that Dorfman has already outlined in his chapter. A key element of the music teacher education course that I lead is reflexivity, which is encouraged at a personal level and "designed" into the university sessions in order to enable individuals to externalize their thinking and to learn from the experiences of their peers who have taught in differing schools. Like Greher, Savage, and Dorfman, I support the novice teachers with whom I work in gaining competence in the use of music hardware and software, and critical engagement and reflection are key to all aspects of the University of Bristol course, including those related to music technology. My concern with school-based teaching routes is whether a novice teachers, independently or with their mentors, will have adequate time for reflection when they are required to become fully immersed in school life very quickly. As Eraut explains, without time to consider, invent, and implement new ideas, there is a danger that even competent new teachers can become "the ossified teachers of tomorrow" (1994, p. 83). Moreover, research indicates that teachers' biographies shape the ways in which they become teachers (Goodson & Mangan, 1992). Given the lack of skills, experience, or interest in music technology displayed by many of the trainee teachers with whom I have worked (Gall, 2013), how prepared would they have been to include music technology in their learning and teaching approaches without the prompting and opportunities they received on the PGCE course? Furthermore, power relationships between novice teachers and their mentors in school settings often result in trainees using and relying upon the pedagogical approaches common within the department and school and, in general, complying with locally situated understandings of all aspects of their work (Ellis, 2010). In relation to technology, might trainees on a school-based teaching route, assigned a training place within a school that chooses to sideline technology within music lessons, feel strongly obliged to follow the departmental lead so as to gain their mentor-assessors' approval?

Thus far, this chapter has focused on music technologies and preservice teachers in England. What of in-service training, which both Greher and Dorfman discuss within the US context? As I have stressed in an earlier publication (Gall, 2013), I see professional development for practicing teachers as inextricably bound up with preservice education: without this, schools will lack good role models for trainees during their practice periods. Like Greher, I have noted ways in which my own PGCE partnership work has acted as a form of professional development for both school and university staff. Indeed, I believe that a "community of practice" (Wenger, 1998) has evolved as a result of the regular meetings held between university music teacher educators and the cohort of school music mentors, and of the ongoing relationships that have developed as colleagues have worked together over a number of years.

During a recent music partnership meeting, which focused upon technology, particularly in relation to classroom work with 11- to 14-year-olds, I asked mentors to complete a questionnaire about their own experiences related to music technology in school and to their work with student teachers. Regarding in-service opportunities for developing learning and teaching skills, including music technology, unlike the majority of UK teachers within Savage's study of 2007–2008 who cited "face to face instruction in a group setting" as their preferred method of professional development regarding technology (Savage, 2010, p. 98), the University of Bristol mentors unanimously expressed a wish to be able to develop ideas within their own school context—so as to enable focus on school-specific aspects of unit design such as available hardware and software, curriculum possibilities and restrictions, physical space, and so on. All those involved in the survey either noted that they were happy to leave "technology work" to more experienced and competent colleagues or cited the need for further opportunities to consider how to use technology effectively and creatively for student learning and for teaching. Both Dorfman and Greher highlight the importance of continual development of thinking and skills related to school music and ICT. Many other music educators across Europe are in agreement, in particular citing the need for further consideration of the pedagogies surrounding the use of music technologies in school classrooms (Gall et al., 2012). This is clearly a global cry and one that needs to be heeded if teachers are to keep abreast with, and capitalize on, the exciting opportunities that advancements in technology can offer school music and, at the same time, are to be in a position to support trainee teachers effectively in their early explorations of technology and musical learning.

References

Armstrong, V. (2011). *Technology and the gendering of music education*. Aldershot, UK: Ashgate.
Crow, B. (2008). Changing conceptions of educational creativity: A study of student teachers' experience of musical creativity. *Music Education Research, 10*(3), 373–388.
Department for Education. (1995). *Music in the national curriculum*. London: HMSO.
Department for Education (2016). *Initial teacher training criteria and supporting advice: Information for accredited initial teacher training providers*. Retrieved January 31, 2017, from https://www.gov.uk/government/uploads/system/uploads/attachment_data/file/279344/ITT_criteria.pdf.
Ellis, V. (2010). Impoverishing experience: The problem of teacher education in England. *Journal of Education for Teaching: International Research and Pedagogy, 36*(1), 105–120.
Ellis, V., & Orchard, J. (2014). *Learning teaching from experience: Multiple perspectives and international contexts*. London: Bloomsbury.

Eraut, M. (1994). The acquisition and use of education theory by beginning teachers. In G. Harvard & P. Hodkinson (Eds.), *Action and reflection in teacher education* (pp. 69–88). Norwood, NJ: Ablex, 69–88.

Gall, M. (2013). Trainee teachers' perceptions: Factors that constrain the use of music technology in teaching placements. *Journal of Music, Technology and Education*, 6(1), 5–27.

Gall, M. (2015, April 14–18). *Interesting challenging times: ICT in music teacher education in England.* Paper presented to symposium Technology within Initial Teacher Education in Music: Cross-continental Perspectives, Ninth International Conference for Research in Music Education, Exeter, UK.

Gall, M. (2017). Music teacher education and TPACK. In A. King, E. Himonides, & S. A. Ruthmann (Eds.), *The Routledge companion to technology and music education* (pp. 305–318). London: Routledge.

Gall, M., de Vugt, A., & Sammer (Eds.). (2012). *European perspectives on music education: New media in the classroom.* Innsbruck: Helbling.

Goodson, I., & Mangan, J. (1992). Subject cultures and the introduction of classroom computers. *British Educational Research Journal*, 21(5), 613–628.

GOV.UK (2011). *Guidance: Teaching standards.* Retrieved January 31, 2017, from https://www.gov.uk/government/publications/teachers-standards.

Green, L. (1997). *Music, gender and education.* Cambridge, UK: Cambridge University Press.

Institute für Music Pädagogik Wien. (n.d.). *Music Education Network (meNet).* Retrieved January 31, 2017, from http://menet.mdw.ac.at/menetsite/english/topics.html?m=2&c=0&lang=en.

Mishra, P., & Koehler, M. (2006). Technological pedagogical content knowledge: A framework for teacher knowledge. *Teachers College Record*, 108(6), 1017–1054.

Office for Standards in Education. (2004). *2004 report: ICT in schools—the impact of government initiatives: Secondary music (HMI 2189).* Retrieved May 5, 2015, from http://dera.ioe.ac.uk/4844/.

Office for Standards in Education. (2012). *Music in schools: Wider still, and wider.* Retrieved January 31, 2017, from http://www.ofsted.gov.uk/resources/music-schools-wider-still-and-wider.

Savage, J. (2010). A survey of ICT usage across English secondary schools, *Music Education Research*, 12(1), 89–104.

Savage, J. (2012). Those who can play; those who can't use music tech: How can teachers knock down the walls between music and music technology? In C. Philpott & G. Spruce (Eds.), *Debates in music teaching* (pp. 169–184). New York: Routledge.

Shulman, L. S. (1987). Knowledge and teaching: Foundations of the new reform. *Harvard Educational Review*, 57(1), 1–22.

Teacher Training Agency. (2002). *National curriculum for using information and communications technology to meet teaching objectives in music initial teacher training: Secondary* (Publication 40/5-99). London: Author.

Wenger, E. (1998). *Communities of practice.* Cambridge, UK: Cambridge University Press.

CHAPTER 54

USING MOBILE TECHNOLOGIES AND PROBLEM-SEEKING PEDAGOGIES TO BRIDGE UNIVERSITIES AND WORKPLACES

JULIE BALLANTYNE

THE task of providing future teachers with the capacity to respond effectively in changing teaching environments and the related issues of retention and job satisfaction of early-career teachers present an ongoing challenge to teacher educators. Many technologies can provide solutions to these issues, and consideration of technologies as a tool for better pedagogical developments in preservice teachers may have multiple benefits to the profession. When technologies are used for other ends, the focus is removed from the technology—much as when using a pencil to write a novel, the content of the novel is the focus, not the pencil. Following this analogy to a logical conclusion, however, the technology is central to the success of the outcome and becomes subsumed into the practices and habits of the user. Technology can arguably change and enhance the ways that early-career teachers arrive at the outcome of effective professional work, enabling greater creativity and fundamentally enhancing their roles as teachers.

In this Further Perspective, I focus predominantly on one typical issue facing music teacher education, and demonstrate how utilizing simple technologies as learning tools can enhance the quality of the teacher education process. The pedagogy I discuss encapsulates a two-way process of facilitating future teachers to become comfortable with new technologies and increasing the likelihood that they will use them effectively in their classrooms. In this way, I will intentionally address one of the Provocation Questions: *how might technology better serve the needs of students and teachers?*

Praxis Shock and the Real-World Divide

It can be very difficult for music education graduates to navigate the divide between the university classroom and first work placements. For students in a variety of disciplines, the divide between tertiary learning and the realities of professional practice looms large (Billett, 2002; Burnard, 2013; Piihl & Philipsen, 2014; Roth, 2010). Music and creative arts undergraduates, in particular, need more opportunities to engage with real-world experiences during their time at university (Bridgestock, 2006). Without such opportunities, graduates can find themselves with simplistic notions of the workplace and lack the skills to successfully engage as professionals. For music teachers, the difficulties associated with "acting as a teacher" from the very first day on the job can be overwhelming and stressful and lead to early-career burnout. Much of this stress is associated with understanding the realities of the job, and prioritizing where attention should be focused (Ballantyne, 2007). Crucially, it is also associated with a failure of tertiary contexts to develop in early-career professionals a realistic understanding of the workplace.

Much research has suggested that case-based work does much to alleviate the "strangeness" of new workplace environments, and assists students to build the skills so as to ascertain what is important, and to problem-solve successfully once graduated. My research (and that of others) has shown that this approach does not go far enough to simulate the complexities of the real-world workplace, leaving graduates unaware of the sheer variety of contexts and issues they are likely to encounter outside a constructed or simulated case study situation (Ballantyne, Harrison, Barrett, & Temmerman, 2009; Ballantyne & Olm, 2013). It is clear that opportunities are needed for students to go out into the workforce, with the sole purpose of practicing real-world problem-solving in situ. In order to accomplish this, preservice teachers must have the opportunity to frame classroom issues, to select and collect relevant information, and, crucially, to "prune" or "ignore" the aspects of the context that, while important, are not relevant to the issue they have identified. Engagement in a process of systematic problem-solving arguably builds the kind of complex understanding of the real-world workplace that inoculates against praxis shock.

To this end, technology aids. An innovative way to collect and frame complex, multifaceted issues is through the use of mobile technologies, such as portable multimedia devices with Internet connectivity. These can be taken into workplace situations and used to collect photos, videos, recordings, interviews, documents, websites, and curriculum documents (whatever might be relevant to the situation identified by the student). The litmus test of whether students have successfully "sought and framed" a problem in the workplace will be seen when they share their issue and the constituent documents with their peers. At this stage, the question arises—can their peers understand the context surrounding the issue, the complexity of the issue, and still work through addressing the issue in a professional and productive manner? When these questions occur, a multifaceted and arguably more "realistic" learning occurs that provides a direct conduit to a deeper and more real understanding of the workplace.

Technology as a Pedagogical Process

The Mobile Technologies Project (MTP) was designed with the aim of to using mobile technologies as pedagogical tools to better serve the needs of students and teachers. The intent has been to use mobile technologies to allow students to construct their own understandings of the challenges faced in the profession, using problem-based learning and reflective practice. The aim of the pedagogical trial has been to increase the relevance of tertiary study to the "real world," linking the reality of teaching with university training. In the trial, preservice teachers have compiled digital portfolio case studies based on interactions with workplaces, narrowing the gaps that exist between research and reality as well as between the university classroom and the broader community. This new curriculum model focused on improving students' learning by embedding discipline interests within a work-integrated approach to problem-solving and community collaboration.

This approach focused on the contextualization of the curriculum through authentic, case-based learning and digital portfolios. The approach required students to move between the university and employment contexts, and was intended to lessen the gap of expectation, which is often to blame for praxis shock (Ballantyne, 2007).

The Mobile Technologies Project

Engagement in the MTP required students to produce their own real-life case studies and to problem-solve with a motivation and realism that is often lacking when fabricated or constructed case study scenarios are used. Case studies have been produced directly by students based on their observations of real-world workplaces, including community groups, schools, festivals, arts councils, arts companies, professional development organizations, music therapy sites, museums, theatres, video game companies, and the broader creative industries. While adaptation to local contexts is crucial to the success of the project, initially all sites involved in the project adhered as closely as possible to the overarching intended approach set out here:

> *Collect rich portfolios using technology*: Students went into the workplace or community to collect, using mobile technologies, evidence or artifacts to contribute to a portfolio (case study) to represent the intricacies of the issues faced in that particular real-world context.

Refine the case study portfolio: Relying on critical analysis to decide whether to include or exclude information and evidence and artifacts in their collections, students produced digital portfolio case studies (created and contained on their mobile technology devices). They could select videos, reflections, discussions, recordings, case files, student works, music compositions, and sharing applications to include in the portfolio; they were limited only by the capabilities of the Internet-connected mobile device and the agreements in place at their case study site.

Frame the selected issue or problem: Students identified problems or issues requiring solving during their engagements, and posed the problems (in writing or video) on their mobile devices. They ensured that their cases were self-contained within their devices (e.g., iPad), and that all relevant information was included.

Collaboratively solve the problems: The mobile device was shared (physically) with another group of students, who then had access to the rich multimedia content compiled in the field by their peers. With the mobile device, students utilized the content included and the framing of the problem to discuss and analyze what they saw and its relationship to their possible futures, suggesting appropriate ways to deal with the issues identified by their peers to solve the problem. This style of "campfire pedagogy" enables students to focus on collaboratively solving problems from the field, as identified by their peers. It is worth noting that it is during the *process* of trying to solve problems that the learning occurs, and that not all problems are solvable.

The Potential Impact on Student Learning

As this innovative project has only recently been piloted, with uptake at various other sites in coming semesters, the ongoing benefits are still theory-driven. However, by equipping graduates with the opportunities to critically examine, identify, frame, and solve intricate problems found in real employment contexts, the project has the potential to mitigate the impact of praxis shock on the profession.

This project's simple premise provides a solution to a complex and ongoing challenge in the higher education sector: how to get students to engage meaningfully with theoretical and practical university work in a way that will enable them to be work-ready upon graduation. When students produce the learning materials themselves while situated directly within real-life contexts, they are applying critical thinking skills similar to those they will require when they become working teachers. This differs fundamentally from the more "formalized" thinking skills they may apply to a constructed case study context. Further, the opportunity to enter the classroom environment before practicum placement provides an "in-between" stage of more objective observation—a learning stage in which students can familiarize themselves with the classroom from the "teaching perspective" prior to having to actively control this context themselves (e.g., as practicum teachers or new graduates). The position of quiet observer is facilitated by the mobile technologies.

From the perspective of the university, the project provides an opportunity to trial, refine, and disseminate a new curriculum model in the arts disciplines, with the potential for future implementation across a wide variety of disciplines. Students involved in this project engage in self-directed analysis, critical thinking, and peer collaboration, and develop skills in selecting and presenting relevant data to facilitate problem-solving. At its foundation, the project aims to support and encourage students in applying theoretical knowledge collaboratively, ultimately increasing their capacities to adapt to a range of professional practice contexts.

By allowing students the responsibility of constructing and shaping the case studies on their mobile technologies (creating their own digital portfolio case studies), the teaching academics are automatically placed in the role of "facilitator" and "guide" in this pedagogy. They are able to access the support of other academics involved in the project through the community of practice, and are also able to learn alongside the students about contexts beyond their existing experiences. The project will have long-term, sustainable benefits as a result of embedding the curriculum model within existing courses, and facilitating partnerships between the academics, the students and workplaces.

Where Is Technology in This, Really?

This approach aligns with Himonides's "new way forward" for music technology in teacher education. Himonides (chapter 59 in this book) asserts:

> The role of technology in education (and, obviously, music education) is central, not complementary, or peripheral.
> Technology can foster the development of teachers' critical thinking, which is . . . the most important ability that an educator should possess.
> Teachers cannot be passive recipients and consequent conduits of predefined educational praxes; they need to be in a position to shape their own and their students' learning, and critically assess what their needs are, as these are being formed.
> It is the social facet of technology that will play a leading role in teachers' competence, credentialing, and professional development.
> Technology is becoming mature enough to allow us to re-rehearse the notion of evidence-based education in a completely novel way.

In conclusion, this project also moves toward the learning community notion espoused by Greher in chapter 51. If the technologies employed in music teacher education are seamlessly integrated and embedded in the pedagogy and curriculum, and they do not seem at odds with the larger aims of music teacher education, then we have reached a point where they are becoming truly useful. This sometimes requires a reconceptualization of the ideal pedagogy for music teacher education—change of any sort can be challenging and should be exciting for teacher educators. Technology provides the opportunity to "shake up" the established norms in our field, to try something new, and to learn about ourselves, our students, and music education in the process.

Acknowledgments

My thanks to the University of Queensland for the provision of a Teaching and Learning Fellowship to enable the piloting of the Mobile Technologies Project. Thanks also to Tammie Olm-Madden, whose work as a research assistant on the project was invaluable.

References

Ballantyne, J. (2007). Integration, contextualization and continuity: Three themes for the development of effective music teacher education programmes. *International Journal of Music Education, 25*, 119–136.

Ballantyne, J., & Olm-Madden, T. (2013). Exploring dialogues in online collaborative contexts with music teachers and pre-service students. In H. Westerlund & H. Gaunt (Eds.), *Collaborative learning in higher music education: Why, what and how?* (pp. 63–76). Burlington: Ashgate.

Ballantyne, J., Kerchner, J. L., & Aróstegui, J. L. (2012). Developing music teacher identities: An international multi-site study. *International Journal of Music Education, 30*(3), 211–226.

Ballantyne, J., Harrison, S., Barrett, M., & Temmerman, N. (2009). *Bridging gaps in music teacher education: Developing exemplary practice models using peer collaboration.* Sydney: Australian Learning and Teaching Council. Retrieved December 30, 2016, from https://www.researchgate.net/publication/43514753_Bridging_gaps_in_music_teacher_education_Developing_exemplary_practice_models_using_peer_collaboration.

Billett, S. (2002). Towards a workplace pedagogy: Guidance, participation and engagement. *Adult Education Quarterly, 53*(1), 22–43.

Bridgestock, R. (2006). Follow your (employable) bliss: The challenge of the Australian applied arts graduate. In *Proceedings AACC06 International Careers Conference*. Sydney. Retrieved January 5, 2017, from https://eprints.qut.edu.au/4013/1/4013_1.pdf.

Burnard, P. (2013). *Developing creativities in higher music education: international perspectives and practices.* London: Routledge.

Piihl, J., Rasmussen, J., & Rowley, J. (2014). Internships as case-based learning for professional practice. In J. Branch, P. Bartholomew, & C. Nygaard (Eds.), *Case based learning in higher education* (pp. 177–196). Faringdon, Australia: Libri.

Roth, M. (2010). Learning in praxis, learning for praxis. In S. Billett (Ed)., *Learning through practice models, traditions, orientations and approaches* (pp. 21–36). Dordrecht: Springer.

CHAPTER 55

APPLICATION OF TECHNOLOGY IN MUSIC EDUCATION IN SELECTED AFRICAN COUNTRIES

BENON KIGOZI

BEING involved with music education practice, theory, and research in Africa, and having taught across primary and secondary schools as well as at university level, I find it interesting to read the Core Perspectives from the United States and England by Dorfman, Greher, and Savage! In most parts of Africa, things are rather different for the simple reason that the way technology is understood in the West is not the way it is perceived in Africa. First of all, technology is not always guaranteed to be available, nor is it "automatic" where it is available, as it is in the West. As I mentioned in chapter 17, "technology," as defined by the Pan African Society for Musical Arts Education, is the "audiovisual aids and tools such as books, systems of musical transmission, aural-oral, mental and other mnemonic aids, [that are] indigenous African, even stories, language and literature, and other aspects of science, the arts and culture" (Pan African Society for Musical Arts Education, 2003a). This certainly means that even if the curricula do not prescribe technology for training teachers, they would find themselves in a position to acquire it through community involvement with music.

TECHNOLOGY IN NONFORMAL MUSIC

During my childhood, I was exposed to music through the neighborhood, and I learned how to sing and play musical instruments during community musical events before I started taking formal lessons. The system of musical transmission was always aural-oral and indigenous African, through poems and stories, language and literature. That is how technological it was in a developing country. In Buganda, the central part of

Uganda where I come from, education involves preparing, training, and transforming a child into a responsible member of society. This process is of utmost importance to the community, and everyone had to be inducted into the heritage of their predecessors, which was understood to be manifested in music education in nonformal situations. Competence in this kind of arrangement is always assured, however, because it is nonformal; the issue of credentials and professional development is not addressed. I agree with Dorfman that music teacher preparation must account for the combination of executive and expressive skills. Even though it is nonformal, music education has a central role in the daily lives of the people, and is treated in a holistic manner with regard to education, involving poetry, dance, drama, storytelling, genealogy, and art.

Dorfman asserts that, in many ways, scholarship in music education echoes the concerns of other subjects. Sometimes, however, it is more challenging in music when macrolevel curriculum design is the order of the day, as it is in various African countries. When that is the case, music education is forced into a weaker position as part of arts education. An example is the Uganda national curriculum, which classifies music under Performing Arts and Physical Education. The government uses nonartistic principles to organize the Performing Arts and Physical Education curriculum to give an impression of unity. The disadvantages of music being part of that curriculum are that the government neglects specific *perception reaction* experiences in favor of a generalized, disembodied "appreciation of the arts." It submerges the character of each individual art by focusing exclusively on family likeness. Because of this, music does not get the attention and support it deserves as a discipline.

Dorfman goes on to point out, discussing teacher education in Boston, that undergraduate students have limited awareness of technologies that support musical learning. This seems to be the trend here, too, where music teachers lack technological skills because the music teacher training programs do not equip prospective teachers with technological skills, augmenting the challenge of initiating sustainable music programs involving technology. Primary school music education is the most appropriate stage to pioneer small-scale changes in the use of music technology; however, primary music educators have yet to assume a more active role in directing the future course of new music technology.

Much of the time, the way technology is viewed in the West is not how it is perceived in Africa. Culturally, Africans feel that African music must be taught in context and through methods that are specific to Africa. Thus far, "African culture" and instructional practices in Africa have not succeeded in incorporating computer-based technology for music education into regular classroom instruction, even at those few schools that can afford it. Computer-based technology must therefore prove to have a generic role of preservation and advancement of the culture if it is to be integrated in music education. Nzewi (1999) warns: "to introduce Africans to modern music learning and appreciation of European music thoughts, contents, practices and pedagogy is a radical, de-culturating process. It continues to produce the crises of cultural inferiority, mental inadequacy, and pervasive, perverse cultural-human identity characterizing the modern African person in modern social, political, educational and cultural pursuits" (p. 72). In 2012, I conducted a survey on the integration and application of information and communication technology in music education. The survey took the approach of direct observation, and included a quantitative as well as qualitative inquiry in the form of interviews. In order to probe participant perspectives, qualitative interviews were conducted. Participants in the survey consisted of students, teachers, and parents. While teachers were positive about the benefits of technology in music education, many of them only used technology outside the classroom for reporting on student performance.

Savage (chapter 52) seems to be concerned with music technology separated from general music teacher education, class music education, and the like. One of my experiences in music classrooms as I travel across the continent doing PASMAE work is that music teachers prefer to teach music as a separate subject from technology, saying that technology is in most cases a time drain on teachers with already busy schedules. In countries across Africa, a vast majority of music educators need assurance that technology will enhance their teaching experience. It is important for music educators to be aware of the full capabilities of these tools to enhance student learning (Forest, 1995; Nolan, 2009; Ofiesh, 1970; Rudolph, 2004). Research shows that when students become active participants in learning, they gain confidence, learn more effectively, and are motivated to continue their studies (Rudolph, Richmond, Mash, Webster, Bauer, & Walls, 2005).

Even though technologies are not always guaranteed to be available, they are currently accessible in some parts of Africa, and in cases where they have been applied they have positively affected the way educators teach. New and used technologies, in the form of musical instruments and computers, have been oversold almost everywhere in Africa as a vehicle for educational reform and educational practices, but the question is whether or not they help teachers teach better (Kigozi, 2013). Greher (chapter 51 in this book) asserts that "technology in the right hands can expand our creative options as musicians." Does technology enhance students' participation, and does it improve students' general performance? Greher notes: "what sets one production apart from the next comes down to the ideas generated by the people at the helm of the technology." In other words, when music educators come up with creative and engaging approaches to integrating technology into music education, then it works. Rudolph confirms that music educators have always searched for better ways to engage students in learning music and that technology has the key (Rudolph et al., 2005).

Application of Technology in Music Education in Other African Countries

Namibia provides access to music education under its Primary Arts Core, a broad and general arts program that also embraces other arts subjects. According to the Ministry of Basic Education and Culture of Namibia (Namibia, 1999) the Primary Arts Core is designed for all learners in primary school and refers to the development of basic knowledge and skills common to all schools in Namibia. The Primary Arts Core program offers an integrated arts approach to the teaching and learning of the arts, which includes music, dance, drama, and visual arts, with no technological components mentioned at all in the curriculum.

In 1985, the Ministry of Education in Ghana introduced an enrichment program to encourage African music in schools. According to Oerhle (1991), the content and methods of education in Ghana had to be adapted to suit local needs. Consequently, the Cultural Studies Program of Ghana, which was developed in 1987 for primary and junior schools, treats music, dance, drama, and folklore as basic components of Ghanaian culture. In 1987, the West African Exam Council, O Level, also included the study of African and Afro-American music (Oerhle, 1988). Music in schools is therefore integrated into, and dispensed under, the Cultural Studies Program. In the mid-1990s, music education stakeholders launched a campaign for a positive approach toward African music, especially indigenous Ghanaian music, to make it stronger in schools. Music education has an important cultural role in helping to ensure schools become more Ghanaian in character (Flolu, 1996).

Malawi legitimized the instruction of music in schools; music is embraced in the National Curriculum is taught via "standards" 1–8. The focus is on skills, including singing, movement, instrumentation, and composition (form, rhythm, and melody). Malawi's motivation is the belief that music is a vehicle of expression, transmits and preserves the culture of Malawi, provides enjoyment, and encourages creativity and imagination. In addition, there is a belief that music is a source of income; it promotes social development and helps to reinforce learning in other subject areas (Malawi, 1991). In all of this there is no mention of technology being integrated with music in education.

In Botswana, music is seen as a subject that provides students with the opportunity to develop their innate music abilities; the Botswana National Curriculum embraces music as one of the subject areas taught in schools. However, because music is treated as an optional subject in the country's education program, the government has not found it necessary to support technological funding for music education. The major aim of music education in Botswana is to contribute to the preservation and transmission of the cultural heritage of Botswana (Botswana, 2000).

Zimbabwe, Nigeria, and Kenya are working on the consolidating of indigenous music education in their systems of education. In my conversation with a few music teachers in Harare, Zimbabwe has had an upsurge of interest in traditional music, but up until 3 years ago no music was prescribed for primary school education. Now that it is there, the integration of technology in music education will probably be the next stage.

Nigeria follows a curriculum which dispenses music education, however, though music is embedded in the educational system, it is not well ranked and is one of the weakest subjects in the curriculum. The emphasis in music has been placed on Western music (Hauptfleisch, 1997, p. 106). Okafor observes that the government and the examination bodies have no radically new and positive approach to the study of music and so do not include relevant music and technology in Nigeria's public education (1991, pp. 61–67).

In Kenya, music education is not administered under the umbrella of arts education. The Presidential National Music Commission of 1984 proposed the enhancement of traditional music in the education curriculum. Following the recommendations by that commission, the revised music syllabus of 1992 reflects the intention to actively involve learners in living cultures through a process of assimilation (Floyd, 1996b, p. 200). However, Kenya's Ministry of Education still needs to fully promote the education of indigenous music in schools. Even though traditional music and dance is an integral part of life of many Kenyan societies, it still needs to be considered an essential element in the education curriculum of the Ministry of Education. Until that is done, the application of technology in music education will remain on paper. Akuno warns: "the exclusion of music from the formal education system, including its use in the classroom, has resulted in the development of a generation of Kenyans that does not know its music" (1995, p. 45).

Conclusion

In order to have an integrated and relevant system of technology in music education, the governments of African countries need to recognize that the future of the music education discipline lies in the development of technology in music education. Governments, through their respective departments of education, must ensure financial support for the development of technology mastery in music educators. Music teacher training programs should be redesigned to empower teachers in training with the knowledge, skills, dispositions, and norms of the discipline of applying technology in music teaching, as well as to connect with other areas of knowledge without organizing music programs on nonmusical principles.

The music curricula should be designed to accommodate the provision of technology for music education while allowing flexibility of content across the continent of Africa and the West. Music education through the curriculum should promote knowledge, enjoyment, and growth through the application of technology. Finally, since music is considered to be a part of the arts, it should not be placed in a weaker position than the other arts subjects in the curriculum: technological funding should cut across all subjects equitably. This will enable learners to demonstrate individual as well as group musicianship in meeting authentic music challenges within diverse music practices.

References

Akuno, E. (1995). *Teaching music in primary school*. Paper presented at the workshop on music education for primary school music teachers, Nairobi.

Botswana. (2000). *Junior secondary music education syllabus draft*. Gaborone, Botswana: Government Printers, Botswana.

Flolu, J. (1996). Music education in Ghana: The way forward. In M. Floyd (Ed.), *World musics in education* (pp. 157–185). Aldershot, UK: Scholar.

Floyd, M. (1996). Promoting traditional music: The Kenyan decision. In M. Floyd (Ed.), *World musics in education* (pp. 185–206). Aldershot, UK: Scholar.

Forest, J. (1995). Music technology helps students succeed. *Music Educators Journal*, 81(5), 35.

Hauptfleisch, S. (1997). *Transforming South African music education: A systems view*. Unpublished D.Mus. University of Pretoria.

Kigozi, B. (2013). Critical impacts of music technology to music education: A restructuring of the teaching and learning of music at secondary schools in Uganda. *Technologii Inforrmatice Si De Communicatie in Domeniul Muzical*.

Malawi. (1991). *Primary school teaching syllabus: Music Standards 1–8*. Lilongwe: Malawi Institute of Education.

Namibia. (1999). *Ministry of basic education and culture*. Arts syllabus—upper primary phase. Windhoek, Namibia: NIED.

Nolan, K. K. (2009). SMARTer music teaching. *General Music Today*, 22(2), 3.

Nzewi, M. (1999). Strategies for music education in Africa: Towards a meaningful progression from tradition to modern. In C. van Niekerk (Ed.), *Conference proceedings of 23rd International Society for Music Education World Conference* (pp. 456–486). Pretoria: Unisa Press.

Oehrle, E. (1988). A guideline for music education in Southern Africa. In C. Lucia (Ed.), *Proceedings of the Second National Music Educators' Conference* (pp. 32–52). Durban: UN.

Oehrle, E. (1991). Emerging music education trends in Africa. *International journal of Music Education*, 18, 23–29.

Ofiesh, G. D. (1970). Technology, the Road to Freedom. Music Educators Journal, 56(7), 45–48.

Okafor, R. C. (1991). Music in Nigerian education. *Bulletin of the Council for Research in Music Education*, 108, 59–68.

Pan African Society for Musical Arts Education. (2003a). *Musical arts in Africa: Theory, practice and education* (Anri Herbst, Meki Nzewi, & Kofi Agawu, Eds.). Unsa Press.

Pan African Society for Musical Arts Education. (2003b, July). Unpublished.

Pan African Society for Musical Arts Education. (2005). *Emerging solutions for musical arts education in Africa* (Anri Herbst, Ed.). Cape Town: African Minds.

Rudolph, A. T. (1991). Technology and music education. *ACM SIGCUE Outlook*, 21(2), 84–86.

Rudolph, T. E. (2004). *Teaching music with technology*. Chicago: GIA.

Rudolph, T., Richmond, F., Mash, D., Webster, P., Bauer, W., & Walls, K. (2005). *Technology strategies for music education*. Milwaukee, WI: Hal Leonard Corporation.

CHAPTER 56

DEFINING AND ACKNOWLEDGING MUSIC EDUCATION TECHNOLOGY IN MUSIC TEACHER TRAINING

LAURI VÄKEVÄ

My Own Background in Music Education Technology

As a child of the 1960s, my association with music technology began rather modestly. In the early 1980s, as an aspiring rock and jazz bass player, I had to develop some rudimentary capabilities in handling amplifiers and recording equipment. However, I never received training in mixing, recording, or producing. Even during a brief spell of working as a radio music journalist in the early 1990s, my technological preparation was restricted to using the broadcast mixer and performing basic audio editing with a razor blade on quarter-inch magnetic tape. Both onstage and in the studio, I was content to let technicians handle technical matters.

My musical studies focused on three areas: instrumental skills, knowledge about music, and music pedagogy. I eventually chose music education as a major subject, and graduated in 1994, the same year I bought my first personal computer, which I used mainly to write texts and musical notation and, later, for online work. While my professional preparation was musically rather extensive, aspects of technology were left to the minimum, notwithstanding the fact that my specialization subject was Afrodiasporic music, in which cutting-edge technology has always been a decisive factor for artistic innovation.

While being aware of the discrepancy between my area of specialization and my own technological dexterity, I only had time to begin to fulfill the blanks at the end of the

1990s, when I was employed full-time as an assistant in a music teacher training program. Subsequently, it was only in the mid-2000s, around my fortieth birthday, that I begun to feel somewhat comfortable with music technology (however, information and communication technology [ICT] was already an integral part of my teaching at that time). Today, my duties as professor of music education mostly relate to supervising and supporting students' research work; however, my interest in music education related technology has grown to a degree that I have found myself participating in several research and development projects that focus on the pedagogical aspects of music technology and ICT in music teacher training. Still, I see myself very much as a digital immigrant, eager to learn more of the ways in which digital natives organize their musical lives and the implications of these ways for music education and music teacher training.

THE ROLE OF TECHNOLOGY IN FINNISH MUSIC EDUCATION

The split between general and performance-oriented music education is deeply ingrained in the Finnish education system, where comprehensive school music is complemented by a nationwide network of music schools (Korpela et al., 2010). Regardless of the institutional context, Finnish music teaching seems to focus principally on rehearsing precomposed and prearranged pieces for performance in and outside the classroom, with some creative working methods tossed in for good measure (Anttila, 2010; Korpela et al., 2010; Juntunen, 2011; Muukkonen, 2011; Partti, 2013). Even if the significance of technology for musical learning has been acknowledged in the national curricula for both comprehensive schools and music schools (*Perusopetuksen opetussuunnitelman perusteet*, 2004, 2014a, 2014b, 2014c; *Taiteen perusopetuksen musiikin laajan oppimäärän opetuksen perusteet*, 2002; *Taiteen perusopetuksen yleisen oppimäärän opetuksen perusteet*, 2005), digital technology seems to be used more as tool for boosting traditional musical activities than as something worthy of being studied as a creative, expressive, artistic, and aesthetic field of its own. That said, there are notable exceptions to this trend, especially in comprehensive schools, where music teachers have a lot of freedom in designing their courses due to an open and pragmatic national curriculum. Thus, I have witnessed schools that do not have any music technology instruction at all, but also schools in which using recording equipment, digital audio workstations, and mobile technology are an integral part of studying.

While it is easy to envisage digital technology as a tool to serve music making, certainly this is not the sole reason for including it in the comprehensive school and music school curricula. Indeed, there is another significant implication of digital technology that has been repeatedly highlighted in Finnish general-level educational discourse:

the power of ICT to transform the conditions of learning. In a country that in the 1990s set out to be a leading developer of information technology (Niemi, 2003), it would seem to be only natural to allocate resources to the rapidly evolving post-Nokia digital economy that emphasizes the synergy of human transactions as the core of meaningful knowledge construction. Certainly, this seems to be a guiding rationale behind the most recent national comprehensive school curriculum (*Perusopetuksen opetussuunnitelman perusteet*, 2014a, 2014b, 2014c). However, in Finnish music teacher training the social aspect of technology appears not to have rewarded much attention yet, even if Finnish music education researchers seem to generally accept that music is social-cultural practice and digital technology has changed the conditions of this practice to a degree that we can talk about participatory music culture, where sharing resources in information networks fills a major part in becoming a musician (see, e.g., Westerlund, 2002, 2004; Väkevä, 2007, 2009, 2010, 2012; Väkevä & Westerlund, 2007; Salavuo, 2008; Partti, 2012; Partti & Westerlund, 2012; Odendaal, Kankkunen, Nikkanen, & Väkevä, 2014). The recognition of the significance of ICT to musical learning was also one of the guiding ideas of the Finnish virtual university project (MOVE—Musiikinopettaja verkossa [Music educator online]), which set out to develop ways to use networked technology in music education in the early 2000s (Ojala, 2006; Salavuo, 2006). At the time of writing this, as both sectors of music education are expecting a new national curriculum, there are several plans for nationwide projects that target developing all music teachers' skills in applying ICT and music technology through in-service training programs.

As things stand, I maintain that is crucial for technology to be addressed in preservice and in-service music teacher preparation from a wider pedagogical perspective than simply acknowledging its power in making traditional ways of music making more efficient. For instance, recognition of the way social media affect artistic communication and aesthetical sensibilities is surely as important as finding out how the latest apps and gadgets can aid singing, playing, and composing in the classroom. It seems that, at least in Finland, we need a rethinking of how preservice and in-service music education technology courses can widen music teachers' visions. Music teacher training programs would also seem to be natural habitats for experimenting with different expressive and creative aspects of digital technology.

Different Aspects of Music Education Technology

In the educational discourse it is possible to recognize narrow and extensive uses of the word "technology." One way to rationalize this difference is to say that the narrow usage refers to technology as a means, whereas the extensive use also encompasses ends. For instance, in the hands of a skillful hip-hop DJ, a system of two turntables and a mixer can be transformed from technical machinery into technological means, in order to attain aesthetic ends. In the narrow understanding, it is only the machinery that represents technology. In the extensive outlook, it is the whole chain of productive activity, from intent to process to outcome to reception, that rewards our technological interest.

There is an obvious difference between these two points of departure for designing technology courses for music teacher education programs. I suggest that music teacher educators should seriously consider looking at the possibilities of the wider approach when designing technology units. One thing that such wide perspective might reveal is that aesthetic uses vary from context to context: the application of technical means to musical ends differs from culture to culture. An important part of a music teacher's technological understanding will be the ability to help her students both to recognize such cultural differences and to shape their own musical expressions based on such recognition.

Here is another possible perspective that could be beneficial to consider when discussing the role of technology in music teacher training. In education conversation, the word "technology" is not infrequently applied to indicate activity that involves specific "technical" knowledge and application of such knowledge to achieve particular goals. Thus, according to the *Oxford English Dictionary* (2014), "technology" can refer to "a branch of knowledge dealing with the mechanical arts and applied sciences," "the application of such knowledge for practical purposes," and "the products of such application." In other words, technology, understood in the Aristotelian sense of techné—namely, skillful activity that produces results other than itself—can be examined from conceptual, practical, or product-based standpoints, where the application of scientific knowledge offers an epistemological point of departure. However, a general understanding of technology does not have to restrict its applications and outcomes to scientifically informed production. While there may be a sense in which one can talk about the "science" of DJ-ing, this does not make a DJ's practice less an art.

I have addressed the distinctions between narrow and extensive understanding of technology on the one hand and between specific and general understandings of technology on the other in order to clarify my own thinking about technology in music teacher training. What is clear from this discussion is that, from both a wide and general understanding, "technology" indicates more than tools. While tools have been at the heart of pedagogical technological interest for a long time, the extensive perspective helps us also to recognize the specific productive relation between means and ends that technology enacts—or, as Ojala (2006) puts it, we can think of technology as a study of habits of action. Moreover, this relationship between means and ends does not have to be restricted to applying scientific knowledge in manufacturing goods, but it can also encompass other goal-specific approaches; the general approach can thus also reveal how the technological means-ends relationship is enacted in musical and musical-pedagogical practices in different contexts.

Organizing Technology Courses in Music Teacher Education: Three Suggestions

Other contributors to this part of this book present important insights into the potential role of technology in music teacher preparation. From my standpoint, these insights are highly relevant and applicable in the kind of music teacher education that is organized in Finland: combined degrees that cover both bachelor (3 years) and master (2 or 2.5 years) levels and in which music, pedagogy, and research studies convene as dimensions of the same music education major subject, producing the full qualification to teach music as a subject teacher. I will close my own contribution by focusing on three suggestions as to how to organize technology education as part of such an extensive music education training.

1. I have discussed above the need to understand technology not merely as a tool but as a system of means-ends activity that, applying a certain body of knowledge, produces certain outcomes that satisfy certain expectations, be they technical, artistic, or other kinds. Such a stance can draw philosophical inspiration from John Dewey's pragmatist philosophy, which situates techné at the heart of all human aspirations (on this interpretation of Dewey's philosophy, see Hickman, 1990). For music teacher educators, such extensive and general understanding of technology as a mediator between means and ends in some context of purpose indicates that it is not enough to present technology as a set of gadgets and apps; however, it also suggests that it may not be enough to focus only on music technology in music educators' professional training. Instead, it might be more beneficial to approach technology as a common denominator for all human transactions that afford diverse social-cultural configurations of musical and music-related meaning-production. This would mean extending the subject matter of the music education technology courses to all technologically mediated forms of "musicking" (Small, 1999), keeping one's focus on all music-related communication.
2. In line with the preceding, it would be as important to trace learning trajectories that lead to musical and music-related communication networks online and offline than to show and tell how technical tools are to be applied in the classroom (Partti, 2012). Such view would open a possibility of building a wide understanding of how technology can provide new aesthetic uses and, thus, novel meanings (to keep with the pragmatist interpretation that identifies meaning with use). Hence, in reference to the example of hip-hop DJ-ing, understanding how the DJ appropriates audio reproducing equipment as a sophisticated musical-expressive device requires knowledge not only of how the turntables can be used in this context and what they are in technical terms but also of how such practices are

developed in the first place: what purposes they serve in the cultural contexts in which they are discovered, and what are the pedagogical lessons to be learned from such practices.

3. As Peters remarks (chapter 63 in this book), "teacher training programs are bursting at the seams." This concerns Finland as much as other countries. Still, in programs that incorporate both graduate and postgraduate degrees it would seem that there is plenty of time to study music education related technology in a variety of forms. Nonetheless, at least at the Sibelius Academy music education program, unless one specializes in music technology, technological preparation is limited to one introductory course that involves familiarizing the students with basic office and music software and the related hardware.

To have only one dedicated technology course for all students is admittedly not much. However, this does not mean that technology is absent from other parts of the curriculum. On the contrary, there seems to be a common understanding in the three Finnish music teacher education programs (and in many other teacher training programs in the country) that technological preparation should be integrated into all coursework. Hence, at Sibelius Academy, the music student teachers are encouraged to experiment with pedagogical uses of technology throughout their studies. We have also lately launched a series of research and development projects that equip students and teachers with mobile apps, with the idea that they report regularly how they find uses for them in their daily transactions and reflect together on the possible professional implications of such technology (Antila & Väkevä, 2013).

While no doubt beneficial, such general integration of technology into music education studies might not suffice unless technology is approached contextually and examined through a wide enough lens. As Greher points out (chapter 51), to buy the latest gadgets and apps does not yet take us far in developing pedagogical approaches that attend to the aesthetic and educational potential of such gadgets and apps. Whatever the apparatus used, it is the ideas that count, and ideas only become ideas when they lead to specific practical applications in certain historical, social, and cultural contexts. Thus, the main technological focus in music teacher training should be on studying what ends technical means serve in the contexts of learning that afford diverse meanings through diverse uses of technical innovations. Such contextual study of technology should not be left to postgraduate studies; it should serve as a guiding ideal throughout music education studies. Of course, this does not mean that we cannot design postgraduate courses to be theoretically more reflective and to involve more research-based views, as Dorfman suggests (chapter 50). However, I believe that the germs of such reflectivity should already be established in the freshman year, with appropriate theoretical preparation.

To conclude, I return to my suggestion above that we should not accept technology as something that only caters to specific course contents for general music teachers or general institutions of learning. All music educators—were they to work in the "performance track" or "general music track"—benefit from understanding how technology helps to construct new meanings together through the use of means in aesthetic ends. A wide and general focus on technology can bring to light means-ends relationships previously unseen and unheard, and thus extend the music teacher's aesthetic sensibility to previously unimagined artistic realms.

REFERENCES

Antila, S., & Väkevä, L. (2013, November 1). *Going mobile, rocking—recent development research projects of applied digital technology in a Nordic music education program.* Paper presented at College Music Society/Association for Technology in Music Instruction conference, Boston.

Anttila, M. (2010). Problems with school music in Finland. *British Journal of Music Education*, 27(3), 241–253.

Hickman, L. (1990). *John Dewey's pragmatic technology.* Bloomington: Indiana University Press.

Juntunen, M. (2011). Musiikki. In S. Laitinen, A. Hilmola, & M.-L. Juntunen (Eds.), *Perusopetuksen musiikin, kuvataiteen ja käsityön oppimistulosten arviointi 9. vuosiluokalla* (pp. 36–95). Koulutuksen seurantaraportit [Educational assessment reports] 2011:1. Helsinki: Opetushallitus.

Korpela, P., Kuoppamäki, A., Laes, T., Miettinen, L., Muhonen, S., Muukkonen, M., Nikkanen, H., Ojala, A., Partti, H., Pihkanen, T., & Rikandi, I. (2010). Music education in Finland. In I. Rikandi (Ed.), *Mapping the common ground: Philosophical perspectives on Finnish music education* (pp. 14–31). Helsinki: BTJ.

Muukkonen, M. (2011). *Monipuolisuuden eetos: Musiikin aineenopettajat artikuloimassa työnsä käytäntöjä.* Unpublished doctoral dissertation, Sibelius Academy, Helsinki, Studia Musica 42.

Niemi, H. (2003). Towards a learning society in Finland: Information and communications technology in teacher education. *Technology, Pedagogy & Education*, 12(1), 85–104.

Odendaal, A., Kankkunen, O., Nikkanen, H., & Väkevä, L. (2014). What's with the K? Exploring the implications of Christopher Small's "musicking" for music education. *Music Education Research*, 16(2), 162–175.

Ojala, J. (2006). Mitä on musiikkikasvatusteknologia? In J. Ojala, M. Salavuo, M. Ruippo, & O. Parkkila (Eds.) *Musiikkikasvatusteknologia* (pp. 15–21). Keuruu, Finland: Suomen musiikkikasvatusteknologian seura.

Partti, H. (2012). *Learning from cosmopolitan digital musicians: Identity, musicianship, and changing values in (in)formal music communities.* Unpublished doctoral dissertation, Sibelius Academy, Helsinki, Studia Musica 50.

Partti, H. (2013). Uudistuva muusikkoushanke tutkii musiikin luovia työtapoja ja säveltämistä kouluissa ja musiikkioppilaitoksissa. *Finnish Journal of Music Education*, 16(1), 47–54.

Partti, H., & Westerlund, H. (2012). Democratic musical learning: How the participatory revolution in new media challenges the culture of music education. In A. Brown (Ed.), *Sound musicianship: Understanding the crafts of music* (pp. 300–312). Newcastle upon Tyne, UK: Cambridge Scholars.

Perusopetuksen opetussuunnitelman perusteet. (2004). Vammala, Finland: Opetushallitus.

Perusopetuksen opetussuunnitelman perusteet: Opetus vuosiluokilla 1–2. (2014a, September 19). Draft. Helsinki, Finland: Opetushallitus.

Perusopetuksen opetussuunnitelman perusteet: Opetus vuosiluokilla 3–6. (2014b, September 19). Draft. Helsinki, Finland: Opetushallitus.

Perusopetuksen opetussuunnitelman perusteet: Opetus vuosiluokilla 7–9. (2014c, September 19). Draft. Helsinki, Finland: Opetushallitus.

Salavuo, M. (2006). MOVE ja musiikkikasvatusteknologian kohdehankkeet. In J. Ojala, M. Salavuo, M. Ruippo, & O. Parkkila (Eds.), *Musiikkikasvatusteknologia* (pp. 363–367). Keuruu, Finland: Suomen musiikkikasvatusteknologian seura.

Salavuo, M. (2008). Social media as an opportunity for pedagogical change in music education. *Journal of Music, Technology & Education, 1*(2-3), 121–136.

Small, C. (1999). *Musicking: The meanings of performing and listening.* Hanover, NH: University Press of New England.

Taiteen perusopetuksen musiikin laajan oppimäärän opetuksen perusteet. (2002). Helsinki: Opetushallitus.

Taiteen perusopetuksen yleisen oppimäärän opetuksen perusteet. (2005). Helsinki: Opetushallitus.

Väkevä, L. (2007). Art education, the art of education and the art of life. Considering the implications of Dewey's later philosophy to art and music education. *Action, Criticism & Theory for Music Education, 6*(1). Retrieved February 2, 2017, from http://act.maydaygroup.org/articles/Vakeva6_1.pdf.

Väkevä, L. (2009). The world well lost, found: Reality and authenticity in Green's "new classroom pedagogy." *Action, Criticism & Theory for Music Education 8*(2). Retrieved February 2, 2017, from http://act.maydaygroup.org/articles/Vakeva8_2.pdf.

Väkevä, L. (2010, March). Garage band or GarageBand®? Remixing musical futures. *British Journal of Music Education, 27,* 59–70.

Väkevä, L. (2012). Digital artistry and mediation: (Re)mixing music education. *National Society for the Study of Education, 111*(1), 177–195.

Väkevä, L., & Westerlund, H. (2007). The "method" of democracy in music education. *Action, Criticism & Theory for Music Education, 7*(4). Retrieved February 2, 2017, from http://act.maydaygroup.org/articles/Vakeva_Westerlund6_4.pdf.

Westerlund, H. (2002). *Bridging experience, action and culture in music education.* Unpublished doctoral dissertation, Sibelius Academy, Helsinki, Studia Musica 16.

Westerlund, H. (2004). Dewey's notion of experience as a tool for music education. *Nordic Research on Music Pedagogy Yearbook, 7,* 37–50.

CHAPTER 57

LEARNER ENGAGEMENT AND TECHNOLOGY INTEGRATION

MICHAEL MEDVINSKY

THE roles of learner engagement and technology integration in education are two of my passions. It has been quite interesting to read the essays by Jay Dorfman, Gena Greher, and Jonathon Savage (chapters 50–52 in this book) in the context of my own perspective. Their discussions of traditions of music education, invisibility, and authenticity in approaching music technology education have sparked my interest in preservice teacher education, supporting and scaffolding music educators, and continuing professional development through becoming a connected educator. They have really made me consider my time with primary and secondary learner-musicians and in-service and preservice music educators. Savage's European perspective has broadened my understanding of the depth and breadth of music technology in the current climate of music education. In this Further Perspective, I build on the ideas Savage, Dorfman, and Greher have presented and contextualize them within my own learning community and practices.

I have been fortunate to teach in public school districts in southeastern Michigan where technology has always been available in my classrooms and where teachers have been supported by innovative district-level technology coordinators. I was able to purchase a class set of iPads a few years ago that have been regularly used by the learner-musicians in my classes throughout their primary and secondary music learning experiences. In my teaching, I find it important to design opportunities for learner-musicians to experience creating music in many ways, including digitally. Savage writes: "for teachers, the key is to find a way to integrate music technology into musical activities, games, curricula, and conversations with their students in a way that facilitates their students' creativity and engagement with music itself." It is my role as a music educator to design relevant and timely musical experiences for learner-musicians in our classroom to engage in these experiences. I found there was a disconnect between my urgency for developing a technologically rich music curriculum and my experiences as an undergraduate music education major.

Teachers of my generation entered into the classroom at a time when personally accessible, mobile technology was first emerging, and it was essential that we be prepared to create experiences for learners supported by that technology; yet we just missed the window of having been prepared for these levels of technology in our undergraduate classrooms. Music educators are not alone. Dorfman writes: "in many ways, scholarship in music education echoes the concerns of other subjects. Technology integration, in particular, is a component of music teacher preparation for which researchers have examined concerns similar to those of our general education counterparts." Technology is constantly changing and being upgraded. More learners are coming to class with mobile devices, excited to use these devices to learn and create. Educators must capitalize on their interest and enthusiasm and integrate this technology into learning and teaching processes.

In my experience of trying to do this, what has been most useful has been the ways the approaches taught in my undergraduate experience lent themselves to these kinds of transformations. I understood that it was my role as teacher to help learners focus on the processes of learning, and that together we constructed understandings of our thought processes and the ways we thought *in* music, rather than *about* music (Wiggins, 2015). The focus of my undergraduate training had not been on how to teach with a guitar or synthesizer; rather, I was taught to teach music in ways that made it easy for me to see how I could easily incorporate these new tools for music making within my vision of curriculum and learning and teaching processes. The world of the classroom (and society) constantly changes. Undergraduate programs need to provide students with the bases for making decisions and adaptations throughout their careers. No teacher education program can teach new teachers everything they will need to know for their entire careers. Continual thinking, planning, and growth are the responsibility of the practicing teacher-professional.

A good teacher education program provides teachers with the understandings they need in order to continue to work innovatively throughout their careers and to reflectively meet the needs of the constantly changing student population. As a music teacher educator in higher education, one of my goals is to help reframe music teachers' understanding of the way today's musicians engage in digital musical processes. Most of the music teachers who enroll in my courses are entry-level music technology users who are quite overwhelmed, at first, with all the possibilities that different technologies offer. This can happen for a number of reasons. In my experience, it may be that these teacher-learners (1) have not had prior experience because of the era in which they became music teachers, (2) did not have enough emphasis on music technology in their methods courses, (3) have difficulty relating to the different approach to musicianship I am presenting, or (4) may be unsure or fearful because of their lack of experience with the technology. Regardless of the backgrounds they bring, I design opportunities for us to consider ways we already function as musicians and transfer those musical processes into digital musicianship. Contemporary performance practices and digital resources may require new operational processes, but authentic musical processes remain at the core (Ruthmann, 2013; Wiggins 2015).

The students and I do not approach learning the technology itself; rather, through the experience of being digital musicians, we perform, we compose and improvise, and we learn mobile technologies, GarageBand and Logic X interfaces, digital recording, and postproduction. After their experiences and reflective dialogue about what they consider to be authentic processes of musicians, how to best support them using digital tools, how technology can launch musicians in ways less familiar to them and bring musicians from different skill levels together, my students become quite excited to implement these tools and techniques in their own classrooms.

Teachers face many constraints when they put themselves in front of eager learners every day while they are being asked to document and show proof of learning to administration. When experienced in-service teachers begin engaging a classroom full of learner-musicians, while excited about music technology, they may fall back into using the traditional techniques they know well and with which they feel comfortable. It is for these reasons that I believe that music technology should not be taught as an independent techniques course. I propose that music technology be integrated within primary-secondary methods courses. This must be done in order for music educators to feel comfortable enough with the technology itself that it becomes transparent to the musical experiences. Dorfman proposes to "distribute technology skill development and pedagogy throughout the larger curriculum where appropriate." I suggest that practical skill development and pedagogy support one another throughout the learning experience. Understanding the interface and what button to press to trigger the appropriate musical response will support the experience design when considering pedagogy. Music educators must have such a rich understanding of their content area that technology becomes a support for authentic processes, as opposed to being an add-on.

Savage argues: "a more coherent approach to the development of music technology skills within a music teacher's pedagogy is urgently needed." I agree and propose that we increase connections between preservice and in-service teachers. The sooner these connections begin, the more contextualized the learning will become. This is such a meaningful time for the observer to experience authentic music teaching and learning and for the in-service teacher to be reflective in practice and explain teaching strategies. These observations should be cyclical between classroom hours debriefing with the in-service teacher, supported by the student's reflective process, and discussion in a college methods course with other student observers and the mentoring professor. The in-service music teacher should model appropriate best practices when implementing music technologies; the student should discuss teacher decisions and learner-musicians' engagement with the in-service teacher, reflect upon those discussions and classroom experiences, and bring her perspectives back to class. This process will provide the preservice teacher with authentic experiences to draw upon when approaching music technology techniques and methods classes. It is an integrated approach such as this that will more adequately equip newly in-service teachers to transparently integrate technology into appropriate musical experiences.

If we use technology to do old things in new ways, we are still doing old things. Technology best serves music educators when they reimagine musicianship and design opportunities to explore nontraditional ways of being a musician. Using loop software such as LoopHD or Loopseque, individual musicians can create multilayered polyphonic pieces using voice, sampled sounds, or a library of sounds. Music educators must be in tune with current musical events to continue to provide relevant experiences. Savage takes issue with Hugill's suggestion that "digital musicians are a discrete class of musician who exhibit certain characteristics that separate them from other musicians who (he acknowledges) may make use of digital technologies on odd occasions within their work." Savage goes on to discuss the problems that stem from characterizing musicians. I, too, believe that musicians should not be characterized as "types"; rather, musicians change performance practices in reaction to current musical trends. Digital musicians exist across so many genres. Some may exclusively use music technology, while others may add a digital means as another voice within their acoustic landscape. Either way, the term "digital musician" covers so many possibilities in today's musical world that it is as difficult to characterize as the category "composer" or "instrumental musician."

This brings to mind an experience I had in 2014, when I saw Jimmy Fallon and Billy Joel on *Late Night with Jimmy Fallon* perform a vocal duet of "The Lion Sleeps Tonight" using the LoopyHD app and a microphone. That evening, I installed LoopyHD on the iPads in my classroom. The next week of classes, we re-created the duet of "The Lion Sleeps Tonight" as a whole class. Their project, to show their understanding of texture change and loops, was to create, perform, and record an original piece of music using loops. They collaborated and rehearsed in small composition groups, performed their pieces for the class, recorded their final performances, and shared their pieces on our classroom's social media streams.

In my experience as a primary and secondary Musicians' Workshop teacher, developed from the concept of Writers' Workshop (Calkins, 1986) and Composers' Workshop (Ruthmann, 2007), learner-musicians want to create the music they hear in their heads. Sometimes it is more complicated than what they can perform themselves; other times, it may be realized on instruments that are not found in traditional music rooms. Technology may support both these types of musician. The GarageBand iOS app has many points of entry through instrument choice, smart instruments, and choosing a scale to play a section of notes that sound good together. For musicians whose music is more complicated than what they can perform at their level, GarageBand has smart instruments and autoplay, which provide opportunities for musicians to manipulate and arrange sound in ways that represent their music. When approaching scales while improvising or composing melody, for an instance, learner-musician who understands the pattern of whole steps and half steps can use the chromatic interface to create melodies, while a learner-musician who may not have that prior knowledge can be scaffolded by the interface by only showing diatonic pitches. The two learner-musicians can perform together at different levels and both feel successful. Savage discusses how "performers can variably relate to each other, they can variably engage with the technologies and instruments before them." This type of experience exemplifies the transparency of music technology.

There will always be emerging technologies, startups, peripherals, and renaissances of old school technology becoming new again. How do music educators stay abreast of all of the new advancements? One approach is to create a personal learning network through social media and become a connected educator. Personal learning networks continuously provide professional development opportunities to improve teaching practices, support innovation, connect like-minded educators, connect the less experienced with professionals in the field, and provide opportunities for learning anytime and anywhere. Educators in this network freely share their experiences, failures, and successes and support each other while taking risks. As new technologies emerge, they are shared within the network via Twitter, Facebook, LinkedIn, and other social media, experimented with in classrooms, reflected upon through blog posts, and shared through video-sharing websites such as YouTube or Vimeo.

It is connectivity that will be the new intelligence. It is not necessarily what you know, it can be what your network knows that matters. A learning network is an integral part of a professional community, which may also include people you interact with in person outside the social digital network. When professional-learners engage in such networks, they become "professional learning communities" (Wenger, 1998). Staying current with advances in current music practices is the foundation for the design of relevant and timely experiences for learner-musicians. When these experiences are designed for musicians to develop, the classroom environment becomes one of experimentation and innovation. In a TED talk, Charlie Reisinger (2013) discusses the outcomes of enabling students in the digital age. He suggests: "When you push the boundaries of conventional practice, you break new ground." His words resonate with me as I continue to develop the environment in my classroom. Learner-musicians who engage in current musical practices in an environment where risk-taking is encouraged and supported are poised to break new ground. My classroom is successful when it is the learner-musicians who are the innovators, creating and performing music in new ways. For this to be the case, I need to continually learn from my professional community of innovators. I engage in learning new things from my personal learning network, while taking musical risks and sharing my written reflections through my blog, so as to model this behavior to the learner-musicians. Together, we cocreate our music curriculum, which, because of today's performance practice, is supported and often enabled by technology. I do not teach technology as a separate class; rather, I transparently integrate digital tools within our daily engagement with music. Reflecting on the ways people think and learn and drawing upon Savage's writing, I support his view that technologies, digital or otherwise, need to firmly be contextualized within music itself. It is music that is at the core of the experience. The ways musicians engage in these experiences change, but the core remains the same.

In my recent career, I have been invited to share my story and the experiences I design for learner-musicians to engage in authentic musical processes transparently supported by technology at conferences, school districts, and various publications. In-service music educators are seeking to develop their understanding and gain the confidence they need to implement technology into their curriculum. I agree with Gall, de Vugt, and Sammer (2012) when they write that both Dorfman and Greher highlight the importance of continual development of thinking and skills related to pedagogies surrounding the use of music technologies in school classrooms. However, we need not focus on the technology itself. Technology constantly moves forward and changes. Educators must embrace a "growth mindset" and discover the power of the word "yet" (Dweck, 2006), envision new possibilities, reach just beyond their grasp, foster divergence, and think differently. Technology will never replace a great educator, but a great educator who understands the possibilities of supporting learning with technology will replace a great educator who does not.

REFERENCES

Calkins, L. (1986). *Art of teaching writing*. Portsmouth, NH: Heinemann.
Dweck, C. (2006). *Mindset: The new psychology of success*. New York: Random House/Ballantine.
Gall, M., de Vugt, A., & Sammer, M. (Eds.) (2012). *European perspectives on music education: New media in the classroom*. Innsbruck, Austria: Helbling Verlagsgesellschaft.
Hugill, A. (2008). *The digital musician*. London: Routledge.
Reisinger, C. (2013, May 30). *Charlie Reisinger: Enabling students in a digital age* (Video). Retrieved January 5, 2017, from https://www.youtube.com/watch?v=f8C037GO2Fc.
Ruthmann, S. A. (2007). The composers' workshop: An approach to composing in the classroom. *Music Educators Journal, 93*(4), 38–43.
Ruthmann, S. A. (2013). Exploring new media musically and creatively. In P. Burnard, & R. Murphy (Eds.), *Teaching music creatively* (pp. 85–97). London: Routledge.
Wenger, E. (1998). *Communities of practice: Learning, meaning, and identity*. New York: Cambridge University Press.
Wiggins, J. (2015). *Teaching for musical understanding* (3rd ed.). New York: Oxford University Press.

4 B

Core Perspectives

CHAPTER 58

FACULTY DEVELOPMENT IN AND THROUGH THE USE OF INFORMATION AND COMMUNICATION TECHNOLOGY

PATRICIA A. GONZÁLEZ-MORENO

In recent years, educational policies around the world have emphasized the use of information and communication technology (ICT) and the development of technological competencies in an attempt to develop better teaching and learning practices (Anderson & Bonefas, 2002; Epper & Bates, 2001; Lowenthal, 2008). Given these conditions, university professors, like music educators at any other educational level, are expected to enhance their academic performance in connection with technological advances. Consequently, higher education institutions have been pushed to respond to a growing demand for faculty development in and through the use of technology for instructional and academic purposes (Epper & Bates, 2001). In Mexico, the Secretariat of Public Education, through diverse national programs (Programa de Fortalecimiento de la Calidad en Instituciones Educativas—Program to Strengthen Quality in Educational Institutions [in higher education]; Programa para el Desarrollo Profesional Docente—PRODEP; Fondo de Modernización para la Educación Superior), has fostered the use of ICT in higher education and of support projects that aim to enhance teachers' development and their "inception into a knowledge society" (Secretaría de Educación Pública, 2014). However, despite governmental and institutional initiatives, these goals are far from being met, particularly in both music and music education programs. In this chapter, I argue that within the Mexican context, faculty resistance is one of the most important factors that inhibit a greater impact of ICT on their professional development. Overcoming restrictive attitudes by helping faculty members to reflectively embrace new technologies may result in greater possibilities for the development of innovative educational practices in postsecondary music education.

This chapter presents an introductory overview of the role of technology in faculty development and the particular challenges for universities and faculty members, followed by a description of a web-based professional development program in the arts (mainly targeting music education topics) initiated at a public university in northern Mexico. The chapter also discusses some possibilities and challenges of overcoming faculty resistance. I conclude with a reflection on what universities and faculty members can do to avoid becoming obsolete, given the technological advancement of music education.

Background on Information and Communication Technology and Faculty Development

In recent years, research on teacher professional development has extensively examined the influence of ICT in the advancement of teachers' competencies (Bowles, 2003; Dede, 2004; Epper & Bates, 2001; Ertmer, Gopalakrishnan, & Ross, 2000; Hammel, 2007; King, 2002; Lowenthal, 2008; Sherer, Shea, & Kristensen, 2003; Triggs & John, 2004; Windschitl & Sahl, 2002). While most of this research has been conducted in general education, a more limited literature exists in the field of music education (Bauer, 1999, 2007; Friedrichs, 2001; Hookey, 2001), how it relates to the impact of ICT on faculty development in post secondary music education (Anderson & Bonefas, 2002), and to what extent university professors have embraced and applied technological advances in order to improve teaching and learning processes. Given the limited research in this area, the following subsections provide a glimpse of current trends in general technology use in higher education, the particular uses observed among music educators in different educational levels, the influence of ICT on faculty's overall professional practice, and the role of faculty development in helping university professors overcome their resistance to technology use and academic obsolescence.

Technology Use in Higher Education

Current trends in technology use suggest that Mexican university students are more intrinsically motivated to use technology than faculty members, who show a slow adoption of current technologies in their educational practices (López de la Madrid, 2007). This can be explained by the fact that younger generations have been highly exposed to new ICT, as digital natives, and university professors have been extrinsically encouraged and sometimes forced to adopt technology and have not easily delivered the expected results imposed by their institutions. According to Hall (2013), a significant proportion of faculty members respond slowly to technological innovations. He argues that some faculty members will never adopt new technologies, despite the nature of the incentives. He found that only a small proportion of early adopters (faculty members) take the initiative to embrace new learning technologies.

Faculty changes do not occur with the expected speed because of lack of experience, the difficulty in changing paradigms and pedagogical practices, and lack of interest in investing additional time and work in alternative educational strategies (Benítez Lima & Ávila Gómez, 2012). As a result, faculty develop an inherent resistance to change that is also characterized by self-imposed barriers to what is unknown and feelings of threatened security. In order to avoid a loss of confidence as a result of experimenting with educational innovations and possibly failing, faculty tend to maintain traditional teaching practices. For instance, faculty resist distance learning because it represents a significant change in teaching preparation. In this educational context, they become facilitators, a role that involves a substantial increase in the time and energy invested in developing a course (Smith, 1996). Both factors, the increased workload and the altered role of the instructor, seem to be the main source of faculty resistance. This could also represent a problematic issue in terms of equity among faculty members in teaching loads and in release time to prepare classes.

In educational environments where technology is available, most faculty members often lack a clear vision of how it can best support educational practices (Ertmer, Gopalakrishnan, & Ross, 2000). Often, faculty's use of technology is limited to email communication, as well as the use of Internet resources and social and professional networks. In a study that involved students and professors from five of the most important universities in Mexico, López de la Madrid and Flores-Guerrero (2010) found that, despite a high level of technological enablement, faculty had not integrated technology systematically in their teaching practices. All faculty members in the sample (N = 65) reported using computers and the Internet, but they mainly used technology for personal purposes rather than as a support for their academic courses, and they reported no meaningful changes in their teaching given the accessibility of new technologies. In a previous study, López de la Madrid (2007) also found that faculty members reported a limited use of ICT in their teaching. These results suggest that while an adequate technological infrastructure is important, faculty perceptions and attitudes may have greater influence on the successful implementation of ICT in higher education teaching. The following section more specifically describes music teachers' common uses and responses to technological demands that characterize the technology use in music education.

Music Teachers' Attitudes toward Technology Use

Literature on music teachers' attitudes toward technology has systematically increased over the years. Bauer (1999) found that the top uses of the Internet among 70 music educators were directly related to professional development, including learning about music and teaching, communicating and networking with colleagues and experts, and being informed about current trends in music education. Bauer, Reese, and McAllister (2003) investigated the impact of one-week technology workshops on the professional development and technology use of music teachers. They found that the frequency of technology use was positively correlated to access to technological resources, but they also noted the importance of extended professional development experiences in order to get teachers to begin and continue to use music technologies.

While these studies inform us about music teachers' responses to and self-expressed interest in technology at lower educational levels, little is known about university professors' attitudes and practices and the role of ICT in their development. In Mexican university programs, advanced technological competencies are more commonly observed among those who are specifically trained in music technology and technology-enhanced music education. However, given the lack of research, it is difficult to generalize about the extent to which ICT has directly impacted faculty development and the uses or misuses of technologies in higher education teaching and overall faculty academic activities.

Given previous research on music teachers' perceptions, it can be hypothesized that a current interest also exists among faculty members to know about technological advances in order to facilitate their academic work. However, although professors' interest and a beginning professional development in technology use are important, these seem to be insufficient to foster such faculty change as to significantly impact the educational context. We must recognize that technology imposes challenges on teaching and may confuse and intimidate learners and users (King, 2002).

While experiences in other academic areas in higher education and the perceptions of music teachers at other educational levels inform us about possible university professors' perceptions and attitudes, a close and deeper examination is still required in postsecondary music education in order to understand the needs for professional and current technological development and ICT integration that can enhance faculty's instructional and academic activities.

Faculty Development, Information and Communication Technology, and Teacher Quality

Undoubtedly, an enormous responsibility lies on university professors to prepare future music teachers and musicians. In this information age, they are expected to be at the cutting edge of technological advances in order to respond to professional needs in music education. Even music and music education faculty with outstanding academic profiles and sufficient technological expertise require continuous opportunities in order to meet the standard professional expectations (DuFour, 2004) and to avoid academic obsolescence. As Camblin and Steger (2000) argue, "the life span for the standard of excellence grows shorter and shorter, and the likelihood that either junior or senior faculty members can maintain distinctive levels of performance without the full support of their college or university is preposterous" (p. 2).

In order to keep faculty up to date with technology innovations, universities usually establish programs to develop faculty members' general pedagogical and technological competencies. These initiatives have been an integral part of higher education institutions' strategy for self-improvement. Unfortunately, institutional programs usually lack the specificity required in each discipline. In "one size fits all" approaches, professional developers are mainly concerned with providing a general education for professors in different disciplines rather than more specific disciplinary and technological expertise (Hubbard & Atkins, 1995). This type of training does not differentiate among levels of expertise, needs, and particular interests, mainly due to a lack of communication among faculty, administrators, and faculty developers (Hammel, 2007).

It is important to acknowledge that the implementation of faculty development programs focused on the use of technology or providing faculty with basic technological tools to enrich their academic activities (e.g., Internet access, computers, projectors, etc.) are not necessarily sufficient as strategies to improve teacher quality. As Epper and Bates (2001) argue, "the most daunting challenge in implementing technology in college teaching is still ahead of us: the development and training of faculty" (p. xv). Schools and educational authorities need to make changes that extend far beyond the mere installation of a network of computers (Ertmer, Gopalakrishnan, & Ross, 2000). For technology to make a difference in teacher quality and student learning, faculty must develop a solid philosophical foundation for the use of technology in teaching, establish a clear vision of the possibilities of ICT in facilitating pedagogical processes in postsecondary music education, and embrace a genuine commitment to their ongoing learning. In addition, faculty require continuous institutional and organizational support for their technology-enhanced professional development.

Another challenge in faculty development is that typical programs follow an approach of "changing the world in a day" (Lowenthal, 2008, p. 350) in order to attract faculty members. Professors seem to prefer these intensive programs, which are commonly held on consecutive days to reduce costs and time investment. However, seeking to effect change in teaching and professional practice in a short period of time is unrealistic. Research suggests that in order to observe substantial differences in teaching practice, professional development programs should be long term, with adequate support and follow-up (Richardson & Placer, 2001). By increasing the time involvement of faculty in professional development programs, faculty developers have aimed to avoid the problem of trying to change the world in a day.

Finally, one of the most difficult challenges for faculty development in higher education is changing values, beliefs, assumptions, and cultures (Richardson & Placier, 2001). As Dede (2004) explains, "altering deeply ingrained and strongly reinforced rituals of schooling takes more than an information interchange of the kind typical in conferences and 'make and take' professional development. Intellectual, emotional, and social support is essential for 'unlearning' and for transformational relearning that can lead to deeper behavioral changes" (p. xii). Within the university environment, it is important to reduce the negative influence of external factors that may impede teachers' positive experiences with ICT in order to favor intrinsic motivation toward realizing institutional technological goals and toward teachers' overall development and teaching practice. In the following example, I describe an initiative that has aimed to support faculty members in their professional development through the implementation of ICT.

A Technology-Enhanced Professional Development Program

In 2011, a series of webinars for professional development in the arts was implemented at the Faculty of Arts, Autonomous University of Chihuahua. The program aimed to use a webinar platform in order to facilitate communication with leaders in diverse music education areas and to a lesser extent with experts in other arts areas. The program gradually developed into an international partnership that included universities and research institutions in Europe, North America, and South America. Participants at each institution joined the virtual lectures as a group and at the hosting institution; an extended postwebinar discussion allowed participants to reflect on and consider the implications of the speakers' research studies and professional practice. The program was initially conceived as providing opportunities for professional development of university professors and graduate students in order for them to actualize their knowledge and improve their expertise about music and arts education by linking research to practice. Given the lukewarm response of the faculty at the hosting institution, it was decided to open up the professional development opportunity to undergraduate students, in-service music teachers, and other music practitioners.

Online learning opportunities such as webinars have been increasingly considered as a way to deliver professional development and offer access to professional training for faculty who otherwise would not be able to attend workshops. While there are some restrictions, such as simultaneous transmission in specific locations within each institution, webinars also provide opportunities for face-to-face interaction among attendees and can facilitate a deeper discussion and understanding, taking into account each particular educational context. It is generally believed that teachers benefit from technology-related professional development not only when it involves the collective participation of teachers from the same school or department and teachers relying on one another in developing technological skills (Desimone, Porter, Garety, Yoon, & Birman, 2002) but also when the opportunities to expand the level of collaboration cross geographical borders and reduce ideological barriers, such as professional isolation and weak collaboration (Bondy, 1997; Marginson & van der Wende, 2007). It was expected that the webinar professional development program at the Autonomous University of Chihuahua would facilitate skill development, increase technology use, and foster cooperation among professors within each institution and across institutions, multiplying the effects and the scope of the program, as previously demonstrated by Camblin and Steger (2000). The following sections describe the scope, strategies, results, and challenges of this initiative.

STRATEGIES TO OVERCOME FACULTY DEVELOPMENT CHALLENGES: BUILDING COMMUNITIES OF PRACTICE

Faculty development is shifting to an approach focusing on a community of practice rather than an individualistic practice. This is the basic feature of the webinar series. Baldwin (1998) suggests that "as technology advances further into academe, it may challenge the 'lone wolf' or 'independent entrepreneur' professorial model. To succeed in a technologically advanced era, professors may need to become more interdependent and mutually supportive" (cited in Sherer, Shea, & Kristensen, 2003, p. 17). The formation of learning communities around computer mediated communication technologies has the strength to assist faculty members in redefining teaching, learning, research, service, and professional development in higher education (Di Petta, 1998; Sherer, Shea, & Kristensen, 2003; Triggs & John, 2004). An "Internet-enhanced" learning community, in situations where face-to-face communication is not possible, has the potential to foster faculty development and upgrade higher education by reducing geographical and financial barriers to knowledge exchange. Forming learning communities and partnerships within institutions and among local, regional, national, and international communities may help higher education to become more socially vibrant and sustainable (Anderson & Bonefas, 2002; Conkling, 2007; Cortese, 2003; Soto, Lum, & Campbell, 2009).

A meaningful approach to technologically enhanced faculty development based on the webinar program is what Pennell and Firestone (1996) describe as a "teachers-teaching-teachers" model. This model is based on interaction and knowledge exchange among empathic peers (either within or across institutions) based on a collaborative culture and trust where faculty members share and teach significant knowledge to colleagues. Since giving and receiving knowledge is at the heart of academic life, university environments and organizational structures should foster quality knowledge sharing through flexible mechanisms of collaboration facilitated by ICT rather than through imposed implementation regulations. The webinar program aimed to foster knowledge-based interactions among academics in more informal contexts where relevant communities of learning and practice emerge, regardless of school constraints. As Antal and Richebé (2009) suggest, the most unsatisfying experiences of knowledge sharing take place in the contexts of academic meetings and formal events, while the most rewarding experiences take place in informal contexts.

Through the webinars, the program has aimed to facilitate the creation of communities of practice and interdisciplinary collaboration (see Barrett, Cappleman, Shoib, & Walsham, 2004; Haythornthwaite, 2006; Hildreth & Kimble, 2003; McKay & Higham, 2012; Sherer, Shea, & Kristensen, 2003; Waldron, 2009). These contexts have allowed faculty members to actively participate and exchange knowledge with peers and foster contextualized discussions within and across audiences, including practitioners beyond the institution. The program has facilitated the creation of a community of practice within the hosting institution in which university professors, in-service teachers, and graduate and undergraduate students share their opinions and expertise and learn vicariously by sharing their diverse experiences. In university-school partnerships, university students interact with in-service teachers in a nonhierarchical way, learning from one another. Similarly, while university professors are responsible for the professional education of preservice and in-service music teachers, teachers in schools also provide relevant and current practice-based knowledge that may inform professors' research agendas. As Conway, Albert, Hibbard, and Hourigan (2005) emphasize, we need to develop collaborative communities of arts teachers who are the real experts and jointly work to solve problems and exchange ideas. Other types of relationships that might develop among music educators, faculty, researchers, advocates, politicians, and the music industry also have the potential to develop into communities of practice (Conkling, 2007; Robbins & Stein, 2005; Soto, Lum, & Campbell, 2009).

While the program was not initially conceived as an international partnership, it has evolved into a faculty development program that facilitates stronger cultural understanding among participant audiences. As a result of the initial resistance from other faculty members in the hosting institution, it was decided to invite colleagues from other institutions in Mexico and other places around the world. During the third series of professional development workshops in February 2012, colleagues from Ball State University (Karin Hendricks and Tawnya Smith, USA), Benedictine University (Allen Legutki, USA) and Universidad Veracruzana (Rafael Toriz, Veracruz, Mexico) agreed to collaborate as coordinators of the program in their own institutions. Following this initiative, the series has welcomed other institutions from Argentina (Instituto Universitario de Arte, Ana Lucía Frega), Brazil (Universidade do Estado de Santa Catarina, Sergio Figueiredo), Canada (Simon Fraser University, Susan O'Neill; University of Western Ontario, Patrick Schmidt), Cyprus (University of Cyprus, Chrysanthi Gregoriou), Greece (Musics R.E.D. Group, Efthymios Papatzikis), Mexico (SEECh, Diana Piñón and Laura Gamboa; Conservatorio de Música de Chihuahua, Eliel Reyes; Instituto de Investigación e Innovación Educativa, Cruz Montes), the United Kingdom (University of Cambridge, Pamela Burnard; University of London, Andrea Creech) and the United States (Gettysburg College, Brent Talbot; Lake Forest College, Scott Edgar; Illinois State University, Kimberly McCord). The communication and collaboration across institutions in these diverse contexts have provided a global perspective strengthened by the multiple possibilities for communication and collaboration among invited speakers, institutions, and audiences.

Given the common technological possibilities, the coordinators of the program have aimed to use strategies consistent with the literature on general professional development (Anderson & Bonefas, 2002; Lowenthal, 2008; Lyons, 2006; Patterson & Rolheiser, 2004; Richardson & Placier, 2001) in order to positively impact teaching and learning in these ways:

1. Fostering school-wide educational changes within participant higher education institutions, as well as in elementary and secondary education through participation and collaboration with in-service teachers
2. Encouraging professional discussion and teamwork at different levels (faculty–university students; researchers–practitioners; preservice teachers, in-service teachers) and fostering reflective engagement
3. Modeling collaborative work by sharing expertise from experienced researchers and practitioners
4. Encouraging collegiality through collaboration among peers
5. Choosing a meaningful focus that is relevant to individuals' needs and will strengthen faculty's, teachers', and students' knowledge base, mainly focused on music education issues but also providing a wide range of topics that address multiple interests
6. Incorporating current content-specific knowledge
7. Developing a strong skill base
8. Building a capacity for sharing
9. Honoring the complexity of teachers' practices
10. Empowering faculty, students, and practitioners

A Research Study

During the first three years of the webinar program, a research study was conducted to examine the impact of the use of ICT on the motivation, attitudes, and values that participants at the hosting institution attributed to their experience in this kind of blended model of distant learning. A mixed methods approach was used to examine motivational and environmental factors that inhibited or fostered the participants' experiences, through surveys, semistructured interviews with invited speakers, and observations during webinar transmissions.

In relation to faculty perceptions, results from the study (González-Moreno, 2012a) showed that faculty members, as well as in-service music teachers and other attendees, reported high levels of interest, usefulness, and importance in connection with this kind of ICT training facilitation (no statistical differences among groups were found). Consistent with previous research, these results show that technology use is a topic of major interest during professional development for music educators, in terms of how it might enhance their teaching strategies and professional practice in music-related content areas and of how technology might provide opportunities for their professional development in general (see also Bauer, 1999, 2007; Bowles, 2003). In addition, faculty members attributed a higher personal cost to participating in traditional learning environments compared to virtual or blended models of learning, like this program in particular (understanding cost as the negative aspects of engaging in the task in terms of a cost versus benefit; see Eccles, 1983).

The perceived level of competence influenced by participation in the series was relatively high (4.3–5.8 on a 7-point Likert scale) and the perceived difficulty of applying the acquired knowledge was in the middle to low range (2.6–3.8 on a 7-point Likert scale). This suggests that the webinar series fostered a perceived sense of competence that might influence participants' teaching practice on a regular basis. While some of the participants interviewed recognized that personal effort is required in order to creatively transfer research-based knowledge into practice, they acquired a higher level of understanding of how research may impact music education. They also expressed their appreciation for how technology has provided them with opportunities to access new knowledge and to facilitate communication with experts from around the world, fostering knowledge mobilization (see González-Moreno, 2014).

Some of the biggest constraints on participation included an inadequate institutional environment, such as a lack of authorization or encouragement to attend when the topics were not directly related to a participant's particular area of professional expertise or teaching workload. In general, participants perceived a lack of institutional support (e.g., time release, incentives) to participate in faculty development opportunities of any kind. They also reported limited opportunities for professional development in music and music education within the university or even through other educational agencies or departments (e.g., the Secretariat of Public Education).

While the results of the study in relation to participants' motivation have been presented elsewhere (González-Moreno, 2012a, 2012b), the focus here is on what has been observed about nonparticipant faculty, a peripheral aspect of webinar observations that requires a deeper analysis and interpretation. Despite the fact that the program was initially conceived to help university professors to gain more knowledge about current topics in arts and music education research and the respective implications for teaching, faculty members in the hosting institution were the most reluctant group of attendees to actively participate in the sessions during the time the webinar series took place (12 semesters, 89 webinars in total). This trend was also noted in other participant institutions, suggesting faculty resistance and lack of intrinsic motivation to actively participate and benefit from professional development opportunities. These trends are consistent with faculty members' slow adoption of and resistance to technology (Dede, 2004; Hall, 2013; López de la Madrid, 2007).

Some of the possible explanations for this faculty resistance range from political contexts to academic jealousy. However, these aspects of faculty members' resistance are difficult to investigate, given the political climate of institutions: limited rapport with other faculty members, lack of confidence to openly express individual opinions about institutional programs and policies, and lack of confidence to express opinions about projects coordinated by other faculty members. Particular institutional environments determined by internal politics and unhealthy rivalry among faculty members may result in professional burnout and a lack of interest in opportunities that might contribute to faculty members' own learning and future development. Previous research supports this notion. For example, Hamman, Daugherty, and Sherbon (1988) found that music professors are continuously exposed to professional burnout in academic environments as a result of job dissatisfaction, workloads that are inconsistent with career goals, and a lack of administrative direction and support. Camblin and Steger (2000) suggest that "vitality is difficult to maintain if the faculty experience a loss of purpose in their work or if a sense of collegiality is supplanted by a counter-productive competitive spirit spurred on by a day-to-day struggle to survive in an environment lacking proper support systems" (p. 4). In order to overcome these challenges, it is evident that social, institutional, and collegial support is essential to transformative learning that may result in significant changes for faculty, motivating them to participate actively in the creation of the next generation of educational practices.

Overall, the webinar series has expanded from a local program to becoming an international partnership. It has positively impacted the profession by providing access to first-class experts in arts and music education, developing international collaborations, and building communities of practice that include faculty members, in-service teachers, university students, and other practitioners. However, it is still important to recognize the challenges inherent in facilitating professional development (technologically enhanced or not) for a large number of faculty members.

Final Considerations

If higher education environments are to continue to be relevant to the music education profession, faculty development programs must evolve, or faculty will become outdated in the rapidly changing world dominated by new technologies. New approaches must take into consideration that faculty members, at various stages of their professional careers, have different professional development objectives that require diverse strategies (Bauer, 2007). In order to implement sustained, long-term faculty development and avoid academic obsolescence, institutions of higher education must design programs that address the personal, developmental, professional, organizational, and technological needs of faculty and encourage them to fully enhance their activities as scholars, academic leaders, researchers, and trainers of future music educators (Camblin & Steger, 2000). Faculty members' learning depends on how well these types of programs match their interests and needs and to what extent they address music-related topics according to the faculty's profiles.

Music and music education professors require continuous training in basic computer utilization and development of instructional materials, in addition to developing flexible skills in managing their own learning and professional growth. They also need to experiment with new modalities of academic work (e.g., teaching, researching, collaborating) using virtual environment tools and to positively enhance their attitudes toward technology use, assessing their instruction in the use of technological tools, and taking into consideration students' own technological abilities and the rapid advance of ICT.

Given the lack of research-based evidence, higher education institutions must take into consideration what effective initiatives for professional development in music and music education look like and what strategies should be implemented in order to address the particular needs, interests, and preferences of faculty in diverse, content-specific programs. DuFour (2004) suggests that the best professional development is situated in the workplace. Therefore, it seems indispensable to create organizational structures that require faculty to work together while at the same time facilitating communication and collaboration with professionals across disciplines and geographical boundaries, developing communities of practice. In this regard, ICT represents the means to accomplish this aim at the local, regional, national, and even international level.

Any comprehensive professional development program should consider the individual needs of faculty members, according to the different stages of their careers. On the one hand music and music education faculty members must have access to diverse and extensive development programs that are content-specific and must be free to select content and learning processes. On the other hand administrators and educational authorities must provide release time, resources, strong leadership, focused goals, and support structures, as well as internal and external evaluations for continuous improvement. Several lessons must be learned and applied to any traditional or technologically enhanced professional development programs. Higher education institutions and faculty learning communities must aim to build a university-wide community through teaching and learning, nourish collaborative work, foster the use of ICT to enhance teaching and its application to student learning, increase communication across disciplines, strengthen a technological infrastructure that supports the learning communities, and develop partnerships with other learning communities with mutual interests at departmental, regional, national, and international levels.

The lack of specific research examining the impact of ICT on faculty development in music teaching and learning in higher education requires a self-reflective examination by faculty members and researchers. Research in this area should identify trends in, and challenges to embracing, technological innovations that take place in higher education and permeate other educational contexts. Results might indicate to what extent universities and faculty evaluate their own performance on keeping up to date with technological advances and developing their pedagogical competencies.

References

Anderson, W., & Bonefas, S. (2002). Technology partnerships for faculty: Case studies and lesson learned. *New Directions for Higher Education, 120*, 47–54.

Antal, A. B., & Richebé, N. (2009). A passion for giving, a passion for sharing: Understanding knowledge sharing as gift exchange in academia. *Journal of Management Inquiry, 18*(1), 78–95.

Baldwin, R. (1998). Technology's impact on faculty life and work. *New Directions for Teaching and Learning, 76*, 7–21.

Barrett, M., Cappleman, S., Shoib, G., & Walsham, G. (2004). Learning in knowledge communities: Managing technology and context. *European Management Journal, 22*(1), 1–11.

Bauer, W. I. (1999). Music educators and the Internet. *Contributions to Music Education, 26*(2), 51–63.

Bauer, W. I. (2007). Research on professional development for experienced music teachers. *Journal of Music Teacher Education, 17*(1), 12–21.

Bauer, W. I., Reese, S., & McAllister, P. A. (2003). Transforming music teaching via technology: The role of professional development. *Journal of Research in Music Education, 51*(4), 289–301.

Benítez Lima, M. G., & Ávila Gómez, J. C. (2012). Los profesores de educación superior y la integración de la tecnología educativa. *Tlatemoani: Revista Académica de Investigación, 10*. Retrieved April 5, 2014, from http://www.eumed.net/rev/tlatemoani/10/blag.pdf.

Bondy, E. (1997). Overcoming barriers to collaboration among partners-in-teaching. *Intervention in School and Clinic, 33*(2), 112–115.

Bowles, C. (2003). The self-expressed professional development needs of music educators. *Update: Application of research in music education, 21*(2). Retrieved January 27, 2014, from http://upd.sagepub.com/content/21/2/35.full.pdf.

Camblin, L., & Steger, J. A. (2000). Rethinking faculty development. *Higher Education, 39*(1), 1–18.

Conkling, S. W. (2007). The possibilities of situated learning for teacher preparation: The professional development partnership. *Music Educators Journal, 93*(3), 44–48.

Conway, S. M., Albert, D., Hibbard, S., & Hourigan, R. (2005). Arts education and professional development. *Arts Education Policy Review, 107*(1), 3–9.

Cortese, A. D. (2003). Critical role of higher education in creating a sustainable future. *Planning for Higher Education, 31*(3), 15–22.

Dede, C. (2004). Preface. In C. Vrasidas & G. V. Glass (Eds.), *Online professional development for teachers: Current perspectives on applied information technologies* (pp. xi–xiv). Greenwich, CT: Information Age.

Desimone, L. M., Porter, A. C., Garet, M. S., Yoon, K. S., & Birman, B. F. (2002). Effects of professional development on teachers' instruction: Results from a three-year longitudinal study. *Educational evaluation and policy analysis, 24*(2), 81–112.

Di Petta, R. (1998). Community online: New professional environments for higher education. In K. Gillespie (Ed.), *The impact of technology on faculty development, life, and work* (New directions for teaching and learning, 76) (pp. 53–66). San Francisco: Jossey-Bass.

DuFour, R. (2004). Leading edge: The best staff development is in the workplace, not in a workshop. *Journal of Staff Development, 25*(2). Retrieved January 26, 2014, from http://theptc.squarespace.com/storage/images/Leading%20edge.pdf.

Eccles, J. S. (1983). Expectancy, values, and academic behaviors. In J. T. Spence (Ed.), *Achievement and achievement motives: Psychological and sociological approaches* (pp. 75–146). San Francisco: Freeman.

Epper, R. M., & Bates, A. W. (2001). *Teaching faculty how to use technology: Best practices from leading institutions*. Westport, CT: American Council on Education and Oryx Press.

Ertmer, P. A., Gopalakrishnan, S., & Ross, E. M. (2000). Technology-using teachers: Comparing perceptions of exemplary technology use to best practice. *Journal of Research on Technology in Education, 33*(5), 1–26.

Friedrichs, C. (2001). The effect of professional growth opportunities as determined by California public high school instrumental music teachers. *Dissertation Abstracts International, 62*(3), 955.

González-Moreno, P. A. (2012a, May). *Desarrollo profesional de profesores mediante el uso de las TIC: Un proyecto de innovación e investigación educativa en las artes*. Paper presented at Primer Congreso Internacional de Educación: Construyendo Inéditos Viables, Chihuahua, Mexico. Retrieved June 11, 2012, from http://cie.uach.mx/cd/docs/area_02/a2p20.pdf

González-Moreno, P. A. (2012b, July). *Impact of a technology-based program for professional development in the arts*. Paper presented at the research seminar of the International Society for Music Education, Thessaloniki, Greece.

González-Moreno, P. A. (2014). Knowledge mobilization through the use of digital technologies: Possibilities and constraints. In S. A. O'Neill (Series Ed. & Vol. Ed.), *Research to Practices: Vol. 6. Music and media infused lives: Music education in a digital age* (pp. 343–360). Waterloo, Ontario: Canadian Music Educators' Association.

Hall, O. P., Jr. (2013). Assessing faculty attitudes toward technological change in graduate management education. *MERLOT Journal of Online Learning and Teaching, 9*(1), 39–51.

Hamman, D. L., Daugherty, E., & Sherbon, J. (1988). Burnout and the college music professor: An investigation of possible indicators of burnout among college music faculty members. *Bulletin of the Council for Research in Music Education, 98*, 1–21.

Hammel, A. M. (2007). Professional development research in general education. *Journal of Music Teacher Education, 17*(1), 22–32.

Haythornthwaite, C. (2006). Learning and knowledge networks in interdisciplinary collaborations. *Journal of the American Society for Information Science and Technology, 57*, 1079–1092. doi: 10.1002/asi.20371.

Hildreth, P., & Kimble, C. (Eds.) (2003). *Knowledge networks: Innovation through communities of practice*. Hershey, PA: Idea Group.

Hookey, M. (2001). Professional development. In R. Colwell & C. P. Richardson (Eds.), *The new handbook of music teaching and learning* (pp. 887–904). New York: Oxford University Press.

Hubbard, G., & Atkins, S. (1995). The professor as a person: The role of faculty well-being in faculty development. *Innovative Higher Education, 20*, 117–128.

King, K. P. (2002). Educational technology professional development as transformative learning opportunities. *Computer & Education, 39*, 283–297. Retrieved January 26, 2014, from http://58.59.135.118:8081/BOOKS%5C026%5C21%5CHXYWPJH263329.pdf.

López de la Madrid, M. C. (2007). Uso de las TIC en la educación superior de México: Un estudio de caso. *Apertura: Revista de innovación educativa, 7*(7), 63–80.

López de la Madrid, M. C., & Flores Guerrero, K. (2010). *Las TIC en la educación superior de México: Políticas y acciones*. Retrieved April 5, 2014, from http://reposital.cuaed.unam.mx:8080/jspui/handle/123456789/1507.

Lowenthal, P. R. (2008). Online faculty development and storytelling: An unlikely solution to improving teacher quality. *MERLOT Journal of Online Learning and Teaching, 4*(3), 349–356.

Lyons, N. (2006). Reflective engagement as professional development in the lives of university teachers. *Teachers and Teaching: Theory and Practice, 12*(2), 151–168.

Marginson, S., & van der Wende, M. (2007). *Globalisation and higher education* (OECD Education Working Papers, No. 8). OECD. Retrieved January 29, 2014, from http://doc.utwente.nl/60264/1/Marginson07globalisation.pdf.

McKay, G., & Higham, B. (2012). Community music: History and current practice, its constructions of "community," digital turns and future soundings: An Arts and Humanities Research Council research review. *International Journal of Community Music, 5*(1), 91–103.

Patterson, D., & Rolheiser, C. (2004). Creating a culture of change: Ten strategies for developing an ethic of teamwork. *The Journal of Staff Development, 25*(2), 1–4.

Pennell, J. R., & Firestone, W. A. (1996). Changing classroom practices through teacher networks: Matching program features with teacher characteristics and circumstances. *Teachers College Record, 98,* 46–76.

Richardson, V., & Placier, P. (2001). Teacher change. In V. Richardson (Ed.), *Handbook of research on teaching* (4th ed.) (pp. 905–947). Washington, DC: American Educational Research Association.

Robbins, J., & Stein, R. (2005). What partnerships must we create, build, or reenergize in K–12 higher and professional education for music teacher education in the future? *Journal of Music Teacher Education, 14*(2), 22–29.

Secretaría de Educación Pública. (2014). *Programa de Fortalecimiento de la Calidad en Instituciones Educativas: Objetivo estratégico del programa.* Retrieved January 25, 2014, from http://pifi.sep.gob.mx/.

Sherer, P. D., Shea, T. P., & Kristensen, E. (2003). Online communities of practice: A catalyst for faculty development. *Innovative Higher Education, 27*(3), 183–194.

Smith, K. L. (1996). *Preparing faculty for instructional technology: From education to development to creative independence.* Paper presented at annual conference of CAUSE. Retrieved January 28, 2014, from http://net.educause.edu/ir/library/pdf/CNC9659.pdf.

Soto, A. C., Lum, C. H., & Campbell, P. S. (2009). A University-School music partnerships for music education majors in a culturally distinctive community. *Journal of Research in Music Education, 56*(4), 338–356.

Triggs, P., & John, P. (2004). From transaction to transformation: Information and communication technology, professional development and the formation of communities of practice. *Journal of Computer Assisted Learning, 20*(6), 426–439.

Waldron, J. (2009). Exploring a virtual music community of practice: Informal music learning on the internet. *Journal of Music, Technology & Education, 2*(2-3), 97–112.

Windschitl, M., & Sahl, K. (2002). Tracing teachers' use of technology in a laptop computer school: The interplay of teacher beliefs, social dynamics, and institutional culture. *American Educational Research Journal, 39*(1), 165–205.

CHAPTER 59

EDUCATORS' ROLES AND PROFESSIONAL DEVELOPMENT

EVANGELOS HIMONIDES

INTRODUCTION AND BACKGROUND

I grew up in a family that has been "producing" teachers for nearly three centuries. Thus, becoming an educator was somewhat of a prescribed resolution for my brother and myself; the same as my father and his sister, their own parents, and so on. My introduction to technology was at a very early stage, around the age of four, spending time in my father's office in the "Informatik" department of Siemens, in Munich, Germany, with an extremely serious job assignment: go through countless boxes of continuous (dot matrix printer) paper printouts and spot the odd blank page, tear it out, and put it in a separate stack for the programming team to use for jotting down programming code. My compensation for this "vital" green work was always twofold; in the office, having the chance to choose from a limited array of "ASCII art"[1] to be printed out on paper for me proudly to take home; and, outside the office, having a special treat visit to the cinema. The usual choice for ASCII art printouts was Astérix le Gaulois, or his friend Obelix, both protagonists in the comic book series written by René Goscinny and illustrated by Albert Uderzo, both characters who have entertained (and educated!) me all the way to adulthood. The cinematic repertoire during my early childhood was amusingly monothematic; it was solely science fiction movies—the ones available in the 1960s and 1970s, involving a lot of glitter, and a lot of fishing line for the suspension of crude, nonconvincing mock space-age objects against black cardboard. (Later in life, I was able to put two and two together and realize that this was probably because of my father's cinematic preferences.)

One could therefore argue that I was somewhat "nurtured" (if not indoctrinated) into becoming friends with technology from an early stage. I was 10 when the IBM personal computer became available, and 10 1/2 when I managed to burn the IBM personal computer's power supply—the first ever imported from the United States into Greece—by plugging it straight into a 220V outlet and powering it on—and just a little before 11 before I managed to experience an interaction with that same computer, once a new power supply had been ordered via Telex[2] and received a couple of months later. At 12, I remember giving my first ever programming lesson to a friend, and I was 14 when I actually managed to get paid by a professional for private lessons in the BASIC[3] programming language, DOS[4] scripting, Edlin,[5] and primitive word processing. As far as my parents are concerned, especially my father, I am, and have always been, a computer scientist. Music came at a later stage, around the age of 20, mainly in order to conduct my own "revolution" and disappoint my extended family, who were convinced that I could not become anything else but a computer scientist and/or mathematician! Gladly, things became calmer once I decided to "settle" within academia and attempted to keep my passion (music) and my "osmotically" as well as academically acquired craft (technology) in a symbiotic relationship.

This somewhat ambiguous self-reflexive ode to my past trajectory will, I hope, start making more sense once I get the chance to present how my craft has helped me develop my passion. It will also help the reader to understand why I am so fervent about the role of technology in learning (and, actually, in most aspects of life, being aware that this may sound overly dramatic at this juncture). Since this personal "revolution," I have managed to study music, and choral conducting, and have tried very hard to overcome the severe handicap of starting late, compared to what might be perceived as "the norm" for a person who pursues a career in music and music education. I can only express my sincere conviction that this journey would have been impossible for me without technology. A "computer" in any of its current incarnations is my first instrument of choice, my first learning tool, and my first outlet, when I need to "understand" something new or complex. I still don't, and honestly don't think that I ever will, possess the ability to "hear" music in my head by looking at dots of ink on a piece of paper (notwithstanding the fact that I believe—and to some extend have managed to demonstrate elsewhere [Himonides, 2012]—that ink on a piece of paper can be as cutting edge in music technology as a computer, an iPad, a synthesizer, or a genetic musical algorithm). I might be extremely envious of the people who are able to perform such feats, but I have gained the confidence to realize after years of disappointment that I am not really a musical child of a lesser god, besides this perceived handicap. Once more, I owe this acquired confidence to technology.

Attitudes

Over the nearly 30 years that I have been involved in training, teaching, educating, and honing and developing people's technological skills (the use of these different terms is intentional, as they are rehearsed differently within different contexts), and in trying to address people's technological needs and problems as an IT professional, I have come to the realization that in almost all of these cases, people's attitudes toward technology has been the one most crucial factor for the genesis of a successful (or not so) engagement. Almost all cases of failure, frustration, and disengagement that I can possibly recall somehow have happened as a result of fear, lack of confidence or courage, and sometimes a lack of drive to address things rationally, engage the "critical brain" instead of the "impulsive brain,"[6] and harness what technology can provide, free from intimidation. These attitudes are highly relevant to Prensky's (2001) notions of digital "natives" and "immigrants" and later work on children and gaming (2006), as well as Savage's elegant "sounding off" in a recent position chapter (2012) wherein he presented potential threats regarding the future of education and the role technology might play in that context, and introduced the notion of a digital "expat." Naturally, this extended personal experience has had a direct impact on my thinking and has branded my beliefs regarding teacher education, teacher credentialing, and further accreditation, which is the focus of this essay. Here, I will present what I believe the key foci should be in such a discourse. I am doing so while fully aware that I am almost certainly not practising what I am usually preaching, by presenting a self-reflexive and somewhat empirically formed conviction instead of reporting the findings of carefully controlled scientific research. This, I believe, is a risk worth taking.

Declarations

It is somewhat paradoxical, if not amusing, that a discussion about something perceived as being as praxial and vocational as the education and certification of music teachers in the area of technology (this section's core focus) could also form a plateau for colorful philosophical debate. With regard to the context itself, one could wonder:

- Are we all in the same "profession" after all?
- Do all "music teachers" have similar education and certification needs?
- Is "the area of technology" a discrete subset, niche, or neighborhood of music teacher education and training?
- Can we infer that the present discussion is sensible and would address similar challenges regardless of the geography and sociopolitical focus?

Naturally, these questions could only be addressed once technology and its "role" have been defined and accepted by those who would be willing to engage in the philosophical debate! Therefore, there should probably be some kind of common understanding about these issues:

- What is technology?
- What is the role of technology in music education and its numerous facets?

- Is this role context specific, context sensitive, or context free? and, most importantly,
- Can this role remain static once defined?

Consequently, once technology and its role(s) have been established (see the second set of questions above) and the context has been sketched out (see the first set), our focus could shift toward the notions of education and certification, more specifically:

- Is music teacher education approached similarly around the world?
- Can technology training be seen as an integral part of music teacher education, or is it simply an "add-on"? What are the parameters, dynamics, challenges, benefits, and potential threats of certification or accreditation of music teachers in technology?
- How can we ensure that certified or accredited teachers become equipped to face future challenges *in* technology, but also *with* technology? What are the roles of the overarching arts, music, and manufacturers' industries in shaping accreditation and certification?

Finally, one cannot forget who the protagonists are or, at least, who they should be! Teachers and their needs cannot be rehearsed in a vacuum; they have to be seen as part of a symbiotic relationship with those of their students. Therefore, we will probably need to discuss what teachers' and students' needs are (if it is possible to quantify those); how universal these might be; what our core aims and objectives for catering to these needs should be; and, last but not least, what the most cost-effective and pragmatic way to address these might be. And what would we expect to see as the result of a systematic debate that would address all (or most) of the foregoing bullet points? Perhaps unsurprisingly, a new, or revamped, educational curriculum is needed. This curriculum, being so appropriately and sufficiently robust as to command respect as the product of systematic and rigorous assessment of current educational needs, will obviously feature components wherein the importance of information and communication technology (ICT)[7] is celebrated. Although I wouldn't necessarily present the above syllogism and suggested framework as utterly futile, I would invite us to consider a novel approach.

A New Way Forward

A synopsis of my core beliefs regarding teachers' development comprises these five points, the last initially appearing to be somewhat controversial:

> The role of technology in education (and, obviously, music education) is central, not complementary, or peripheral.
> Technology can foster the development of teachers' critical thinking, which is . . . the most important ability that an educator should possess.

- Teachers cannot be passive recipients and consequent conduits of predefined educational praxes; they need to be in a position to shape their own and their students' learning, and critically assess what their needs are, as these are being formed.
- It is the social facet of technology that will play a leading role in teachers' competence, credentialing, and professional development.
- Technology is becoming mature enough to allow us to re-rehearse the notion of evidence-based education in a completely novel way.

What follows is a somewhat more extensive unpacking of these five points.

Activity and Proactivity

In the concluding part of a creative presentation on the value of what he calls intermediate technologies, Purves (2012) suggests that a shift is needed away from the parochial view that teachers should be passive recipients of education and knowledge and toward a new model wherein the teachers themselves are responsible for shaping the future (i.e., the development of their curricula, their assessment, and their accreditation). Assessment and accreditation cannot exist in isolation from learning and development. Assessment and accreditation should be seen as vital parts of an overarching learning and development framework. Therefore, a proposed future model of teacher assessment and accreditation is one that has teachers as the driving force of their own development. This is in line with—and also celebrates—our current understanding of how learning and development work, placing the learner in the center of the learning context, if we agree to view the educators as "senior learners" or "lifelong learners" and not as masters responsible for molding apprentices' minds.[8]

Teacher Education

Thinking about music teacher education, we also need to take into account teacher education in general (i.e., not necessarily remaining solely within the world of music). For that, we need to decide and assess what our praxial, philosophical, as well as moral stance is, and what an educator's actual role is, or should be. There is, for example, a shift in England in the past two decades towards teacher education that is mainly or wholly school based. I believe that this is conceptually flawed, certainly in terms of the understanding of technology in pedagogy, and will result in teachers becoming less equipped in the long run to deal with future challenges effectively (a concern that is highly relevant to the present discussion). University-led support in teacher development is essential for the advancement of teachers' critical thinking and ability to face new challenges as they emerge, because this is likely to be where would-be teachers access the latest research in their subject areas, in pedagogy, and in the pedagogical power of technology.

The ability to adapt in the face of new challenges is something much broader and requisite for an effective educator, as opposed to becoming seasoned or honing one's skills to deal with challenges within very particular existing contexts (i.e., in the real-world school praxis, constrained by the pressing reality and logistics).

Obviously, we are not advocating that teachers should be nurtured inside a theoretical, abstract, utopian, or "non-real-world" environment; on the contrary! At least in England, preservice teachers have always been required to undergo school placements and slowly become critically immersed in the school culture. But formerly they were required to do so whilst developing their critical thinking at university, and whilst being mentored and supported by academics and expert teachers experienced in both the academic as well as the school contexts. This duality needs to be sustained and to continue forming the cornerstone of teacher development. Taking the whole framework of teacher education and placing it in the school is something that shifts things toward the far side of the preservice continuum. Whilst this is likely to solve the need for "producing" adequate numbers of teachers in the interim, unfortunately it is bound to bring major repercussions for the overall quality of teaching, as well as the future of education and the development of students in the long term. Teacher development will be shaped primarily by the local context, rather than by a broader vision in which new developments, insights, and initiatives are accessible to all. Trailblazing educational systems around the world support this argument. Preservice teachers in Finland, for example, are required to conduct rigorous postgraduate research in order to become qualified. Disregarding similar exemplars and moving toward the "vocationalization" of teacher education (Department for Education, 2014) is a myopic and highly insular approach that may have a catastrophic impact on future society.

The Teacher's Role

When we are interested in nurturing and developing critical thinking for teachers, we need to also have a vision about their greater role within educational and developmental ecologies. This role, on many occasions and within many different educational contexts around the world, appears to be quite controversial. We need to address whether there is—or should be—an intellectual hierarchy, in the sense that certain careers and pathways are valued or esteemed more than others, and we need to assess what the teachers' placement is within this hierarchy. Having subscribed to a model of intellectual hierarchy, we will consequently need to assess what teachers' role(s) are within the hierarchical structures and how these structures might compare to (or map against) internationally established educational contexts (e.g., the often-discussed Finnish example; see Kupiainen, Hautamäki, & Karjalainen, 2009; Laukkanen, 2007; Rinne, 2000).

Most of us serving in the overarching education, higher education, and research fields are well aware of what I consider the highly insulting aphorism "[T]hose who can't, teach" (Fuller, 1969, p. 208). We often read about young professionals' aspirations, especially young artists' (Triantafyllaki, 2010), but also their "contingency plans" to settle on teaching (Creech et al., 2008a, 2008b) should an aspired-to career in performance not come to fruition (Welch, Purves, Hargreaves, & Marshall, 2011; Welch et al., 2008). One of the interesting realizations when looking closer at the Finnish paradigm is that teachers are highly valued in Finland by all sides of the social spectrum (Simola, 2005) and certainly aren't perceived as second-rate artists or scientists who failed to become trailblazers and succeed in their preferred career choices and pathways. Finnish teachers are highly respected and perceived to be catalytic in forging knowledge and student development in such a successful way (i.e., Finland consistently being found to have one of the leading educational systems worldwide).

With England having one of the youngest teaching workforces in the Organization for Economic Co-operation and Development (OECD) countries, especially at the primary level, and the latest statistics suggesting that a massive 31% of teachers are below age 30 (Weston, 2013), in tandem with published evidence (Barmby, 2006) that teacher retention is very challenging in England, and with mature teaching professionals citing stress, workload, and bureaucracy as a reason to leave their profession, we are sadly experiencing the reality that teachers are far from being celebrated, developed, championed, fostered, and mentored here. Under these circumstances, one could have strong doubts about whether teachers in England are in a position to even aspire to the continual development of their critical thinking; they merely try to survive in an intellectually, physically, and emotionally hostile milieu that sees them as commodities. This conservative notion of an expendable workforce is almost certainly not in line with the Finnish (and greater Scandinavian) paradigm.

We will therefore need to address these concerns, and I believe that the role of technology in doing so is, once again, vital. Interestingly, this role is somewhat indirect and not really centered on technology providing tools and heuristic remedies for certain curricular and learning objectives. It is rather for technology to become the conduit through which awareness, discourse, dialogue, pluralism, and intellectual democracy can thrive. I therefore believe that, once more, the emphasis should be placed on the social aspects of technology.

The Multiple Facets of Technology

This discussion naturally leads to the portrayal of the multiple "facets" of technology and its role, but also to the immense power the social facet of technology possesses. We really don't need to be that intuitive in order to realize how much of a difference social technologies have made in influencing and developing current practices (Schlager & Fusco, 2003), such as the private studio, the classroom, the school, the local, the global, the *glocal* (Himonides & Purves, 2010). For example, findings from a recently concluded European collaborative project, where students and teachers from Italy, Denmark, Spain, and the United Kingdom formed online synergies in music making and music production, suggested that the development of skills other than musical ones was also fostered through online collaborative practice. Participants in the OpenSoundS project reported that in engaging with a collaborative online platform, and participating in music-making projects with other participants from around Europe, they benefited musically but also gained skills and confidence in project management and communication (Fiocchetta, Himonides, Nicollet, & Rodá, 2013). What was particularly interesting in participants' reports was the empowerment that shared leadership and shared ownership brought to the learning context, supported by and working within a democratic technological platform. Therefore, I would like to argue that harnessing this kind of potential in order to voice, communicate, organize, network, structure, and frame context-specific and context-sensitive policy, where each individual maintains a dynamic role, is (or at least should be) a part of a new organic system, and is close to a democratic model that would have been perceived as simply utopian not long ago. In addition, we are continually acquiring more tools and modalities to support our achieving this somewhat "utopian" state.

Therefore, it is important to focus on the overarching role of technology within education and music education when considering further, continual development of educators and their students. We refer to this overarching role, as opposed to the very specific role technology plays in providing modern tools to the classroom, because of the need to engage in a discourse concerning the social benefits, channels, and modalities that technology introduces. It is the ways that technology enables us, but also empowers us, to access, communicate, assess, discuss, debate, and develop further in a novel, democratic, and accessible way that we should be continually critical about. Interestingly, this is not necessarily the facet of technology on which educational policy-makers have been focusing. This is somewhat understandable, as this facet is much more difficult to define, document, and map within a set structured curricular framework.

Continual Professional Development

The need for teachers to learn continually in order to develop their knowledge and skill and to adapt and develop their roles, especially through classroom enquiry (Sutherland, 2009), should be recognized and supported (see, among others, Welch, 2007). The Teaching and Learning Research Programme (2008) claims that this theme appeared consistently throughout the span of its research enquiry. Similar findings were presented in an England-wide study performed by myself and my colleagues (Himonides, Saunders, Papageorgi, & Welch, 2011) wherein an almost perfectly linear relationship was found to exist between practitioners' and teachers' confidence in teaching challenging subjects and the number of "continuous professional development" sessions that they had participated in.

Continuous professional development is also raised as a vital issue in an overwhelming proportion of globally available relevant literature, in certain contexts even having been rendered compulsory. (A plethora of examples exists in the legal and medical professions; see, e.g., the Dental Council and their policy for continuing professional development, http://www.gdc-uk.org, or the Solicitors' Regulation Authority in the United Kingdom, http://www.sra.org.uk.) It is important, therefore, for professional development facilitators to appreciate that both their own and their learners' development are inevitably ongoing and that the advent of new technologies can only be seen as positively challenging and stimulating. Teachers can harness a new technology in tandem with learners, offering their experiences in being critical reflectors and assessors of the learning process, without necessarily being heuristic experts in the new tools. A novel approach is, therefore, necessary, with core values and remit being to keep the role of technology central in the learning dyad's (and learning communities') developmental ecology(ies). In addition, in tandem with past and current face-to-face learner development, the introduction of fresh learning experiences online promises to challenge stereotypical educational hierarchies, promote equality, and foster the democratization of continual education and development. Slowly emerging examples of multiplayer games for learning "agile ways of working and creative problem solving through social drama" (AltoGame, 2014) introduce new tools that facilitate development and critical discourse and are not necessarily focused on the transmission of knowledge in a vacuum (something that a large proportion of "further training" and "further development" programs unfortunately still do). They promote "active learning opportunities" (Desimone, Porter, Garet, Yoon, & Birman, 2002), which have been found to stimulate further development.

EVIDENCE-BASED EDUCATION

In supporting the argument about technology's vital role in redefining education and educational practice, in redefining teacher and student development, and in redefining assessment, it is necessary to at least attempt to reintroduce the notion of "evidence-based education" (the somewhat controversial fifth bullet point in my earlier synopsis). Of course, one has to be particularly careful when doing so, because numerous colorful failures of the past could be presented as reasons for caution when trying to implement evidence-based education. But I believe that this is exactly where technology can now shine: by providing novel realizations of evidence-based education, free from past problems. I could not provide a better example than massive open online courses (MOOCs) to support my argument. This is the first juncture in the history of educational practice, in the history of "learning and learner development" if you prefer, where we can easily rehearse sound educational (as well as assessment) hypotheses, theories, notions, procedures, methods, and practices, and research their effectiveness among vast cohorts of learner samples. This can be performed within very narrow time windows, with a very diverse geographical, as well as socioeconomic, spread and representation, and in a highly rigorous as well as extensible manner. I believe that this is one of the most powerful possibilities that educational research has ever had for shaping our understanding efficiently and systematically and rendering educational and assessment practice that is based on robust datasets and their analyses. If, for example, we wanted to rehearse two different accreditation approaches, or two different learning pathways, for a music technology thematic, we could conduct an appropriate randomized controlled trial using different MOOC cohorts as empirical research participants. We have never had the opportunity to do this in the past at a similar magnitude. Therefore, some researchers' concerns regarding the potential "sterility" of such a clinical approach within a highly qualitative context become less of a challenge, given the vast numbers of participants involved, as well as the immense potential to perform qualitative research in tandem with the core quantitative approach. I believe that this is something that is not to be taken lightly; this is a unique opportunity to develop sound and robust practice from scrutinized theory in a novel way. It is technology that enables us to do so.

Novel technology is now beginning to mature and allow us not only to address design, interface, and human–computer interaction issues and challenges but also to further celebrate the importance of the semantic rationalization, interrogation, distillation, archival, and retrieval of information in a critical way, thus laying some of the crucial first building blocks toward the meaningful exploitation of large, nonconventional, real-life, educational datasets (i.e., big data). This fairly sketches my stance regarding the provision of support to educators and learners, as well as policy-makers, organizations, and governments, for experiencing a more meaningful, sound, robust, socially inclusive, and systematic form of education (Bridges, Smeyers, & Smith, 2008). In line with Goldacre's beliefs (Goldacre, 2013), I too believe that this will empower educators to become more confident in identifying best practice, based on the intelligent (and creative) mining of data for evidence. This new praxis will also take into account the (admitted) failures of the previous foci that the U.S. government maintained (Coalition for Evidence-Based Policy, 2003) for educational research funding (i.e., adopting a sterile model that ineffectively tried to copy medical research), as is also highlighted by Whitty (2013) in a recent online discourse. Teacher credentialing and professional development should consequently focus primarily on the research and development of context-sensitive, blended (i.e., deterministic and stochastic in conjunction with qualitative) systems that will be continually exploiting big educational datasets, whilst being highly sensitive and adaptive to qualitative educational measures and annotations and the semantic metainterpretation of the datasets.

Agile Thinking and Practice

One last point concerning what I believe to be essential for the development of competent future music educators (or any educator, in any field, for that matter) is the importance of developing teachers for an uncertain future, what Fisher, Higgins, and Loveless (2006) very intuitively portray as the "renaissance in teacher development," as opposed to focusing on the advancement of particular competencies that might prove to be ephemera. Savage augments this argument with great success (Savage, 2012), voicing an additional concern about the potential intellectual "stasis" into which one might come, even as a gratified "digital native."

Savage (2012) focuses on the need to be able to initiate and sustain meaningful change; I would echo this argument and add that "properly qualified" future educators will also need not only to initiate and perform sustained critical thinking in initiating change but also to manage change and develop novel strategies for driving their own and their students' development forward. This is a concept that modern software and general technological development has been following successfully for some time (Martin, 2003). I believe that education, teacher development, and teacher accreditation can benefit immensely from adopting a similar model. Adopting such a model will, naturally, have implications for curriculum design and assessment, both of which are currently quite reliant on a linear understanding of development. Technology will have to play an important role once more in fostering this new praxis. Social technologies, in particular, can assist us in safeguarding a democratic modus operandi such that each step of this new agile approach will be open to scrutiny, and open to change, according to the needs of the participating groups. Even being able to study the evolution of such a model harbors vast potential for understanding how our needs develop, interact, and evolve, and technology, once again, can play a crucial role in facilitating this study.

Coda

In conclusion, and in line with Purves (2012), I wish to close this brief contribution to the valuable discourse presented within this book by reminding ourselves once more that educators can and should proactively work toward shaping their own future. A competent educator who is full of credentials is not an educator who has mastered a set of shiny ephemeral tools (either soft of hard tools). Competent educators are critical thinkers who are ready to polish old tools, forge new ones, creatively abuse existing tools, methods, and processes, and harness everything available to them creatively in order to facilitate learning and development—their own and their students'. I see technology, particularly "social technology," as the competent educator's most powerful ally in this endeavor. All of us within the overarching field of education appear to have reached a state of critical mass. We can either remain passive, or we can become proactive and shape our futures and those of our children and the generations to come. This cannot be performed by clicking on a thumbs-up, like, or thumbs-down button; this can only happen if we harness technology and voice our needs, challenges, and aspirations, and form praxial synergies with other colleagues in our social circles. We have open access to the technology to do so very effectively.

Notes

1. http://en.wikipedia.org/wiki/ASCII_art
2. http://en.wikipedia.org/wiki/Telex
3. http://en.wikipedia.org/wiki/BASIC
4. http://en.wikipedia.org/wiki/DOS
5. http://en.wikipedia.org/wiki/Edlin
6. As rehearsed by Nobel laureate Daniel Kahneman (2012) in his bestselling monograph *Thinking Fast and Slow*.
7. A once noble and very promising notion, but nowadays somewhat meaningless synonym of "anything that is battery or mains operated" in the United Kingdom, and fully interchangeable with what people call "IT" everywhere else in the world.
8. See Himonides (2013).

References

AltoGame. (2014). AltoGame. Retrieved March 28, 2014, from http://www.altogame.com/.

Barmby, P. W. (2006). Improving teacher recruitment and retention: The importance of workload and pupil behaviour. *Educational Research, 48*(3), 247–265.

Bridges, D., Smeyers, P., & Smith, R. (2008). Educational research and the practical judgment of policy makers. *Journal of Philosophy of Education, 42*(1), 5–14.

Coalition for Evidence-Based Policy. (2003). *Identifying and implementing educational practices supported by rigorous evidence: A user friendly guide* (Practice Guide No. NCEE EB2003). Institute of Education Sciences. Retrieved from http://ies.ed.gov/ncee/pdf/evidence_based.pdf. Retrieved on January 5, 2017.

Creech, A., Papageorgi, I., Duffy, C., Morton, F., Haddon, E., Potter, J., de Bezenac, C., Whyton, T., Himonides, E., & Welch, G. (2008a). From music student to professional: The process of transition. *British Journal of Music Education, 25*(3), 315–331. doi: 10.1017/S0265051708008127.

Creech, A., Papageorgi, I., Duffy, C., Morton, F., Haddon, E., Potter, J., de Bezenac, C., Whyton, T., Himonides, E., & Welch, G. (2008b). Investigating musical performance: Commonality and diversity amongst classical and non-classical musicians. *Music Education Research, 10*(2), 215–234.

Department for Education. (2014, March 26). School-based training—Get into teaching—Department for Education. Retrieved March 28, 2014, from http://www.education.gov.uk/get-into-teaching/teacher-training-options/school-based-training.aspx?sc_lang=en-GB.

Desimone, L. M., Porter, A. C., Garet, M. S., Yoon, K. S., & Birman, B. F. (2002). Effects of professional development on teachers' instruction: Results from a three-year longitudinal study. *Educational Evaluation and Policy Analysis, 24*(2), 81–112. doi: 10.3102/01623737024002081.

Fiocchetta, G., Himonides, E., Nicollet, Q., & Rodá, A. (2013). OPEN SoundS: main results of the testing activities. In Gemma Fiocchetta (Ed.), *OPEN SoundS: Peer education on the internet for social sounds* (pp. 195–254). Rome: Editoriale Anicia Srl.

Fisher, T., Higgins, C., & Loveless, A. (2006). *Teachers learning with digital technologies: A review of research and projects* (Vol. 14). Bristol, UK: Futurelab Bristol.

Fuller, F. F. (1969). Concerns of teachers: A developmental conceptualization. *American Educational Research Journal, 6*(2), 207–226.

Goldacre, B. (2013). *Building evidence into education* (No. 110313). Department for Education. Retrieved January 5, 2017, from http://www.education.gov.uk/inthenews/inthenews/a00222740/building-evidence-into-education.

Himonides, E. (2012). The misunderstanding of music-technology-education: A meta-perspective. In *The Oxford handbook of music education* (Vol. 2, pp. 433–456). New York: Oxford University Press.

Himonides, E., & Purves, R. (2010). The role of technology. In S. Hallam & A. Creech (Eds.), *Music education in the 21st century in the United Kingdom: Achievements, analysis and aspirations* (pp. 123–140). London: IOE.

Himonides, E., Saunders, J., Papageorgi, I., & Welch, G. F. (2011). *Researching Sing Up's workforce development: Main findings from the first three years (practitioners' singing self-efficacy and knowledge about singing)*. London: Institute of Education.

Himonides, E. (2013). Technology enhanced learning in the 21st century: The ethos of OPEN SoundS. In Gemma Fiocchetta (Ed.), *OPEN SoundS: Peer education on the internet for social sounds* (pp. 285–292). Rome: Editoriale Anicia Srl.

Kahneman, D. (2012). *Thinking, fast and slow*. London: Penguin.

Kupiainen, S., Hautamäki, J., & Karjalainen, T. (2009). *The Finnish education system and PISA*. Helsinki, Finland: Ministry of Education Finland.

Laukkanen, R. (2007). Finnish strategy for high-level education for all. In *Governance and performance of education systems* (pp. 305–324). London: Springer.

Martin, R. C. (2003). *Agile software development: Principles, patterns, and practices*. Upper Saddle Brook, NJ: Prentice Hall.

Prensky, M. (2001). Digital natives, digital immigrants—part 1. *On the Horizon, 9*(5), 1–6.

Prensky, M. (2006). *Don't bother me, Mom, I'm learning! How computer and video games are preparing your kids for 21st century success and how you can help!* New York: Paragon House.

Purves, R. (2012). Technology and the educator. In *The Oxford handbook of music education* (Vol. 2, pp. 457–475). New York: Oxford University Press.

Rinne, R. (2000). The globalisation of education: Finnish education on the doorstep of the new EU millennium. *Educational review, 52*(2), 131–142.

Savage, J. (2012). Driving forward technology's imprint on music education. In *The Oxford handbook of music education* (Vol. 2, pp. 492–511). New York: Oxford University Press.

Schlager, M. S., & Fusco, J. (2003). Teacher professional development, technology, and communities of practice: Are we putting the cart before the horse? *The Information Society, 19*(3), 203–220. doi: 10.1080/01972240309464.

Simola, H. (2005). The Finnish miracle of PISA: Historical and sociological remarks on teaching and teacher education. *Comparative Education, 41*(4), 455–470.

Sutherland, R. (Ed.). (2009). *Improving classroom learning with ICT*. New York: Routledge.

Teaching and Learning Research Programme. (2008). *New teachers as learners: A model of early professional development* (Research Briefing No. 56). London: Author. Retrieved January 5, 2017, from http://www.tlrp.org/pub/documents/Mcnally56final.pdf.

Triantafyllaki, A. (2010). Performance teachers' identity and professional knowledge in advanced music teaching. *Music Education Research, 12*(1), 71–87. doi: 10.1080/14613800903568254.

Welch, G. F. (2007). Addressing the multifaceted nature of music education: An activity theory research perspective. *Research Studies in Music Education, 28*(1), 23–37. doi: 10.1177/1321103X070280010203.

Welch, G., Papageorgi, I., Haddon, L., Creech, A., Morton, F., de Bézenac, C., Duffy, C., Potter, J., Whyton, T., & Himonides, E. (2008). Musical genre and gender as factors in higher education learning in music. *Research Papers in Education, 23*(2), 203–217. doi: 10.1080/02671520802048752.

Welch, G., Purves, R., Hargreaves, D., & Marshall, N. (2011). Early career challenges in secondary school music teaching. *British Educational Research Journal, 37*(2), 285–315. doi: 10.1080/01411921003596903.

Weston, D. (2013, August 1). Why are teachers leaving education? *Guardian.* Retrieved April 13, 2013, from http://www.theguardian.com/teacher-network/2013/aug/01/why-are-teachers-leaving-education.

Whitty, G. (2013, March 19). *CERP—Evidence-informed policy and practice—we should welcome it, but also be realistic!* Centre for Education Research and Policy. Retrieved April 13, 2013, from https://cerp.aqa.org.uk/perspectives/evidence-informed-policy-practice.

CHAPTER 60

MUSIC TECHNOLOGY PEDAGOGY AND CURRICULA

DAVID A. WILLIAMS

Adapting Technology

Taken as a whole, the music education profession in the United States has not embraced technology. We are good at not welcoming new things. Certainly there are music teachers who have embraced technologies in their teaching and learning environments, but this simply isn't true for the vast majority of our profession. Jazz was more than 50 years old before it interested the profession to any great extent. Rock music still isn't widely accepted, and only in the past few years have we recognized the guitar as a real instrument by including it among the musical options in some schools. There is some irony in the fact that only when the guitar began to lose its grip as the dominant musical instrument in popular music did our profession start to pay attention to it. Prevalent in popular music since the 1980s, digital music making is still not a part of very many school programs. Reese and Rimington (2000), Taylor and Deal (2000), Noxon (2003), and Dorfman (2008) all reported that music teachers tended to use technology more as an administrative tool than as a part of instruction or for musical applications.

As a university supervisor of student teachers, I have seen evidence of technology as administrative tool first-hand. By the late 1990s, it was not unusual to find several computers in secondary school music rehearsal rooms. It was unusual, however, to see students actually using computers for musical purposes. As years passed and these computers became antiquated, many of them disappeared from the music rooms, not to be replaced. Today, outside select schools that have established a music computer lab, I see very few band, choir, and orchestra rooms that have computers available for student use. I have observed, as did Dorfman (2008), that teachers tend to use technology more than students do in classroom situations.

Bauer, Reese, and McAllister (2003) and Gall (2013) reported that, generally speaking, music teachers have not been adequately trained to use music technology in their classrooms. Training music teachers, and preservice teachers, in the use of technology is complicated. Music teachers tend to be busy people, and finding time to learn new things is difficult. A one-time workshop or in-service event is not normally enough to provide teachers with skills necessary to bring technology into the classroom, and few teachers find it possible to devote more time to learning about new technologies. While preservice teachers may have more opportunities to learn about music technology, their college professors often lack the expertise required to provide adequate instruction. It is quite possible that the music education field in the United States will need to wait until a new generation of teachers and professors enters the workforce before we will witness the kind of training required for the majority of music teachers to feel comfortable with music technology. This is a massive sea change that has only recently begun.

Training teachers to use new technologies is further complicated by how fast change in music technology happens. It is difficult for even early adopters of technology to keep up with this fast pace of change. New technologies, instruments, and techniques are often replaced before they can get old. It can be frustrating for teachers to learn a new technology only to see it replaced or changed before they can really make good use of it. Again, it is possible that we must wait until we have a mostly new generation of teachers, and teachers of teachers, before the profession as a whole can adapt to the pace of change. As practitioners become more comfortable with new technologies, they will no doubt also learn to be more acclimated to quick change.

Perhaps the most important issue for the profession to overcome has to do with the natural reluctance many teachers (and students) have toward change. Change can be troublesome, and it is difficult to change something we've done for a very long time. We find it hard to trust a change to time-honored traditions, and the music education profession has a lot of them. Music teachers (and again, students) tend to have strong convictions about what we have been doing and often go out of the way to ensure that nothing changes. This might be the most difficult challenge related to the use of new technologies in music education. In addition to the need for a new generation of teachers and professors in the workforce, we may also need a new type of music education student. I'll return to this idea later.

While preparing teachers and preservice teachers to use music technologies is obviously a matter for concern within the music education profession, I want to suggest there is an even more important issue affecting the use of music technologies in schools. This has to do with the pedagogy we employ in our classrooms. The music education profession in the United States remains closely tied to a particular pedagogical technique borrowed from the eighteenth-century orchestral model. In this model, there is one person in control. The teacher (normally referred to as the director or conductor) plans the entirety of the student's musical experience. This typically includes both long- and short-range educational and musical goals. The teacher will generally have a plan for the number of concerts to be performed and will make a schedule for them. The teacher will characteristically select previously composed music to be rehearsed and performed and will plan the rehearsal schedule. During rehearsals the teacher will guide students' learning based on a predetermined goal. This will include making the great majority of musical decisions for the students while helping them understand their individual roles within the ensemble. This is a teacher-centered pedagogy that places the teacher in a central position within the classroom from which the teacher controls all aspects of the environment.

In this model, students are arranged in rows within large groups and normally have predetermined places to sit or stand. After receiving music from the teacher, their role is to realize in sound the teacher's musical conception. Many times the teacher will assess the preparation of the student's musical part through performance tests.

The end result of classroom work is almost always a formal concert. In most concerts the teacher will conduct as the students perform. Students will often dress in a uniform manner in concert settings, and they normally interact with audience members only through their group performance. At the end of the pieces, the audience will show their appreciation through applause, and the teacher will normally recognize the audience with a bow, sometimes asking the students, most often as a group, to do so as well. After a concert the cycle will normally begin again with a fresh set of pieces and another formal concert date for which to prepare.

This pedagogical model has been very successful. School ensembles have achieved a high standard of performance that has continued to improve over the years. The model is predictable and allows for effective classroom management. Students know what to expect and, with at least a moderate amount of effort, generally do well. This teacher-centered model that is fixated on performance in very particular ways has also proven to be resistant to change. Calls for change within the music education profession, in the United States, began as early as the 1960s. Incorporation of comprehensive musicianship was encouraged by the Contemporary Music Project for Creativity in Music Education. Results from The Yale Seminar of 1963 included suggestions for making the development of musicality the main goal of the K–12 music curriculum, broadening the school music repertory, including listening instruction, and incorporating additional performance ensemble opportunities (even though "technology" certainly wasn't the focus at the time!).

An even stronger call for change resulted from the Manhattanville Music Curriculum Program, which began in 1965. The objectives of the program included the development of a sequential music curriculum for grades 3–12. This extensive program was realized in three phases that included student learning, curriculum reform, classroom procedures, curriculum creation with curriculum guides, plans for teacher retraining, and the development of an assessment instrument. The resulting curriculum was based on a creative approach in which children encounter stages of exploration and creativity (spiral curriculum), where they have multiple opportunities for composing, performing, conducting, listening, and evaluating. Ronald Thomas, the director of the program, was critical of traditional music education values and practices. Among other things, he found that in-service teachers were too attached to skill development and performance, were too method-oriented to work in different teaching environments, and lacked significant prior experiences to be comfortable with creative atmospheres (Thomas, 1970).

Since the 1960s in the United States, interest in change within the profession has continued. There have been attempts to increase the use of world musics (multicultural education) in music classrooms, as well as sustained interest in the incorporation of popular and folk musics. The 1990s welcomed National Standards in music that called for the incorporation of more creative activities, such as composition and improvisation. While these efforts have resulted in some change in specific schools, we have witnessed little overall difference in how music education is practiced, especially at the secondary school levels. New National Standards appeared in early 2014 with an even stronger call for the incorporation of creative activities in school music classes, and more recently a report from a College Music Society task force, with interests in changing college and university level music instruction, "takes the position that improvisation and composition provide a stronger basis for educating musicians today than the prevailing model of training performers in the interpretation of older works" (College Music Society, 2014, p. 2). It is yet to be seen if these recent attempts to change music instruction in schools will have any significant effect. Other fields of education have been experimenting with alternate modes of pedagogy for several years. Many of these efforts have been guided by more student-centered, constructionist approaches; however, these teaching strategies have failed to take root in our teacher-centered model.

We are a profession with a rich tradition of remaining true to tradition. I realize I am generalizing and lumping the vast majority of our profession into a very big whole. Certainly there are individuals, sometimes even small groups of individuals, who are exceptions to the rule. There are teachers who are doing amazing work with new technologies and different pedagogical approaches, but these exceptions tend to be few and very far between. Accepting change, particularly within secondary performance classes, has been difficult for the vast majority of music teachers. It is apparent that working within accepted musical styles, with set instrumentations, within a prescribed methodology, and with established goals has become our "traditional model." It is no wonder that mixing new technologies into this setting, especially electronic and digital instruments, has not proven very effective in the great majority of programs. I suggest we might address potential change in the profession from two different directions: changes within this traditional model, and changes via different types of music classes that are perhaps better suited to address the needs of new technologies. While aspects of the following discussion about pedagogy could certainly be implemented in the former, I am primarily thinking of the latter—separate courses that involve an entirely different approach that is better suited to address the needs of new technologies.

Pedagogical Change

There are certainly many possibilities for the structure of a new genre of music classes. In addition to looking at implications for curricula, I will use the remainder of this chapter to describe one pedagogical model that may be flexible enough to support a wealth of alternatives. It is a pedagogy that could incorporate almost any combination of instruments (including the voice) and would be especially suited to music technology. This model shares much in common with the ways individuals make music in popular music settings outside schools (Green, 2002). It is a cooperative, interactive, and democratic process of music making and differs from the traditional school music model in at least five important ways:

1 Ensemble Size

The guiding principle behind the large ensemble model is "the bigger the better." Ensemble directors normally report increases in class size with great jubilation. This way of thinking is not shared across other areas of education, however. Most high school math teachers, for example, would complain if they were given a class of 80 students. It would be hard to find a middle school science class filled with 60 students. We know that small class sizes are educationally sound and allow for a better learning environment, yet the music education profession is consumed with a craving for bigger.

A pedagogical model better suited to music technology would incorporate class sizes similar to other academic classes in a given school. Students would then work independently and in small group settings within the class. These small groups (perhaps two to six in a group) would function in democratic and interactive tasks, realizing solutions to musical problems presented by their teacher. It would be important that each student have opportunities to provide significant input in the process. Student groupings should also be flexible over time so that it would be feasible for students to move from group to group as needs arise.

2 Musician Autonomy

While the traditional large ensemble model gives almost complete control to the teacher, a different pedagogical model would require the teacher to give up the bulk of control to the students in a class. Such autonomy would incorporate at least three areas, including instrument choice, musical styles, and creative decision-making. First, students would be free to select the instruments and equipment they would like to use in the class. Choice would come from any instrument the school could make available or to which students might have access. Obviously this would extend beyond the traditional band and orchestral instruments that presently dominate music rooms to include any and all electronic and digital technologies. It would also be possible (perhaps preferable) for each student to use a variety of instruments (and voice) over the course of a semester or year. Whereas students in the traditional model normally focus on one instrument, students in this model would be free to learn a variety of instruments that best fit the musical needs of what they are doing.

Second, students would have opportunities to select the musical styles with which they wish to work and study. Again, this would include styles and genres outside the music for traditional band, chorus, and orchestra. It would certainly allow students to dig deeper into musical styles that interest and hold significant meaning for them. If there is anything we know about education it is that students learn best when they are working with material that interests them. Most any popular music styles that students were to select would have significant opportunities to include electronic and digital technologies.

And finally, students would take over the responsibility of making musical and artistic decisions as they work. This should happen with students working both individually and in collaboration with other students. This process isn't normally fast, and it isn't always clean; however, students can gain tremendous musical understanding by working as creative artists.

Giving students autonomy in these three ways might occur in a class where the end product is not a performance of preselected standard pieces but rather a performance of pieces selected or originally created by students that fulfill broad guidelines and requirements set forth by the teacher. For example, an assignment might ask students to select one or more "classical" music pieces and to arrange them in a new performance piece. To realize the assignment one group of students might take the theme from Beethoven's "Für Elise," add lyrics, and work it into an electronic dubstep piece performed on various digital instruments. At the same time, another group of students could take a little more traditional approach with themes from Haydn's Symphony no. 94, performing their own arrangement on rock band instruments and a tuba.

3 Aural and Oral Process

Traditional school ensembles tend to be tied to conventional staff notation as they prepare music, written by others, for their specific type of ensemble. This use of notation is closely related to ensemble size and lack of individual autonomy, which together almost necessitate the use of notation. An alternative music education model, especially one involving music technology, would almost certainly emphasize aural transmission over written notation. Being bound to staff notation would limit the musical possibilities from which students could explore. One potential process would be developed through a cooperative aural and oral process, with individuals making musical suggestions and responding to the suggestions of others. This should be a highly interactive experience, with everyone taking part in a democratic system. Such a process would open up worlds of musical possibilities for students, who would no longer be tied to only the music they could find in notated form. Encouraging students to become better aural musicians could be the single most important step in helping develop lifelong musical skills, especially for students interested in working with music technology.

4 Concerts

Traditional school concerts, especially at the secondary level, typically emulate the eighteenth-century orchestral model. This concert experience is almost totally devoted to the sonic properties of the music being performed. The students focus on making the music, and the audience sits quietly and takes in the performance. This type of audience behavior is foreign to most people who are involved primarily with popular culture, and many find it difficult to conform to the limitations imposed in classically oriented concerts. An alternate model of music education centered on music technology would create an excellent opportunity for presenting concerts more like those found in mainstream popular culture—concerts that audience members and performers would possibly enjoy more and find more meaningful.

Such a concert could resemble musical shows or productions more than classically orientated concerts do. There is almost no limit to what could be included in a concert. A wide range of musical styles could be involved, such as rap, heavy metal, African drumming, electronic, disco, dubstep, classic rock, blues, gospel, and yes, even classical. Individual songs could include both vocals and instruments at the same time, something not found in most school concerts yet the norm in musical cultures outside the schools. There could be as much to see as to hear, including lighting effects, flexible and adjustable staging, haze machines, and multiple video screens with constantly changing projections.

Depending on the piece being performed, the music itself might not always be the only focus. On some pieces dancers could use movement, staging, and lighting to convey motion and interest, and other pieces might include actors playing out roles with action and dialogue, all connected to the music being performed. Visual artists and poets could be used on other pieces. In addition, the audience could be made to feel welcome and a part of the show. They could be allowed to socialize and urged to sing along with songs, clap, move, and dance. Audience members could become an integral part of the event. The main limitation for staging such productions is our own imaginations, but when we allow our students to have active roles in planning and producing concerts our imaginations expand exponentially.

5 The Role of the Teacher

Obviously the teacher's role in a class like this would be very different from the traditional model wherein the teacher takes on the central role in just about every aspect of the learning environment. Instead of being "the sage on the stage" the teacher would become more of "the guide on the side," assisting students in realizing their own interests and needs. This would require the teachers to have an understanding of a wider range of musical instruments, the ability to keep up with technological advancements in digital instruments, and the capacity to help students develop skills with instruments the teachers don't play themselves. It would also call for a deeper understanding of how to assist students who are helping each other, and it would require the teacher to take a deep breath and remain quiet at times when they normally would step in and take over a teaching and learning situation. It would also require teachers to address classroom management situations that are very different from those typically encountered in the traditional music classroom.

A New Curriculum

The suggestions here represent a radical change for a music education profession that has been mired in doing music in one form for a very long time, and doing it with a very limited vocabulary of musical styles. There are serious ramifications for music teacher training and education programs if they allow such change to take place. American composer Libby Larson forecast these changes in her 1997 plenary address at the National Association of Schools of Music National Convention:

> I posit again: it is probable in our time that the larger population will come to prefer produced sound over acoustic sound. If this is true, then the questions raised about how we teach music, what we teach, and the expected outcomes of our degree programs, are enormous. We will need to vigorously expand our teaching knowledge of professions in music to include not only our traditional professions but also producers, sound technicians, environmental sound scientists and recording engineers and managers. We will need to expand our teaching techniques to train our students in recording studio performance practice techniques. And some point in time, we will need to come to grips with the microphone. It is a new instrument. Without it performers such as Bobby McFerrin could not exist. It is central to the produced sound core. It permeates the ensembles, the music, the notational systems and the compositions of those who work in produced sound. To ignore the microphone and the recording studio in teaching music to young students is, in my opinion, a grave error. Ladies and gentlemen, I am suggesting that we now have alongside of the core of classical music education, another core, and that is the core of produced sound. We need to develop a rigorous course of study around this core. This course of study must have the highest standards and demand discipline and extraordinary craftsmanship from its students. (Larson, 1997)

Future music teachers will require time to develop skills in at least six new areas of study. The first is alternate pedagogies that lend themselves to new music technologies. This may be the most important piece of the puzzle for bringing these technologies into music classrooms. Understanding the technology is simply not sufficient in itself. Methods of teaching must change in order to take full advantage of new musical possibilities brought by new music technology, and preservice students will need extended experience and practice with new classroom pedagogies, perhaps similar to that described above for face-to-face classrooms. An additional pedagogical approach that will be increasingly important for preservice music teachers is in the area of distance learning. Having the ability to teach music in online environments will soon become as important as teaching face-to-face. College music education majors need guidance in best practices associated with distance education while they experience online courses from the vantage point of both the student and the teacher. Ignoring the possibilities and potentials of online learning in music teacher preparation programs is a serious mistake and an injustice to preservice teachers, many of whom will still be teaching through the middle of the twenty-first century, when online learning could well be the prevailing model of education.

The second area of study for future music teachers involves the art of digital performance. One of the most important changes technology has created is the continued blurring of what have previously been cleanly divided areas of music involvement. With most new music technologies, composing, improvising, performing, and listening are no longer separate categories, and students will require as much maturity in these different types of musical experiences as they have had with traditional performance. This will necessitate a fuller understanding of composition, improvisation, beat development, and songwriting (both individually and collaboratively) than is now customary. It will also call for a broad understanding of a wide array of musical instruments, including the guitar, electric guitar, digital keyboards and workstations, drum set, drum machines, microphones, turntables, computers, digital controllers, mobile devices, and so on.

Third, and closely related to performance, is an understanding of audio recording, editing, and mixing. In the blurring of musical fields, these areas could certainly qualify as new performance mediums. This would include recording both digital instruments and traditional acoustic instruments, computer-based editing and sequencing, and mixing for both live performance and recording. A general knowledge of music distribution would also be important, as the concert hall is no longer the most appropriate venue for many music performances.

A fourth area of study would involve app development. Digital music is progressively becoming portable and new handheld devices provide more and more possibilities for music teachers. Just as the ability to create webpages has enhanced potential learning, it will be important for preservice music teachers to establish skills in the creation of new apps for both music making and for educational materials that could further enrich the learning environment for students.

Music made with technology is increasingly mixed with various media, and this is a fifth area of study for preservice teachers. New curricula for future music educators should include opportunities where students work with a variety of media, including video and lighting. Video is so much a part of music technology that the two are often inseparable, especially in youth culture, where most K–12 students live. Lighting (including the use of video) plays an important part of music in most live performance settings outside traditional large ensembles, especially with music involving technology. Preservice music teachers need an understanding of the intersections between these various media and how to help their future students use them.

Sixth, and often associated with video and lighting, preservice teachers will need to understand relationships with music across various art forms, including dance, theater, visual art, and architecture. Musics outside the western European classical tradition, especially music produced with new technologies, are increasingly involved with sensory information beyond the sonic properties of the music itself. Music making with technology often involves movement, acting, visual stimulation, staging, and audience interaction, and future music teachers will benefit from a sensitivity to how music interacts with related art forms.

Taken together, new classroom and online pedagogies, digital performance, audio recording, editing and mixing, app development, multimedia, and arts education total a significant change for music education. What I am suggesting here is not a simple addition of a new course or two. It could potentially involve an overhaul of existing curriculum. However, adding new coursework to prevailing undergraduate music education programs is problematic. Between music core classes, state mandated education courses, and college and university general education requirements, most music teacher preparation programs are already busting at the seams with credit hour requirements.

I propose that, if we are serious about preparing preservice music education majors to reach a broader range of students with music technology (mainly those students uninvolved in traditional large ensembles), we must seriously rethink both curriculum and pedagogy. By seriously, I mean approaching the teaching of technology with the same commitment we have historically given to traditional music education. We have developed an entire curriculum around the preparation of music teachers to direct bands, choirs, and orchestras in K–12 schools, and it would be a disservice to our students to approach the teaching of music with technology with any less zeal.

I expect that many college and university programs will continue to approach the teaching of music technology in the form of a single course, wherein preservice music teachers learn some basic skills with various hardware and software. Many may also attempt to embed various knowledge and skills into an assortment of existing classes. Both these approaches are fine as far as they go, however it is doubtful that this will be enough to allow music technology to ever reach its true potential as a serious course of study for those who will be teachers in future elementary and secondary school classrooms.

I also expect that there will be daring individuals at certain colleges and universities who will take dramatic steps to ensure that their preservice teachers are ready to greet their own future students with a variety of new courses that involve music technology—courses that, certainly, large numbers of K–12 students are ready and willing to dive into. Some of these college programs might develop a new music education track that lives alongside current traditional tracks. Others might abandon the traditional and concentrate on the preparation of a new type of music teacher.

There are certainly many different approaches to the necessary curricular changes that would need to take place in order to bring music technology to the forefront of our profession. One such approach might begin with questioning the common music core associated with most music education programs. This core stems from a traditional music curriculum wherein students with various degree emphases have a common set of classes and experiences. The thinking is that no matter what a music major plans to do after graduation, musical "basics" exist in which every music student must be proficient. These tend to include performance skills on a principal instrument, developed through both private study and ensemble experience, knowledge of note reading and music theory, and an understanding of the history of Western music. Most often other experiences are also involved, including keyboarding skills and basic conducting competence. All of these—performance, notation, theory, history, keyboarding, and conducting, tend to be based on a particular perspective from a classical musical tradition with roots in eighteenth-century western Europe.

Questioning this core does not necessarily mean disputing the importance of specific material but rather questioning the organization of the core itself. The structure of the traditional core might be applicable for students interested in classical performance careers but perhaps not so much for students eager to begin teaching in twenty-first-century K–12 schools, especially with music technology. One alternative would be to separate music education students out and have them experience a core of material that is specifically tailored for twenty-first-century teachers. Again, there are a plethora of possibilities, but rethinking the musical core for music education majors could open up the degree program in ways such that digital performance, audio recording, editing and mixing, app development, multimedia, and arts education could be included within new classroom and distance learning pedagogies.

Such a model could take on aspects of comprehensive musicianship wherein various learnings are used together in practical ways. Perhaps taken by students over a series of four to six semesters, the core of musical learning for music education majors might take place in musicianship for music educator courses. These courses could focus on students working in various small, collaborative groups where appropriate music theory and history would be learned while developing musicianship skills through performing, listening, composing, arranging, songwriting, copying, improvising, and conducting. These musicianship courses would serve as a central hub experience, perhaps replacing many of the traditional academic classes, ensembles, and private study normally taken by music education students. Skills and understandings developed in all other music education courses would tie directly back to these hub experiences. Regardless of the design, a rethinking of the common music core for undergraduate music majors may be vital to the success of future music teachers who are interested in taking advantage of music technology in serious ways.

Finally, I want to return to my previous reference to the type of student normally admitted to music education degree programs. This typical student usually looks a lot like other music majors, as they all tend to come from the same classically oriented, large ensemble model in high schools. Such students probably excelled in the traditional system and are now looking forward to teaching in the same type of program, and in the same ways their teachers taught them. Many of these students are, unsurprisingly, not very interested in exploring different music-making models or using unfamiliar pedagogical approaches. They tend to enjoy using music technology for certain things, such as listening to recordings, but typically not as musical instruments. In addition to requiring a new generation of teachers and professors in the workforce, we may need a new type of music education student. As I suggested before, this might be the most difficult challenge related to the use of new technologies in music education. We need to start admitting students with more diverse musical backgrounds to teacher education programs—students who play in rock bands, DJ with turntables, and perform hip-hop need opportunities to become music teachers—and we need them to be music teachers. Bernard (2012) suggests:

> individuals with non-traditional backgrounds, because they may not be tied to the way things have always been done in music education, may be able to think outside the box in terms of repertoire, musical activities, teaching strategies, and performance practices, bringing their unique musical experiences and perspectives to the ways that they structure their classroom practice. They may be able to incorporate different musical styles into their teaching—musical styles that may be more interesting and meaningful to their students. They may thereby be able to increase the relevance of the field of music education by forging stronger connections between the music that their students listen to outside of school and the music that their students encounter in music class or rehearsal. (p. 6)

In addition to rethinking the core of music studies for music education majors, we must also reconsider admission requirements for students interested in teaching. These two changes actually fit together quite nicely. A reimagined core that places high value on music technology, creative thinking through composition and songwriting, and small group collaborative work would be well suited to students from a wide range of popular musical involvements. These students, with this type of musical training in music education programs, would create a welcome change for a profession that is in desperate need of just such a change.

References

Bauer, W., Reese, S., & McAllister, P. (2003). Transforming music teaching via technology: The role of professional development. *Journal of Research in Music Education*, 51(4), 289–301.

Bernard, R. (2012). Finding a place in music education: The lived experiences of music educators with "non-traditional" backgrounds. *Visions of Research in Music Education, 22*. Retrieved October 29, 2013, from http://www.rider.edu/~vrme/.

College Music Society. (2014). *Transforming music study from its foundations: A manifesto for progressive change in the undergraduate preparation of music majors* (Report of the Task Force on the Undergraduate Music Major, David Myers, Task Force Chair). Retrieved January 30, 2017, from http://www.mtosmt.org/issues/mto.16.22.1/manifesto.pdf.

Dorfman, J. (2008). Technology in Ohio's school music programs: An exploratory study of teacher use and integration. *Contributions to Music Education*, 35, 23–46.

Gall, M. (2103). Trainee teachers' perceptions: Factors that constrain the use of music technology in teaching placements. *Journal of Music, Technology and Education*, 6(1), 5–27.

Green, L. (2002). *How popular musicians learn: A way ahead for music education*. Aldershot, UK: Ashgate.

Hoeprich, E. (2008). *The clarinet*. New Haven, CT: Yale University Press.

Holmes, T. (2012). *Electronic and experimental music: Technology, music, and culture*. Hoboken, NJ: Taylor & Francis.

Larson, L. (1997, November 23). *The role of the musician in the 21st century: Rethinking the core*. Plenary address at the National Association of Schools of Music National Convention, San Diego, CA. Retrieved January 5, 2014, from http://libbylarsen.com/as_the-role-of-the-musician/

Millard, A. J. (2004). *The electric guitar: A history of an American icon*. Baltimore: Johns Hopkins University Press.

Noxon, J. 2003. Music technology as a team sport. *Journal of Technology in Music Learning*, 2(2), 56–61.

Reese, S., & Rimington, J. (2000). Music technology in Illinois public schools. *Update: Applications of Research in Music Education*, 18(2), 27–32.

Taylor, J., & Deal, J. (2000, November 2). *Integrating technology into the K–12 music curriculum: A national survey of music teachers*. Poster session presented at the annual meeting of the Association for Technology in Music Instruction, Toronto.

Thomas, R. B. (1970). *Manhattanville Music Curriculum Program. Final report*. Purchase, NY: Manhattanville College.

4 B

Further Perspectives

CHAPTER 61

WHY ISN'T MUSIC EDUCATION IN THE UNITED STATES MORE TWENTY-FIRST-CENTURY PC?

BARBARA FREEDMAN

TRAINING preservice teachers in technology and technology pedagogy as we would train them on brass, woodwinds, percussion, and piano is vital for future educators to be fully prepared for all possibilities of employment and to serve a greater number of students in the twenty-first century. In his Core Perspective, David Williams states: "training teachers to use new technologies is further complicated by how fast change in music technology happens" (chapter 60 in this book). Having taught high school music technology classes for the last 14 years, my initial response to this statement is to wonder whether the fast changing technology is an issue for the college students or for the professors! This is probably true for both.

PREPARING NEW TEACHERS

Even in this day and age of pervasive technology, I completely agree that we need to address a population of students who may not be savvy users of music technology. However, a growing number of students are graduating from schools such as mine who have taken extensive music technology and audio engineering classes in high school. They participate in these classes much as their peers participate in performing ensembles, gaining 3 or 4 years of ongoing experience and training in high school. Many of these students who might venture to become music educators will have no issue teaching with technology, since they have experience and have had teaching with technology modeled in their own educations. We need to nurture these students' desired expression of music pedagogy, the use of music technology as a valid form of music creation and expression, and, in agreement with Williams, separate and apart from the current and standard course offerings of music performance ensembles.

I do agree with Williams that technology is changing very quickly and that this creates many challenges for the university and the continuing education educator *if* we are training teachers on a specific device or software. Students learn what we present to them. College teachers need to sift through the vast amount and variety of technology to decide what is most useful and covers a broad need for their students. It is possible to use the current technologies as examples of types of technologies that are available. An iPad or a computer is a technology tool. Granted, I cannot train teachers on all of the apps and software available, but I can use some of them as examples of apps or software for creation, ear training, notation reading, and so on. University professors can teach that these apps and software exist, that they share some common traits, and how to evaluate apps and software.

The opportunity of the university or continuing education setting is to expose preservice and in-service educators to basic common technology tools useful in music education and encourage them to learn on their own and explore. We can educate preservice and in-service educators in such a way that they do not need to be technology experts to use technology. Current technology makes self-learning easier and more readily available through YouTube and professional learning networks such as the Music Teachers Facebook Group and the Facebook subgroups that have specialties, such as "I'm a Music Technology Teacher." Furthermore, I am in complete agreement with Himonides's declaration (chapter 59): "teachers can harness a new technology in tandem with the learners, offering their experiences in being critical reflectors and assessors of the learning process, without necessarily being heuristic experts in the new tools." Although Himonides's statement refers to the preservice or in-service educator, university professors can also view themselves as in-service educators. Removing the notion of the "sage on stage," where educators think they need to be the experts on everything in their classrooms, allows for students to contribute to their own education, their peers, and the willing professor.

In the classroom, music teachers will always be the music experts, regardless of the tools they use to teach music. My motto, "Teach music. The technology will follow," helps focus the educator with little background in using or teaching with technology as a music teacher, not a technology instructor. Technology is the tool. Trained woodwind teachers may understand the basic concepts of playing brass instruments, having learned this in a brass pedagogy class in college or grad school. They are not the expert brass player, nor do they need to be. Experienced band teachers have no problem asking their advanced trumpet player students to help the beginning trumpet player students, calling on the students as experts and partners in the classroom. Why don't we use this model with technology? Experienced music technology teachers and teachers integrating technology into their classrooms do this all the time. This is a best practice for education that is not specific to music or technology.

My personal experience has shown that I do not need to be an absolute expert in a specific technology to use it in the classroom. Sometimes bringing in a new device and handing it to an enthusiastic student can be a great learning experience for the student and teacher. That student can learn the device and then teach me or, better yet, other students. In addition, if the software I use in my classroom is professional level and sophisticated, I simply cannot learn everything every time there is a major upgrade to it. I have had no problem telling my students that I can show them the basic framework of the software and that there are things that we will learn and discover together. They have no problem with that because they know that the technology is the tool and I am the music teacher and they came to my classroom to learn music. We deal with the tool together until I do actually become the expert! Inevitably, there is always a student who knows more than I do about the technology tool, and I always learn from that student. Good for us all!

An Honest Conversation about Music Education in the United States

For Williams, "perhaps the most important issue for the profession to overcome has to do with the natural reluctance many teachers (and students) have toward change" (p. 634). This might even be truer of many university music education programs. I would ask university professors to consider how much change there has been in their university preservice teacher training to address the current needs of the K–12 educational environment both from an educational and socioeconomic perspective. Williams suggests a new curriculum for preservice teachers (and I'll conjecture that he also includes continuing education for in-service teachers) to "develop skills in at least six new areas of study" (p. 640). He further challenges higher education, asserting: "we have developed an entire curriculum around the preparation of music teachers to direct bands, choirs, and orchestras in K–12 schools, and it would be a disservice to our students to approach the teaching of music with technology with any less zeal" (p. 642). What Williams is suggesting is that kindergarten through university music education aims to address mostly the needs of traditional performing ensembles, which serves, according to Elpus and Abril (2011), on average, only 20% of a school's population. We are clearly not serving a majority of the students in public education.

I'd like to take this one step further and offer an extremely controversial perspective, if for no other purpose than to stimulate a larger conversation. Let me preface this by saying that there are many teachers bringing nontraditional music education opportunities for learning music outside the traditional performing ensembles of band, orchestra, and chorus into their classrooms through technology (iPad ensembles, music recording and composition classes), ethnic performing ensembles (mariachi, steel pan, etc.), new and diverse nontechnological instruments (ukulele, Boomwhackers, kazoos, harmonicas, etc.), and classroom music listening and singing and playing "world" and "folk" music. These teachers bring opportunities to a diverse population of students who might not otherwise want or have the means to participate in traditional performance ensembles. This might be due to personal preferences or the economics of the individual or the school, given that it is expensive to run an instrumental program like a band or orchestra. However, I suggest that we have institutionalized music education in such a way as to primarily serve those who have performance ensembles as a tradition or cultural background and to serve the economically advantaged—those who can afford instruments and lessons, individually or institutionally. It should be noted that American football as a tradition is a great influencer of music culture in U.S. schools, but this chapter does not allow me the space to elaborate on the impact American football has had as a driving force behind U.S. band programs. It could be argued that the American music education system, K–university, is implicitly skewed toward the majority middle to upper economic class culture and, by nature, disinherits those that seek to engage in other musical cultures.

Research by Dammers (2012) indicates that the percentage of students in "Technology-Based Music Classes" from economically disadvantaged schools exceeds that of students from advantaged schools. This research does not indicate the data for the schools' performing ensembles. After teaching music technology classes for the past 14 years, I have accumulated a great deal of data with regard to my students' socioeconomic status (e.g., students on "Free or Reduced Lunch") and their ethnicity (declared minority population tracked in schools and sometimes linked to socioeconomically disadvantaged). My music technology classes have a considerably higher percentage of students who are of the region's minority than the performing ensembles. At my school, there have been more students on free and reduced lunch in my music technology classes than are in all of my school's bands, orchestras, and choruses combined. Clearly, my school and others serve students outside the traditional performing ensembles with hands-on applied music learning and other engaging music-creating opportunities. This levels the playing field for all students, regardless of cultural preferences or socioeconomic background. It is one example of an attempt to bridge the achievement gap.

Again, these statistics and data are not true of *all* music programs K–12 or university music education programs in the United States, and a great number of individual teachers K–university are driving the creation of new opportunities for music to be learned and expressed with and without technology. Dammers's (2012) research indicates that 59% of music technology teachers in U.S. high schools have 11 to 20-plus years of teaching experience. Teachers with fewer than 10 years of teaching only constitute 9% of those teaching a music technology class in the United States. If it is primarily the experienced teacher who creates and teaches music technology classes in the United States, can it be that preservice music education does not properly prepare new teachers for this challenge? Williams might agree.

Can Technology Impact the Economically Disadvantaged Individual and School?

A high school band director from a nearby town with a solid reputation in music education and a successful band program greeted me recently with the comment "I really need to learn what you do or I won't have a job." This was not the first ensemble director in recent years to tell me this. In certain parts of the United States, ensemble programs are increasingly under jeopardy, and music teachers are scurrying to save their employment. Given the deep roots of instrumental music ensemble performance programs in the United States, why would so many ensemble directors fear for their job security? Why would so many music educators recognize that turning toward teaching with technology might be a solution to their problem? It's a matter of economics. Although a single band, orchestra, or chorus director can have extremely large classes and make their student-to-teacher ratio more cost-effective than some other subjects, running instrumental ensembles is extremely expensive.

Furthermore, an ensemble director in a high school may teach more than one ensemble with multiple overlapping students, for example, concert band, jazz band, and marching band all drawing from the same pool of students. That teacher's time (full-time teacher equivalent or FTE 1.0) is spent teaching many of the same students. In addition, a strong high school ensemble program requires students to repeat the course throughout their high school careers—again, one teacher with overlapping students. I taught "general music" in a New York City public school and had 50 high school students in a class. Fifty students in "General Music" is not education, it's a credit mill, and the administration knew it. They can put 50 kids through a music class, get them their credit in one semester and then put another 50 the next semester. Multiply this by a single teacher's class load (typically, in New York state, for example, five classes, or 0.2 FTE per class, per semester), and you can give 500 *unique* students a music credit in one year with one teacher.

In a school or district that is facing economic hardships, how teachers are utilized becomes critical to survival. When people or institutions are in survival mode, only what are considered the basic necessities are included. Typically, art and music are the first to go. If music is required for graduation, then you need to give those students credit in the most economical way possible. This is exactly what happened in my classroom, in the state of California in 2010 and in Yonkers, New York in 2011 (Dev, 2010).

Technology costs have plummeted in recent years. Computing devices are as ubiquitous as the smartphones in your students' back pockets. Like them or not, initiatives like "No Child Left Behind" and "Race to the Top" have provided funding for technology but not for tubas (U.S. Department of Education, 2009). Time and time again, I have heard teachers who have started music classes that incorporate technology as the primary learning and music creation tool tell me how many students want to take the class and want to take more advanced classes. They tell me, and I have had the same experience, that these are new music students. They are not, for the most part, students who are already engaged in band, orchestra, or chorus. These new classes increase enrollment in the music program, and more enrollment in music classes means job security for music teachers. Technology is already available to teachers and students, and teachers would do well to utilize it for their career longevity and to help serve the larger student populations of their schools.

While I agree with Williams in principle, I might, given my perspective, change one of his sentences slightly: let's leave the traditional large ensemble model alone to do what it does best, and use "an entirely different approach in other music classes that is better suited to address the needs" *of all our students through the use* "of new technologies."

References

Dammers, R. J. (2012). Technology-based music classes in high schools in the United States. *Bulletin for the Council of Research in Music Education, 194*, 73–90.

Dev, R. (2010, June 12). *Survey shows how California schools are coping with budget cut pains.* Retrieved January 15, 2015, from http://newamericamedia.org/2010/06/survey-shows-how-california-schools-are-coping-with-budget-cut-pains.php.

Elpus, K., & Abril, C. (2011). High school music students in the United States: A demographic profile. *Journal of Research in Music Education, 59*(2), 128–145.

U.S. Department of Education (2009, July 24). *President Obama, U.S. Secretary of Education Duncan Announce National Competition to Advance School Reform: Obama Administration Starts $4.35 Billion "Race to the Top" Competition, Pledges a Total of $10 Billion for Reforms* [Press release]. Retrieved from http://www2.ed.gov/news/pressreleases/2009/07/07242009.html

CHAPTER 62

GENERATING INTERSECTIONS BETWEEN MUSIC AND TECHNOLOGY

MATTHEW HITCHCOCK

Contextual Background

I currently hold the position of director of music technology programs at the Queensland Conservatorium, Griffith University, in Brisbane, Australia. I came to academia after having spent some 30 years of my teenage and adult life as a professional performing musician, recording studio owner, record producer, recording engineer, software programmer, and music teacher. I am still active in these domains. Further, I have also acted as a music education curriculum consultant for some secondary schools around Brisbane. I am therefore familiar with secondary school contexts as well as tertiary contexts.

Technology has played a pivotal role in my musical endeavors, and not always in obvious ways. Music technologies were certainly central to performance and studio craft; however, in the 1980s computer languages such as HTML, PHP and MySQL were integral to many successful musical careers, including my own. The Internet became a natural extension of music creation, collaboration, and dissemination.

Subsequently I entered academia at the tertiary level in the early 2000s in a dual (50/50) role as a music technology lecturer and an e-learning specialist. I held this dual role for 3 years before the music technology role took over. The e-learning specialist role was established to provide the Conservatorium with someone who could navigate the divide between the music faculty and IT. The role was largely divided between interpreting, translating, and recontextualizing the competing educator and institutional needs of IT specialists and educators, and assisting academic staff with pedagogical, technological, and philosophical aspects of using technology in music education.

Consequently, I was embedded in IT almost as much as in music and pedagogy. I have continued to maintain leadership roles in technological advancement and implementation in learning and teaching since the official title expired. Herein, the dual specializations afford me insights into both extremes of a technological-pedagogical continuum, as well as some ways that tensions between the two can be mediated for successful outcomes.

In my local environs, technology and teaching were not easy bedfellows. In the mid-2000s the technological-pedagogical continuum typically manifested itself as technocentric thinking (Papert, 1987) with little to no regard for pedagogical needs at one end of the continuum and stubborn refusal to consider alternatives to historically accepted teaching practices at the other.

Over the last 10 years, however, the extent of the divide has narrowed, with far less IT department drive to implement technology in order to justify their employment, and many more academics embracing technologies in their pedagogical and musical practices. This change has come about both from intrinsic motivation and as a result of changes in leadership and philosophical stances on technology in education.

As the divide has narrowed, there has been exponential success in the uptake of technologies in teaching. This supports observations raised by Himonides and Gonzalez that "people's attitudes toward technology has been the one most critical factor for the genesis of a successful (or not so engagement" (chapter 59 in this book).

In the Australian context, there are three primary modes of MUSIC teaching in schools: instrumental, ensemble, and classroom. Instrumental lessons can range from one-to-one to small groups of around four students. If the school has an instrumental program it will typically also have an ensemble program. Some states in Australia have very limited in-school music programs, largely confined to classroom music education. Queensland is very fortunate to have extensive and long-standing school-based music programs extending beyond classroom music, embedding instrumental and ensemble teachers in most schools.

In all three contexts, it is extremely rare to see music technologies used within the school curriculum. (In the few instances where music technologies are core to the school curriculum they are used extensively and to great effect.) Over the last 12 years, incoming Conservatorium music technology students have been surveyed about their secondary school music technology experiences. With few exceptions, these students say that their previous music technology learning occurred outside the school, occurred as self-driven engagements, and that they felt they were typically more adept with the basics of musical technologies than their music teachers.

FURTHER PERSPECTIVES: EDUCATIONAL, ADMINISTRATIVE, SOCIAL, AND *MUSIC* TECHNOLOGIES

Here it is important to further differentiate between different categories of technologies used in music education, as variously raised by González-Moreno, Himonides, and Williams.

I separate these into four broad categories: educational, administrative, social, and music technologies. While these categories are not mutually exclusive, educational and administrative technologies can be differentiated from social and musical technologies philosophically, because typically any mention of technology as a means to delivery and management puts the technology in the control of the teacher rather than the students. This aspect is highlighted by Williams in particular. These four categories are briefly unpacked below.

Educational Technologies

Broadly, this is where the use of technology is principled and facilitates learning and assessment processes—including quizzes, tutorials, learning engagements, feedback, and assessments. Theses sorts of technology implementations include the use of content management systems, learning management systems, discussion boards, virtual classrooms, game environments, collaborative tools, learning labs, networks, and digital repositories.

González-Moreno, in particular, discusses the impactful use of ICT technologies for virtual or blended models of knowledge access, sharing, and mobilization. She raises the importance of collaboration, sense of community, and ownership in professional development achievable through communities of practice (chapter 58). Clearly depicted is a use of technology wherein sharing and contributing to distributed knowledge networks has assisted in establishing richer experiences than the initial "top-down" webinar created.

Administrative Technologies

Broadly, this is where the use of technology supports managing people, circumstances, and resources. The typical interfaces again include content management systems, learning management systems, digital repositories, and extend to resource booking systems, student administration portals, finance systems, and so on.

Social Technologies

There is little need to expound on definitions of social technologies; however, for the purposes of clarity I propose the following as a useful broad definition: "any technology that facilitates social interactions and is enabled by a communications capability . . . [and is] targeted at and enable[s] social interactions" (Social Technologies, 2013).

Music Technologies

Here I will limit discussion of music technologies to any technology used in the creation of music that requires electricity to operate. It is important to make this distinction because while all acoustic instruments are also examples of technological innovation, they are not under scrutiny in this context. Consequently, relevant examples of secondary and tertiary musical endeavors and their music technologies can, and I propose *should*, to varying extents, include:

- Music engraving (e.g., MakeMusic's Finale and Avid's Sibelius)
- Audio capture and manipulation (e.g., digital audio workstations, audio file editors, portable hardware recorders)
- Sound design and synthesis (physical, virtual, and blended)
- Performance using electronic technologies (e.g., synthesizers, tablets, phones, laptops with performance software such as Ableton Live, and human–computer interfaces and controllers)

Critical Intersections between Music and Technology

One valuable provocation to institutionalized music education at any level is that music is a global phenomenon proliferating at an unprecedented rate. Accessing music and the tools to create music has never been easier. Further, access and ownership are rapidly converging, such that the democratization of technology means that those who 10 years ago would have been mere recipients of music are now able to engage in the creation of it.

This has fostered a new generation of individuals who can create music without the need for formal "instruction." This has also fostered unprecedented levels of cross-pollination of ideas, cultures, inspiration, and rewards. Far from *negating* the need for music curricula and teachers to engage in these music practices, however, I make the case that the roles of educators and institutions in these musical activities are increasingly critical. Musical and cultural history, music theory, genre studies, and self-directed learning skills all contribute to the creation of a more resilient and skilled musician, artist, educator, and practitioner.

It is an axiom outside academia that DIY music culture is saturated by copy-and-paste music making. Consequently, and in contrast to the above, large amounts of modern music genres do not in reality benefit from healthy cross-pollination but are instead heavily derivative. This lack of diversity is desensitizing; however, it also means that musicians with deeper musical backgrounds and broader skills and knowledge are more likely to be heard above the noise.

While the democratization of technology has caused major disruptions to multinational corporations (the major record companies), which situated themselves as the "gatekeepers" of music, it has also empowered every modern musician. The aggregated power of niche markets is seen to be overwhelming the once-omnipotent monopolies, and this is a space that modern musicians and educators need to be engaging in to remain relevant to the current realities of music making.

The Role of Technology in Music and Artistic Practice

At the core of the provocations raised by González-Moreno, Himonides, and Williams is the question *what are we educating our students for?* This may range from fostering a deep appreciation of music to providing a grounding in musical language and developing competencies that might lead students to undertake careers in music or continue high-level artistic practice alongside a different career.

Notably, one of the key aspects to engagement in music, be it performance, education, creation, or dissemination, is music technology. There seems little sense in overlooking this; rather, we must work to embrace it before being outmoded by it. It is therefore up to us as institutions to correctly manage our inheritance of hundreds of years of music making. This inheritance includes a growing importance on technology in music making, and for some contexts, technology as core, not as "add-on."

This is not about casual awareness of social media, administrative technologies, and online strategies either but is an educated, forward thinking, and sophisticated engagement with intersections of music and technology. This needs to be done with the same level of critical engagement and "cognitive apprenticeship" (Collins, Brown & Newman, 1987) that, for example, musicians undertaking a performance pathway are engaged in while learning to "be" and "become" (Brown & Duguid, 2000; Wenger, 1998).

Curriculum Relevance

In moving forward, it is central to the relevance of music education that educators remain acutely aware of what the marketplace is doing. Williams draws on this concept to form a central tenet of his discussions, provoking thoughts on moving away from any sole reliance on eighteenth-century western European musical practices, toward alternative approaches to the *massification* of education through large ensembles, and toward framing this proposition using a series of core activities that respond to the realities of twenty-first-century music making.

Here, it is important to recognize that twenty-first-century music making is not limited to any style or genre, the term instead indicating any form of engagement in music making in current contexts (Hugill, 2012).

At the Conservatorium, alumni from traditional music departments who are seeking careers in traditional music pathways are communicating with us about the importance of technological skills in their career successes; for instance, composers' ability to work on virtual instrument mockups to "win" gigs, or the ability of performers on traditional instruments to work effectively in recording contexts (live and studio).

These graduates are telling us that they are "getting the gigs" over other equally musical musicians simply because comfort and skills with technology in music making are regarded by many employers as a must-have. Similarly, performing musicians who have expertise with digital audio recording and production technologies are saying that they are able to enhance their performance opportunities, or to create financially accessible opportunities to produce high quality recorded product for their fan and employer bases, or to make themselves appear relevant to their twenty-first-century peers, collaborators, and employers. The general consensus appears to be that technology skills, and specifically *music* technology skills, are no longer an add-on but are increasingly a core requirement.

Teacher Training

Catch-22: Perpetual Traditionalism

Many people in leadership positions in secondary and tertiary education have been teaching in excess of 15 years and therefore received their teacher education prior to the advent of now ubiquitous music technologies. Consequently, in my experiences, it is rare to find teachers, or teachers of teachers, who understand the potential of music technologies in developing core musical skills, and even rarer to find those who can use technology with the ease and fluidity that precipitates the embedding of music technologies as core to curricula and pedagogies.

This is certainly not to suggest these teachers are disengaged in any way. The music teachers I know work many hours in excess of what they are required. They are highly trained, engaged, forward thinking people who are good communicators and excellent teachers. For many, their tertiary experience was 20–30 years ago. They rely on incoming young teachers to bring expertise and skills with technology to bear in music departments.

However, it should be recognized that newly trained music teachers with technology skills are stepping into a landscape of an unmoving curriculum with limited power to affect change on anything but a surface level. The leaders who interpret curriculum outcomes and how to achieve the state-based pedagogical goals need guidance and leadership from more than "fresh blood." The self-perpetuating cycle of nontechnical traditionalist tertiary music educators creating yet another generation of technologically nescient educators needs to be disrupted in a positive and constructive way.

A Suggestion for a Music Technology Enabled Curriculum

Electronic music technologies such as those typically found in studios are no longer just the domain of recording engineers and producers. With the democratization of technology, the expectations placed on musicians in everyday life have broadened to include technology as a core skill set. We are witnessing the collapsing of boundaries between different types of artistic media and a growth in blurring of distinctions between professional, prosumer (Toffler, 1980), and technologically enabled consumer.

What separates people is not the technology but the ideas. It has been my experience that music education in Australia has demonstrable strengths in teaching ideas, but mainly in traditional contexts. Technologically-enabled musical ideas and negotiations should also be taught in schools. The democratization of technology now means that, in conjunction with a music education, there may be little obstacle in the ability of an artistic individual to output original or innovative music. To ignore or, worse, to resist the teaching of music through ubiquitous music technologies seems indefensible.

Further, many of the music technologies such as computers, recording and editing devices, signal processors, and reproduction technologies have played central roles in high-level music making for over half a century. Surely it is time to actively engage in the last 50 years of music making with more than a casual musicological interest and to embrace these technologies in areas such as genre studies, analysis, music theory, music reading, aural skills, composition, performance, and creation.

A Curriculum Approach

Music, like any creative pursuit, is arguably one of the few disciplines that can foster a purposeful fusion of intellect, know-how, and emotion. This is truest where students are required to create something new, an act of thoughtful and intuitive origination. Intuition in music creation, the underlying tenet of improvisation, does not occur quickly or without significant intellectual, emotional, and skills training.

The importance of immersive experiential training for musicians cannot be overstated. Artistic outcomes are simply not forthcoming if the immersive and intensive learning engagements are shortened and simplified. However, music technologies can change the threshold to successful entry into musical engagements.

I propose that learning music comprises a range of processes: deconstruction, reconstruction, construction, and the peak activity of origination. At the highest level of origination there is innovation, however, for this context I choose to have the finer distinction of innovation subsumed within the term origination. Simultaneously, an array of progressively complex philosophical, intellectual, and creative engagements occurs. I propose these as copying, emulation, inspiration, and (again) origination.

Similarities can be drawn to Bloom's revised taxonomy (Sosniak, 1994) wherein creation is placed at the higher-order level of the cognitive processes dimension. It should be recognized, however, that the last three levels I outline below all refer to various levels of creation; hence I use the term "origination" as a distinct form of creation. Whether undertaken by a new learner or an experienced musician, the levels remain relevant. These are:

1. Imitation (the starting point) (deconstruction/analysis) (largely genesis—understanding how others think and feel; copying to learn, to understand—to "become")

2. Emulation (level 2) (reconstruction and recontextualization)
    Bringing the "you" into the equation but still derivative—reinterpreting, enhancing, bringing a personal "artistic vocabulary" to bear—to be "like" someone else, to demonstrate an understanding of someone else's approaches to creative endeavors
3. Inspiration (Level 3) (sophisticated concepts of construction)
    You—but triggered and inspired by external forces; greater levels of intuition and imagination than emulation and imitation
4. Origination (Level 4)—completely new and from within. The peak form of origination is innovation.

It is beyond the scope of this chapter to enter into the complex details of such a taxonomy. However, broadly speaking, there is a range of free and cheap software titles and accessible hardware devices that can enable the learning of music to occur in complex and interactive ways, such that learning from books can be replaced or enhanced by, for example, using exemplar MIDI files that can be deconstructed, reconstructed, and recontextualized. Compositional and arrangement activities can similarly take advantage of technologies that afford us the ability not only to see but to hear the results of our musical decisions in real time. Extending on this, collaboration and negotiation of ideas and actions are fostered by such use of technology.

This means that traditional music curriculum outcomes can still remain, but the resources used to achieve these are modified to concurrently create new learning outcomes. The engagement with technology to learn traditional aspects of music establishes a familiarity with technology in a musical setting such that when the more sophisticated and complex musical engagements of creation and origination are enacted, the range of musical encounters is richer and holds more relevance to twenty-first-century contexts.

Conclusion

Moving forward requires taking risks—the sort of risks that involve discomfort and disruption. History tells us the "normal" condition is one of chaos and entropy occasionally resolving into brief periods of stasis; however, music education in Australia, for all its forward thinking proponents and positive outcomes, appears to me to be paused on the stasis part of its evolution. If nothing is at risk, then nothing new is created, and the arts and education are both in high-risk categories at the moment in Australia. Embracing music technologies as a means to rich and meaningful learning engagements in music requires action at the point where music teachers are taught to be musicians and teachers, and is therefore equally critical in the development and administration of curriculum and teacher training.

References

Brown, J. S., & Duguid, P. (2000). *The social life of information*. Cambridge, MA: Harvard Business School.

Collins, A., Brown, J. S., & Newman, S. E. (1987). *Cognitive apprenticeship: teaching the craft of reading, writing, and mathematics*. Cambridge, MA: University of Illinois at Urbana-Champaign, Center for the Study of Reading.

Hugill, A. (2012). *The digital musician* (2nd ed.). New York: Routledge.

Papert, S. (1987). Computer criticism vs. technocentric thinking. *Educational Researcher*, 16(1), 22–30.

Social Technologies. (2013). In *Gartner IT Glossary*. Retrieved January 5, 2015, from http://www.gartner.com/it-glossary/social-technologies.

Sosniak, L. A. (1994). *Bloom's taxonomy* (L. W. Anderson, Ed.). Chicago: University of Chicago Press.

Toffler, A. (1981). *The third wave*. New York: Bantam Books.

Wenger, E. (1998). *Communities of practice: learning, meaning, and identity*. Cambridge, UK; New York, NY: Cambridge University Press.

CHAPTER 63

PREPARING FOR CHANGE AND UNCERTAINTY

VALERIE PETERS

CHANGE

CURRICULAR expectations for teacher education programs are bursting at the seams as we try to prepare students for the diverse contexts of music teaching in schools (Blondin, Peters, & Fournier, 2010). For every new trend in education we seem to propose a new class or a new teacher education seminar. Fundamentally, universities are conservative places (especially faculties of music) where change happens at a snail's pace. Change is often viewed as negative and opposed to the belief that traditions from the past should be preserved and cherished in the institution. This position is often justified by the argument that if we do not preserve the past, who will? This makes me wonder whether we are museums or institutions of higher learning. All institutions must evolve and change or perish. Information and communication technologies challenge the existing traditions in higher learning and are often deemed the "enemy," with professors elaborating on the negative impact of technology on society. In my experience, this polarized debate seems to go nowhere, offering only reactionary commentary rather than creating a collaborative space in which to imagine creative solutions and new and innovative ways of integrating technology into the curriculum. In terms of technology and music education, it seems ludicrous to ignore the incredible changes in our culture that impact the ways our students in schools conceptualize music learning. In the coming years, I and my colleagues hope to provide a detailed description of youth conceptions of artistic learning in a digital age. Hopefully these data will contribute to a transformation of music teacher curricula in higher learning.

González-Moreno (chapter 58 in this book) discusses the resistance to change by professors at universities and characterizes students of the current generation as more intrinsically motivated to use technology. The younger generation is often described as digital natives while the older generation, including university professors, is described as digital immigrants, not having grown up with technology and being forced to integrate it into their teaching practice by their institutions. While many university students seem comfortable with certain specific types of technology (social networks, the Internet, etc.), I have found that many music education students are not necessarily at ease with technologies that could transform their future teaching practices. In addition, I know many university professors who have embraced technology and use it extensively in their teaching.

I believe that the most important deterrent to faculty technology development and the integration of technology into the university curriculum is time! Even for young professors beginning in the profession, the incredible demands of the first years of university teaching can deter the most passionate and innovative person. Universities need to be proactive and provide infrastructures for the development and the continued support of their workforce if they expect these people to evolve and embrace the many changes that will continue to challenge university teaching in the future. As González-Moreno indicates, a community of practice model is the future for professional development in our field. We need to create a mutually supportive environment where it is acceptable to say "I don't know everything." However, there are few graduate programs that prepare their students to work in this type of paradigm. As González-Moreno states, the "lone wolf" or "independent entrepreneur" professorial model continues to exist in most music education doctoral programs and in most faculties. We can no longer sustain this model if we wish to be at the cutting edge of teaching, research, and technology integration in our institutions. Working alone is no longer acceptable and will not move us forward as a profession.

Technology in Music Teacher Education

A collaborative model with faculty working together should be part of the solution to preparing undergraduate students for the complexity of music teaching in the twenty-first century. However, I continually feel frustration about how much ground I must cover with my music education students in 4 years. It is a colossal job to prepare future music teachers, and the job seems to become more and more complex with each passing year. Do I ignore the state of music education in local schools? What about all the "traditional" programs that exist? How far can I push an undergraduate student toward a more radical approach to teaching and learning that promotes music education for all students rather than a select few? Can both coexist in a school? And where does technology fit in? Should students take a required music education technology class or should it be integrated throughout undergraduate studies?

While I push toward creating seamless technology integration throughout the curriculum, there continues to be pushback from the students themselves. As a professor who had no technology during her undergraduate education (and had to pursue her professional development outside the university), I find it difficult to accept that students do not use the available notation technology to submit an assignment. I wonder who is really the digital native in this situation. Information and communication technology integration is a professional competency, and I want to make it a part of every class our students take. I believe that technology should be an active player in every learning situation, whether it is through audio recording, providing an instrument for a student who cannot play a traditional acoustic instrument in an ensemble, or giving students opportunities to create music in new and innovative ways.

As David Williams remarks (chapter 60), adding extra courses to undergraduate music programs is extremely problematic. When we question the common core of music programs, it is as if we are questioning the foundational sacred beliefs and value systems of many music professors. However, if we want our students to change teaching and learning practices in schools, they must experience these changes at the university level as well. For example, instead of music education students participating in the traditional large ensembles that replicate the performance model, could students attend "participative music-making events" during their studies for credit as a large ensemble? Could a tablet ensemble be considered for credit in a music education program? Again, when we propose such changes or options for our students, it is understood by certain colleagues that we are "questioning" the foundation of music programs in our institutions. We are challenging our institutions to stay current, to accept that music is changing and evolving in society and that we must also embrace these changes so that we do not become obsolete, as González-Moreno warns.

Attitudes and Core Beliefs

Fundamentally, attitudes toward technology and core beliefs about technology are key to people's engagement, according to Himonides (chapter 59). In addition to fear, the lack of confidence, and the lack of drive that may lead to disengagement with technology, I would add that lack of time and strategies for insightful integration of technology into accreditation programs can also lead to disengagement. In addition, practicing teachers often find themselves in situations where technology and technological support are not available to them. Teaching conditions in most places in the world are less than satisfactory, and teaching materials to facilitate technology integration in music classrooms would certainly contribute to the development of positive attitudes and engagement. On a more personal note, I was privileged to be able to attend a doctoral school in one of the most digitally up-to-date universities in the world. As a student, I had access to technology support any time I needed it. When I became a new faculty member at another university, I experienced just the opposite. After 10 years, our faculty still does not have adequate technology support for professors. Imagine my frustration going from being a well-served doctoral student to a faculty member with no resources.

As I stated in my Core Perspective (chapter 26), there is little philosophical debate about music technology integration, the nature and value of technology, and its role in music education. Himonides also mentions this, stating that we need a common understanding about technology and its role in order to subsequently discuss contexts of integration and notions of education and certification around the world. Of course not all countries have the same access to technology in their school systems, and this impacts teacher preparation. As already mentioned, I do not believe that technology education should simply be an "add-on" to existing certification programs. Rather, technology must be an integral part of all music teacher education, so that future music teachers view it organically, as a natural part of what it means to be a musician and a music teacher in the twenty-first century. In addition, Himonides states that the challenge is to prepare music teachers for the unknown, for the continued mutations and changes that will be present in the future. How do we certify people to embrace change? How do we encourage an open, evolving attitude and flexibility that allows music teachers to flow with the times and imagine new and creative music teaching with current technologies?

Himonides enumerates five core beliefs regarding teachers' development and proceeds to develop these in the last part of his chapter:

> The role of technology in education (and, obviously, music education) is central, not complementary, or peripheral.
> Technology can foster the development of teachers' critical thinking, which is . . . the most important ability that an educator should possess.
> Teachers cannot be passive recipients and consequent conduits of predefined educational praxes; they need to be in a position to shape their own and their students' learning, and critically assess what their needs are, as these are being formed.
> It is the social facet of technology that will play a leading role in teachers' competence, credentialing, and professional development.
> Technology is becoming mature enough to allow us to re-rehearse the notion of evidence-based education in a completely novel way.

As I have already affirmed, the role of technology in education is central, not peripheral. In addition, music teachers must own their continued professional development, especially as it pertains to technology. Given the rate of technological change, it is impossible to "prepare" teachers for what the future will be; the idea of lifelong learning in connection with the development of technology competency is not just a nice idea, it is a necessity! Teachers understand their students, and they are in touch with their learning needs. We cannot impose technology development from "above"; it must be based on responding to teachers' need to create curriculum for their students and to engage them in significant projects. Himonides describes how teaching education in England during the past two decades has been mostly school-based. It seems that university–school partnerships are essential to the development of a critical stance regarding technology in schools by making available informed and current research on technology pedagogy and integration.

Himonides touches on the role of teachers in society and the social facets of technology. Given that teachers are not highly valued in most societies (Finland being an exception), technology might be a way forward to educate our populations and create awareness about the importance of music education and the value of music teaching. Perhaps developing technology competency in music teachers could create a perception of "added value" in the general population. We have not come close to harnessing technology's social potential in music education. Himonides affirms: "it is the ways that technology enables us, but also empowers us to access, communicate, assess, discuss, debate, and develop further in a novel, democratic, and accessible way. Interestingly, this is not necessarily the facet of technology on which educational policy-makers have been focusing." This novel, democratic, and accessible way can and should already be part of our music teacher education programs, a model of how the social facets of technology can allow our community to discuss and debate different pedagogical and societal issues.

Perhaps technology will allow us to reengage the debate about evidence-based education by allowing us in the future to identify best practices concerning technology pedagogy. Our access to large data sets through online learning systems opens a window on this new learning context. I personally believe that blended systems in music teacher education are very powerful, allowing us the combine the uniqueness of face-to-face instruction with online tasks and discussion forums. Himonides states that research on these new blended systems needs to address design, interface, and human–computer interactions.

Finally, Himonides affirms that we must prepare teachers for an uncertain future rather than developing particular competencies that are of no lasting significance. How do we do this? Music teacher education programs have long lists of professional competencies that must be developed. We have worked hard to create observable indicators for each professional competency so that cooperating teachers and music education students are able to "know it when they see it." We often say to students that indicators of competency development must be observable and measurable to be evaluated. How would we observe and measure the competency "Ready for an uncertain future" or "Initiates and sustains meaningful change"? These are difficult questions. What is clear is that "qualified music teachers" are able to sustain critical thinking in order to manage change. I agree with Himonides: "competent educators are critical thinkers who are ready to polish old tools, forge new ones, creatively abuse existing tools, methods and processes, and harness everything available to them creatively in order to facilitate learning and development—their own and their students."

Conclusion

What are the ramifications of technology and technological change in teacher education programs? New collaborative models (e.g., communities of practice) must become the new norm. Digital natives and digital immigrants must work together and be mutually supportive of each other in a new educational paradigm that does not conform to traditional ideas about teaching and learning. Music teacher education programs must foster openness to uncertainty and the acceptance of a changed educational context. If we want to change music-teaching practice in schools (nontraditional ensembles, audio recording, tablet integration, etc.), students must experience these changes in their classes during their university studies. Music teacher education programs must reflect the best of the past while looking solidly toward the future. Whose responsibility is it for professional development and certification? It is a responsibility shared among teacher educators, students, and schools. Technology integration is central and must take place throughout the music education program, in core courses and in pedagogy courses. Future music teacher competencies for certification must include "Being able to live with constant change." By initiating and sustaining change in their own local contexts, music teachers will own their own professional development and be able to critically evaluate the place of music education in a modern, technological society.

Reference

Blondin, D., Peters, V., & Fournier, H. (2010). L'orientation identitaire de l'étudiant en formation à l'enseignement de la musique. *Recherche en éducation musicale, 28*, 185–208. Québec: Université Laval.

INDEX

Page numbers followed by *f* indicate figures.

Ableton Live, 195
  Follow Actions, 387
  Launch modes in, 387
  quantization in, 387, 390
  scenes in, 387
  session view, 387
Ableton Push, 387–88
Abril, C., 651
access issues, music technology and, 283–84
Acevedo, P., 298
  music program of, 295–96
ACID Pro (program), 195
acoustemology, 512
acoustic instruments, 72, 458, 468
acoustic knowing, 512
active intervention, 451, 455
active thinking, 145
activity, 623–26
Adams, J., 388
adaptive experts, 169
Adé, K. S., 330
administrative technologies, 657
Adorno, T., 19
  on repetition, 388
Advanced Placement music theory exams, U.S., 299
affordances, 244
Africa. *See also specific countries*
  instruments in, 178
  Internet in, 177, 186–87
  music education in, 177
  music technology in, 178
  oral-aural tradition in, 177–78
  rural, 177
  surveys in, 179–85
  technology in, 177, 185–87, 250–51, 583–85
  urban areas in, 185–86
"Africa Stop Ebola" (song), 333
African culture, technology in, 582–83
African diaspora, 388
African music, 177
Afrobeats, 335
agency, 360–61, 455
  in music education, 168
Aime, G. B., 406, 407
AirPlay, 471
AirServer, 471
Akai, 208, 388
Akuno, E., 3
Alber, D., 610
algebra, 145
alienation, 19
Allsup, R. E., 542
  on child-centered education, 549
Amadou & Mariam, 333
American Federal Radio Commission, 21
Amplified Elephants, 456–57, 460
amps, 205
analog recording, 186
analytical listening, 392
Andersson, A. P., 245
Anhuali Community Center, 411
Animoog, 238, 390
anonymous diversity, 266
Antal, A. B., 609
Antares, 234–35, 390
anxiety, 158
  technology and, 318
APC, 388
Apple (company), 23–24, 75–76, 99, 546
Apple, M., 25, 205

Apple TV, 471
apprenticeship, linear model of, 159–60
apps, 115, 318. *See also specific apps*
 development of, 641
 on iPad, 360
 as scaffold, 468–69
 teachers and, 641
Arendt, H., 150
Aristotle, 590
 on objects, 138
armchair ethnomusicology, 58
Armstrong, L., 21
Armstrong, V., 285
 on gender, 571
Arnold, M., 341
arpeggios, 368
arrangements, 66
 Tuneblocks, 198–99
art, Dissanayake on, 77
Articulator, 441
artifacts, 66
ASCII art, 619
assimilation
 differences in degrees of, 313
 distinction and, 314
 Hayles on, 340
 in practice, 294–95
 of technology, 293–94
assistive music technology, 284
Associated Board of Royal Schools of
  Music, 134
Atanasoff-Berry Computer, 73
Atari, 35, 555
at-risk learners, 324
Attali, J., 18, 23
attitudes, 620–21, 667–69
Audacity, 158
audiation, 426
AudioCubes, 498–99
aural awareness, 559
Aural Books, 134–35
aural processes, 638
Auralia (program), 133
Australasia, 84
Australia, 89, 454
 education in, 92
 ICT in, 92
 music education in, 98–99, 656

authentic music technology, 563–65
authenticity, 226
authoritative complex, 212, 227–28
autism spectrum, 439–40
automobile, as technology, 83–84
Autonomous University of Chihuahua, 608
autonomy, of musicians, 637–38
Auto-Tune, 234–35, 390
Awadi, Didier, 333
awareness, 300

Babbitt, M., 495
Bach, J. S., 379, 399
Baldwin, R., on teachers, 314
Ball State University, 610
Ballantyne, J., 7
Bamberger, J., 198
 on students, 326
Band Aid, 332–33
Bandhub, 467
"Bangla-desh" (Harrison), 335
banking model, 341
Barrett, J. R., 303
BASIC, 620
battles (rap), 207
Bauer, W. I., 424, 633
 on music education, 549
Bauerlein, M., 153
Bauman, Z., 359, 360, 365
BBC, 33–34
Beastie Boys, 304
beatboxing, 209
the Beatles, 384
Bebot (app), 441, 443*f*
bedroom producers, 234, 422–23
Beethoven, 72, 378*f*, 638
 music theory and, 379
Bell, 391
belonging, 455
Benchley, R., 555–56
Benjamin, W., 19
Bennett, H. S., 205
Bentley Test, 34
Berber music, 235
Berkovitz, J., 3–4
Berlin Phonogramm-Archiv, 57–58
Bernard, R., 644
Bernhard, H. C., 424

Bernstein, B., 345
 on resources, 349
Berry, A., 524, 528
"Beware the Dandelions," 305
Bieber, J., 424
binary mindset, 549
Biophilia, 320
Bjork, 320
Blackboard (management system), 530
Bledsoe, R., 6
blended learning, 221
blogs, 467
Bluetooth, 167
BMI Award, 207
Boehm, C., on music technology, 324
Bonde, L. O., 246
Bongo Flava, 335
Bono, 330, 332
Border Crossing, 409
Botswana, music education in, 584
Bourdieu, P., 21
Bowers, J.
 on variable engagement and interactivity, 561–62
 on variable sociality, 560
Boyette, R. B., 209–12
Boys and Girls Club, 194–95
Breeze, N., 40
Briggs, D., 284
bring-your-own-device (BYOD), 424, 472
British Empire, 490
Bruner, J., 339
BTEC. *See* Business and Technology Education Council
Buckingham, D., 250, 261
 on education, 262
 on mediatization, 263
 on participation, 267
Bulgarian bagpipers, 69
Bullmershe College, 105
Burnard, P., 281
Bush, J.-M., 3
Business and Technology Education Council (BTEC), 38
 First, 557
 National, 557
Bwana, J., 409
BYOD. *See* bring-your-own-device

Caillois, R., 247
Cakewalk Sonar (program), 195
CAL. *See* computer-assisted learning
Camblin, L., 608
 on faculty, 613
Campbell, P. S., 205
Candy Crush Saga, 318
Canvas (management system), 530
capitalism, 263
Cappelen, B., 245
captured sound, 20, 22
Caputo, V., 300
Carcassi, M., 399
Carr, N., 279
Carroll, L., 149
Caruso, E., 73
case-based work, 576
 framing, 578
 learning and, 578–79
 MTP and, 577–78
 problem-solving, 578
 refining, 578
 technology and, 579–80
Casella, M., 480
Catholic Church, 370
CDs, 205
Celemony, 320
cerebral palsy, 246–47
change
 in aural processes, 638
 in concert structure, 639
 degrees of, 294–95
 in faculty, 605
 Kigozi on, 237
 learning and, 649–51
 in music curriculum, 640–44
 in music education, 342–43, 371
 in oral processes, 638
 positive, 294
 Savage on, 629
 students and, 634
 teachers and, 634, 639–40
 of teacher's roles, 401–2
 technology and, 146, 292, 297–99
 Williams, D. A., on, 651
Chaplin, C., 18
Chen, J., 3
Chen-Edmund, J.-J., 408

Cher, 390
Chernoff, J. M., 388
Chez Panisse, 78
Chicago House, 207
Chicago Public Library, 505
chiffers, 476
child-centered education, 542
    Allsup on, 549
    Westerlund on, 549
China Conservatory, 406, 411
Chistakis, D., 422
choral music, 129
chords, 368
    in GarageBand, 469
Christendom, 370
Christensen, M., music program of, 295
Chrysostomou, S., 3–5, 123, 125, 161–62
    on ICT, 124, 137, 157
    on technology, 124, 139
Churchill, W., 154
cicadas, 512
Circuitbenders Forum, 197
citizenship, 267
    digital, 269
clarinet, 81–82
Clark, F. E., 20
classical guitar, 399
classical music, 638
classroom
    creativity in, 497
    flipped, 91, 377
    innovation in, 599
    iPad in, 319
    knowledge and, 318
    music technology in, 133
    structured time in, 497
    technology in, 463, 467–68, 547–49
Clay, M., 313
Clayton, J., 235
Clean Bandit, 332
Clifford, J., 59
cloud-based services, 91
cognitive apprenticeship, 659
collaboration, 364
collective music making, byproducts of, 455
College Music Society Advisory Committee for Education, 205
collegiality, 611

Columbia University, 61, 325
commerce
    education and, 147–48
    technology and, 147–48
commodification, 20, 53–54
    music education and, 21–22
Common Core, in United States, 542
communication, 364
communities, 364. *See also* mobile learning communities
    Greher on, 579–80
    of learners, 549–51
    nonformal learning, 505
    professional learning, 599
communities of practice, 141n2, 208, 573
    faculty and, 609–11
    Lave on, 228
    online music communities as, 228
    Wenger on, 228, 549
communities of response, 141n2
    formation of, 208–9
    Pignato on, 228–29
community music
    active intervention approach to, 451
    leader in, 460
    technology and, 456
comparative ethnomusicology, 58
composition, 244
    defining, 425
    digital, 242–43
    mobile learning communities and, 166–67
    music technology for, 73–74
    music theory and, 379
    in secondary school, 287
    teachers and, 321
    technology and, 317–18, 423–24
Computer Clubhouse Network, 504, 505
computer-assisted learning (CAL), 100
computers
    human-computer interaction, 628
    labs, 132
    languages, 655
    as media, 386
    Turkle on, 242–43, 323
    in Western schools, 245
concerts
    augmentation of experience, 433–35
    change in structure of, 639

formal, 635
  in pedagogical models, 639
Concordia University, 74
conspicuous consumption, 21
constructivist movement, 384, 544
consumption
  conspicuous, 21
  of music, 317
contemporary ethnomusicology, 58
Contemporary Music Project for Creativity in Music Education, 635
content knowledge, 305
content providers, technology as, 541–44
context, 249
  technology, 250–51
context-specific experiences, 325–27
contingency plans, 624–25
contingent practical configurations, 560–61
continual professional development, 626–27
controllerism, 388, 436
Conway, S. M., 610
cool factor, 252
copyright, 263
core beliefs, 667–69
  Himonides on, 668
Countryman, J., 425
Coupé Dècalé, 335
Coyne, R., 516
Create Digital Music, 197
Creative Arts Laboratory, 325
Creative Commons, 115, 291
creative turn, 192
creativity, 222, 244
  in classroom, 497
  iPad and, 545–46
  music technology and, 320
  participation and, 259
  products and, 416
  rhythmic video games and, 236
  risk-taking and, 499
  technology and, 147, 245, 410–11
Creativity Labs, 192
Cristofori, B., 72
critical listening, 193
  learning and, 235–36
critical literacy, 341
critical pedagogy, 258

critical thinking, in music curriculum, 286–88
criticality, 516
crosscultural understanding, 228
Cuban, L., on laptops, 545
Cubase (program), 35, 384, 571
cultural awareness, 246
  technology and, 407
culture, 144
  addressing, through praxis, 302–5
  knowledge of, 559
  music and, 59
  music education and, 195–96
  participatory, 259
  practices, 169
  reproduction of, 30
  resources, 168
  teachers on, 299–302
  technology and, 303
Cunha, R., 455
Cycling 74, 436
Cyprus
  music education in, 114
  teachers in, 114

D12, 332
Dammers, R. J., 282, 285
  on students, 652
Damrosch, W., 21
dance music, 394
  language of, 425
"The Dance of Eternity" (Dream Theater), 210
Dar es Salaam, 178
Darling-Hammond, L., 531
Darrow, A., 324
Daugherty, E., 613
Davidson, C. N., 263
Davis, M., 78
Davis, N., 386
DAW. See digital audio workstation
Day, G., 410
de Vugt, A., on pedagogical models, 599–600
Deal, J., 633
Dean, E., 389
decentralization, 137–38
declarations, 621–22
deconstruction, 661

Deleuze, G., 4, 206, 365
democratization, 21
   Kigozi on, 222
   MOOCs and, 222
   of technology, 285, 658, 661
Destinn, E., 73
determinism. *See* human determinism; technological determinism
deterritorialized experience, 206
DeVito, D., 5
Dewey, J., 20, 549, 591
   on education, 263, 268
   on ignorance, 264
DH. *See* digital humanities
Diamante, L.
   on distinction, 298
   music program of, 296–97
Digital Advancement, 109
Digital Agenda, 105–6, 109
digital arts pedagogy, 415
   challenges for, 416
digital audio workstation (DAW), 5, 207, 317, 385, 465, 503
   notation in, 386
   quantization in, 390
digital citizenship, 269, 341–42
digital composition, 242–43
digital divide, 94
digital expats, 621
digital habitat experiences, 257
digital humanities (DH), technology and, 60–62
digital immigrants, 262, 670
digital literacy, 90, 105–6, 138, 317, 319
digital musicians
   challenging notion of, 558–60
   Hugill on, 598
digital natives, 108, 260, 424, 670
   defining, 477
   pedagogical models and, 319
   Prensky on, 621
   rise of, 477–80
   Savage on, 629
   Turkle on, 479
digital pedagogy, 169
digital performance, 641
digital recording, 186

Digital School, 110
   national initiative, 113
digital studio
   language and, 394
   moving beyond, 392–93
   musicianship and, 385–90
   pedagogical models of, 390–92
   Ruthmann on, 392
digital technology
   Howell on, 233
   knowledge of, 533
Digital 21 Strategy, 169
digital vernissage, 245
Digital Youth Network, 505
Dillon, S., 384
directors, 507
disability, 241
Disquiet Junto, 392
disruptive innovation, 150, 237–38
Dissanayake, E., on art, 77
distinction
   assimilation and, 314
   Diamante on, 298
   Hayles on, 340
   in practice, 294–95
   of technology, 293–94
distraction, 237–38
distribution, 209
   Dorfman on, 597
   through Nimbit, 402
   Peppler on, 228
   through SoundCloud, 402
DIY. *See* do-it-yourself
DJ Hero, 94
DJ Spooky, 304
Djenet, C., 235
DJing, 299, 425–26, 590, 644
DJs, 207
"Do They Know It's Christmas" (song), 332–33
Doering, A., 547
do-it-yourself (DIY), 24–25, 658
dopamine, 246
Dorfman, J., 5, 7, 421
   on distribution, 597
   on ICT, 573
   on preservice teachers, 570
   on scholarship, 582

on teachers, 633
on technology, 570
DOS, 20, 105
downloading, illegal, 237
Downton, M., 194
Dream Theater, 210
drum circles, 406
dub reggae, 320
Duckworth, E., 539
DuFour, R., 614
The Dumbest Generation: How the Digital Age Stupefies Young Americans and Jeopardizes Our Future (Bauerlein), 152
Dykema, P., 20

ease of use, 546–47
EBD. *See* emotional and behavioral difficulties
Ebola
 education about, 330
 songs about, 331–33
 spread of, 329–30
"Ebola in Town" (Shaddy, D12, and Kuzzy), 332
"Ebola is Real" (HOTT FM and UNICEF Liberia), 332
Ebony Village Studios, 186
eBooks, 109–10, 113–14, 133, 545, 546
ecological issues, music technology and, 283
economic issues, music technology and, 283
Edexcel Board, 35, 38
edit windows, 374, 376*f*, 385*f*
Edmodo (management system), 530
educated hope, 269–70
education. *See also* higher education; music education
 Arendt on, 150
 in Australia, 92
 banking model, 341
 Buckingham on, 262
 child-centered, 542, 549
 commerce and, 147–48
 defining, 222–23
 Dewey on, 263, 268
 about Ebola, 330
 evidence-based, 627–28
 formal, 221
 Freire on, 270
 goals of, 659
 in Greece, 109–12
 hierarchy, 540
 ICT and, 92–93, 94–95
 language and, 146–47
 laptops in, 545
 liberal, 154
 mathematics and, 144–45
 methods, 525
 Morozov on, 150–51
 music and, 145–46
 philosophy of, 150, 464–65
 privatization of, 137–38
 Rancière on, 347–48
 reform centers, 540
 in Singapore, 92–93
 standards, 542–43
 Talbert on, 151–52
 of teachers, 127, 314–15, 523–24, 533–35, 623–24, 666–67
 techniques, 525
 technology and, 106, 115–16, 144–45
 tertiary, 221, 576
 traditional, of teachers, 572
 traditions of, 524–28
 trust in, 152
 in Uganda, 582
 in United Kingdom, 151
 visual arts and, 146–47
 Williams, D. A., on, 659
Education 3.0 paradigm, 94–95
Education and Manpower Bureau, 165, 166, 168
Education for All, 90
educational technologies, 657
edutainment, 492
eighth notes, 373–74
Einstein, A., 101
Eisner, E., 52
e-learning, 93*f*, 167, 168, 221, 655
electrical music/media equipment (EME), 179
electroacoustic music, 560
electronica, 197
Ellington, D., 21
Elliott, D. J., 402
Elliott, Missy, 388
Ellis, A., 57, 58
Ellsey, D., 323, 325, 327

Elpus, K., 651
embouchure, 160
EME. *See* electrical music/media equipment
Emmeleia, 112
Emmerson, S., 562
emotional and behavioral difficulties (EBD), 409
emulation, 662
endeere, 237
end-users, 540
English National Curriculum, 42
English schools
  information communications technology in, 36–39
  music technology in, 33–36
engoma, 237
ensembles, 652
  augmentation of, 435–36
  high school, 653
  size of, 637
Ephpheta Paul VI Pontifical Institute, 406
episteme, 17–18
equity, 472–73
Ertmer, P. A., on technology, 315
ethnomusicology
  armchair, 58
  comparative, 58
  contemporary, 58
  music technology and, 59
  origins of, 57–59
Etopia, E., 400
European Association for Music in Schools, 39
European Union, 105
Euterpe, 113
evidence-based education, 627–28
exemplars, teachers as, 362–63
experience
  concert, 433–35
  context-specific, 325–27
  deterritorialized, 206
  digital habitat, 257
  multilocated, 206
  perception reaction, 582
  technology, 525–26
Experience Music Project Museum, 281
explorers, 507

Facebook, 73–74, 197, 208, 467, 599
facilitator, 363
  in pedagogical models, 579
  technology and, 459–60
faculty
  Camblin on, 613
  changes in, 605
  communities of practice and, 609–11
  development, 604–11, 614
  ICT and, 604–8, 614
  roles of, 530–33
  Steger on, 613
failing, 319
Faith, P., 332
Fakoly, T. J., 333
Fallon, J., 598
Faulkner, R., 546
Feenberg, A., 26
Feld, S, 59
  on place, 512
Fenélon, 347, 348
Figure (program), 238, 391
Finale (program), 24, 133
Finland, 124
  ICT in, 589
  music education in, 588–89
  music technology in, 125
  teachers in, 625
  technology in, 588–89
Finney, J., 205
Firestone, W. A., 608
Fisher, K., 107
Fisher, T., 628
FL Studio, 133, 195, 335
flipped classroom, 91, 377
flow channel, 247
fluency, 68–69. *See also* literacy
  defining, 367–68
  music theory and, 376–77
fluxus, 320
folk music, 652
Folkways, 58
Follow Actions (Ableton), 387
Ford, S., 67
formal concerts, 635
formal music, 423
Foucault, M., 16

fourth environment, 241
four-track tape recorder, 269–70, 325, 367
Frankfurt School, 19
Franklin, J., 213
Freedman, B., 5, 8–9, 400, 402, 421
   on learners, 413
freedom, 241
   Heidegger on, 24
FreeStyle (program), 367
Freire, P., 264, 341
   on education, 270
Fuse ODG, 333
Future Shock (Toffler), 19

Gall, M., 2–3, 7, 40, 546–47
   on pedagogical models, 599–600
   on teachers, 548
game-based learning, 167
GarageBand (program), 23–24, 132–33, 167, 179, 197
   chord progressions in, 469
   instruments in, 469
   learning and, 528
   loops in, 389
   points of entry to, 598
Gardner, H., 327
Garvey, G., 332
gatekeepers, 658
Gauntlett, D. D., 262
GCSE. *See* General Certificate of Secondary Education
Geertz, C., 59
Geldof, B., 332
Gelineck, S., 387
gender, 300
   Armstrong, V., on, 571
   inequalities, 298
   music technology and, 285
   technology and, 571–72
General Certificate of Secondary Education (GCSE), 556–57
General Guitar Gadgets, 197
Ghana, 179, 283
   Cultural Studies Program, 584
   Ministry of Education in, 584
   music education in, 584
Gil, G., 291

Gill, A., 406
Gillespie, T., 25
"Girls" (Beastie Boys), 304
Giroux, H., 268
Gjertsen, L., 466
glitch (music), 320
glitches, 107
Global Information Technology Report, 89
globalization, 126
   ICT and, 90, 137
   Leong on, 140
   of music, 97–98
   music education and, 101
   technology and, 89, 348
glocalization, 54, 625
Goldacre, B., 628
Goldberg, D. T., 263
Goldberg Variations (Gould), 22
GoldieBlox, 304
González-Moreno, P. A., 8, 665–67
   on technology, 656, 657
Goodall, H., 369
Google Hangout, 405
Gopalakrishnan, S., on technology, 315
Goscinny, R., 619
GoTalk, 407
Gould, G., 78
Goulding, E., 332
government, 101
graduate music education, 532
Grammys, 207
Grateful Dead, 384
Greece, 108, 124, 221, 620
   education and technology in, 109–12
   ICT in, 111–12
   Internet in, 111
   language in, 112
   music education in, 112–15
   music technology in, 112–15
   teachers in, 114
Greek National Digital Learning Repositories, 110
Greek National Educational Content Aggregator Photodentro, 110
Greek Observatory of Digital Society, 109
Green, J., 67, 251, 362
   on musicians, 220

Green, L., 85, 193, 346, 476
 on informal learning, 482
Greene, M., 54
Greenhow, C., 267
Greenwich High School, 367
Greenwood, J., 386
Greher, G., 7, 242, 497
 on communities, 579–80
 on ICT, 573
 on pedagogical models, 592
 on preservice teachers, 570
 on technology, 583
groove, 425
Group Laiengee, 406
growth mindset, 600
Guattari, F., 365
Guinea, 329
Guitar Hero (game), 23, 94, 194, 417
guitars, 193
Gunbord, 242–43

Haiti, 407–8
Hamlen, K., 400
Hamman, D. L., 613
Hammoud, L. N., 406–7
Harare, 178
hardware
 evolution of, 65–67
 incompatibilities, 220
harmony, 369–70
Harrison, G., 335
Hay, K., 194
Haydn, 638
Hayles, N. K.
 on assimilation, 340
 on distinction, 340
 on technogenesis, 340
 on technology, 293
health, music and, 246–47
Heidegger, M., 3, 16
 on freedom, 24
 on *poiësis*, 17–18
 on *techné*, 17–18
 on technology, 17
Hein, E., 4–5, 421–22, 425, 427
Heines, J. M., 497
Hepp, A., 257
Herbst, A., 178

Hibbard, S., 610
Higgins, C., 628
Higgins, L., 451
higher education, 539, 624–25
 technology in, 604–5
 in United States, 540
Himonides, E., 6, 24, 579, 650
 on core beliefs, 668
 on teachers, 669
 on technology, 656
Hipco, 335
hip-hop, 197, 305, 384, 644
 beatboxing in, 209
 music technology and, 320
 ostinati in, 379
Hit Record, 208
Hitchcock, M., 8–9
holistic transformation, 95
Hong Kong, 89, 93, 124, 165
 ICT and, 167–69
 music education in, 99, 167–69
Honoré, C., 76
Hooper, S., 358
Hopscotch, 319
Horan, Niall, 332
Horkheimer, M., 19
Hornbostel, E., 57–58
hosho, 237
HOTT FM, 332
Hourigan, R., 610
How Music Works (documentary), 369–70
How Popular Musicians Learn (Green, L.), 476
Howell, G., 4–6, 491
 on digital technology, 233
 on music education, 238
 on technology, 233
HTML, 655
Huffman, D., 547
Hughes, J. E., 547
Hugill, A., 7, 436
 on digital musicians, 598
 on musicians, 558, 559
 on pitch, 558
 on rhythm, 558–59
 on timbre, 559
Huizinga, J., 247
Hullick, J., 456
human determinism

in music education, 314
  technological determinism and, 311–12
human needs, technology and, 140
human subjectivity
  music and, 19
  technology and, 18–23
human-computer interaction, 628
humanity
  Pignato on, 249
  technology and, 101, 278–80
  Turkle on, 152
human-techno future, 99–101
Humberstone, J., 5
hybrid genres, 130
Hypercard, 15–16
hyperconnectivity, 95
Hyperscore, 324

IBM, 620
iCompositions, 197
ICT. *See* information communications technology
identification, 331
identity
  individual and, 361–62
  music and, 34–35
  music education and, 161
  YouTube and, 361
ignorance, Dewey on, 264
The Ignorant Schoolmaster: Five Lessons in Intellectual Emancipation (Rancière), 347
Ihde, D., 26n3
illegal downloading, 237
imitation, 661
immediacy, 453–57
immersion, 193–94
Impromptu (program), 198
improvisation, 177, 253, 560–63
  in secondary school, 288
Improvising Machines (Bowers), 560
In Search of Music Education (Jorgensen), 205
inadequacy, 158
Indaba Music, 208, 467
indigenous music
  education, 584
  in Kenya, 130–31
individualism, 263
  technology and, 361

Indonesia, 450–51
inequality, 265–66
informal learning, 263, 346
  defining, 483
  Green, L., on, 482
  technology and, 251–53, 423
informality, 249
information age, knowledge in, 90–91
information communications technology (ICT), 159, 175
  accessibility of, 101
  in Australia, 92
  Chrysostomou on, 124, 137, 157
  Dorfman on, 573
  ecology of, 168f
  education and, 92–93, 94–95
  in English schools, 36–39
  in European countries, 39–40
  faculty development and, 604–8, 614
  in Finland, 589
  globalization and, 90, 137
  in Greece, 111–12
  Greher on, 573
  in Hong Kong, 167–69
  implementation of, 161–63
  learning and, 161
  Leong on, 124, 137, 157
  in Mexico, 603
  in music curriculum, 282
  National Curriculum for, 570
  network readiness and, 89
  progress of, 312–13
  UNESCO on, 101
information technology (IT), 655–56
in-group, 229
innovation, 249
  play and, 253–54
  technology and, 150, 253–54, 466
in-person interactions, 161
in-service training, in United States, 572–73
inspiration, 662
Instagram, 208, 467
instantaneity, 358
instructionism, 54
instrumentalists, 237
instruments, 360
  acoustic, 72, 458, 468
  in Africa, 178

instruments (*Cont.*)
  building, 451–52
  designing, 466
  in GarageBand, 469
  technology and, 81–82, 95, 399–400
  traditional, 82
  traditional indigenous musical, 179
  Varese on, 108
integrated learning, 93
integration, 342
intellectual property, 237
interactive textbooks, 113–14
interdisciplinary collaboration, 610
interface, 628
International Baccalaureate, 179, 299
International Classification of Function, Disability and Health, 246
International General Certificate, 178
International School of Uganda, 185
International Society for Music Education, 113, 408–9
Internet
  access to, 318
  in Africa, 177, 186–87
  cafes, 177
  connectedness to, 318
  evolution of, 65–67
  in Greece, 111
  growth of, 477
  Kigozi on, 227
  in music education, 107
  music education and, 39, 91
  radio, 132
  roboyette on, 212
  voice over Internet protocol, 208
Interpretation of Cultures (Geertz), 59
interpretive turn, 59–60
INXS, 393
i-Orchestra, 418
Iowa Tests of Music Literacy, 400
iPad, 41, 179, 208, 362, 410, 411
  apps on, 360
  in classroom, 319
  as controller, 320, 470
  creativity and, 545–46
  learning with, 134–35, 153, 440
  motivation and, 165–66
  notation on, 167

  social inclusion and, 440
iPod, 23–24, 73, 179
Ipswich, 561
iRadio, 208
Irish traditional music, 226, 475, 476
isolation, 161
IT. *See* information technology
iTunes, 73
iTunes store, 153
Ivory Coast, 335

Jackson, M., 335, 384
Jackson, Tiny, 207, 210
Jackson State University, 406
Jacotot, J., 347
jam sessions, 460
Jam2Jam, prosumers and, 414–15
James Madison University, 406
JamLink, 208
JamPod, 38
Java, 278
J-DISC, 61
Jenkins, H., 67
Jerusalem, 406–7
Jiaxing, X., 411
Joel, B., 598
Johnson, M., 339
joint enterprise, 228
Jordan, B., 206
Jorgensen, E., 205
Journal of Music, Technology and Education, 1
joy, in learning, 153

kakalo, 452
Kampala, 178
Kampala International School, 179, 185
Kanam, B., 333
Kandinsky, W. W., 243
Kanté, M., 333
Kasiwukira, 178
Katz, M., 22
Keita, S., 333
Kenny, A., 4–5
Kent, C., 319
Kenya, 179, 335, 584
  indigenous music in, 130–31
  music education in, 129–35, 585
  music in, 130–31

music technology in, 131–35
public school in, 130
Kenya Music Festival, 129
Kenyatta University, 130
Khan, A., 207
Khan Academy, 151
Kidjo, A., 332, 333
Kigozi, B., 4, 7, 349
    on change, 237
    on democratization, 222
    on Internet, 227
    on music education, 238
    on music technology, 219, 220
    on place, 225
    on poverty, 236
    on smartphones, 237
    on technology, 233, 249, 250
Kisumu, 176
knowledge. *See also* technological pedagogical
        and content knowledge
    classroom and, 318
    content, 305
    cultural, 559
    of digital technology, 533
    in information age, 90–91
    official, 25, 205
    Papert in, 547
    pedagogical, 528–30
    pedagogical content, 569
    personally situated, 319
    technology and, 140, 533
    transference of, 312–13
Koby Nanopads, 409
Kodable, 319
Kodály method, 194, 295
Koehler, M., 533
Koh, A., 60–62
Kolinski, M., 62
Kora, K., 333
Kriaras, E., 222
Kuhn, T., 342
Kuhn, W., 375
Kuzzy, 332
Kwaito, 335
Kwami, R., 176, 177

Lady Gaga, 423
Lakoff, G., 339
Lambert, A. D., 527
Lang, F., 18
language, 77, 339
    computer, 655
    of dance music, 425
    digital studio and, 394
    education and, 146–47
    in Greece, 112
    music as, 380
    music education and, 340
    remixes as, 393
    technology and, 146–47, 317, 340
laptops. *See also* computers
    Cuban on, 545
    in education, 545
Larson, L., 640
Launch modes (Ableton Live), 387
Launchpad, 388
Laurillard, D., 319
Lave, J., 141, 208, 211
    on communities of practice, 228
Lawler, P., on screens, 153
leaders, in community music, 460
learners
    at-risk, 324
    community of, 549–51
    engagement of, 595
    with exceptionalities, 440, 444
    Freedman on, 413
    Lum on, 413
    prosumers as, 413–15
    scaffolding, 464–65
    twenty-first century, 422–23
    universal design for, 439
    video games and, 417
learning. *See also* education; mobile learning
        communities
    augmentation of, 436–37
    blended, 221
    case-based work and, 578–79
    change and, 649–51
    computer-assisted, 100
    control of, 221
    critical listening and, 235–36
    e-learning, 93f, 167, 168, 221, 655
    game-based, 167
    GarageBand, 528
    ICT and, 161

learning (*Cont.*)
  informal, 251–53, 263, 346, 423, 482–83
  integrated, 93
  with iPad, 134–35, 153, 440
  joy in, 153
  Logic Audio, 528
  mediatization and, 258–59
  MIDI, 528
  Mixcraft, 528
  nonformal learning communities, 505
  online, 608
  out-of-school music, 504–5
  participation and, 263–64
  place and, 515–16
  Pro Tools, 528
  professional development programs and, 611
  professional learning communities, 599
  rhythmic video games and, 195–96
  self-directed, 167, 252
  self-learning, 650
  situating, 211
  sound and, 515–16
  technology and, 18, 221
  trajectories, 591
  universal design for, 439
  YouTube and, 160, 476
Learning in an Online World: The School Action Plan for the Information Economy, 92
Learning Labs, 505
Learning Nation, 93
Learning Through Producing, 126
Leclercq, H., 423
Lenhar, A., 480
Lennon, John, 330
Leong, S., 3, 5, 123–25, 161–62
  on globalization, 140
  on ICT, 124, 137, 157
  on music education, 481–82
  on technology, 124
Leonhard, G., 357
Levitan, D., 379
liberal education, 154
liberalization, 137–38
liberation, technology and, 238
Liberia, 329, 335

Lievrouw, L., 66
Lifelong Kindergarten, 238
Lightman, A., 75–76
Lil Wayne, 235
Limerick University, 406
Lindsay, E., 194
linguistic relativity, 339
LinkedIn, 599
Lipscomb, S., on music education, 549
listening
  analytical, 392
  critical, 193, 235–36
  defining, 425
  mobile learning communities and, 165–66
  in music curriculum, 169
literacy
  critical, 341
  digital, 90, 105–6, 138, 317, 319
  musical, 342, 368–70, 400–401
Logic Audio, 185, 197, 399, 409, 571
  learning and, 528
Logic Pro X, 370–71
Lomax, A., 62
loneliness, 152
long-distance courses, 221
LoopHD, 598
loops, 388
  in GarageBand, 389
Loopseque, 598
López de la Madrid, M. C., 605
Lorenzino, L., 455
Lospalos, 451
Loveless, A., 628
Lum, C.-H., 3, 5, 75, 79, 403
  on learners, 413
  on teachers, 402
  on technology, 433
  on YouTube, 434–35

machine music, 560
Machover, T., 324
MacJams, 197
Mack, J., 61
Madden Foundation, 61
Madeon, 423, 426
MadPad, 442–43, 469
Maetzel, J., 72

Maghrebi music, 235
Magnusson, T., 391
Malawey, V., 389
Malawi, music education in, 584
Malaysia, 305
Malik, Z., 332
Manhattanville Music Curriculum
    Program, 635
Manifold (studio), 78
Mansfield, J., 25, 27n14
Mantie, R., 2–3
  on music technology, 83
  on technology, 138
Marcus, G. E., 59
Marcuse, H., 19
market forces, technology and, 140
Markus, 333
Marrington, M., 386
Marsh, K., 252
Marshall, W.
  on pedagogical models, 303–4
  on remixes, 393
Martin, C., 332, 411
Martinez, S. L., 497
Marx, K., 19
mashups, 393
massification, 659
massive open online courses (MOOCs), 158,
    627, 628
  completion of, 222
  democratization and, 222
  growth of, 221
mathematics
  education and, 144–45
  technology and, 144–45
Max/MSP, 320, 436
mbira, 237
McAllister, P., 633
McCartney, P., 236
McClary, S., 388
McDonald's, 76–77
McGrail, E., 547
McLuhan, M., 19
MDI, 95
meaning, 455
Médecins Sans Frontières (MSF), 333
media, music and, 33

media culture, 257
mediated music, 16
mediation, 75
mediatization
  Buckingham on, 263
  learning and, 258–59
  pedagogical models and, 262–65
Medvinsky, M., 6, 7, 491
Meintjes, L., 60
melisma, 235
Melodica (app), 443*f*
Melodisia, 113
melody, 369–70
Melodyne, 320
"The Menace of Mechanical Music" (Sousa), 19, 23
MENC Handbook, 1
metaeducators, 143
metamodernism, 427
Metaphors We Live By (Lakoff and Johnson), 339
Metropolis (Lang), 18
Metropolitan Opera House, 73
Mexico, ICT in, 603
Middle Ages, 159–60
MIDI, 16, 22–23, 356, 436, 465
  display, 380*f*
  files, 378*f*
  learning, 528
  notation, 193, 371, 375
  sequencing, 371, 383
MIDI keyboard, 242
Miller, A. L., 527
Miller, P. D., 304–5
Miller, R., 490
Mills, N., 318
minimalism, 379
Ministry of Education, 112–13
mirroring devices, 471–72
Mishra, P., 533
MIT, 324
Mixcraft (program), learning and, 528
MLPP. *See* Music Learning Profiles Project
mobile apps, 97, 132
mobile labs, 418
mobile learning communities, 163
  composition and, 166–67
  expectations of, 166*f*
  listening and, 165–66

mobile phones, 41, 331, 596
Mobile Technologies Project (MTP), 7
　case-based work and, 577–78
Modern Drummer (magazine), 210
Modern Times (Chaplin), 18
Modugno, A., 367
Moha, 246–47
Mokobé, 333
Monkey Drum, 238
Monson, I., 388
Montclair State University, 384
MOOCs. *See* massive open online courses
Moodle (management system), 530
Moog, 390
"Moonlight Sonata" (Beethoven), 378f
Moore, B., 425
More Music, 410
Morozov, E., 75–76, 152
　on education, 150–51
Moser, P., 410
MOTU, 367
Moyer, M., 411
Moylan, W., 392
Mozart, 72, 304
MPC (Akai), 208
MSF. *See* Médecins Sans Frontières
MTP. *See* Mobile Technologies Project
Mullen, P., 408, 411
multilocated experience, 206
　Rodman on, 213–14
multimedia-based instruction, 100
　teachers and, 641–42
multisensory feedback, 458–59
multitrack recording, 317
Munich, 619
Murs, O., 332
music. *See also specific types*
　appreciation, 21
　augmentation of, 432–33
　Berber, 235
　consumption of, 317
　culture and, 59
　daily experience of, 204
　defining, 222–23
　education and, 145–46
　elements of, 425
　formal, 423
　globalization of, 97–98
　health and, 246–47
　human subjectivity and, 19
　identity and, 34–35
　industry, 101
　in Kenya, 130–31
　as language, 380
　locating, 205
　media and, 33
　mediated, 16
　without music education, 233–34
　nonformal, 581–83
　place and, 226
　public health and, 330–31, 335
　public policy and, 335
　social engagement and, 304–5
　society and, 32–33
　standards in, 313
　technology and, 145–46, 658
　timing of, 497–98
　written forms, 193
Music and Movement (radio program), 33–34
music curriculum, 499–500
　approaches to, 661–62
　change in, 640–44, 665–66
　critical thinking in, 286–88
　ICT in, 282
　listening in, 169
　music technology and, 282
　performance in, 169
　relevance of, 659–60
　sequential, 635
　software and, 391
　technology-enhanced, 417–18, 660
　vocational, 557
　Williams, D. A., on, 667
music education, 1
　in Africa, 177
　agency in, 168
　in Australia, 98–99, 656
　barriers to, 221
　Bauer on, 549
　boredom and, 359
　in Botswana, 584
　change in, 342–43, 371
　commodification and, 21–22
　context of, 564

in Cyprus, 114
daily practices, 125–27
defining, 175–76, 222–23
doorways to, 192–96
in Finland, 588–89
flow-centric view of, 385
formal, 153, 227, 384
formal concerts, 635
in Ghana, 584
globalization and, 101
graduate, 532
in Greece, 112–15
in Hong Kong, 99, 167–69
Howell on, 238
human determinism in, 314
identity and, 161
indigenous, 584
informal, 192, 193, 227
innovation and, 83
Internet and, 39, 91
Internet in, 107
in Kenya, 129–35, 585
Kigozi on, 238
language and, 340
legitimacy of, 265–68
Leong on, 481–82
literature, 481–83
in Malawi, 584
music culture and, 195–96
music technology and, 81–82, 84
music without, 233–34
in Nigeria, 584
nonformal, 175–76
nuance in, 313–14
Nzewi on, 582–83
opportunities for, 191–92
pedagogical models and, 265–68
Pignato on, 251
poietic spaces in, 203–4
principles, 500–501
in public school, 653
radical pedagogy in, 268–69
rhythmic video games and, 235
Rudolph on, 583
scholarship in, 582
Skype and, 408
slow food and, 76–77

social reality in, 126–27
specialization in, 116n2
structures in, 168
tablets in, 545
technological determinism and, 300
technology and, 95–99, 107, 125–27, 149–50, 176–85, 227, 292, 300, 305–6, 342–43, 407–9, 556–58, 564, 589–90
technology without, 206–11
in Texas, 69
traditional, 235
in twenty-first century, 422–23
in Uganda, 184
in United Kingdom, 546–47, 556–58
in United States, 69, 82, 98, 191, 546–47, 636, 651–52
Music Education for the Digital Age, 74, 76
Music Educators Journal, 213
Music Learning Profiles Project (MLPP), 206, 212, 213
Music Mark (journal), 409
Music National Curriculum, 37
    technology in, 570
music program
    of Acevedo, 295–96
    of Christensen, 295
    of Diamante, 296–97
music technologists, 401
music technology, 32–33, 71–72, 280–85
    access issues and, 283–84
    addressing, 394
    in Africa, 178
    applications of, 220
    authentic, 563–65
    Boehm on, 324
    in classroom, 133
    for composition, 73–74
    concentration in, 325
    concerns about, 74
    costs of, 236–37
    creativity and, 320
    defining, 66, 657–58
    development of, 81–82
    ecological issues and, 283
    economic issues and, 283
    in English schools, 33–36
    ethnomusicology and, 59

music technology (*Cont.*)
  in Finland, 125
  gender and, 285
  in Greece, 112–15
  hip-hop and, 320
  integration of, 597
  in Kenya, 131–35
  Kigozi on, 219, 220
  Mantie on, 83
  music curriculum and, 282
  music education and, 81–82, 84
  negative impacts of, 236–38
  pedagogical models and, 84, 85
  for performance, 73–74
  potentialities of, 286
  pushback against, 74–76
  for recording, 73–74
  resistance to, 85
  Savage on, 583, 597
  social structures and, 160–62
  society and, 280–81
  sociocultural issues, 282–83
  specialization in, 325
  suspicion of, 135
  unexpected uses, 234–35
  video and, 641
Music Technology, 35–36
music technology lab, in secondary school, 287
music theory, 368
  Beethoven and, 379
  composition and, 379
  fluency and, 376–77
  instructional materials, 378*f*
  teaching, 375–80
music writing software, 132–33
musical beginners, teachers and, 244–45
musical engagement, augmentation of, 432–33
Musical Futures program, 506, 507
musical literacy, 342, 368–70
  new, 400–401
  pedagogical models and, 400
musical tools, musicking and, 244–45
musicality
  institutionalized, 359
  technology and, 356, 409–10
musicians, 141n1, 203

autonomy of, 637–38
defining, 425
digital, 558–60
Green, J., on, 220
Hugill on, 558, 559
Rock Band and, 196
Savage on, 598
Small, 213
vulnerability of, 204–5
Webster on, 211
YouTube and, 207, 234
Musicians' Workshop, 463, 598
musicianship, 342
  digital studio and, 385–90
  reimagining, 597–98
  technology and, 409–10
musicking, 5, 241–42, 425
  musical tools and, 244–45
  technology and, 541
music-making, technology and, 197–99
Musition (program), 133
Mussorgsky, M., 37
mutual engagement, 228
MySpace, 197, 208
MySQL, 655

NAfME. *See* National Association for Music Education
Nairobi, 178
narcissism, 492
NASM. *See* National Association of Schools of Music
National Association for Music Education (NAfME), 542, 544
National Association of Schools of Music (NASM), 543–44
National Association of Schools of Music, 640
National Core Arts Standards, 543
National Core Music Standards for Music Education, 542–43
National Curriculum for Music, 112
National Education Technology Plan, 92
National Goals for Schooling in the Twenty First Century, 92
National Music Council, 43n10
National Standards for Music Education, 542–43, 636

"Nauru Elegies" (Miller, P. D.), 304
neoliberalism, 137–38
Netflix, 73–74
networked technology, 476–77
New School, 109
New York, 207
New York Philharmonic Orchestra, 417
New York University, 384
Newton, A., 300–302
Nigeria, 179, 335
  music education in, 584
Nilsson, B., 4–5
Nimbit, distribution through, 402
"No Child Left Behind," 653–54
noise (music), 320
Nonesuch, 58
nonformal learning communities, 505
nonformal music, technology in, 581–83
NotateMe, 167
notation
  in DAWs, 386
  grafts, 372
  on iPad, 167
  MIDI, 193, 371, 375
  in Rock Band, 194–95
  standard, 370–72
  Western, 369
Noteflight (program), 24
Noteworthy (program), 133
Notre Maison, 406, 407–8
Novation, 388
novice musicians, 97
Noxon, J., 633
NuMu, 39–40, 97–98, 348
NUMU, 506
Nzewi, M., on music education, 582–83

Oakley, G., 546
Objective C., 319
objects, Aristotle on, 138
obsolescence, of technology, 319
Ocarina and Guitar!, 414
O'Connor, S., 332
OECD. *See* Organisation for Economic Co-operation and Development
Oehler, E., 584
official knowledge, 25, 205

Ojala, J., on technology, 590
On Becoming a Rock Musician (Bennett), 205
O'Neill, S., 279, 282, 284
online collaboration, 476
online learning, 608
online music communities, 127, 482
  as communities of practice, 228
  place and, 229
  recombinance in, 228
  research into, 227
  sharing through, 197–98
Onwenu, O., 330
Open Educational Resources, 91
Open Sound Control (OSC), 436
open source software, 482
OpenSoundS project, 625–26
opera companies, 346
Opportunity-to-Learn Standards for Music Technology, 543
Ora, R., 332
oral processes, 638
oral-aural tradition, in Africa, 177–78
orchestras, 346
Orff training, 295, 406
Orff xylophone, 390
ORFI, 245
Organisation for Economic Co-operation and Development (OECD), 105, 106, 116n1, 138, 625
origination, 661, 662
OSC. *See* Open Sound Control
ostinati, 379
out-of-school music learning, 504–5
oxytocin, 246

Pan African Society for Musical Arts Education (PASMAE), 176, 180, 581, 583
  on technology, 177
Papert, S.
  on knowledge, 547
  on technology, 541–42, 544
Parikka, M., 125
participation, 66–67
  Buckingham on, 267
  creativity and, 259
  gap, 266, 341

participation (*Cont.*)
  learning and, 263–64
  Partti on, 341
  peripheral, 212
participatory culture, 259
Partti, H., 3, 246, 320, 325
  on participation, 341
  on pedagogical models, 340–41
PASMAE. *See* Pan African Society for Musical Arts Education
passeur culturel, 341
pause tape productions, 234
Payne, L., 332
pedagogical content knowledge, 569
pedagogical models, 312
  alternate, 640–41
  aural processes in, 638
  autonomy in, 637–38
  change in, 636–40
  concerts in, 639
  courses on, 544
  critical, 258
  de Vugt on, 599–600
  development of, 526
  digital, 169
  digital arts, 415–16
  of digital studio, 390–92
  emerging, 528–30
  ensemble size in, 637
  facilitator in, 579
  fundamentalist, 258, 260–62, 340–41
  Gall on, 599–600
  Greher on, 592
  knowledge, 528–30
  Marshall on, 303–4
  mashups, 304
  mediatization and, 262–65
  music education and, 265–68
  music technology and, 84, 85
  musical literacy and, 400
  oral processes in, 638
  Partti on, 340–41
  populist, 258, 260–62, 340–41
  radical, 258
  Sammer on, 599–600
  sound, 544
  teacher-centered, 635, 639–40
  technological determinism and, 261–62
  technology and, 204, 346–47, 481
  traditional, 636
pedagogical processes, technology as, 577
Pegrum, M., 546
Peluso, D. C. C., 282, 284
Pennell, J. R., 608
Peppler, K., 4, 6, 219, 235
  on distribution, 228
  on place, 225
  on play, 252–53
  on privilege, 250
  on technology, 229, 249
perception reaction experiences, 582
performance, 596
  augmentation of, 435–36
  defining, 425
  digital, 641
  doorways to, 192–96
  in music curriculum, 169
  music technology for, 73–74
  paradigms of, 313
  Rock Band and, 195
  telematic, 209
peripheral participation, 212
personal computers, 145
personalization, 161
personally situated knowledge, 319
Pestano, C., 408–9
Peters, V., 340, 341, 592
  on teachers, 592
Petrini, C., 76–77
Pew Internet surveys, 480
PGCE. *See* Postgraduate Certificate in Education
Philips (company), 129
philosophy, 110–11
  of education, 150, 464–65
Phonogramm-Archiv, 58
phonograph, Reimer on, 369
Photodentro, 114
Photodentro EduVideo, 110
Photodentro Repository, 110
Photodetro LOR, 110
PHP, 655
phronesis, 139
piano, 356
piano keyboard, 22–23, 129
pickups, 205

Pignato, J. M., 3, 219
  on communities of response, 228–29
  on humanity, 249
  on music education, 251
  on place, 225
  on privilege, 250
  on technology, 227, 234, 249
Pinkard, N., 506
Pinterest, 467
PISA. *See* Programme for International Student Assessment
pitch, Hugill on, 558
place
  defining, 511–12
  Feld on, 512
  Kigozi on, 225
  learning and, 515–16
  literature on, 513–14
  music and, 226
  online music communities and, 229
  Peppler on, 225
  Pignato on, 225
  sound and, 512–14
  teaching and, 515–16
  tuning of, 514–15
Plato, 19
play, 457–59
  innovation and, 253–54
  Peppler on, 252–53
  technology and, 251–53
pleasure, 77
pluralism, 421
plurality, 285
Pogonowski, L., 325
poiësis, 139
  Heidegger on, 17–18
poietic spaces, in music education, 203–4
policy, technology and, 140
Poll Everywhere, 465
Popplet, 465
popular culture, 237, 241
popular music, 84
Porsdam, H., 60–62
Portnoy, M., 210–11
Postgraduate Certificate in Education (PGCE), 569–70
  trainees in, 571

posthumanism, 19
poverty, 330
  Kigozi on, 236
practices, 66
pragmatism, 384, 591
praxis, 139
  addressing social and cultural issues through, 302–5
  informing, 303
  shock, 576–77
Prensky, M., 424, 491
  on digital natives, 621
preservice teachers, 527–28, 532
  Dorfman on, 570
  Greher on, 570
  technology and, 634–35
  in United Kingdom, 571
pressure points, 158–59
Price, D., 422
Prince, 388
privatization, of education, 137–38
privilege
  Peppler on, 250
  Pignato on, 250
Pro Tools (program), 133, 185, 356, 384
  learning and, 528
proactivity, 623–26
producing, 385
products, creativity and, 416
professional development programs, 628
  continual, 626–27
  learning and, 611
  strategies, 609–11
  teaching and, 611
  technology and, 608–9
professional learning communities, 599
Programme for International Student Assessment (PISA), 106, 111, 115
progression pathways, 506–8
progressive rock, 210
Propellerhead, 195, 391
prosumers, 25, 660
  defining, 414
  Jam2Jam and, 414–15
  as learners, 413–15
  SoundCloud and, 414
  Spotify and, 414
  YouTube and, 414

Provocation Questions, 16, 137, 138, 146
public health, music and, 330–31, 335
public policy, music and, 335
public school
  in Kenya, 130
  music education in, 653
publishing, on YouTube, 209
punk, 34
Purves, R., 24, 491, 623, 629
  on teachers, 623
putonghua, 165

Q-Tip, 234–35
quantization, 373–75
  in Ableton Live, 387, 390
  in DAWs, 390
quarter notes, 373–74
Quebec, 277–78
"Queen of the Night" (Mozart), 304
Queen's University, 74
Queensland Conservatorium, 655, 659
Queensland University, 406

"Race to the Top," 653–54
radical pedagogy, 258
  in music education, 268–69
radio, 33
  in Uganda, 184
  in West Africa, 331
Radio Luxembourg, 34
Radiohead Remix (Reich), 426
raï, 235
Rainbow International School, 185
Rainie, L., 479, 480
Rancière, J., 345
  on education, 347–48
rap, 197. *See also* hip-hop
  battles, 207
Rasinen, A., 125
Rathbone, M., 409
real-world divide, 576–77
Reason (program), 195, 384
Rebirth (program), 384
recombinance, in online music communities, 228
reconstruction, 661

recording
  analog, 186
  digital, 186
  music technology for, 73–74
  teaching, 367, 641
  techniques, 367
Rediffusion, 355
Reese, S., 423, 633
reference groups, 360
reflection
  encouraging, 531
  of students, 531
  YouTube in, 531
reflection-in-action, 531
refusers, 506–7
Regelski, T.
  on *techné*, 139
  on technology, 139
Reich, S., 379, 388, 426
Reimer, B., 368, 378*f*
  on phonograph, 369
  on technology, 371
Reisinger, C., 599
remixes
  as language, 393
  Marshall on, 393
  Weidenbaum on, 393
repetition, Adorno on, 388
Resnick, M., 391, 504
resources, 472–73
  Bernstein on, 349
Re-thinking Pedagogy for a Digital Age (Laurillard), 319
rhizomatic structure, 365
  technology and, 433
rhythm, 145–46, 373–75
  defining, 374
  Hugill on, 558–59
rhythmic video games, 192–93, 505
  creativity and, 236
  learning and, 195–96
  music education and, 235
Rice, T., 69
Richebé, N., 609
Richter, M., 426
Rieber, L. P., 358
riffs, 388

Rimington, J., 633
risk-taking
　creativity and, 499
　teachers, 499
Ritchie, L., 335
ritual ceremonies, 77
Robelia, B., 267
robotic moment, 152
roboyette, 209–11
　on Internet, 212
Rock Band (game), 23, 94, 236, 252, 508
　musician response to, 196
　notation in, 194–95
　performance and, 195
Rodman, M. C., 206
　on multilocated experience, 213–14
Roland Music, 99
Rose, L. S., 425
Rosenbaum, E., 391, 550
Roskilde Festival, 334
Ross, E. M., on technology, 315
routine experts, 169
Routledge Companion to Music, Technology, and Education, 1
Rozendal, M., 400
Rudolph, T. E., 312
　on music education, 583
Runfala, M., 400
Ruthmann, A., 383
　on digital studio, 392
Ruud, E., 246

Sammer, G., 40
　on pedagogical models, 599–600
Sandé, E., 332
Sangaré, O., 333
Sapir, E., 339
Sapir-Whorf hypothesis, 339
Satie, E., 434
Savage, J., 7, 24, 75, 79
　on change, 629
　on digital natives, 629
　on music technology, 583, 597
　on musicians, 598
　on teachers, 548, 573
　on technology, 545–46
　on United Kingdom, 571

savoring, 52
scales, 368
Scandinavia, 84
scenes (Ableton Live), 387
Schafer, R. M., 6
　on sound, 515, 516
scholarship
　Dorfman on, 582
　in music education, 582
screens, Lawler on, 153–54
Seal, 332
Second Life, 152
secondary school
　composition in, 287
　improvisation in, 288
　music technology lab in, 287–88
Seeger Melograph, 62
self-directed learning, 167, 252
self-evaluation, 39
self-expression, of students, 498
self-learning, technology and, 650
self-publishing, 208
Sennett, R., 364
sequencer software, 185
sequencing, MIDI, 371, 383
Serafin, S., 387
session view (Ableton Live), 387
Seuss Band, 441
Shaddy, 332
shared repertoire, 228
Sherbon, J., 613
Shulman, L., 7, 529
Sibelius (program), 24, 133, 386, 399, 592
Sidney Lanier Center School, 405–6, 408, 410
Siemens, 619
Sierra Leone, 329
sight-reading, 145–46
silence, 75
Silverman, M., 402
Simon, P., 384
simulation, 199
Sing & See, 96
Singapore, 53–55
　education in, 92–93
Singing Fingers, 441
sixteenth notes, 373–74

Skype, 208, 405
  music education and, 408
slow food, 76–77
slow music
  evidence of, 77–78
  students and, 78–79
Small, C., 5, 20
  on musicians, 213
  on technology, 213
Smalley, D., 562
SMART boards, 132, 548–49
SmartMusic (program), 24, 96, 98
smartphone, 115, 318, 653–54
  Kigozi on, 237
Smith, J., 214, 229
Smule, 414
snapping to grid, 374
social bonding, 455
social constitution, of technology, 294
social engagement
  music and, 304–5
  technology and, 304–5
social inclusion, 439
  iPad and, 440
social issues
  addressing, through praxis, 302–5
  songs and, 330
  teachers on, 299–302
  technology and, 303
social networks, 91, 208, 241
social structures, music technology and, 160–62
social technologies, 492, 629
  defining, 657
sociality, variable, 561
society
  music and, 32–33
  music technology and, 280–81
sociocultural frameworks, mobilizing, 302
sociocultural issues, music technology, 282–83
socioeconomic status
  of students, 652
  technology and, 653–54
Socrative, 465
software, 96. *See also specific software*
  creation of, 143–44
  evolution of, 65–67
  incompatibilities, 220
  music curriculum and, 391
  music writing, 132–33
  open source, 482
  sequencers, 185
Sole, G., 164
Solicitors' Regulation Authority, 627
Solis, G., 3
songs
  about Ebola, 331–33
  social issues and, 330
Songs for You and Me (iBook), 444
Sor, F., 399
Sor, L. D., 355
sound
  blocks, 392
  curation of, 515
  learning and, 515–16
  objects, 498
  pedagogy, 544
  place and, 512–14
  Schafer on, 515, 516
  before sight, 377, 402
  teaching and, 515–16
  technology and, 320, 514–15
  visualizing, 372
Sound Forge (program), 195
Sound of Africa (Meintjes), 60
sound systems, 205
Soundation, 467
Soundbrush, 441
SoundCloud, 197, 208, 348
  distribution through, 402
  prosumers and, 414
Soundrop (app), 441, 443*f*
Sounds of Intent, 407
Soundtrap, 467
Sousa, J. P., 19, 23
South Africa, 179, 335
Special Education Centre, 406
speed, 75
Spotify, 208
  prosumers and, 414
spreadability, 66
staff, 373*f*
Stager, G. S., 497
standard notation, 370–72
standing reserves, 18

Steger, J. A., 608
    on faculty, 613
STEM, 540
Stige, B., 456
Strand, K., 423
stress, of teachers, 576
String Project, 325, 327
Striphas, T., 25, 28n19
structuralism, 339
The Structure of Scientific Revolutions
    (Kuhn, T.), 342
structures, in music education, 168
student teachers, 570
students. *See also* learners
    Bamberger on, 326
    change and, 634
    Dammers on, 652
    defining, 643–44
    exploration of, 496–97
    reflection of, 531
    self-expression of, 498
    slow music and, 78–79
    socioeconomic status of, 652
    surveys, 181–82
studio. *See* digital studio
Stumpf, C., 57–58
Suárez, E. R., 406
subjectivity, of technology, 210
Suoranta, J., 270
"SuperEverything," 305
sustainability, 450–53
Sweden, 39
Swift, T., 424
swing, 375
switches, 244
syllabi, 134–35
symbols, rethinking, 372
syntax, 562
synthesizers, 245, 278
Syracuse University, 406
tablets, 115
    in music education, 545
    in Western schools, 245
Talbert, R., on education, 151–52
Tanganyika, 179
Tanzania, 179, 335
tapes, 205
Taylor, J., 633

TBMCs. *See* technology-based music classes
teachers
    acting as, 576
    adaptation of, 127
    ages of, 625
    apps and, 641
    Baldwin on, 314
    biographies of, 572
    change and, 634, 639–40
    changing roles of, 401–2
    competence, 572
    composition and, 321
    confidence with technology, 495–96
    constraints of, 597
    cooperation among, 314
    credentialing, 628
    on cultural issues, 299–302
    in Cyprus, 114
    development, 628–29
    Dorfman on, 633
    education of, 127, 314–15, 523–24, 533–35,
        623–24, 666–67
    as exemplars, 362–63
    in Finland, 625
    Gall on, 548
    in Greece, 114
    Himonides on, 669
    lone wolf model, 666
    Lum on, 402
    methods, 525
    multimedia-based instruction and, 641–42
    music theory, 375–80
    musical beginners and, 244–45
    pedagogical models centered on,
        635, 639–40
    perceptions of difficulty of, 423–24
    Peters on, 592
    preservice, 527–28, 532, 570, 571, 634–35
    professional development of, 314–15
    Purves on, 623
    recording and, 367, 641
    reflection-in-action, 531
    risk-taking, 499
    roles of, 624–25, 639–40
    Savage on, 548, 573
    on social issues, 299–302
    stress of, 576
    student, 570

teachers (Cont.)
  surveys, 182–84
  techniques, 525
  technology challenges, 184–85, 358
  technology used by, 107, 633–34
  traditional education of, 572
  traditions of, 524–28
  training, 614, 634–35, 660
  in United Kingdom, 625
teaching, 68
  augmentation of, 436–37
  content in, 551
  context, 169
  place and, 515–16
  professional development programs and, 611
  recording, 367, 641
  sound and, 515–16
  technology in, 124–25
Teaching and Learning Research Programme, 626
Teaching Music through Composition: A Curriculum Using Technology (Freedman), 371
TeachMeets, 408
techné, 16, 138, 300, 590
  Heidegger on, 17–18
  Regelski on, 139
technical skills, 559
technogenesis, 293, 311
  Hayles on, 340
technological capital, 259
technological determinism, 18–19, 293
  human determinism and, 311–12
  music education and, 300
  pedagogical models and, 261–62
  Turkle on, 478
technological pedagogical and content knowledge (TPACK), 7, 424, 528, 532
  rise of, 534–35
  theory of, 529
technology, 2–3, 32–33. *See also* information communications technology; music technology
  acceleration of, 466
  access and, 356
  adapting, 633–36
  administrative, 657
  affordances and constraints on, 312–14
  in Africa, 177, 250–51, 583–85
  in African culture, 582–83
  anxiety and, 318
  assessment of, 139
  assimilation of, 293–94
  authoritative complex of, 212, 227–28
  automobile as, 83–84
  availability of, 312, 583
  barriers to, 192, 221
  brain friendly, 221
  case-based work and, 579–80
  change and, 146, 292, 297–99
  Chrysostomou on, 124, 139
  in classroom, 463, 467–68, 547–49
  commerce and, 147–48
  community music and, 456
  competency with, 533–34
  composition and, 317–18, 423–24
  concerns about, 220
  as constraint, 515
  as content provider, 541–44
  context, 250–51
  costs of, 653–54
  creativity and, 147, 245, 410–11
  cultural awareness and, 407
  cultural issues and, 303
  cynicism towards, 422
  dangers of, 313
  defining, 65–66, 175–76, 222–23, 452, 492–93, 590, 621–22
  delegation of, 564–65
  democratization of, 285, 658, 661
  digital, 233, 533
  digital humanities and, 60–62
  distinction of, 293–94
  Dorfman on, 570
  education and, 106, 115–16, 144–45, 527
  educational, 657
  embedding of, 177
  entry points, 399–400
  eras of, 477–80
  Ermer on, 315
  experiences, 525–26
  facets of, 625–26
  facilitator and, 459–60

in Finland, 588–89
fluidity of, 357–58
gender and, 571–72
globalization and, 89, 348
González-Moreno on, 656, 657
Gopalakrishnan on, 315
in Greece, 109–12
Greher on, 583
Hayles on, 293
Heidegger on, 17
in higher education, 604–5
Himonides on, 656
Howell on, 233
human needs and, 140
human subjectivity and, 18–23
humanity and, 101, 278–80
impact of, 134–35
implementation, 500
individualism and, 361
informal learning and, 251–53, 423
initiation of, 564–65
innovation in, 150, 253–54, 466
instruments and, 81–82, 95, 399–400
integration of, 463, 468–69
intervention, 564–65
invisible, 544
Kigozi on, 233, 249, 250
knowledge of, 140, 533
language and, 146–47, 317, 340
learning and, 18, 221
learning community, 550*f*
legislative priorities and, 204
Leong on, 124
liberation and, 238
limitations of, 449
Lum on, 433
maintenance of, 237
Mantie on, 138
market forces and, 140
mathematics and, 144–45
music and, 145–46, 658
within music education, 204–6
without music education, 206–11
music education and, 95–99, 107, 125–27, 149–50, 176–85, 199–200, 211–12, 227, 292, 300, 305–6, 342–43, 407–9, 556–58, 564
in Music National Curriculum, 570

musicality and, 356, 409–10
musicianship and, 409–10
musicking and, 541
music-making and, 197–99
networked, 476–77
of 1990s, 16
in nonformal music, 581–83
obsolescence of, 319
Ojala on, 590
Papert on, 541–42, 544
PASMAE on, 177
pedagogical models and, 204, 346–47, 481
as pedagogical process, 577
Peppler on, 229, 249
Pignato on, 227, 234, 249
play and, 251–53
policy and, 140
portfolio collection with, 577–78
positive impacts of, 185–86
potentials of, 465–67
power of, 359–61
preservice teachers and, 634–35
prevalence of, 544
principles, 500–501
productive work with, 564
professional development programs and, 608–9
progress of, 294
Regelski on, 139
Reimer on, 371
rhizomatic structure and, 433
roles of, 621–22
roots of, 16
Ross on, 315
Savage on, 545–46
in schools, 98
self-learning and, 650
skill development, 526–28, 570
Small on, 213
social, 492, 629, 657
social constitution of, 294
social engagement and, 304–5
social issues and, 303
socioeconomic status and, 653–54
sound and, 320, 514–15
space of flows of, 480
spread of, 314

technology (Cont.)
  standards, 543
  stewards, 214, 228, 234
  subjective-social interaction complex of, 212
  subjectivity of, 210
  supervision of, 564–65
  suspicion of, 135
  teacher challenges in, 184–85, 358
  teacher confidence with, 495–96
  teacher use of, 633–34
  in teaching, 124–25
  as thing, 65–67, 491
  thinking with, 108
  training and, 356, 527
  Turkle on, 477
  ubiquity of, 540
  unexpected uses, 234–35
  in United Kingdom, 570
  unpacking, 65–66
  Upitis on, 306
  used by teachers, 107
  users of, 134, 140
  visual arts and, 146–47
  Wajcman on, 292
  Webster on, 358
  in Western schools, 582
  Williams, D. A., on, 649–50
Technology and the Gendering of Music Education (Armstrong, V.), 285, 571
Technology Institute for Music Education, 533
technology-based music classes (TBMCs), 280, 283
technology-enhanced curriculum, 417–18, 660–61
technomusicking, 414
technomusicology, 303
TED talks, 599
Télémaque (Fenélon), 347, 348
telematic performances, 209
"Terra Nova: Sinfonia Antarctica" (Miller, P. D.), 304
tertiary education, 221, 576
Texas, music education in, 69
theory. *See* music theory
Thibeault, M. D., 386
things, technology as, 65–67

thinking
  active, 145
  agile, 628–29
  critical, 286–88
  with technology, 108
Thinking Schools, 93
third environment, 241
Thomas, R., 635
Thorpe, C., 466
Three Ring, 465
3D printing, 418
Tiemann, M., 78
timbre, Hugill on, 559
TIMI. *See* traditional indigenous musical instruments
Timor-Leste, 450–51
tinkering, 426
To Save Everything, Click Here (Morozov), 150–51
Tobias, E., 5, 340
Toca Band, 238
Toffler, A., 19
Tolno, S., 333
Tompkins, M., 466
tool-oriented approach, 125
totally pedagogized society, 345
totally technologized society, 345–46
touch-typing, 376–77
TPACK. *See* technological pedagogical and content knowledge
T-Pain, 414
traditional indigenous musical instruments (TIMI), 179
traditional instruments, 82
traditional skills, 40
traditionalism, 162n1
  perpetual, 660
trainees, in PGCE, 571
training
  in-service, 572–73
  teachers, 614, 660
  technology and, 356, 527
treble clef, 373f
Triple Revolution, 479
trumpet, 160
trust, in education, 152
Tuneblocks, arrangement, 198–99

The Tuning of Place (Coyne), 516
Turkle, S., 6, 19, 75–76
 on computers, 242–43, 323
 on digital natives, 479
 on humanity, 152
 on technological determinism, 478
 on technology, 477
 on virtuality, 478
The Twilite Tone, 210
Twitter, 434, 467, 599
TwitterBand, 198
Uderzo, A., 619
Uganda, 179, 226
 education in, 582
 music education in, 184
 radio in, 184
ukulele, 360
Understanding and Providing a Developmental Approach to Technology Education, 126
UNESCO, 90, 101
 on ICT, 101
UNICEF Liberia, 332
United Kingdom, 84
 education in, 151
 music education in, 546–47, 556–58
 preservice teachers in, 571
 Savage on, 571
 teachers in, 625
 technology skill development in, 570
United States, 89
 Advanced Placement music theory exams, 299
 Common Core in, 542
 higher education in, 540
 in-service training in, 572–73
 music education in, 69, 82, 98, 191, 546–47, 636, 651–52
 National Standards in, 542–43, 636
 urban areas in, 225–26
universal design for learning, 439
Universal Music France, 334
Universidad de Londrina, 406
Universidad Veracruzana, 610
University of Bristol, 572
University of London, 406
University of Minneapolis-Duluth, 406

University of South Florida, 406
University of Windsor, 406
Upitis, R., 3, 302
 on technology, 306
urban areas
 in Africa, 185–86
 in United States, 225–26
Ure, Midge, 332
use, ease of, 546–47
utopia, 626
Väkevä, L., 7
van den Akker, R., 426–27
VanderLinde, D., 5
Varese, E., 32
 on instruments, 108
variable engagement and interactivity, 561–62
variable sociality, 561
Veblen, T., 21
Veenema, S., 327
Vermuelen, T., 426–27
Victorian synthesizers, 496
victory narratives, 227
video, music technology and, 641
video feedback, 327
video games. *See also* rhythmic video games
 collaborative, 482
 learners and, 417
 mods, 415
Vidrhythm, 469
Vietnam war, 330
Villa-Lobos, H., 399
Vimeo, 599
vinyl records, 78
Virtual Classroom Disabilities Project, 406
virtual contact, 279
virtuality, Turkle on, 478
virtuosity, 22
visual arts
 education and, 146–47
 technology and, 146–47
visual feedback voice training, 96
vitality, 455
Vivaldi Recomposed (Richter), 426
vocational curriculum, 557
vocationalization, 624

voice over Internet protocol (VoIP), 208
WAI. *See* Western acoustic musical instruments
Wajcman, J., on technology, 292
Waldron, J., 3, 6, 489
Waters, A., 78
Wavebot (app), 443*f*
waverers, 507
"We Are the World" (Jackson, M., and Ritchie), 335
We Got Tickets, 409
web browsers, 144
web-based instruction, 100
Weber, M., 24
webinars, 608, 610
    perceptions and, 612
    research study, 611–13
Webster, P., 15–16, 244, 325
    on musicians, 211
    on technology, 358
Weidenbaum, M., 392
    on remixes, 393
Welch, G., 407
well-being, 455
Wellington, J., 281
Wellman, B., 479, 480
Wenger, E., 141, 208, 211, 214, 229
    on communities of practice, 228, 549
West, K., 235
the West, 176–77
West Africa, 329
West African Exam Council, 584
Westerlund, H., 542
    on child-centered education, 549
Western acoustic musical instruments (WAI), 179
Western art music, 68–69
Western music systems, 53–54
Western notation, 369
Western schools
    computers in, 245
    tablets in, 245
    technology in, 582
Whistlecroft, L., 426

White, N., 214, 229
White, P., 369
Whitty, G., 628
WHO. *See* World Health Organization
Whorf, B. L., 339
Wi-Fi, 167
Wikipedia, 71
Williams, D. A., 3, 8, 654
    on change, 651
    on education, 659
    on music curriculum, 667
    on technology, 649–50
Williams, D. B., 211, 325
Wired (magazine), 291
Wise, S., 386
The Wizard of Oz (movie), 512
Woosup, L, 61
"The Work of Art in the Age of Mechanical Reproduction" (Benjamin), 19
workplace, new, 576
World Health Organization (WHO), 333
world music, 651–52

Yahoo, 73–74
Yale Seminar, 635
YOUmedia, 50, 505, 506
Youth Music, 409
Youth Orchestra, 325, 327
YouTube, 39, 52, 132, 599, 650
    identity and, 361
    learning and, 160, 476
    Lum on, 434–35
    musicians and, 207, 234
    prosumers and, 414
    publishing on, 209
    in reflection process, 531
    tutorials, 318

Zambia, 179, 226
Zambia-Ireland Teacher Education Partnership Program, 226
Zap Mama, 291
Zeichner, K., 526, 531, 533
Zimbabwe, 179, 584